Arithmetic Operations:

$$ab+ac = a(b+c)$$

$$\frac{a}{b} + \frac{c}{d} = \frac{ad+bc}{bd}$$

$$\frac{a+b}{c} = \frac{a}{c} + \frac{b}{c}$$

$$\frac{\left(\frac{a}{b}\right)}{\left(\frac{c}{d}\right)} = \frac{ad}{bc}$$

$$a\left(\frac{b}{c}\right) = \frac{ab}{c}$$

$$\frac{a-b}{c-d} = \frac{b-a}{d-c}$$

$$\frac{ab+ac}{a} = b+c$$

$$\frac{\left(\frac{a}{b}\right)}{c} = \frac{a}{bc}$$

$$\frac{a}{\left(\frac{b}{c}\right)} = \frac{ac}{b}$$

Exponents and Radicals:

$$a^0 = 1 \ (a \neq 0)$$

$$\frac{a^x}{a^y} = a^{x-y}$$

$$\left(\frac{a}{b}\right)^x = \frac{a^x}{b^x}$$

$$\sqrt[n]{a^m} = a^{m/n} = (\sqrt[n]{a})^m$$

$$a^{-x} = \frac{1}{a^x}$$

$$(a^x)^y = a^{xy}$$

$$\sqrt{a} = a^{1/2}$$

$$\sqrt[n]{ab} = \sqrt[n]{a}\,\sqrt[n]{b}$$

$$a^x a^y = a^{x+y}$$

$$(ab)^x = a^x b^x$$

$$\sqrt[n]{a} = a^{1/n}$$

$$\sqrt[n]{\left(\frac{a}{b}\right)} = \frac{\sqrt[n]{a}}{\sqrt[n]{b}}$$

Algebraic Errors to Avoid:

$\dfrac{a}{x+b} \neq \dfrac{a}{x} + \dfrac{a}{b}$ (To see this error, let $a = b = x = 1$.)

$\sqrt{x^2+a^2} \neq x + a$ (To see this error, let $x = 3$ and $a = 4$.)

$a-b(x-1) \neq a-bx-b$ (Remember to distribute negative signs. The equation should be $a-b(x-1) = a-bx+b$.)

$\dfrac{\left(\frac{x}{a}\right)}{b} \neq \dfrac{bx}{a}$ (To divide fractions, invert and multiply. The equation should be

$$\frac{\frac{x}{a}}{b} = \frac{\frac{x}{a}}{\frac{b}{1}} = \left(\frac{x}{a}\right)\left(\frac{1}{b}\right) = \frac{x}{ab}.)$$

$\sqrt{-x^2+a^2} \neq -\sqrt{x^2-a^2}$ (We can't factor a negative sign outside of the square root.)

$\dfrac{\cancel{a}+bx}{\cancel{a}} \neq 1+bx$ (This is one of many examples of incorrect cancellation. The equation should be $\dfrac{a+bx}{a} = \dfrac{a}{a} + \dfrac{bx}{a} = 1 + \dfrac{bx}{a}$.)

$\dfrac{1}{x^{1/2}-x^{1/3}} \neq x^{-1/2}-x^{-1/3}$ (This error is a sophisticated version of the first error.)

$(x^2)^3 \neq x^5$ (The equation should be $(x^2)^3 = x^2 x^2 x^2 = x^6$.)

Conversion Table:

1 centimeter = 0.394 inches	1 joule = 0.738 foot-pounds	1 mile = 1.609 kilometers
1 meter = 39.370 inches	1 gram = 0.035 ounces	1 gallon = 3.785 liters
= 3.281 feet	1 kilogram = 2.205 pounds	1 pound = 4.448 newtons
1 kilometer = 0.621 miles	1 inch = 2.540 centimeters	1 foot-lb = 1.356 joules
1 liter = 0.264 gallons	1 foot = 30.480 centimeters	1 ounce = 28.350 grams
1 newton = 0.225 pounds	= 0.305 meters	1 pound = 0.454 kilograms

COLLEGE
ALGEBRA

COLLEGE ALGEBRA

Roland E. Larson
Robert P. Hostetler
THE PENNSYLVANIA STATE UNIVERSITY
THE BEHREND COLLEGE

With the assistance of
David E. Heyd
THE PENNSYLVANIA STATE UNIVERSITY
THE BEHREND COLLEGE

D. C. Heath and Company
Lexington, Massachusetts / Toronto

PREFACE

Success in college level mathematics courses for those interested in any one of a variety of disciplines such as computer science, engineering, management, statistics, or one of the natural sciences begins with a firm understanding of algebraic concepts. The goal of our textbook is to further the preparation of students, who have completed two years of high school algebra, in such important areas as graphical techniques, functions, and analytic geometry. These are some of the fundamental elements used in the calculus and other mathematical endeavors that many students pursue.

The features of our book have been designed to create a comprehensive teaching instrument that employs effective pedagogical techniques.

• *Order of Topics.* Chapter 1 provides a thorough review of the concepts of algebra, including complex numbers. With this early coverage of complex numbers, the algebra of polynomial and rational functions (Chapters 1 through 5) can be brought to a logical conclusion with a discussion of the Fundamental Theorem of Algebra. Then, in Chapters 6 through 9 coverage is given to additional topics in algebra: exponential and logarithmic functions, systems of linear equations, matrices, sequences, series, and probability.

• *Algebra of Calculus.* Special emphasis has been given to the *algebra of calculus*. Many examples and exercises consist of algebra problems that arise in the study of calculus. These examples are clearly identified.

• *Examples.* The text contains over 600 examples, each carefully chosen to illustrate a particular concept or problem-solving technique. Each example is titled for quick reference and many examples include side comments (set in color) to justify or explain the steps in the solution.

• *Exercises.* Over 3000 exercises are included that are designed to build competence, skill, and understanding. Each exercise set is graded in difficulty to allow students to gain confidence and understanding in the use of algebra. To help prepare students for calculus, we stress a graphical approach in many sections and have included numerous graphs in the exercises.

• *Graphics.* The ability to visualize a problem is a critical part of a student's ability to solve a problem. This text includes over 600 figures.

• *Applications.* Throughout the textual material we have included numerous word problems that give students concrete ideas about the usefulness of the topics included.

• *Calculators.* Although we do not require the use of calculators in any section, techniques for calculator use are provided at appropriate places throughout the text. In addition, calculators have allowed us to include many realistic applications that are often excluded because of lengthy or tedious computations. Exercises meant to be solved with the help of a calculator are clearly indicated.

• *Supplements.* For the student, the *Study and Solutions Guide* by Dianna L. Zook is available. This guide includes detailed steps of solutions to nearly half of the odd-numbered exercises. The guide also includes a review of important concepts for each chapter as well as practice chapter tests. For the instructor, the *Instructor's Guide* by Meredythe M. Burrows is available and it includes answers to the even-numbered exercises as well as sample tests for each chapter.

We would like to thank the many people who have helped us at various stages of this project. Their encouragement, criticisms, and suggestions have been invaluable to us. The following reviewers offered many excellent ideas: Ben P. Bockstage, Broward Community College; Daniel D. Bonar, Denison University; H. Eugene Hall, DeKalb Community College; William B. Jones, University of Colorado; Jimmie D. Lawson, Louisiana State University; Jerome L. Paul, University of Cincinnati; and Shirley C. Sorensen, University of Maryland.

The mathematicians listed below completed a survey conducted by D. C. Heath and Company in 1983 which helped outline our topical coverage: Stan Adamski, University of Toledo; Daniel D. Anderson, University of Iowa; James E. Arnold, University of Wisconsin; Prem N. Bajaj, Wichita State University; Imogene C. Beckemeyer, Southern Illinois University; Bruce Blake, Clemson University; Dale E. Boye, Schoolcraft College; Sarah W. Bradsher, Danville Community College; John H. Brevit, Western Kentucky University; Milo F. Bryn, South Dakota State University; Gary G. Carlson, Brigham Young University; Louis J. Chatterley, Brigham Young University; Mary Clarke, Cerritos College; Lee G. Corbin, College of the Canyons; Milton D. Cox, Miami University; Robert G. Cromie, St. Lawrence University; Bettyann Daley, University of Delaware; Clinton O. Davis, Brevard Community College; Karen R. Dougan, University of Florida; Richard B. Duncan, Tidewater Community College; Don Duttenhoeffer, Brevard Community College; Bruce R. Ebanks, Texas Technological University; Susan L. Ehlers, St. Louis Community College; Delvis A. Fernandez, Chabot College; Leslee Francis, Brigham Young University; August J. Garver, University of Missouri; Douglas W. Hall, Michigan State University; James E. Hall, University of Wisconsin; Nancy Harbour, Brevard Community College; Ferdinand Haring, North Dakota State University; Cecilia Holt, Calhoun Community College; James Howard, Ferris State College; Don Jefferies, Orange

Coast College; William B. Jones, University of Colorado; Eugene F. Krause, University of Michigan; Richard Langlie, North Hennerin Community College; Jimmie D. Lawson, Louisiana State University; John Linnen, Ferris State College; Joseph T. Mathis, William Jewell College; Walter M. Potter, University of Wisconsin; Sandra M. Powers, College of Charleston; Nancy J. Poxon, California State University; George C. Ragland, St. Louis Community College; Ralph J. Redman, University of Southern Colorado; Emilio O. Roxin, University of Rhode Island; Charles I. Sherrill, University of Colorado; Donald R. Snow, Brigham Young University; William F. Stearns, University of Maine; Don L. Stevens, Eastern Oregon State College; Warren Strickland, Del Mar College; Billy J. Taylor, Gainesville Junior College; Henry Tjoelker, California State University; Richard G. Vinson, University of Southern Alabama; Glorya Welch, Cerritos College; Dennis Weltman, North Harris County College; Larry G. Williams, Schoolcraft College; and Rey Ysais, Cerritos College.

We would also like to give special thanks to our publisher, D. C. Heath and Company, and in particular, the following people: Tom Flaherty, editorial director; Mary Lu Walsh, mathematics editor; Mary LeQuesne, developmental editor; Cathy Cantin, production editor; Nancy Blodget, designer; Carolyn Johnson, editorial assistant; and Mike O'Dea, manufacturing supervisor.

Several of our colleagues also worked on this project with us. David E. Heyd assisted us with the text; Dianna Zook wrote the *Study and Solutions Guide;* and Meredythe Burrows wrote the *Instructor's Guide*. Three students helped with the computer graphics and accuracy checks: Wendy Hafenmaier, Timothy Larson, and A. David Salvia. Linda Matta spent many hours carefully typing the instructor's manual and proofreading the galleys and page-proofs. Deanna Larson had the enormous job of typing the entire manuscript.

On a personal level, we are grateful to our children for their interest and support during the three years the book was being written and produced, and to our wives, Deanna Larson and Eloise Hostetler, for their love, patience, and understanding.

If you have suggestions for improving this text, please feel free to write us. Over the past several years we have received many useful comments from both instructors and students and we value this very much.

Roland E. Larson
Robert P. Hostetler

THE LARSON AND HOSTETLER PRECALCULUS SERIES

To accommodate the different methods of teaching college algebra, trigonometry, and analytic geometry, we have prepared four volumes. These separate titles are described below.

COLLEGE ALGEBRA

A text designed for a one-term course covering standard topics such as algebraic functions, exponential and logarithmic functions, matrices, determinants, probability, sequences, and series.

TRIGONOMETRY

This text is used in a one-term course covering the trigonometric functions, exponential and logarithmic functions, and analytical geometry.

ALGEBRA AND TRIGONOMETRY

This title combines the content of the two texts mentioned above (with the exception of analytic geometry). It is comprehensive enough for two terms of courses or may be covered, with careful selection, in one term.

PRECALCULUS

With this book, students cover the algebraic and trigonometric functions, and analytic geometry in preparation for a course in calculus. This may be used in a one- or two-term course.

CONTENTS

Introduction to Calculators   xv

Review of Fundamental
Concepts of Algebra

Chapter 1
1.1   Algebra: Its Nature and Use   1
1.2   The Real Number System   3
1.3   The Real Number Line: Order and Absolute Value   11
1.4   Integral Exponents   17
1.5   Rational Exponents and Radicals   24
1.6   Complex Numbers   34
1.7   Polynomials and Special Products   40
1.8   Factoring   48
1.9   Rational Expressions   56
1.10 Algebraic Errors and the Algebra of Calculus   62

      Review Exercises   68

Algebraic Equations
and Inequalities

Chapter 2
2.1   Linear Equations   71
2.2   Formulas and Applications   78
2.3   Quadratic Equations   90
2.4   The Quadratic Formula and Applications   95
2.5   Other Types of Equations   104
2.6   Solving Inequalities   111

      Review Exercises   120

Functions and Graphs

Chapter 3
3.1   The Cartesian Plane and the Distance Formula   121
3.2   Graphs of Equations   127
3.3   Functions   136
3.4   Graphs of Functions   145

3.5   Linear Functions and Lines in the Plane   156
3.6   Composite and Inverse Functions   166
3.7   Variation and Mathematical Models   172

      Review Exercises   177

Polynomial Functions:
    Graphs and Zeros   Chapter 4

4.1   Quadratic Functions and their Graphs   179
4.2   Graphs of Polynomial Functions of Higher Degree   188
4.3   Polynomial Division and Synthetic Division   196
4.4   Rational Zeros of Polynomial Functions   204
4.5   Complex Zeros and the Fundamental Theorem of Algebra   211
4.6   Approximation Techniques for Zeros of Polynomials
      [Optional]   217

      Review Exercises   221

Rational Functions and
    Conic Sections   Chapter 5

5.1   Rational Functions and Their Graphs   222
5.2   Partial Fractions   234
5.3   Conic Sections   241
5.4   Conic Sections and Translation   251

      Review Exercises   256

Exponential and
Logarithmic Functions   Chapter 6

6.1   Exponential Functions   258
6.2   Logarithms   268
6.3   Logarithmic Functions   279
6.4   Exponential and Logarithmic Equations   286

      Review Exercises   291

Systems of Equations
    and Inequalities   Chapter 7

7.1   Systems of Equations   294
7.2   Systems of Linear Equations in Two Variables   303
7.3   Systems of Linear Equations in More Than Two Variables   310
7.4   Systems of Inequalities and Linear Programming   319

      Review Exercises   328

Matrices and
    Determinants   Chapter 8

8.1   Matrix Solutions of Systems of Linear Equations   330
8.2   The Algebra of Matrices   339

8.3   The Inverse of a Matrix   348
8.4   Determinants   355
8.5   Properties of Determinants   362
8.6   Cramer's Rule   369

      Review Exercises   375

Sequences, Series
and Probability   Chapter 9

9.1   Sequences and Summation Notation   377
9.2   Arithmetic Sequences and Series   386
9.3   Geometric Sequences and Series   392
9.4   Mathematical Induction   402
9.5   The Binomial Theorem   408
9.6   Counting Principles, Permutations, and Combinations   413
9.7   Probability   424

      Review Exercises   431

Appendix A   Exponential Tables   A1
Appendix B   Natural Logarithmic Tables   A3
Appendix C   Common Logarithmic Tables   A6
Appendix D   Trigonometric Tables   A9
Appendix E   Table of Square Roots and Cube Roots   A12

Answers to Odd-Numbered Exercises   **A15**

Index   **A45**

INTRODUCTION TO CALCULATORS

This text includes some examples and exercises that make use of a scientific calculator. A calculator can assist you in both learning and applying mathematics. Moreover, a calculator can significantly extend the range of practical applications. Instructions in the use of a calculator will be given as we encounter new functions and applications. Of necessity, the instructions that we provide are somewhat general and may not agree precisely with the steps required by your calculator.

One of the basic differences in calculators is their internal hierarchy (priority) of operations. For use with this text, we recommend a calculator with the following features.

1. (At least) 8-digit display with scientific notation
2. Four arithmetic operations: $\boxed{+}$, $\boxed{-}$, $\boxed{\times}$, $\boxed{\div}$
3. Exponential keys: $\boxed{y^x}$ or $\boxed{a^x}$, $\boxed{e^x}$ or $\boxed{\text{INV}}$ $\boxed{\ln x}$
4. Natural logarithm: $\boxed{\ln x}$
5. Pi: $\boxed{\pi}$
6. Inverse, reciprocal, square root: $\boxed{\text{INV}}$, $\boxed{1/x}$, $\boxed{\sqrt{}}$
7. Trigonometric functions: $\boxed{\sin}$, $\boxed{\cos}$, $\boxed{\tan}$
8. Memory: $\boxed{M}$ or $\boxed{\text{STO}}$
9. Parentheses: $\boxed{(}$, $\boxed{)}$
10. Change sign key: $\boxed{+/-}$ (Note that this is not the subtraction key. It is used to enter negative numbers into the calculator.)

In this text, all calculator steps will be given with *algebraic logic,* that is, the calculator logic using the normal algebraic order of operations. For example, the calculation 4.69[5 + 2(6.87 − 3.042)] can be performed with the following sequence of steps:

4.69 $\boxed{\times}$ $\boxed{(}$ 5 $\boxed{+}$ 2 $\boxed{\times}$ $\boxed{(}$ 6.87 $\boxed{-}$ 3.042 $\boxed{)}$ $\boxed{)}$ $\boxed{=}$

which should yield the value 59.35664. Without parentheses, we would work from the inside out with the following sequence to obtain the same result:

6.87 $\boxed{-}$ 3.042 $\boxed{=}$ $\boxed{\times}$ 2 $\boxed{+}$ 5 $\boxed{=}$ $\boxed{\times}$ 4.69 $\boxed{=}$

When rounding off decimals, we use the following rules:

1. Determine the number of positions you wish to keep. The digit in the last position you keep is called the rounding digit, and the digit in the first position you discard is called the decision digit.

2. If the decision digit is 5 or greater (≥ 5), round up by adding 1 to the rounding digit.

3. If the decision digit is 4 or less (≤ 4), round down by leaving the rounding digit unchanged.

4. Keep the decimal point in the same place.

We cannot control the internal round-off that occurs in calculators. What does your calculator display when you compute 2 ÷ 3? Some calculators simply truncate (drop) the digits that exceed their display range (of eight digits) and display .66666666. Others have an internal round-off subroutine and display .66666667. Although the second display is more accurate, *both* of these decimal representations of $\frac{2}{3}$ contain a round-off error. One of the best ways to minimize error due to round-off is to *leave numbers in your calculator* until your calculations are complete. If you want to save a number for future use, store it in your calculator memory.

1

REVIEW OF FUNDAMENTAL CONCEPTS OF ALGEBRA

Algebra: Its Nature and Use

1.1

What Is Algebra?

Whatever your algebraic background, you already have some concept of what algebra is. First and foremost, algebra is a *useful language for solving practical* (real-life) *problems*. Second, algebra is a *convenient tool for writing generalizations* of specific statements involving the operations of arithmetic. For instance, the equation

$$a + b = b + a$$

is used to indicate that the operation of addition is commutative on the set of real numbers. Algebra, in this sense, provides a kind of short-hand version of the rules of arithmetic. Finally, knowledge of algebra is a *prerequisite for the study of advanced courses in mathematics*.

Since real-life problems do not come to us expressed in algebraic form, one truly challenging part of algebra is the translation of such word problems into algebraic problems. The algebraic solution is then translated back as the solution to the word problem. This process has the following scheme:

REAL-LIFE PROBLEM → VERBAL STATEMENT → translate → ALGEBRAIC STATEMENT → solve → ALGEBRAIC SOLUTION → translate back → REAL-LIFE SOLUTION

For Those Planning to Take Calculus

To prepare for calculus, you will need to spend time learning to manipulate or rewrite algebraic expressions. For example, consider the following expression:

$$\frac{9 - 2x}{x^2 + x - 6}$$

which has the equivalent partial fractions form

$$\frac{9 - 2x}{x^2 + x - 6} = \frac{1}{x - 2} - \frac{3}{x + 3}$$

Integration, a basic operation in calculus, is difficult to carry out on the left-hand expression, but is rather simple to do on the right-hand expression. Throughout the text, we will refer to the algebra used in calculus as the **algebra of calculus.** In many instances, this algebra of calculus will seem backwards—the reverse of our regular algebra. For instance, adding and subtracting fractions would be a regular use of algebra—but in the partial fractions example just cited, we used algebra to *rewrite* a given fraction as the difference of two simpler fractions. Study the following chart for a preview of what algebra can do for us.

HOW DO WE USE ALGEBRA?

A. TO SYMBOLIZE REAL-LIFE PROBLEMS

Verbal Statement	Algebraic Statement
• Sixteen is 45% of what number?	• x = Number $16 = (0.45)x$
• Ed is now twice as old as Cindy. Eight years ago he was 18 years older than Cindy was. How old is each now?	•

	Now	8 Years Ago
Cindy's Age:	x	$x - 8$
Ed's Age:	$2x$	$2x - 8$
		$(2x - 8) = 18 + (x - 8)$

Verbal Statement	Algebraic Statement
• A certain amount is invested at 12% compounded annually and grows to $840.00 at the end of one year. How much was the original investment?	• x = Original Investment $x + (0.12)x = 840$

B. TO WRITE GENERAL STATEMENTS OF ARITHMETIC PROPERTIES

Arithmetic Equation

- $3 + 2 = 2 + 3$

  $\dfrac{1}{2} + \dfrac{3}{4} = \dfrac{3}{4} + \dfrac{1}{2}$

- $\sqrt[3]{8} = \sqrt[3]{2^3} = 2$

  $\sqrt{4^3} = (\sqrt{4})^3 = 2^3 = 8$

- $\dfrac{1}{2} + \dfrac{3}{5} = \dfrac{5 + 6}{10}$

- $13(49) = 13(50 - 1)$

  $= 650 - 13 = 637$

Algebraic Rule

- $a + b = b + a$

- $\sqrt[n]{a^m} = (\sqrt[n]{a})^m = a^{m/n}$

- $\dfrac{a}{b} + \dfrac{c}{d} = \dfrac{ad + bc}{bd}$

- $a(b - c) = ab - ac$

C. TO SIMPLIFY THE OPERATIONS OF CALCULUS

Algebraic Expression

- $\sqrt[3]{(2x - x^2)^5}$

- $\dfrac{5}{(2x + 5)^4}$

- $\ln \dfrac{\sqrt{x}}{(x + 2)^2}$

- $\dfrac{x^3 - 3x^2 + 5x}{x^3}$

- $\dfrac{1}{1 + e^{-x}}$

- $\dfrac{3x^2 - x - 2}{x^3 + 2x^2}$

Rewritten for Calculus

- $(2x - x^2)^{5/3}$

- $5(2x + 5)^{-4}$

- $\dfrac{1}{2} \ln x - 2 \ln(x + 2)$

- $1 - \dfrac{3}{x} + 5x^{-2}$

- $\dfrac{e^x}{e^x + 1}$

- $\dfrac{3}{x + 2} - \dfrac{1}{x^2}$

The Real Number System   1.2

We begin our study of algebra with a look at the **real number system.** We use real numbers every day to describe quantities like age, miles per gallon, container size, population, and so on. To represent real numbers we use symbols such as

$$9, \quad -5, \quad \sqrt{2}, \quad \pi, \quad \tfrac{4}{3}, \quad 0.6666\ldots,$$
$$28.21, \quad 0, \quad \text{and} \quad \sqrt[3]{-32}$$

There are four fundamental operations on the real numbers: **addition, subtraction, multiplication,** and **division,** denoted by the symbols $+$, $-$, $\times$ (or $\cdot$), and $\div$. The set of real numbers is **closed** relative to these four operations (with the exception that division by zero is undefined). This means that the *sum, difference, product, and quotient of two real numbers are also real numbers.*

The set of real numbers is made up of the following five subsets:

Natural Numbers	$\{1, 2, 3, 4, \ldots\}$
Whole Numbers	$\{0, 1, 2, 3, \ldots\}$
Integers	$\{\ldots, -3, -2, -1, 0, 1, 2, 3, \ldots\}$
Rational Numbers	$\{$all numbers of the form $p/q\}$* or $\{$all terminating or repeating decimals$\}$
Irrational Numbers	$\{$all nonrepeating, nonterminating decimals$\}$

Note in Figure 1.1 that if we begin with the natural numbers we have closure for addition and multiplication. However, we do not obtain closure for subtraction until we get to the integers and we do not obtain closure for division until we get to the rational numbers.

For the purpose of this text, we consider addition and multiplication as the two basic operations of arithmetic and we use these operations without formally defining them. Later, we will define subtraction and division in terms of addition and multiplication, respectively. The following list summarizes the properties of real numbers under the two basic operations.

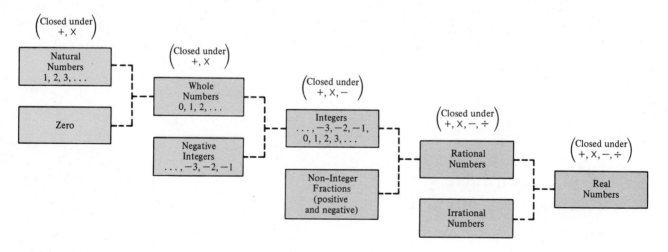

FIGURE 1.1   *Subsets of the Real Numbers*

*Rational numbers can be expressed as the ratio of two integers; that is, they can be written in the form p/q, where p and q are integers with $q \neq 0$.

PROPERTIES OF REAL NUMBERS

For all real numbers a, b, and c:

Property	**Addition**	**Multiplication**
1. **Closure:**	$a + b$ is a real number.	$a \cdot b$ is a real number.
2. **Commutative:**	$a + b = b + a$	$a \cdot b = b \cdot a$
3. **Associative:**	$(a + b) + c = a + (b + c)$	$(a \cdot b) \cdot c = a \cdot (b \cdot c)$
4. **Identity:**	0 is the identity. $a + 0 = a = 0 + a$	1 is the identity. $a \cdot 1 = a = 1 \cdot a$
5. **Inverse:**	$-a$ is the inverse of a. $a + (-a) = 0 = (-a) + a$	$\dfrac{1}{a}$ is the inverse of a. $a\left(\dfrac{1}{a}\right) = 1 = \left(\dfrac{1}{a}\right)a, \qquad a \neq 0$
6. **Distributive:**	$a(b + c) = a \cdot b + a \cdot c$ $(a + b)c = a \cdot c + b \cdot c$	*(left)* *(right)*

EXAMPLE 1
Properties of Real Numbers

Identify the property illustrated in each of the given equations.

(a) $5 + 4 = 4 + 5$

(b) $(3 + 7)2 = 3{\cdot}2 + 7{\cdot}2$

(c) $4x\left(\dfrac{1}{4x}\right) = 1, \qquad x \neq 0$

(d) $(x + 6) + 8 = x + (6 + 8)$

Solution:

(a) $5 + 4 = 4 + 5$ — — — — — — — — — — — *Commutative*

(b) $(3 + 7)2 = 3 \cdot 2 + 7 \cdot 2$ — — — — — *Distributive*

(c) $4x\left(\dfrac{1}{4x}\right) = 1, \qquad x \neq 0$ — — — *Inverse*

(d) $(x + 6) + 8 = x + (6 + 8)$ — — — — *Associative*

Remark: When working with the additive inverse (Property 5), don't confuse the *negative of a number* with a *negative number*. If a number b is already negative, then its additive inverse, $-b$, is positive. For instance, if $b = -5$, then $-b = -(-5) = 5$.

PROPERTIES OF NEGATIVES

For all real numbers a and b:

Properties	**Examples**
1. $(-1)a = -a$	$(-1)7 = -7$
2. $-(-a) = a$	$-(-6) = 6$

3. $(-a)b = -(ab) = a(-b)$ $(-5)3 = -(5 \cdot 3) = 5(-3)$

4. $(-a)(-b) = ab$ $(-2)(-6) = 2 \cdot 6 = 12$

5. $-(a + b) = (-a) + (-b)$ $-(3 + 8) = (-3) + (-8)$

A Word About Equations

An **equation** is a statement of equality between two expressions. For example, in algebra the statement

$$a = b$$

(read "a equals b") means that a and b represent the same number. Therefore, since $7 - 2$ and $3 + 2$ both represent the number 5, we write

$$7 - 2 = 3 + 2$$

We make frequent use of the following properties of equality.

PROPERTIES OF EQUALITY

For all real numbers a, b, and c:

Properties	Examples
1. **Reflexive:** $a = a$	$6 = 6$
2. **Symmetric:** If $a = b$, then $b = a$.	If $a + 2 = 5$, then $5 = a + 2$.
3. **Transitive:** If $a = b$ and $b = c$, then $a = c$.	If $a = b$ and $b = 4$, then $a = 4$.
4. **Substitution:** If $a = b$, then a can be replaced by b in any statement involving a or b.	If $a = b$ and $a + 2 = 5$, then $b + 2 = 5$.

Two important consequences of the substitution property are the following rules:

1. If $a = b$, then $a + c = b + c$.

2. If $a = b$, then $ac = bc$.

The first rule allows us to add the same number to both sides of an equation, while the second allows us to multiply both sides of an equation by the same number. The converses of these two rules are called the **Cancellation Laws** for addition and multiplication.

1. If $a + c = b + c$, then $a = b$.
2. If $\quad ac = bc$, $\quad$ then $a = b$, $\quad c \neq 0$.

You may recall from arithmetic that it is possible to define the operations of subtraction and division in terms of addition and multiplication, respectively.

DEFINITION OF SUBTRACTION AND DIVISION

For all real numbers a and b:

Subtraction: $\quad a - b = a + (-b)$

Division: $\quad$ If $b \neq 0$, then $a \div b = a\left(\dfrac{1}{b}\right) = \dfrac{a}{b}$

Remark: In the definition of subtraction, $-b$ is called the **negative** (or additive inverse) of b. In the definition of division, $1/b$ is called the **reciprocal** (or multiplicative inverse) of b. Furthermore, we often use the fraction symbol a/b in place of $a \div b$. In this fractional form, a is called the **numerator** and b is called the **denominator.**

For many students, fractions and properties of zero are two of the hardest topics in arithmetic. To help you achieve success with these topics, we offer the following piece of advice. *Memorize* the rules and try to gain an *intuitive feeling* for their validity.

PROPERTIES OF FRACTIONS

For all fractions a/b and c/d where $b \neq 0$ and $d \neq 0$:

1. Equivalent Fractions: $\dfrac{a}{b} = \dfrac{c}{d}$ if and only if $ad = bc$

2. Rule of Signs: $-\dfrac{a}{b} = \dfrac{-a}{b} = \dfrac{a}{-b}$

3. Generate Equivalent Fractions: $\dfrac{a}{b} = \dfrac{ac}{bc}$, $c \neq 0$

4. Add with Like Denominators: $\dfrac{a}{b} \pm \dfrac{c}{b} = \dfrac{a \pm c}{b}$

   Add with Unlike Denominators: $\dfrac{a}{b} \pm \dfrac{c}{d} = \dfrac{ad \pm bc}{bd}$

5. Multiply Fractions: $\dfrac{a}{b} \cdot \dfrac{c}{d} = \dfrac{ac}{bd}$

6. Divide Fractions: $\dfrac{a}{b} \div \dfrac{c}{d} = \dfrac{a}{b} \cdot \dfrac{d}{c} = \dfrac{ad}{bc}$, $c \neq 0$

Remark: The ± symbol means that both addition and subtraction satisfy the stated property. Moreover, when no symbol appears between two letters (as in Property 1, where $ad = bc$), multiplication is implied.

EXAMPLE 2
Properties of Fractions

Evaluate the following, using the properties of fractions.

(a) $\dfrac{a}{2} \cdot \dfrac{3}{5}$

(b) $\dfrac{a}{2} \div \dfrac{3}{5}$

(c) $\dfrac{a}{2} - \dfrac{3}{5}$

(d) $\dfrac{a}{(1/2) + (3/5)}$

Solution:

(a) $\dfrac{a}{2} \cdot \dfrac{3}{5} = \dfrac{(a)(3)}{(2)(5)} = \dfrac{3a}{10}$

(b) $\dfrac{a}{2} \div \dfrac{3}{5} = \dfrac{a}{2} \cdot \dfrac{5}{3} = \dfrac{5a}{6}$

(c) $\dfrac{a}{2} - \dfrac{3}{5} = \dfrac{(a)(5) - (2)(3)}{(2)(5)} = \dfrac{5a - 6}{10}$

(d) The denominator is

$$\frac{1}{2} + \frac{3}{5} = \frac{(1)(5) + (2)(3)}{10} = \frac{11}{10}$$

Therefore

$$\frac{a}{(1/2) + (3/5)} = \frac{a}{11/10} = \frac{a}{1} \cdot \frac{10}{11} = \frac{10a}{11}$$

(Note that we can write a as $(a/1)$.)

For adding or subtracting fractions, Property 4 is often overlooked because of too great a dependence upon the **lowest common denominator** (LCD) method. For instance, to add $\frac{3}{10}$ and $\frac{1}{12}$, we can use Property 4 to get

$$\frac{3}{10} + \frac{1}{12} = \frac{36 + 10}{120} = \frac{46}{120} = \frac{23}{60}$$

Using the LCD method, we must first find the LCD (sometimes this is quite time consuming). In this case, it is 60 and we write

$$\frac{3}{10} + \frac{1}{12} = \frac{3 \cdot 6}{10 \cdot 6} + \frac{1 \cdot 5}{12 \cdot 5} = \frac{18}{60} + \frac{5}{60} = \frac{23}{60}$$

We suggest that you try to strike an effective balance between these two methods. For adding *two* fractions, Property 4 is often more convenient. For *three* or *more* fractions, the LCD method is usually better.

EXAMPLE 3
The LCD Method of Adding or Subtracting Fractions

Evaluate

$$\frac{2}{15} - \frac{5}{9} + \frac{4}{5}$$

Solution:
By prime factoring the denominators

$$15 = 3 \cdot 5, \quad 9 = 3 \cdot 3, \quad \text{and} \quad 5 = 5$$

we see that the LCD is

$$3 \cdot 3 \cdot 5 = 45$$

and it follows that

$$\frac{2}{15} - \frac{5}{9} + \frac{4}{5} = \frac{2(3)}{45} - \frac{5(5)}{45} + \frac{4(9)}{45}$$

$$= \frac{6 - 25 + 36}{45} = \frac{17}{45}$$

Remark: In Example 3, recall that a **prime number** is a positive integer that has precisely two factors: itself and 1. For example, 2, 3, 5, 7, and 11 are prime numbers, whereas 1, 4, 6, 8, and 10 are not prime.

PROPERTIES OF ZERO

For real numbers a and b:

1. $a \pm 0 = a$ 2. $a \cdot 0 = 0$ 3. $0 \div a = \frac{0}{a} = 0, \quad a \neq 0$

4. Factorization Law: If $ab = 0$, then $a = 0$ or $b = 0$.

5. Division by Zero: $\frac{a}{0}$ is undefined.

EXAMPLE 4
Properties of Zero

Evaluate the following:

(a) $3 - \frac{0}{5}$ (b) $-4(9 - 11 + 2)$ (c) $\frac{7}{3 - (8 - 5)}$

Solution:

(a) $3 - \frac{0}{5} = 3 - 0 = 3$

(b) $-4(9 - 11 + 2) = -4(-2 + 2) = -4(0) = 0$

(c) This expression is *undefined* since its denominator is zero.

Denominator: $3 - (8 - 5) = 3 - 3 = 0$

Section Exercises 1.2

In Exercises 1–20, identify the property (or properties) of real numbers (closure, commutative, associative, distributive, identity, inverse) illustrated in each of the given equations.

1. $3 + 4 = 4 + 3$
2. $x + 9 = 9 + x$
3. $2(x + 3) = 2x + 6$
4. $18 - 6 = 12$
5. $\frac{70}{25} = 2.8$
6. $-15 + 15 = 0$
7. $\frac{1}{(h + 6)}(h + 6) = 1$
8. $57 \cdot 1 = 57$
9. $h + 0 = h$
10. $(5 + 11) \cdot 6 = 5 \cdot 6 + 11 \cdot 6$
11. $6 + (7 + 8) = (6 + 7) + 8$
12. $x + (y + 10) = (x + y) + 10$
13. $\dfrac{\sqrt{21}}{\sqrt{7}} = \sqrt{3}$
14. $(x + y) - (x + y) = 0$
15. $2x + 2y = 2(x + y)$
16. $\frac{1}{7}(7 \cdot 12) = (\frac{1}{7} \cdot 7)12 = 1 \cdot 12 = 12$
17. $x(3y) = (x \cdot 3)y = (3x)y$
18. $\frac{1}{3}(x + 3) = \frac{1}{3}x + \frac{1}{3} \cdot 3 = \frac{1}{3}x + 1$
19. $x(y + 0) = xy + x \cdot 0 = xy$
20. $-10 + (8 + 10) = -10 + (10 + 8)$
$$= (-10 + 10) + 8 = 0 + 8 = 8$$

In Exercises 21–30, evaluate each expression.

21. $(8 - 17) + 3$
22. $3 - (-6)$
23. $10 - 6 - 2$
24. $-3(5 - 2)$
25. $(4 - 7)(-2)$
26. $(-5)(-8)$
27. $2\left(\dfrac{77}{-11}\right)$
28. $\dfrac{27 - 35}{4}$
29. $\dfrac{-10}{-3 + (-2)}$
30. $8(-6)(-2)$

In Exercises 31–40, perform the indicated operations and reduce each fraction to lowest terms.

31. $\frac{3}{16} + \frac{5}{16}$
32. $\frac{6}{7} - \frac{4}{7}$
33. $6(-\frac{3}{8})$
34. $\frac{10}{11} + \frac{6}{33} - \frac{13}{66}$
35. $\frac{2}{5} \cdot \frac{7}{8}$
36. $4 \cdot \frac{3}{4}$
37. $\frac{4}{5} \cdot \frac{1}{2} \cdot \frac{3}{4}$
38. $\frac{4}{5} - \frac{1}{2}$
39. $\dfrac{x}{y} - \dfrac{x}{2}$
40. $\dfrac{x}{(1/2) - (1/6)}$

In Exercises 41–44, evaluate (if possible) each expression using the properties of zero.

41. $\dfrac{81 - (90 - 9)}{5}$
42. $\dfrac{(x + 2) - (x + 2)}{x - x}$
43. $10(23 - 30 + 7)$
44. $\dfrac{8}{-9 + (6 + 3)}$

45. Let m and n be any two integers, then $2m$ and $2n$ are even integers and $(2m + 1)$ and $(2n + 1)$ are odd integers. Show that
    (a) the sum of two even integers is even.
    (b) the sum of two odd integers is even.
    (c) the product of an even integer with any integer is even.
46. Using the properties of real numbers, show that if $a = -a$ then $a = 0$.
47. One worker can assemble a component in 7 days and a second worker can do the same task in 5 days. If they work together, what fraction of the component can they assemble in 2 days?
48. One foot of copper wire weighs 1 ounce ($\frac{1}{16}$ pound). What is the weight of $\frac{5}{8}$ mile of this wire?

Calculator Exercises

49. In this section you learned that a rational number expressed in decimal form is a terminating decimal or a nonterminating decimal with a repeating pattern. Use your calculator to find the decimal form of each of the following rational numbers. If it is a nonterminating decimal, give the repeating pattern
    (a) $\frac{5}{8}$ (b) $\frac{1}{3}$ (c) $\frac{41}{333}$ (d) $\frac{6}{11}$ (e) $\frac{85}{750}$
    (Note: For some rational numbers one must have more decimal places than on a typical calculator before observing the repeating pattern. For example, find the decimal representation of 4/23.)
50. Complete the following table to observe that $5/n$ increases without bound as we let n approach zero.

n	10	1	0.5	0.01	0.0001	0.000001
$\dfrac{5}{n}$	$\dfrac{1}{2}$	5				

51. When $h = 0$, $\dfrac{\sqrt{4 + h} - 2}{h}$

    is undefined. Complete the following table to demonstrate that this expression approaches the value of 0.25 as h assumes real numbers progressively closer to zero.

h	10	1	0.5	0.01	0.0001	0.000001
$\dfrac{\sqrt{4 + h} - 2}{h}$	0.174	0.236				

The Real Number Line: Order and Absolute Value

1.3

The model we use to represent the real number system is called the **real number line.** It consists of a horizontal line with an arbitrary point (the **origin**) labeled 0. Positive units are measured to the right from the origin and negative units are measured to the left, as shown in Figure 1.2.

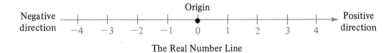

The Real Number Line

FIGURE 1.2

The importance of the real number line lies in the fact that *each point on the line corresponds to one and only one real number* and *each real number corresponds to one and only one point on the real line*. This type of relationship is called a **one-to-one correspondence.** (See Figure 1.3.)

Every real number corresponds to a point on the real line.

Every point on the real line corresponds to a real number.

One-to-One Correspondence

FIGURE 1.3

The number associated with a point on the real line is called the **coordinate** of the point. For example, in Figure 1.3, $-\frac{5}{3}$ is the coordinate of the left-most point and $\sqrt{2}$ is the coordinate of the right-most point.

Ordering the Real Numbers

The real number line is useful in demonstrating the order property of real numbers. We say that the real number a **is less than** the real number b if a lies to the left of b on the real number line. Symbolically, we denote this relationship by

   $a < b$ *(a is less than b)*

or equivalently

   $b > a$ *(b is greater than a)*

Arithmetically, we have

   $a < b$ if and only if $b - a > 0$ *(b − a is positive)*
   $a > b$ if and only if $b - a < 0$ *(b − a is negative)*

Remark: In mathematics we use the phrase "if and only if" as a means of stating two implications in one sentence. For instance, the statement $a < b$ if and only if $b - a > 0$ means

$$\text{if } a < b \text{ then } b - a > 0 \qquad and \qquad \text{if } b - a > 0 \text{ then } a < b$$

The symbols $<$ and $>$ are referred to as **inequality signs.** They are sometimes combined with an equal sign as follows:

$a \leq b$ *(a is less than or equal to b)*

$b \geq a$ *(b is greater than or equal to a)*

Inequalities are useful in denoting subsets of the real numbers. For instance,

Inequality		Subset of Reals
$x \leq 2$	denotes	All real numbers less than or equal to 2.
$-2 \leq x < 3$	denotes	All real numbers between -2 and 3, including -2 (but not including 3).
$x > -5$	denotes	All real numbers greater than -5.

EXAMPLE 1
Interpreting Inequalities

Use inequality notation to describe each of the following:

(a) c is nonnegative
(b) b is at most 5
(c) d is negative and at least -3
(d) x is positive but not more than 6
(e) c is at most 2 and b is between 2 and 3
(f) x is greater than y and both lie between -3 and -5

Solution:

(a) "nonnegative" means a number is greater than or equal to zero:

$$c \geq 0$$

(b) "at most" means $\leq$:

$$b \leq 5$$

(c) "at least" means $\geq$:

$$d < 0 \quad \text{and} \quad d \geq -3 \quad \Rightarrow \quad -3 \leq d < 0$$

(d) "not more than" means $\leq$:

$$0 < x \quad \text{and} \quad x \leq 6 \quad \Rightarrow \quad 0 < x \leq 6$$

(e) $c \leq 2 \quad \text{and} \quad 2 < b < 3 \quad \Rightarrow \quad c \leq 2 < b < 3$
(f) $-5 < y < x < -3$

Subsets of real numbers are sometimes expressed in the **interval** forms shown in Table 1.1. (We use the symbols ∞ and $-\infty$ to denote positive and negative infinity.)

TABLE 1.1
Intervals on the Real Line

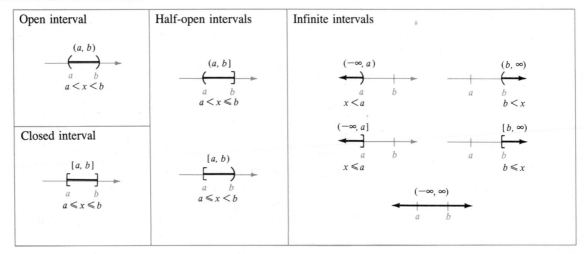

Open interval (a, b) $a < x < b$	Half-open intervals $(a, b]$ $a < x \leq b$	Infinite intervals $(-\infty, a)$ $x < a$ (b, ∞) $b < x$
Closed interval $[a, b]$ $a \leq x \leq b$	$[a, b)$ $a \leq x < b$	$(-\infty, a]$ $x \leq a$ $[b, \infty)$ $b \leq x$ $(-\infty, \infty)$

EXAMPLE 2
Intervals and Inequalities

Write an inequality to represent each of the following intervals:

(a) $(-3, 5]$ (b) $(-3, \infty)$ (c) $[0, 2]$

Solution:

(a) $(-3, 5]$ corresponds to $-3 < x \leq 5$.
(b) $(-3, \infty)$ corresponds to $-3 < x$.
(c) $[0, 2]$ corresponds to $0 \leq x \leq 2$.

It should be clear from our discussion of order that for any two real numbers a and b *exactly one* of the following is true:

$$a = b, \quad a < b, \quad \text{or} \quad a > b$$

This is referred to as the **law of trichotomy.** Some other important properties of inequalities are summarized as follows.

PROPERTIES OF INEQUALITIES

For all real numbers a, b, c, and d:

Property

1. **Transitive:**

   $a < b$ and $b < c \Rightarrow a < c$

2. **Addition of Inequalities:**

   $a < b$ and $c < d \Rightarrow a + c < b + d$

3. **Addition of a Constant:**

   $a < b \Rightarrow a + c < b + c$

4. **Multiplying by a Constant:**

   (i) For $c > 0$,

       $a < b \Rightarrow ac < bc$

   (ii) For $c < 0$,

       $a < b \Rightarrow ac > bc$

Example

Since $-2 < 5$ and $5 < 7$,
then $-2 < 7$.

Since $2 < 4$ and $3 < 5$,
then $2 + 3 < 4 + 5$.

Since $-3 < 7$,
then $-3 + 2 < 7 + 2$.

Since $5 > 0$ and $3 < 9$,
then $3(5) < 9(5)$.

Since $-5 < 0$ and $3 < 9$,
then $3(-5) > 9(-5)$.

Remark: Note in Property 4(ii) that we *reverse* the inequality when multiplying by a negative number. (This property is true for division as well. That is, an inequality is preserved in division by a positive number and is reversed in division by a negative number.) Note also that each of the four properties is true if $<$ is replaced by $\leq$.

We have compared two numbers using the order relations $<$ and $>$. Two numbers can also be compared using their **absolute value.** By the absolute value of a number, we mean its *magnitude (its value disregarding its sign).* We denote the absolute value of a by $|a|$. Thus, the absolute value of -5 is

$$|-5| = 5$$

DEFINITION OF ABSOLUTE VALUE

For any real number a:

$$|a| = \begin{cases} a, & \text{if } a \geq 0 \\ -a, & \text{if } a < 0 \end{cases}$$

The absolute value of a number can never be negative, for if a is negative ($a < 0$), then $-a$ is actually positive. For instance, let $a = -5$, then

$$|a| = |-5| = -a = -(-5) = 5$$

Absolute value is useful in finding the distance between two numbers on the real number line. To see how this is done, consider the numbers -3 and 4 shown in Figure 1.4.

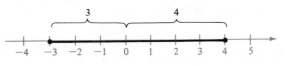

The distance between −3 and 4 is 7.

FIGURE 1.4

To find the distance between these two points, we subtract *either* number from the other and then take the absolute value of the resulting difference. Thus, using absolute value, we have

$$\text{distance} = |4 - (-3)| = |4 + 3| = |7| = 7$$

or

$$\text{distance} = |-3 - 4| = |-7| = 7$$

DISTANCE BETWEEN TWO POINTS ON THE REAL LINE

For any real numbers a and b:
The **distance between a and b** is

$$\text{distance} = d(a, b) = |b - a| = |a - b|$$

EXAMPLE 3
Distance and Absolute Value

Use absolute value to denote each of the following.

(a) The distance between $\sqrt{7}$ and 4
(b) The distance between c and -2 is at least 7
(c) The distance between x and 2.3 is less than 1
(d) x is closer to 0 than to -4

Solution:

(a) $d(\sqrt{7}, 4) = |4 - \sqrt{7}| = 4 - \sqrt{7} \approx 4 - 2.646 \approx 1.354$
(b) Since $d(c, -2) = |c + 2|$, we have

$$|c + 2| \geq 7$$

(c) Since $d(x, 2.3) = |x - 2.3|$, we have

$$|x - 2.3| < 1$$

(d) Since $d(x, 0) = |x - 0| = |x|$ and $d(x, -4) = |x - (-4)| = |x + 4|$, we have

$$|x| < |x + 4|$$

Section Exercises 1.3

In Exercises 1–10, locate the two real numbers on the real number line and place the appropriate inequality sign ($<$ or $>$) between them.

1. $\frac{3}{2}$, 7
2. -3.5, 1
3. π, -6
4. -8, $-\frac{25}{2}$
5. -4, -8
6. 1, $-\frac{16}{3}$
7. $\frac{5}{6}$, $\frac{2}{3}$
8. $-\frac{8}{7}$, $-\frac{3}{7}$
9. -1.75, -2.5
10. $-\frac{5}{4}$, 6

In Exercises 11–16, use the additive property of inequalities to combine each pair of inequalities into one inequality.

11. $-3 < 4$, $3 < 5$
12. $-\frac{7}{2} < -2$, $-4 < 2$
13. $3 > 0$, $2.5 > -2.5$
14. $-4 > -10$, $18 > 15$
15. $-2 < x$, $6 \leq y$
16. $-5 \geq b$, $-3 \geq c$

In Exercises 17–22, multiply each inequality by the given constant to produce a new inequality. (Be careful to reverse the inequality sign if the constant multiple is negative.)

17. $-6 < 3$;   2
18. $-6 > -10$; $\frac{1}{2}$
19. $-5 < x$; -3
20. $y > -6$; $-\frac{1}{3}$
21. $b > -5$; $-\frac{1}{2}$
22. $-\frac{2}{3} < a$; -6

In Exercises 23–30, complete the two missing descriptions of the given interval.

Interval Notation	Inequality Notation	Graph
23. _____	_____	
24. $(-\infty, -4]$	_____	_____
25. _____	$3 \leq x \leq \frac{11}{2}$	_____
26. $(-1, 7)$	_____	_____
27. _____	_____	
28. _____	$10 < x < \infty$	_____
29. $(\sqrt{2}, 8]$	_____	_____
30. _____	$\frac{1}{3} < x \leq \frac{22}{7}$	_____

In Exercises 31–36, use inequality notations to rewrite the given expression.

31. x is negative.

32. y is greater than 5 and less than or equal to 12.
33. Burt's age, A, is at least 30.
34. The yield, Y, is estimated to be no more than 45 bushels per acre.
35. The monthly rate of inflation, R, is expected to range from $R = \frac{1}{3}\%$ to $R = 1\%$.
36. The price, P, of unleaded gasoline is not expected to go above \$1.50 per gallon during the coming year.

In Exercises 37–44, find the distance between a and b.

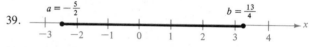

37.

38.

39.

40.

41. $a = 126$, $b = 75$
42. $a = -126$, $b = -75$
43. $a = 9.34$, $b = -5.65$
44. $a = \frac{16}{5}$, $b = \frac{112}{75}$

In Exercises 45–50, use absolute value notation to describe the given expression.

45. The distance between x and 5 is no more than 3.
46. The distance between x and -10 is at least 6.
47. The distance between z and $\frac{3}{2}$ is greater than 1.
48. The distance between z and 0 is less than 8.
49. y is closer to 0 than to 8.
50. y is at most 2 units from a.

Calculator Exercises

51. (a) Use a calculator to order the following real numbers starting with the smallest:

$$\frac{7071}{5000}, \quad \frac{584}{413}, \quad \sqrt{2}, \quad \frac{47}{33}, \quad \frac{127}{90}$$

(b) Which of the rational numbers in part (a) is closest to $\sqrt{2}$?

52. Use a calculator to order the following real numbers starting with the smallest:

$$\frac{26}{15}, \quad \sqrt{3}, \quad 1.73\overline{20}, \quad \frac{381}{220}, \quad \sqrt{10} - \sqrt{2}$$

Integral Exponents

1.4

In arithmetic, multiplication is sometimes described as repeated addition. That is,

$$5 \cdot 4 = 4 + 4 + 4 + 4 + 4 \qquad \text{or} \qquad 3 \cdot x = x + x + x$$

In this section, we look at *repeated* multiplications like

$$7 \cdot 7, \qquad a \cdot a \cdot a \cdot a \cdot a \cdot a, \qquad (-4)(-4)(-4),$$
$$\text{or} \qquad (2x)(2x)(2x)(2x)$$

The standard short forms for these multiplications are

$$7^2 = 7 \cdot 7$$
$$a^6 = a \cdot a \cdot a \cdot a \cdot a \cdot a$$
$$(-4)^3 = (-4)(-4)(-4)$$
$$(2x)^4 = (2x)(2x)(2x)(2x)$$

This short way to represent repeated multiplication is called **exponential form.** In general, we have

$$a^n = \underbrace{a \cdot a \cdot a \cdots a}_{n \text{ factors}}, \qquad\qquad n \text{ is a positive integer}$$

where n is called the **exponent** (or power) and a is called the **base.** We read a^n as ''a to the nth power'' or simply ''a to the n.'' For example, in 3^5, 5 is the exponent and 3 is the base. Thus, $3 \cdot 3 \cdot 3 \cdot 3 \cdot 3$ is 3^5 (3 to the fifth power).

Remark: It is important to recognize the difference between exponential forms like $(-2)^4$ and -2^4. In $(-2)^4$, the parentheses indicate that the power applies to the negative sign as well as the 2, but in $-2^4 = -(2^4)$, the power applies only to the 2. Similarly, in $(5x)^3$, the parentheses indicate that the power applies to the 5 as well as to the x, whereas in $5x^3 = 5(x^3)$, the power applies only to the x.

Multiplying Exponentials with Like Bases

To see what happens when we multiply two exponential expressions having the same (nonzero) base, note the following:

$$2^2 \cdot 2^5 = \underbrace{(2 \cdot 2)}_{2 \text{ factors}} \cdot \underbrace{(2 \cdot 2 \cdot 2 \cdot 2 \cdot 2)}_{5 \text{ factors}} = \underbrace{2 \cdot 2 \cdot 2 \cdot 2 \cdot 2 \cdot 2 \cdot 2}_{7 \text{ factors}}$$

that is,

$$2^2 \cdot 2^5 = 2^{2+5} = 2^7$$

This suggests that when multiplying exponential expressions with the same base, we *add* powers. In general, we have

$$a^m \cdot a^n = a^{m+n}$$

Dividing Exponentials with Like Bases

If we divide 2^5 by 2^2, we get

$$\frac{2^5}{2^2} = \frac{2 \cdot 2 \cdot 2 \cdot 2 \cdot 2}{2 \cdot 2} = 2 \cdot 2 \cdot 2 = 2^3 = 2^{5-2}$$

which suggests that when dividing exponential expressions, we *subtract* powers. That is, for $m > n$, we have

$$\frac{a^m}{a^n} = a^{m-n}, \qquad a \neq 0$$

Two special cases arise from this rule for dividing exponential expressions. First, if $m = n$, then

$$\frac{a^m}{a^n} = a^{m-n} = a^0 = 1, \qquad a \neq 0$$

and we say that *any nonzero number raised to the zero power is 1*. Second, if the power of the denominator is larger than the power of the numerator, then we have

$$\frac{2^2}{2^5} = \frac{2 \cdot 2}{2 \cdot 2 \cdot 2 \cdot 2 \cdot 2} = \frac{1}{2^3} = 2^{2-5} = 2^{-3}$$

This suggests that if n is a positive integer, then

$$\frac{1}{a^n} = a^{-n}, \qquad a \neq 0$$

These relationships can be summarized as follows. A formal proof of each rule can be given using the Principle of Mathematical Induction (see Section 12.4).

PROPERTIES OF EXPONENTS

For real numbers a, b, x, and y and for integers m and n (assuming that all bases and denominators are nonzero):

Property	Example
1. $a^m a^n = a^{m+n}$	$3^2 \cdot 3^4 = 3^{2+4} = 3^6$
2. $\dfrac{a^m}{a^n} = a^{m-n}$	$\dfrac{x^7}{x^4} = x^{7-4} = x^3$
3. $\dfrac{1}{a^n} = a^{-n}$	$\dfrac{1}{y^4} = y^{-4}$

4. $a^0 = 1$ $(x^2 + 1)^0 = 1$

5. $(ab)^m = a^m b^m$ $(5x)^4 = 5^4 x^4$

6. $(a^m)^n = a^{mn}$ $(y^3)^{-4} = y^{3(-4)} = y^{-12} = \dfrac{1}{y^{12}}$

7. $\left(\dfrac{a}{b}\right)^m = \dfrac{a^m}{b^m}$ $\left(\dfrac{2}{x}\right)^3 = \dfrac{2^3}{x^3}$

The preceding rules hold for *all* integers m and n (not just positive ones). For instance, by Property 2, we have

$$\frac{3^4}{3^{-5}} = 3^{4-(-5)} = 3^{4+5} = 3^9$$

As another example, we can use Properties 6 and 7 followed by Property 3 to obtain

$$\left(\frac{4x^{-2}y^3}{5}\right)^2 = \frac{4^2 x^{-4} y^6}{5^2} = \frac{16y^6}{25x^4}$$

When evaluating expressions having negative exponents, you can avoid errors by converting to positive exponents before evaluating.

EXAMPLE 1
Simplifying with
Negative Exponents

Rewrite each of the following with positive exponents and simplify.

(a) $(-3b^4)(5b^2)\left(\dfrac{1}{9} b^{-7}\right)$ (b) $(-2a^2 b^{-1})^3$

(c) $\dfrac{(2x^3)^{-1}(7x^{-4})^0}{4x^{-2}}$ (d) $\left(\dfrac{y^{-1}}{3x^{-2}}\right)\left(\dfrac{3x^2}{y}\right)^{-2}$

Solution:

(a) $(-3b^4)(5b^2)\left(\dfrac{1}{9} b^{-7}\right) = \left(-\dfrac{15}{9}\right)(b^4)(b^2)(b^{-7})$ *Multiply coefficients*

$$= -\frac{15}{9} b^{4+2-7}$$ *Property 1*

$$= -\frac{15}{9} b^{-1}$$

$$= -\frac{15}{9b} = -\frac{5}{3b}$$ *Property 3*

(b) $(-2a^2 b^{-1})^3 = (-2)^3 (a^2)^3 (b^{-1})^3$ *Property 5*

$$= -8a^6 b^{-3}$$ *Property 6*

$$= -\frac{8a^6}{b^3}$$ *Property 3*

(c) Note how negative exponents can be converted to positive exponents by simply shifting the *factors* with negative exponents from numerator to denominator or vice versa.

$$\frac{(2x^3)^{-1}(7x^{-4})^0}{4x^{-2}} = \frac{(2x^3)^{-1}(1)}{4x^{-2}} = \frac{x^2}{4(2x^3)} \qquad \textit{Shift factors}$$

$$= \frac{x^2}{8x^3} = \frac{1}{8x} \qquad \textit{Subtract exponents}$$

(d) $\left(\dfrac{y^{-1}}{3x^{-2}}\right)\left(\dfrac{3x^2}{y}\right)^{-2} = \left(\dfrac{x^2}{3y}\right)\left(\dfrac{y}{3x^2}\right)^2$

$$= \left(\dfrac{x^2}{3y}\right)\left(\dfrac{y^2}{9x^4}\right) = \dfrac{x^2y^2}{27x^4y} = \dfrac{y}{27x^2}$$

Remark: Notice that in part (d) of Example 1 *fractions* raised to negative powers were simplified by inverting the fraction and changing the sign of the exponent.

Scientific Notation

Exponents provide an efficient way of writing and computing with the very large (or very small) numbers used in science. For instance, a drop of water contains more than 33 billion billion molecules. That's 33 followed by 18 zeros:

33,000,000,000,000,000,000

It is convenient to write such numbers in **scientific notation.** This notation has the form

$$c \times 10^n$$

where $1 \le c < 10$ and n is an integer. Thus, the number of molecules in a drop of water can be conveniently written in scientific notation as

$$3.3 \times 10,000,000,000,000,000,000 = 3.3 \times 10^{19}$$

The *positive* exponent 19 indicates that the number is very *large* and that the decimal point should be moved 19 places to the *right*. A *negative* exponent in scientific notation indicates that the number is *small* (< 1) and that the decimal point should be moved to the *left*. For instance, the mass (in grams) of one electron is approximately

$$9 \times 10^{-28} = 0.00000 \cdots 09$$

$$\underbrace{\qquad\qquad\qquad}_{\text{28 decimal places}}$$

EXAMPLE 2
Scientific Notation

Convert the following numbers to the indicated form.

Decimal Form	**Scientific Form**
(a) [　　　　　]	1.345×10^2
(b) 0.0000782	[　　　　　]
(c) [　　　　　]	9.36×10^{-6}
(d) 836,100,000	[　　　　　]

Solution:

(a) Since the exponent is positive, we move the decimal point to the right to get

$$1.345 \times 10^2 = 134.5$$

2 places

(b) The number 0.0000782 is very small. Hence the exponent is negative and we write

$$0.0000782 = 7.82 \times 10^{-5}$$

5 places

(c) Here again the negative exponent indicates a small number and we write

$$9.36 \times 10^{-6} = 0.00000936$$

6 places

(d) In this case, $c = 8.361$ and we write

$$836,100,000 = 8.361 \times 10^8$$

8 places

Exponents and the Calculator

Most scientific calculators automatically switch their display to scientific notation when computing with large (or small) numbers that exceed the display range. Try multiplying

$$98,900,000 \times 500$$

If your calculator follows standard conventions, its display should be

$$\boxed{4.945 \quad 10}$$

This means that $c = 4.945$ and the exponent is $n = 10$.

To allow you to *enter* numbers in scientific notation, your calculator should have an exponential entry key labeled $\boxed{EE}$ or $\boxed{EXP}$. If you wanted to perform the preceding multiplication using scientific notation, you could begin by writing

$$98,900,000 \times 500 = (9.89 \times 10^7)(5.0 \times 10^2)$$

and then enter

9.89 $\boxed{\text{EE}}$ 7 $\boxed{\times}$ 5 $\boxed{\text{EE}}$ 2 $\boxed{=}$

Remark: A number such as 10^{15} is entered

1 $\boxed{\text{EE}}$ 15

and *not* just

$\boxed{\text{EE}}$ 15

This latter sequence would be interpreted as 0×10^{15}, which is zero.

The Exponential Key

Scientific calculators are capable of evaluating exponential expressions via the keys $\boxed{y^x}$ and $\boxed{e^x}$. The second of these two keys will be discussed in Chapter 6. To use the first key, remember that y is the *base* and x is the *exponent*. Thus, to calculate 3^6, we can enter

	Display
3 $\boxed{y^x}$ 6 $\boxed{=}$	729

Similarly, to calculate

$$\left(1 + \frac{0.09}{12}\right)^{12}$$

we enter

	Display
.09 $\boxed{\div}$ 12 $\boxed{+}$ 1 $\boxed{=}$ $\boxed{y^x}$ 12 $\boxed{=}$	1.0938069

base   exponent

Negative exponents are entered into a calculator by pressing the change-sign key $\boxed{+/-}$ immediately after entering the exponent. For instance, to enter the numbers

$$3.753 \times 10^{-4} \quad \text{and} \quad 27^{-6}$$

we use the following sequences

	Display
3.753 $\boxed{\text{EE}}$ 4 $\boxed{+/-}$ $\boxed{=}$	3.753 -04

and

	Display
27 $\boxed{y^x}$ 6 $\boxed{+/-}$ $\boxed{=}$	2.5812 -09

Section Exercises 1.4

In Exercises 1–10, fill in the missing descriptions.

Exponential Form	Base	Exponent	Factored Form
1. 4^3	___	___	_____
2. $(x-1)^2$	___	___	_____
3. ___	5	4	_____
4. ___	___	___	$(-5)(-5)(-5)(-5)(-5)$
5. $(2x)^4$	___	___	_____
6. $2x^4$	___	___	_____
7. ___	___	___	$(x^2+4)(x^2+4)(x^2+4)$
8. ___	$-x$	3	_____
9. ___	___	___	$\left(\dfrac{x}{y}\right)\left(\dfrac{x}{y}\right)\left(\dfrac{x}{y}\right)\left(\dfrac{x}{y}\right)\left(\dfrac{x}{y}\right)\left(\dfrac{x}{y}\right)$
10. $\left(\dfrac{8h}{3}\right)^4$	___	___	_____

In Exercises 11–16, evaluate each expression for the given value of x.

11. $-3x^3 \quad (x=2)$

12. $\dfrac{x^2}{2} \quad (x=6)$

13. $4x^{-3} \quad (x=2)$

14. $7x^{-2} \quad (x=4)$

15. $6x^0 - (6x)^0 \quad (x=10)$

16. $5(-x)^3 \quad (x=3)$

In Exercises 17–40, simplify the given expression.

17. $5x^4(x^2)$

18. $(8x^4)(2x^3)$

19. $6y^2(2y^4)^2$

20. $z^{-3}(3z^4)$

21. $10(x^2)^2$

22. $(4x^3)^2$

23. $\dfrac{7x^2}{x^{-3}}$

24. $\dfrac{r^4}{r^6}$

25. $\dfrac{12(x+y)^3}{9(x+y)}$

26. $(2x^2yz^5)^0$

27. $(-2x^2)^3(4x^3)^{-1}$

28. $\dfrac{(4y^{-2})(8y^4)}{6y^2}$

29. $\left(\dfrac{3z^2}{x}\right)^{-2}$

30. $\left(\dfrac{x^{-3}y^4}{5}\right)^{-3}$

31. $(4a^{-2}b^3)^{-3}$

32. $(5x^2y^4z^6)^3(5x^2y^4z^6)^{-3}$

33. $(2x^2+y^2)^4(2x^2+y^2)^{-4}$

34. $[(x^2y^{-2})^{-1}]^{-1}$

35. $\left(\left(\dfrac{y^2}{x^2}\right)^{-1}\right)^2$

36. $\left(\dfrac{4x^{-2}}{3}\right)^{-3}$

37. $\left(\dfrac{a^{-2}}{b^{-2}}\right)\left(\dfrac{b}{a}\right)^3$

38. $(-3x^2)(4x^{-3})(\tfrac{1}{6}x)$

39. $\left(\dfrac{10x^{-2}y^3}{x^4}\right)^0$

40. $(-2r^2s^3u^{-1})^{-3}$

41. Change each of the following from decimal form to scientific notation.
   (a) 93,000,000   (b) 900,000,000   (c) 0.00000435

42. Change each of the following from decimal form to scientific notation.
   (a) 0.000087   (b) 6.87   (c) 0.004392

43. Change each of the following from scientific notation to decimal form.
   (a) 1.91×10^6 (b) 2.345×10^{11}
   (c) 6.21×10^0 (d) 8.52×10^{-3}
   (e) 7.021×10^{-5} (f) 3.798×10^{-8}

44. Change each of the following from scientific notation to decimal form.
   (a) 2.65×10^7 (b) 9.4675×10^4
   (c) 3.0025×10^8 (d) 1.0909×10^{-4}
   (e) 3.2×10^{-7} (f) 4.6666×10^{-5}

Calculator Exercises

In Exercises 45–51, use a calculator to perform the indicated calculations.

45. (a) $2400(1 + 0.06)^{20}$

   (b) $750\left(1 + \dfrac{0.11}{365}\right)^{800}$

   (c) $\dfrac{(2.414 \times 10^4)^6}{(1.68 \times 10^5)^5}$

   (d) $(9.3 \times 10^6)^3(6.1 \times 10^{-4})^4$

46. (a) $\dfrac{3000}{[1 + (0.05/4)]^4}$

   (b) $\dfrac{4 - 1.25^6}{1 - 0.625^4}$

   (c) $\dfrac{(3.28 \times 10^{-6})^{10}}{(5.34 \times 10^{-3})^{25}}$

   (d) $(2.52 \times 10^4)^5(1.63 \times 10^{-3})^7$

47. (a) $(0.000345)(8,980,000,000)$

   (b) $\dfrac{67,000,000 + 93,000,000}{0.0052}$

48. (a) $\dfrac{848,000,000}{1,624,000}$

   (b) $\dfrac{0.0000928 - 0.0000021}{0.0061}$

49. The speed of light is 11,160,000 miles per minute. The distance from the sun to the earth is 93,000,000 miles. Find the time it takes for light to travel from the sun to the earth.

50. The *per capita public debt* is defined as the gross debt divided by the population. Find the per capita debt of the United States in 1980 if the gross debt was 839.2 billion dollars and the population was 220 million.

51. The amount A after t years in a savings account earning r percent interest compounded n times per year is

$$A = P\left(1 + \frac{r}{n}\right)^{nt}$$

where P is the (original) principal. Complete the following table for $500 deposited in an account earning 12% compounded daily. (Note that 12% interest implies that $r = 0.12$.)

t	5	10	20	30	40	50
A						

(The key sequence for programming a Texas Instruments programmable calculator is

| LRN | × | 365 | = |

| STO | 1.12 | ÷ | 365 | = |

| + | 1 | = | y^x | RCL | 1 | × | 500 | = |

| R/S | RST | LRN | RST |

If you enter the time in years and press the run/start key | R/S |, the calculator will display the value for A.)

Rational Exponents and Radicals   1.5

Many problems in science, business, and economics involve equations with exponential forms like

$$r^2 = 49, \qquad s^3 = 125, \qquad \text{or} \qquad (1 + r)^{48} = 97$$

Solutions to such equations involve the finding of **roots** of numbers. For example, we say:

If $x^2 = 49$,   then x is the **square root** of 49.

If $x^3 = 125$, then x is the **cube root** of 125.

In general, we have the following definition of an nth root of a number.

DEFINITION OF nth ROOT

> For nonnegative real numbers a and b and positive integer n:
> a is an nth root of b if $a^n = b$

Remark: In the special case where n is an *odd* positive integer, we extend the definition of nth root to include negative values for a and b.

In denoting roots of numbers, we use the **radical** form $\sqrt[n]{}$ except for $n = 2$, which is simply written as $\sqrt{}$.

DEFINITION OF $\sqrt[n]{b}$

> For a and b nonnegative and n a positive integer *or* for a and b negative and n an odd positive integer:
>
> $$a = \sqrt[n]{b} \quad \text{if and only if} \quad a^n = b$$
>
> The radical, $\sqrt[n]{b}$, is called the **principal nth root of b,** n is called the **index** of the radical, and b is called the **radicand.**

Remark: It is important that you understand the significance of the restrictions placed on the numbers a, b, and n in this definition. For instance, if $b = -9$ ($b < 0$) and $n = 2$, then there is *no real* number a such that $a^2 = -9$, since the square of every real number is nonnegative. (In Section 1.6 we show how even roots of negative numbers fit into the set of *complex* numbers.)

Odd roots of negative numbers are real numbers. For example,

$$\sqrt[3]{-64} = -4 \text{ because } (-4)^3 = (-4)(-4)(-4) = -64$$

$$\sqrt[5]{-\frac{1}{32}} = -\frac{1}{2} \text{ because } \left(-\frac{1}{2}\right)^5 = \left(-\frac{1}{2}\right)\left(-\frac{1}{2}\right)\left(-\frac{1}{2}\right)\left(-\frac{1}{2}\right)\left(-\frac{1}{2}\right)$$

$$= \left(-\frac{1}{32}\right)$$

For even roots of positive numbers, we obtain more than one answer. For example, since $5^2 = 25$ and $(-5)^2 = (-5)(-5) = 25$, we conclude that *both* $5 = \sqrt{25}$ and $-5 = -\sqrt{25}$ are *square roots* of 25. In general, if n is even and b positive, then there are *two* real nth roots of b, namely

$$a = \sqrt[n]{b} \qquad \qquad \text{(principal nth root of b)}$$

and

$$-a = -\sqrt[n]{b} \qquad \qquad \text{(negative nth root of b)}$$

In summary, we have the following situation with respect to nth roots of a number.

REAL ROOTS OF $a^n = b$

Condition	Roots	Examples
n even, $b > 0$	$a = \sqrt[n]{b},\ -a = -\sqrt[n]{b}$	$a^2 = 5 \Rightarrow a = \sqrt{5},\ -a = -\sqrt{5}$
n even, $b < 0$	no real roots	$a^4 = -10 \Rightarrow a$ is not real
n odd, $b > 0$	$a = \sqrt[n]{b} > 0$	$a^3 = 27 \Rightarrow a = \sqrt[3]{27} = 3$
n odd, $b < 0$	$a = \sqrt[n]{b} < 0$	$a^3 = -27 \Rightarrow a = \sqrt[3]{-27} = -3$

EXAMPLE 1
Roots and Radicals

(a) $\sqrt{81} = 9$ because $81 = 9^2$.

(b) $\sqrt{81} \neq -9$ because $\sqrt{81}$ denotes the *principal* square root, which is always nonnegative.

(c) $-\sqrt{81} = -9$ because $(-9)^2 = 81$.

(d) $\sqrt[3]{-125} = -5$ because $(-5)^3 = -125$.

(e) $\sqrt[4]{-16} \neq -2$ because $(-2)^4 = 16 \neq -16$.

(f) $-\sqrt[3]{-125} = 5$ because $-\sqrt[3]{-125} = -(-5) = 5$.

Remark: It is *incorrect* to write

$$\sqrt{81} = \pm 9 \qquad\qquad \textit{Common mistake}$$

The radical $\sqrt{81}$ denotes only the principal (positive) square root of 81. We use $-\sqrt{81}$ to denote the negative square root of 81.

EXAMPLE 2
Evaluating Radicals

Evaluate the following radicals.

(a) $\sqrt[5]{-32}$ (b) $\sqrt[3]{\dfrac{125}{64}}$ (c) $-\sqrt{\dfrac{1}{121}}$

Solution:

(a) $\sqrt[5]{-32} = -2$ because $(-2)^5 = -32$.

(b) $\sqrt[3]{\dfrac{125}{64}} = \dfrac{5}{4}$ because $\left(\dfrac{5}{4}\right)^3 = \dfrac{5^3}{4^3} = \dfrac{125}{64}$.

(c) $-\sqrt{\dfrac{1}{121}} = -\dfrac{1}{11}$ because $-\left(\dfrac{1}{11}\right)^2 = -\left(\dfrac{1}{121}\right)$.

Rational Exponents

Some roots of numbers raised to powers, like

$$\sqrt[3]{2^6} \qquad \text{and} \qquad \sqrt[6]{8^4}$$

can be evaluated more easily by writing the radicals in exponential form using *rational exponents*.

DEFINITION OF RATIONAL EXPONENTS

> For integer m, natural number n, and real number b such that $\sqrt[n]{b}$ exists:
>
> $$b^{m/n} = (\sqrt[n]{b})^m = \sqrt[n]{b^m}$$

Remark: Note that in this definition the denominator is the *index* for the corresponding radical form and the numerator is the *power* of the radical (or radicand).

$$\overbrace{\underbrace{b^{m/n}}_{} = (\sqrt[n]{b})^m = \sqrt[n]{b^m}}$$

with labels: power, index

Two important special cases arise from the definition of rational exponents.

1. If $m = 1$, then
$$b^{1/n} = \sqrt[n]{b} = \text{principal } n\text{th root of } b$$

2. If $m = n$, then
$$b^{n/n} = \sqrt[n]{b^n} = b \qquad\qquad (n \text{ is odd})$$
$$b^{n/n} = \sqrt[n]{b^n} = |b| \qquad\qquad (n \text{ is even})$$

Note the difference between the parts of the second case, particularly the occurrence of $|b|$ when n is even. For instance,

$$\sqrt[4]{(-6)^4} = |-6| = 6 \qquad \text{and} \qquad \sqrt{x^2} = |x|$$

EXAMPLE 3
Changing from Radical to Exponential Form

Write each of the following in exponential form.

(a) $\sqrt{(3xy)^5}$ (b) $\sqrt[3]{2a - 5b}$ (c) $2x\sqrt[4]{x^3}$ (d) $\dfrac{1}{\sqrt[5]{a^4}}$

Solution:

(a) In this case the base is $(3xy)$, the power is 5, and the index is 2. Thus
$$\sqrt{(3xy)^5} = \sqrt[2]{(3xy)^5} = (3xy)^{5/2}$$

(b) $\sqrt[3]{2a - 5b} = \sqrt[3]{(2a - 5b)^1} = (2a - 5b)^{1/3}$

(c) $2x\sqrt[4]{x^3} = (2x)(x^{3/4}) = 2x^{1+(3/4)} = 2x^{7/4}$

(d) $\dfrac{1}{\sqrt[5]{a^4}} = \dfrac{1}{a^{4/5}} = a^{-4/5}$

EXAMPLE 4
Changing from Exponential to Radical Form

Write each of the following in radical form.

(a) $(x^2 + y^2)^{3/2}$ (b) $2y^{3/4}z^{1/4}$ (c) $a^{-3/2}$ (d) $(4x^2y^{-3})^{2/3}$

Solution:

(a) In this case the index is 2 and the power is 3, so we write
$$(x^2 + y^2)^{3/2} = (\sqrt{x^2 + y^2})^3 = \sqrt{(x^2 + y^2)^3}$$

(b) Note that the factor 2 is not part of the base, so we write
$$2y^{3/4}z^{1/4} = 2(y^3z)^{1/4} = 2\sqrt[4]{y^3z}$$

(c) We first convert to positive exponents, then to radical form as follows:

$$a^{-3/2} = \frac{1}{a^{3/2}} = \frac{1}{\sqrt{a^3}}$$

(d) $(4x^2y^{-3})^{2/3} = \sqrt[3]{(4x^2y^{-3})^2} = \sqrt[3]{4^2x^4y^{-6}} = \sqrt[3]{\dfrac{16x^4}{y^6}}$

Since radicals can be written in exponential form using rational exponents, it follows that radicals possess properties similar to those of exponents. Compare the following list of properties with those given in the previous section.

PROPERTIES OF RADICALS

For integer m, natural number n, and real numbers a and b such that $\sqrt[n]{a}$ and $\sqrt[n]{b}$ are real:

Property	Example
1. $\sqrt[n]{a^m} = (\sqrt[n]{a})^m = a^{m/n}$	$\sqrt[3]{8^2} = (\sqrt[3]{8})^2 = 8^{2/3}$
2. $\sqrt[n]{a} \cdot \sqrt[n]{b} = \sqrt[n]{ab}$	$\sqrt{5} \cdot \sqrt{7} = \sqrt{5 \cdot 7} = \sqrt{35}$
3. $\dfrac{\sqrt[n]{a}}{\sqrt[n]{b}} = \sqrt[n]{\dfrac{a}{b}} \quad (b \neq 0)$	$\dfrac{\sqrt[4]{27}}{\sqrt[4]{9}} = \sqrt[4]{\dfrac{27}{9}} = \sqrt[4]{3}$
4. $\sqrt[m]{\sqrt[n]{a}} = \sqrt[mn]{a}$	$\sqrt[3]{\sqrt{10}} = \sqrt[6]{10}$
5. $(\sqrt[n]{a})^n = a$	$(\sqrt[6]{15})^6 = 15$
6. For n even, $\sqrt[n]{a^n} = |a|$	$\sqrt{(-12)^2} = |-12| = 12$
For n odd, $\sqrt[n]{a^n} = a$	$\sqrt[3]{(-12)^3} = -12$

Simplifying Radicals

An expression involving radicals is in **simplest form** when the following conditions are satisfied.

1. All possible factors have been removed from under the radical sign.
2. The index for the radical has been reduced as far as possible.
3. All fractions have radical-free denominators (accomplished by a process called *rationalizing the denominator*).

EXAMPLE 5
Removing Factors from Under a Radical Sign

Simplify the following radical expressions.

(a) $\sqrt{75x^3}$ (b) $\sqrt[3]{24a^4c^8}$ (c) $\sqrt[4]{(5x^2y^{-3})^4}$

(d) $\sqrt[5]{\dfrac{4r^3}{s^4}} \sqrt[5]{\dfrac{16r^6}{s^3}}$ (e) $\sqrt{\sqrt[3]{a^{10}b}}$

Solution:

(a) $\sqrt{75x^3} = \sqrt{25(x^2)(3x)}$ *Find largest square factors*

$\qquad\qquad = \sqrt{25}\sqrt{x^2}\sqrt{3x}$ *Property 2 of radicals*

$\qquad\qquad = 5x\sqrt{3x}$ *Find roots of perfect squares*

(b) $\sqrt[3]{24a^4c^8} = \sqrt[3]{(8)(a^3)(c^6)(3ac^2)}$ *Find largest cube factors*

$\qquad\qquad = \sqrt[3]{8}\sqrt[3]{a^3}\sqrt[3]{c^6}\sqrt[3]{3ac^2}$ *Property of 2 radicals*

$\qquad\qquad = 2ac^2\sqrt[3]{3ac^2}$ *Find roots of perfect cubes*

(c) Since n is even, we have (by Property 6)

$$\sqrt[4]{(5x^2y^{-3})^4} = |5x^2y^{-3}| = \left|\frac{5x^2}{y^3}\right|$$

(d) $\sqrt[5]{\dfrac{4r^3}{s^4}}\ \sqrt[5]{\dfrac{16r^6}{s^3}} = \sqrt[5]{\dfrac{64r^9}{s^7}} = \sqrt[5]{\left(\dfrac{32r^5}{s^5}\right)\left(\dfrac{2r^4}{s^2}\right)} = \dfrac{2r}{s}\sqrt[5]{\dfrac{2r^4}{s^2}}$

(e) $\sqrt{\sqrt[3]{a^{10}b}} = \sqrt[6]{a^{10}b} = \sqrt[6]{(a^6)(a^4b)} = |a|\sqrt[6]{a^4b}$

EXAMPLE 6
Reducing the Index of a Radical

Simplify the following radicals, reducing the index when possible.

(a) $\sqrt[6]{a^4}$ (b) $\sqrt[3]{\sqrt{125}}$ (c) $\sqrt[4]{27xy^3}\ \sqrt[4]{\dfrac{3y^3}{x^3}}$

Solution:

(a) $\sqrt[6]{a^4} = a^{4/6}$ *Write with rational exponents*

$\qquad = a^{2/3}$ *Reduce exponent*

$\qquad = \sqrt[3]{a^2}$ *Rewrite in radical form*

(b) $\sqrt[3]{\sqrt{125}} = \sqrt[6]{125} = \sqrt[6]{5^3} = 5^{3/6} = 5^{1/2} = \sqrt{5}$

(c) $\sqrt[4]{27xy^3}\ \sqrt[4]{\dfrac{3y^3}{x^3}} = \sqrt[4]{\dfrac{81xy^6}{x^3}}$ *Multiply radicals*

$\qquad\qquad\qquad = \sqrt[4]{(3^4y^4)\dfrac{y^2}{x^2}}$ *Find largest 4th powers*

$\qquad\qquad\qquad = 3y\sqrt[4]{\dfrac{y^2}{x^2}}$ *Find 4th root*

$\qquad\qquad\qquad = (3y)\dfrac{y^{1/2}}{x^{1/2}}$ *Write with fractional exponents*

$\qquad\qquad\qquad = 3y\sqrt{\dfrac{y}{x}}$ *Rewrite in radical form*

Rationalizing Radicals

Our third simplification technique involves a *rationalizing* procedure that removes radicals from either the numerator or the denominator of a fraction. In algebra we usually emphasize rationalizing the denominator. However, in calculus it is also helpful to be able to rationalize the numerator. In both instances, we make use of the form $a + b\sqrt{m}$ and its **conjugate** $a - b\sqrt{m}$. Note that the product of this conjugate pair has no radical:

$$(a + b\sqrt{m})(a - b\sqrt{m}) = a^2 - b^2m$$

EXAMPLE 7
Rationalizing the Denominator

Eliminate the radical(s) in the denominator of each of the following.

(a) $\dfrac{5}{2\sqrt{3}}$ (b) $\dfrac{2}{\sqrt[3]{5x}}$ (c) $\dfrac{\sqrt{3} + \sqrt{x}}{\sqrt{3} - \sqrt{x}}$ (d) $\sqrt{\dfrac{5y^3}{6x^5}}$

Solution:

(a) $\dfrac{5}{2\sqrt{3}} = \dfrac{5}{2\sqrt{3}} \cdot \dfrac{\sqrt{3}}{\sqrt{3}}$ *Multiply by $\sqrt{3}/\sqrt{3}$*

$\quad = \dfrac{5\sqrt{3}}{2(3)} = \dfrac{5\sqrt{3}}{6}$

(b) $\dfrac{2}{\sqrt[3]{5x}} = \dfrac{2}{\sqrt[3]{5x}} \cdot \dfrac{\sqrt[3]{(5x)^2}}{\sqrt[3]{(5x)^2}} = \dfrac{2\sqrt[3]{(5x)^2}}{\sqrt[3]{(5x)^3}} = \dfrac{2\sqrt[3]{25x^2}}{5x}$

(c) $\dfrac{\sqrt{3} + \sqrt{x}}{\sqrt{3} - \sqrt{x}} = \dfrac{\sqrt{3} + \sqrt{x}}{\sqrt{3} - \sqrt{x}} \cdot \dfrac{\sqrt{3} + \sqrt{x}}{\sqrt{3} + \sqrt{x}}$ *Multiply by conjugate*

$\quad = \dfrac{(\sqrt{3})^2 + 2\sqrt{3}\sqrt{x} + (\sqrt{x})^2}{(\sqrt{3})^2 - (\sqrt{x})^2}$

$\quad = \dfrac{3 + 2\sqrt{3x} + x}{3 - x}$

(d) $\sqrt{\dfrac{5y^3}{6x^5}} = \sqrt{\dfrac{5y^3}{6x^5} \cdot \dfrac{6x}{6x}}$ *Create square in denominator*

$\quad = \sqrt{\dfrac{(30xy)(y^2)}{36x^6}}$

$\quad = \dfrac{y\sqrt{30xy}}{6x^3}$ *Find square roots*

EXAMPLE 8
Rationalizing the Numerator

Eliminate the radicals in the numerator of each of the following.

(a) $\dfrac{\sqrt{5} - \sqrt{7}}{\sqrt{4}}$ (b) $\dfrac{\sqrt{x + h} - \sqrt{x}}{h}$

Solution:

(a) $\dfrac{\sqrt{5} - \sqrt{7}}{\sqrt{4}} = \dfrac{\sqrt{5} - \sqrt{7}}{\sqrt{4}} \cdot \dfrac{\sqrt{5} + \sqrt{7}}{\sqrt{5} + \sqrt{7}}$ *Multiply by conjugate*

$$= \dfrac{5 - 7}{\sqrt{4}(\sqrt{5} + \sqrt{7})}$$

$$= \dfrac{-2}{2(\sqrt{5} + \sqrt{7})}$$ *Simplify*

$$= \dfrac{-1}{\sqrt{5} + \sqrt{7}}$$

(b) $\dfrac{\sqrt{x + h} - \sqrt{x}}{h} = \dfrac{\sqrt{x + h} - \sqrt{x}}{h} \cdot \dfrac{\sqrt{x + h} + \sqrt{x}}{\sqrt{x + h} + \sqrt{x}}$

$$= \dfrac{(x + h) - (x)}{h(\sqrt{x + h} + \sqrt{x})} = \dfrac{h}{h(\sqrt{x + h} + \sqrt{x})}$$

$$= \dfrac{1}{\sqrt{x + h} + \sqrt{x}}$$

Although the sum or difference of two radicals can often be simplified by multiplying by its conjugate, it is not so easy to simplify a sum or difference that is part of a single radicand. In particular, note that

$$\sqrt{a + b} \quad \text{DOES NOT EQUAL} \quad \sqrt{a} + \sqrt{b}$$

For example,

$$\sqrt{16 + 9} = \sqrt{25} = 5$$

whereas

$$\sqrt{16} + \sqrt{9} = 4 + 3 = 7$$

Watch out for the following version of this common error:

$$\sqrt{x^2 + y^2} \quad \text{DOES NOT EQUAL} \quad x + y$$

EXAMPLE 9
Combining Radicals

Simplify and combine terms.

(a) $\sqrt{12} - 3\sqrt{27} + 2\sqrt{48}$ (b) $\sqrt[3]{2xyz} + 2\sqrt[3]{54xyz} - \sqrt[3]{\dfrac{xyz}{4}}$

Solution:

(a) $\sqrt{12} - 3\sqrt{27} + 2\sqrt{48}$

$\quad = \sqrt{4 \cdot 3} - 3\sqrt{9 \cdot 3} + 2\sqrt{16 \cdot 3}$ *Find square factors*

$\quad = 2\sqrt{3} - 9\sqrt{3} + 8\sqrt{3}$ *Find square roots*

$\quad = (2 - 9 + 8)\sqrt{3} = \sqrt{3}$ *Combine like terms*

(b) $\sqrt[3]{2xyz} + 2\sqrt[3]{54xyz} - \sqrt[3]{\dfrac{xyz}{4}} = \sqrt[3]{2xyz} + 2\sqrt[3]{27(2xyz)} - \sqrt[3]{\dfrac{2xyz}{8}}$

$$= \sqrt[3]{2xyz} + 6\sqrt[3]{2xyz} - \frac{1}{2}\sqrt[3]{2xyz}$$

$$= \frac{13}{2}\sqrt[3]{2xyz}$$

Radicals and Calculators

There are two methods of evaluating radicals on most calculators. For square roots, you use the *square root key* $\boxed{\sqrt{\ }}$. For other roots, you should first convert the radical to exponential form and then use the *exponential key* $\boxed{y^x}$.

EXAMPLE 10
Evaluating Radicals
with a Calculator

Use a calculator to evaluate each of the following radicals.

(a) $\sqrt[3]{56}$ (b) $\sqrt[3]{-4}$ (c) $(1.2)^{-1/6}$

Solution:

(a) First, we write $\sqrt[3]{56}$ in exponential form:

$$\sqrt[3]{56} = 56^{1/3}$$

Now, there are several options:

Calculator Steps

(i) $1\ \boxed{\div}\ 3\ \boxed{=}\ \boxed{\text{STO}}\ 56\ \boxed{y^x}\ \boxed{\text{RCL}}\ \boxed{=}$ *Use memory key*

(ii) $56\ \boxed{y^x}\ \boxed{(}\ 1\ \boxed{\div}\ 3\ \boxed{)}\ \boxed{=}$ *Use parentheses*

(iii) $56\ \boxed{y^x}\ 3\ \boxed{1/x}\ \boxed{=}$ *Use reciprocal key*

For each of these three keystroke sequences, the answer is

$$\sqrt[3]{56} \approx 3.8258624$$

(b) Since

$$\sqrt[3]{-4} = \sqrt[3]{(-1)(4)} = \sqrt[3]{-1}\cdot\sqrt[3]{4} = -\sqrt[3]{4}$$

we can attach the negative sign of the radicand at the end of the keystroke sequence as follows:

Calculator Steps   **Display**

$4\ \boxed{y^x}\ \boxed{(}\ 1\ \boxed{\div}\ 3\ \boxed{)}\ \boxed{=}\ \boxed{+/-}$ -1.5874011

(c) **Calculator Steps**   **Display**

$1.2\ \boxed{y^x}\ \boxed{(}\ 1\ \boxed{\div}\ 6\ \boxed{+/-}\ \boxed{)}\ \boxed{=}$ $.97007012$

Section Exercises 1.5

In Exercises 1–12, complete the two missing descriptions.

$\sqrt[n]{b^m} = a$	$b^{m/n} = a$	$a^{n/m} = b$
1. $\quad\sqrt{9} = 3$	_____	_____
2. $\quad\sqrt[3]{64} = 4$	_____	_____
3. _____	$32^{1/5} = 2$	_____
4. _____	$-(144^{1/2}) = -12$	_____
5. _____	_____	$14^2 = 196$
6. _____	_____	$8.5^3 = 614.125$
7. $\sqrt[3]{-216} = -6$	_____	_____
8. _____	_____	$-3^5 = -243$
9. _____	$27^{2/3} = 9$	_____
10. $(\sqrt[4]{81})^3 = 27$	_____	_____
11. $\sqrt[4]{81^3} = 27$	_____	_____
12. _____	_____	$16^{5/4} = 32$

In Exercises 13–20, evaluate each radical without using a calculator.

13. (a) $\sqrt{\frac{9}{4}}$
    (b) $\sqrt[3]{\frac{27}{8}}$
14. (a) $-\sqrt{25}$
    (b) $-\sqrt[3]{-27}$
15. (a) $16^{3/2}$
    (b) $8^{2/3}$
16. (a) $(\sqrt[4]{16})^3$
    (b) $(\sqrt[3]{-125})^3$
17. (a) $(\sqrt[6]{326})^6$
    (b) $\sqrt[4]{562^4}$
18. (a) $121^{-1/2}$
    (b) $32^{-3/5}$
19. (a) $64^{-2/3}$
    (b) $(\frac{9}{4})^{-1/2}$
20. (a) $(-\frac{27}{8})^{-1/3}$
    (b) $-1/(144^{-1/2})$

In Exercises 21–26, simplify by removing all possible factors from under the radical. (Assume $x > 0$, $y > 0$.)

21. (a) $\sqrt{8}$
    (b) $\sqrt{18}$
22. (a) $\sqrt[3]{\frac{16}{27}}$
    (b) $\sqrt[3]{\frac{24}{125}}$
23. (a) $\sqrt[3]{16x^5}$
    (b) $\sqrt[4]{32x^4z^5}$
24. (a) $\sqrt[4]{(3x^2y^3)^4}$
    (b) $\sqrt[3]{54x^7}$
25. (a) $\sqrt{75x^2y^{-4}}$
    (b) $\sqrt{5(x-y)^3}$
26. (a) $\sqrt[5]{96b^6c^3}$
    (b) $\sqrt{72(x+1)^4}$

In Exercises 27–30, simplify by reducing the index of the radical as far as possible. (Assume $x > 0$, $y > 0$.)

27. (a) $\sqrt[4]{x^2}$
    (b) $\sqrt[6]{x^2}$
28. (a) $\sqrt[6]{(x+y)^3}$
    (b) $\sqrt[6]{(x+y)^4}$
29. (a) $\sqrt[8]{(3x^2y^3)^2}$
    (b) $\sqrt[10]{(6x^2y^4)^5}$
30. (a) $\sqrt[6]{9x^2y^2}$
    (b) $\sqrt[6]{16(x-y)^4}$

In Exercises 31–38, simplify each radical by rationalizing the denominator and reducing the resulting fraction to lowest terms.

31. (a) $\dfrac{1}{\sqrt{3}}$
    (b) $\dfrac{3}{\sqrt{21}}$
32. (a) $\dfrac{5}{\sqrt{10}}$
    (b) $\dfrac{21}{\sqrt{7}}$
33. (a) $\dfrac{8}{\sqrt[3]{2}}$
    (b) $\dfrac{1}{\sqrt[3]{12}}$
34. (a) $\dfrac{5}{\sqrt[3]{(5x)^2}}$
    (b) $\dfrac{3}{\sqrt[5]{(3x)^3}}$
35. (a) $\dfrac{2x}{5 - \sqrt{3}}$
    (b) $\dfrac{16}{6 + \sqrt{10}}$
36. (a) $\dfrac{5}{\sqrt{15} - 2}$
    (b) $\dfrac{x}{\sqrt{2} + \sqrt{3}}$
37. (a) $\dfrac{3}{\sqrt{x} + \sqrt{y}}$
    (b) $\dfrac{8}{\sqrt{2} - 2\sqrt{3}}$
38. (a) $\dfrac{5}{3\sqrt{x} - 5}$
    (b) $\dfrac{34}{5\sqrt{2} - 4}$

In Exercises 39–42, simplify each radical by rationalizing the numerator and reducing the resulting fraction to lowest terms.

39. (a) $\dfrac{\sqrt{13}}{2}$
    (b) $\dfrac{\sqrt{2}}{3}$
40. (a) $\dfrac{\sqrt{26}}{2}$
    (b) $\dfrac{\sqrt{y}}{6y}$
41. (a) $\dfrac{\sqrt{3} - \sqrt{2}}{x}$
    (b) $\dfrac{\sqrt{15} + 3}{12}$
42. (a) $\dfrac{\sqrt{x+2} - \sqrt{x}}{2}$
    (b) $\dfrac{\sqrt{a-b} + \sqrt{a}}{b}$

In Exercises 43–50, combine and/or simplify the given radicals.

43. (a) $5\sqrt{x} - 3\sqrt{x}$
    (b) $6\sqrt{2} + 7\sqrt{2}$
44. (a) $4\sqrt{27} - \sqrt{75}$
    (b) $5\sqrt[3]{2} + \sqrt[3]{54}$
45. (a) $2\sqrt{4xy} - 2\sqrt{9xy} + 10\sqrt{xy}$
    (b) $3\sqrt{ab/2} + 5\sqrt{2ab}$
46. (a) $2\sqrt{80} + \sqrt{125} - \sqrt{500}$
    (b) $\sqrt{\dfrac{25x}{y}} + 3\sqrt{\dfrac{x}{36y}} - \sqrt{\dfrac{32x}{50y}}$
47. (a) $\sqrt{5x^2y}\,\sqrt{3y}$
    (b) $\dfrac{\sqrt{54a^2}}{\sqrt{2a^4}}$
48. (a) $\sqrt[3]{\dfrac{4z^2}{y^5}}\,\sqrt[3]{\dfrac{2z}{y}}$
    (b) $\dfrac{\sqrt[4]{16b}}{\sqrt[4]{2b^5}}$

49. (a) $\sqrt{\sqrt{32}}$

    (b) $\sqrt{\sqrt[3]{10a^7b}}$

50. (a) $\sqrt{50}\,\sqrt[3]{2}$

    (b) $\sqrt{50}\,\sqrt[3]{10}$

Calculator Exercises

51. Use a calculator to approximate each of the following to four decimal places.

    (a) $\sqrt{57}$

    (b) $\sqrt[5]{562}$

    (c) $\sqrt[3]{45^2}$

    (d) $(-10)^{4/5}$

    (e) $(15.25)^{-1.4}$

    (f) $(9.42 \times 10^5)^{2/3}$

52. Use a calculator to approximate each of the following to four decimal places.

    (a) $\sqrt[6]{125}$

    (b) $\sqrt[5]{-65}$

    (c) $\sqrt{75 + 3\sqrt{8}}$

    (d) $225^{-2/3}$

    (e) $(2.65 \times 10^{-4})^{1/3}$

    (f) $(9.3 \times 10^7)^2$

53. To have uniform depreciation by the declining balance method, we use the formula

$$R = N\left[1 - \left(\frac{S}{C}\right)^{1/N}\right]$$

where R is the percentage depreciation each year, N is the useful life of the item, C is the original cost, and S is the salvage value. Calculate R (to two decimal places) for each of the following.

    (a) $N = 8$, $C = \$10,400$, $S = \$1,500$

    (b) $N = 4$, $C = \$11,200$, $S = \$3,200$

54. The time, t, it takes a funnel to empty when filled with water to a height of h is given by

$$t = 0.03[12^{5/2} - (12 - h)^{5/2}], \qquad 0 \le h \le 12$$

where t is measured in seconds. Find t (to two decimal places) for $h = 7$ centimeters.

Complex Numbers   1.6

So far in our review, we have dealt only with the set of real numbers. Although the real number system is sufficient for most applications of algebra and trigonometry, it does have a serious inadequacy related to even roots of negative numbers. For instance, the square of a real number is always non-negative. Hence, roots like

$$\sqrt{-1} \qquad \text{and} \qquad \sqrt{-4}$$

cannot be real numbers since their respective squares are negative:

$$[(-1)^{1/2}]^2 = (-1)^1 = -1 \qquad \text{and} \qquad [(-4)^{1/2}]^2 = (-4)^1 = -4$$

To resolve this problem, mathematicians created an expanded system of numbers based upon the **imaginary** (nonreal) **number i,** whose square is defined to be -1.

DEFINITION OF THE IMAGINARY NUMBER i

The **imaginary number i** is defined by

$$i = \sqrt{-1}$$

where $i^2 = -1$.

With this single addition to the real number system, we can develop the system of **complex numbers.**

DEFINITION OF A COMPLEX NUMBER

If a and b are real numbers, then the number

$a + bi$

is called a **complex number,** where a is called the **real part** and b the **imaginary part** of the number.

Remark: The form $a + bi$ is called the **standard form** of a complex number.

We can see that the set of real numbers is a subset of the set of complex numbers, since every real number a can be written as a complex number using $b = 0$. That is, for every real number a,

$a = a + 0i$

Powers of i

Using the rules for powers of real numbers, we see that

$$i = \sqrt{-1}$$
$$i^2 = -1$$
$$i^3 = i^2 \cdot i = -i$$
$$i^4 = i^2 \cdot i^2 = (-1)(-1) = 1$$
$$i^5 = i^4 \cdot i = i$$

The pattern begins to repeat after the 4th power. Therefore, to compute the value of i^n for any natural number n, we simply factor out the multiples of 4 in the exponent and compute the remaining portion. For instance,

$$i^{38} = i^{36} \cdot i^2 = (i^4)^9 \cdot i^2 = (1)^9(-1) = -1$$

Operations with Complex Numbers

Since a complex number consists of a real part plus a multiple of i, we define operations with complex numbers in a manner consistent with the rules for real numbers and the imaginary unit.

EQUALITY OF TWO COMPLEX NUMBERS

If $a + bi$ and $c + di$ are two complex numbers written in standard form, then

$a + bi = c + di$

if and only if $a = c$ and $b = d$.

To add (or subtract) two complex numbers, we add (or subtract) the real and imaginary parts of the numbers separately.

ADDITION AND SUBTRACTION OF COMPLEX NUMBERS

If $a + bi$ and $c + di$ are two complex numbers written in standard form, then their sum and difference are

Sum: $(a + bi) + (c + di) = (a + c) + (b + d)i$

Difference: $(a + bi) - (c + di) = (a - c) + (b - d)i$

The **additive identity** in the complex number system is zero (the same as in the real number system). Furthermore, the **additive inverse** of the complex number $a + bi$ is given by

Given Number	Additive Inverse
$a + bi$	$-(a + bi) = -a - bi$

Thus, we have

$$(a + bi) + (-a - bi) = 0 + 0i = 0$$

EXAMPLE 1
Adding and Subtracting Complex Numbers

Write the following sums and differences in standard form.

(a) $(3 - i) + (2 + 3i)$ (b) $2i + (-4 - 2i)$
(c) $7 - (5 - i) + (-2 + 3i)$

Solution:

(a) $(3 - i) + (2 + 3i) = 3 - i + 2 + 3i$
$$= 3 + 2 - i + 3i$$
$$= (3 + 2) + (-1 + 3)i$$
$$= 5 + 2i$$

(b) $2i + (-4 - 2i) = 0 + 2i - 4 - 2i$
$$= -4 + 2i - 2i$$
$$= -4$$

(c) $7 - (5 - i) + (-2 + 3i) = 7 + 0i - 5 + i - 2 + 3i$
$$= 7 - 5 - 2 + i + 3i$$
$$= 4i$$

Remark: Note in part (b) of Example 1 that the sum of two complex numbers can be a real number.

Many of the properties of real numbers are valid for complex numbers as well. Some such properties are

Associative Property of Addition and Multiplication

Commutative Property of Addition and Multiplication

Distributive Property of Multiplication over Addition

Notice how these properties come into play when two complex numbers are multiplied.

$$
\begin{aligned}
(a + bi)(c + di) &= a(c + di) + bi(c + di) && \textit{Distributive Law}\\
&= ac + (ad)i + (bc)i + (bd)i^2 && \textit{Distributive Law}\\
&= ac + (ad)i + (bc)i + (bd)(-1) && \textit{Definition of i}\\
&= ac - bd + (ad)i + (bc)i && \textit{Commutative Law}\\
&= (ac - bd) + (ad + bc)i && \textit{Associative Law}
\end{aligned}
$$

We summarize this result as follows.

MULTIPLICATION OF COMPLEX NUMBERS

If $a + bi$ and $c + di$ are two complex numbers written in standard form, then their product is

$$(a + bi)(c + di) = (ac - bd) + (ad + bc)i$$

Remark: Rather than trying to memorize this rule, you may just want to remember how the Distributive Law is used to derive it.

EXAMPLE 2
Multiplying Complex Numbers

Find the following products.

(a) $(i)(-3i)$ (b) $(2 - i)(4 + 3i)$ (c) $(3 + 2i)(3 - 2i)$

Solution:

(a) $(i)(-3i) = -3i^2 = -3(-1) = 3$

(b) $(2 - i)(4 + 3i) = 8 + 6i - 4i - 3i^2$
$$= 8 + 6i - 4i - 3(-1)$$
$$= 8 + 3 + 6i - 4i$$
$$= 11 + 2i$$

(c) $(3 + 2i)(3 - 2i) = 9 - 6i + 6i - 4i^2$
$$= 9 - 4(-1) = 9 + 4$$
$$= 13$$

Complex Conjugates

Pairs of complex numbers of the forms

$$(a + bi) \quad \text{and} \quad (a - bi)$$

are called **complex conjugates.** That is, the conjugate of $(3 + 2i)$ is $(3 - 2i)$. As illustrated in part (c) of Example 2, the product of two complex conjugates is a real number:

$$\begin{aligned} (a + bi)(a - bi) &= a^2 - abi + abi - b^2i^2 \\ &= a^2 - b^2(-1) \\ &= a^2 + b^2 \end{aligned}$$

We can use complex conjugates to perform division in the complex number system. That is, if we want to find the quotient

$$\frac{a + bi}{c + di}$$

we simply multiply numerator and denominator by the conjugate of the denominator as follows:

$$\begin{aligned} \frac{a + bi}{c + di} &= \frac{a + bi}{c + di}\left(\frac{c - di}{c - di}\right) \\ &= \frac{(a + bi)(c - di)}{c^2 + d^2} \end{aligned}$$

EXAMPLE 3
Dividing Complex Numbers

Write the following in standard form.

(a) $\dfrac{1}{1 + i}$

(b) $\dfrac{2 + 3i}{4 - 2i}$

Solution:

(a) $\dfrac{1}{1 + i} = \dfrac{1}{1 + i}\left(\dfrac{1 - i}{1 - i}\right)$

$\qquad = \dfrac{1 - i}{1 - (-1)} = \dfrac{1 - i}{2} = \dfrac{1}{2} - \dfrac{1}{2}i$

(b) $\dfrac{2 + 3i}{4 - 2i} = \dfrac{2 + 3i}{4 - 2i}\left(\dfrac{4 + 2i}{4 + 2i}\right)$

$\qquad = \dfrac{(2 + 3i)(4 + 2i)}{16 + (-2)^2} = \dfrac{1}{20}(8 + 12i + 4i + 6i^2)$

$\qquad = \dfrac{1}{20}(8 - 6 + 12i + 4i) = \dfrac{1}{20}(2 + 16i) = \dfrac{1}{10} + \dfrac{4}{5}i$

In algebra it is common to obtain a result such as $\sqrt{-3}$. In standard form this complex number can be written as

$$\sqrt{-3} = \sqrt{(3)(-1)} = \sqrt{3}\sqrt{-1} = \sqrt{3}\,i$$

In general, we have the following rule for writing the square root of a negative number in standard form.

STANDARD FORM OF SQUARE ROOT OF A NEGATIVE NUMBER

If $a > 0$, then
$$\sqrt{-a} = \sqrt{a}\,i$$

Thus,

$$\sqrt{-16} = \sqrt{16}\,i = 4i$$

and

$$\frac{-2 - \sqrt{-2}}{2} = \frac{-2}{2} - \frac{\sqrt{2}\,i}{2} = -1 - \frac{\sqrt{2}}{2}\,i$$

Remark: When working with square roots of negative numbers, be sure to convert to standard form *before* multiplying. For instance, consider the following:

$$Correct: \quad \sqrt{-1}\sqrt{-1} = i \cdot i = i^2 = -1$$
$$Incorrect: \quad \sqrt{-1}\sqrt{-1} = \sqrt{(-1)(-1)} = \sqrt{1} = 1$$

EXAMPLE 4
Manipulating Expressions with Radicals

Expand the following expression.

$$(-1 + \sqrt{-3})^2 + 2(-1 + \sqrt{-3}) + 4$$

Solution:

$$(-1 + \sqrt{-3})^2 + 2(-1 + \sqrt{-3}) + 4$$
$$= (-1 + \sqrt{3}\,i)^2 + 2(-1 + \sqrt{3}\,i) + 4$$
$$= (-1)^2 - 2\sqrt{3}\,i + (\sqrt{3})^2(i^2) - 2 + 2\sqrt{3}\,i + 4$$
$$= 1 + 3(-1) - 2 + 4 - 2\sqrt{3}\,i + 2\sqrt{3}\,i$$
$$= (1 - 3 - 2 + 4) + (-2\sqrt{3} + 2\sqrt{3})i$$
$$= 0$$

Section Exercises 1.6

1. Write out the first 20 positive integer powers of i (i.e., i, i^2, i^3, ..., i^{20}), and express each in the simplest form of i, $-i$, 1, and -1. What is the pattern of the simplest form?

In Exercises 2–20, write the complex number in standard form and give the complex conjugate.

2. $3 + \sqrt{-16}$

3. $4 + \sqrt{-9}$

4. $-10 - \sqrt{-15}$

5. $-7 - \sqrt{-64}$

6. $\sqrt{-4} - 5$

7. $\sqrt{-81} - 7$

8. $1 + \sqrt{-8}$

9. $2 - \sqrt{-27}$

10. 45

11. $\sqrt{-75}$

12. $4i^2 - 2i^3$

13. $-6i + i^2$

14. $(-i)^3$

15. $8 + (-\sqrt{-7})^3$

16. $(\sqrt{-4})^2 - 5$

17. 8

18. $2 + \sqrt{3}$

19. $5 - \sqrt{27}$

20. $-5i^5$

In Exercises 21–60, perform the indicated operation and write the result in standard form.

21. $(5 + i) + (6 - 2i)$

22. $(13 - 2i) + (-5 + 6i)$

23. $(8 - i) - (4 - i)$

24. $(3 + 2i) - (6 + 13i)$

25. $6i - (5 + 3i)$

26. $6 - (6 - 3i)$

27. $(-2 + \sqrt{-8}) + (5 - \sqrt{50})$

28. $(8 + \sqrt{-18}) - (4 + 3\sqrt{2}\,i)$

29. $-(\frac{3}{2} + \frac{5}{2}i) + (\frac{5}{3} + \frac{11}{3}i)$

30. $(1.6 + 3.2i) + (-5.8 + 4.3i)$

31. $(1 + i)(3 - 2i)$

32. $(6 - 2i)(2 - 3i)$

33. $(4 + 5i)(4 - 5i)$

34. $(6 + 7i)(6 - 7i)$

35. $6i(5 - 2i)$

36. $-8i(9 + 4i)$

37. $(15i)(6i)$

38. $4(3 - 2i)$

39. $-8(5 - 2i)$

40. $(\sqrt{5} - \sqrt{3}\,i)(\sqrt{5} + \sqrt{3}\,i)$

41. $(\sqrt{14} + \sqrt{10}\,i)(\sqrt{14} - \sqrt{10}\,i)$

42. $(3 + \sqrt{-5})(7 - \sqrt{-10})$

43. $(4 + 5i)^2$

44. $(2 - 3i)^3$

45. $(4 + 5i)^3$

46. $\dfrac{1}{2 + 3i}$

47. $\dfrac{4}{4 - 5i}$

48. $\dfrac{3}{1 - i}$

49. $\dfrac{2 + i}{2 - i}$

50. $\dfrac{8 - 7i}{1 - 2i}$

51. $\dfrac{6 - 7i}{i}$

52. $\dfrac{8 + 20i}{2i}$

53. $(12 + 7i) \div (2 + i)$

54. $(3 + 2i) \div (3 - 2i)$

55. $\dfrac{1}{(2i)^3}$

56. $\dfrac{1}{(4 - 5i)^2}$

57. $\dfrac{5}{(1 + i)^3}$

58. $\dfrac{(2 - 3i)(5i)}{2 + 3i}$

59. $\dfrac{(21 - 7i)(4 + 3i)}{2 - 5i}$

60. $\dfrac{1}{i^3}$

In Exercises 61–64, find real numbers a and b so that the equation is true.

61. $a + bi = -10 + 6i$

62. $a + bi = 13 + 4i$

63. $(a - 1) + (b + 3)i = 5 + 8i$

64. $(a + 6) + 2bi = 6 - 5i$

65. Show that the sum of a complex number and its conjugate is a real number.

66. Show that the difference of a complex number and its conjugate is an imaginary number.

67. Show that the product of a complex number and its conjugate is a real number.

68. Show that the conjugate of the product of two complex numbers is the product of their conjugates.

69. Show that the conjugate of the sum of two complex numbers is the sum of their conjugates.

Polynomials and Special Products   1.7

In this section we will be working with *literal* (algebraic) expressions and statements instead of strictly *numerical* ones. We will use letters (called **variables**) *and* numbers in making up the expressions and statements with which we operate.

An **algebraic expression** is a collection of variables and real numbers (called **constants**) organized in some manner by using additions, subtractions, multiplications, divisions, or radicals. Some examples are

$$3x - 7, \qquad \frac{5xy^2 - 2x^2}{x + 2}, \qquad \sqrt{x^2 - y^2}, \qquad 5x^3 - 7x^2 + x - 11$$

A **term** of an algebraic expression is any product of a constant and one or more variables raised to powers. For instance,

Expressions	**Terms**
$3x - 7$	$3x, 7$
$\dfrac{5xy^2 - 2x^2}{x + 2}$	$5xy^2, 2x^2, x, 2$
$5x^3 - 7x^2 + x - 11$	$5x^3, 7x^2, x, 11$

Note that *an algebraic expression has no equal sign in it*. In Chapter 2 we deal with algebraic statements called equations that *do* have equal signs. This parallels the study of a language where an expression is just *part* of a complete statement (or sentence). This distinction is crucial, since some of the ways we manipulate equations are not permissible with algebraic expressions. As we work with algebraic expressions in this chapter, we will use equal signs *only* to indicate that the new form of the expression is algebraically **equivalent*** to the old (original) form. Note the use of the equal sign in simplifying the expression $3x + 5(x - 2)$.

$$3x + 5(x - 2) = 3x + 5x + 5(-2) \quad \textit{Distributive Law}$$
$$= 8x - 10 \quad \textit{Combine like terms}$$

One of the simplest kinds of algebraic expressions is the **polynomial.** Some examples are

$$2x + 5, \qquad 3x^4 - 7x^2 + 2x + 4, \qquad 5x^2y^2 - xy + 3$$

The first two are *polynomials in x* and the last one is a *polynomial in x and y*. The terms of a polynomial in x have the form cx^k, where c is called the **coefficient** and k the **degree** of the term.

DEFINITION OF A POLYNOMIAL IN x

For real numbers $a_0, a_1, a_2, \ldots, a_n$ and nonnegative integer n:
A **polynomial in x** is an expression of the form

$$a_n x^n + a_{n-1} x^{n-1} + \cdots + a_1 x + a_0$$

where $a_n \neq 0$. The polynomial is of **degree** n. The numbers $a_0, a_1, a_2, \ldots, a_n$ are called the **coefficients** and a_n is called the **leading coefficient.**

Remark: Polynomials with one, two, or three terms are called **monomials, binomials,** or **trinomials,** respectively.

Note in the above definition that the polynomial is written in *order* with decreasing powers of x. This is referred to as **standard form.** Note also that the polynomial is written as a *sum*. Consequently the coefficients take on the sign between terms. For instance, the polynomial

---

*Two expressions are algebraically equivalent if they yield the same values for all x-values for which the expressions are defined.

$$2x^3 - 5x^2 - 3x + 1 = 2x^3 + (-5)x^2 + (-3)x + 1$$

has coefficients 2, -5, -3, and 1.

Some examples of polynomials rewritten in standard form, together with terms, degrees, and coefficients, are

Polynomial	Standard Form	Degree	Leading Coefficient	Terms	Coefficients
$4x^2 - 5x^7 - 2 + 3x$	$-5x^7 + 4x^2 + 3x - 2$	7	-5	$-5x^7, 4x^2, 3x, -2$	$-5, 4, 3, -2$
$4 - 9x^2$	$-9x^2 + 4$	2	-9	$-9x^2, 4$	$-9, 4$
8	8	0	8	8	8

$$(8 = 8x^0)$$

For polynomials in more than one variable, the degree of a *term* is the sum of the powers of the variables in the term. The degree of the *polynomial* is the highest degree of all its terms. For instance, the polynomial

$$5x^3y - x^2y^2 + 2xy - 5$$

has two terms of degree 4, one term of degree 2, and one term of degree 0. The degree of the polynomial is 4.

Operations with Polynomials

We can add and subtract polynomials in much the same way that we add and subtract real numbers. We simply add or subtract the coefficients of *like terms* (terms having the same variables to the same powers).

EXAMPLE 1
Sums and Differences
of Polynomials

(a) Add $5x^3 - 7x^2 - 3$ and $x^3 + 2x^2 - x + 8$.
(b) Subtract $3x^4 - 4x^2 + 3x$ from $7x^4 - x^2 - 4x + 2$.
(c) Subtract $4x^2 + 2xy - 3$ from $x^2 - 3xy + 2$.

Solution:

(a) $(5x^3 - 7x^2 - 3) + (x^3 + 2x^2 - x + 8)$

$\qquad = (5x^3 + x^3) + (2x^2 - 7x^2) - x + (8 - 3)$ *Group like terms*

$\qquad = (5 + 1)x^3 + (2 - 7)x^2 - x + 5$ *Distributive Law*

$\qquad = 6x^3 - 5x^2 - x + 5$ *Combine like terms*

(b) $(7x^4 - x^2 - 4x + 2) - (3x^4 - 4x^2 + 3x)$ *Distribute minus sign*

$\qquad = 7x^4 - x^2 - 4x + 2 - 3x^4 + 4x^2 - 3x$

$\qquad = (7x^4 - 3x^4) + (4x^2 - x^2) + (-3x - 4x) + 2$ *Group like terms*

$\qquad = 4x^4 + 3x^2 - 7x + 2$ *Combine like terms*

(c) $(x^2 - 3xy + 2) - (4x^2 + 2xy - 3)$

$$= x^2 - 3xy + 2 - 4x^2 - 2xy + 3$$
$$= (x^2 - 4x^2) + (-3xy - 2xy) + (2 + 3)$$
$$= -3x^2 - 5xy + 5$$

Remark: A very common mistake is to fail to change the sign of *each* term inside parentheses preceded by a minus sign.

$$-(3x^2 - 7xy + 3) = -3x^2 + 7xy - 3$$
$$\neq -3x^2 - 7xy + 3 \qquad \text{. } Common\ mistake$$

In finding the product of two polynomials, the Distributive Laws

$$a(b + c) = ab + ac \qquad \text{and} \qquad (a + b)c = ac + bc$$

are useful. For example, if we treat $(3x - 2)$ as a *single* quantity in the Distributive Law, we have

$$(3x - 2)(5x + 7)$$

$= (3x - 2)(5x) + (3x - 2)(7)$	*Left Distributive Law*
$= (3x)(5x) - 2(5x) + (3x)(7) - 2(7)$	*Right Distributive Law*
$= 15x^2 - 10x + 21x - 14$	*Law of Exponents*
$= 15x^2 + 11x - 14$	

When multiplying two polynomials, you must take care to see that each term of one polynomial is multiplied by each term of the other one. The following vertical arrangement works well for such multiplications.

$$
\begin{array}{r}
2x^2 - 3x + 5 \\
\underline{x - 4} \\
2x^3 - 3x^2 + 5x \qquad \Longleftarrow \qquad x(2x^2 - 3x + 5) \\
\underline{-8x^2 + 12x - 20} \qquad \Longleftarrow \qquad -4(2x^2 - 3x + 5) \\
2x^3 - 11x^2 + 17x - 20 \qquad\qquad\qquad Combine\ like\ terms
\end{array}
$$

EXAMPLE 2
Multiplying Polynomials

(a) Multiply $(x^2 + 3)(5x^3 - x + 6)$.
(b) Multiply $(x^2 - 2x + 2)(x^2 + 2x + 2)$.

Solution:

(a) Using the Distributive Law, we have

$$(x^2 + 3)(5x^3 - x + 6)$$
$$= (x^2 + 3)(5x^3) + (x^2 + 3)(-x) + (x^2 + 3)(6)$$
$$= 5x^5 + 15x^3 - x^3 - 3x + 6x^2 + 18$$
$$= 5x^5 + (15x^3 - x^3) + 6x^2 - 3x + 18$$
$$= 5x^5 + 14x^3 + 6x^2 - 3x + 18$$

(b) For two trinomials, the vertical method is simpler.

$$x^2 - 2x + 2$$
$$\underline{x^2 + 2x + 2}$$
$$x^4 - 2x^3 + 2x^2 \qquad\qquad \Leftarrow \quad x^2(x^2 - 2x + 2)$$
$$2x^3 - 4x^2 + 4x \qquad \Leftarrow \quad 2x(x^2 - 2x + 2)$$
$$\underline{\qquad\qquad 2x^2 - 4x + 4} \quad \Leftarrow \quad 2(x^2 - 2x + 2)$$
$$x^4 + 0x^3 + 0x^2 - 0x + 4 = x^4 + 4$$

This is a surprising answer, isn't it?

Special Products

We list next some special binomial products that should be memorized so that the factoring process in the next section will go smoothly.

SPECIAL BINOMIAL PRODUCTS

Form	Product	Example
(Binomial)(Binomial)	$(ax + b)(cx + d)$ $= acx^2 + adx + bcx + bd$ $= acx^2 + (ad + bc)x + bd$ $= mx^2 + nx + r$	$(3x - 2)(x + 5) = 3x^2 + 15x - 2x - 10$ $= 3x^2 + 13x - 10$
(Binomial)^2	$(u + v)^2 = u^2 + 2uv + v^2$ $(u - v)^2 = u^2 - 2uv + v^2$	$(3x - 2)^2 = (3x)^2 + 2(3x)(-2) + (-2)^2$ $= 9x^2 - 12x + 4$
(Sum)(Difference)	$(u + v)(u - v) = u^2 - v^2$	$(7x + 4)(7x - 4) = (7x)^2 - (4)^2$ $= 49x^2 - 16$
(Binomial)^3	$(u + v)^3 = u^3 + 3u^2v + 3uv^2 + v^3$	$(x + 2)^3 = x^3 + 3x^2(2) + 3x(2)^2 + (2)^3$ $= x^3 + 6x^2 + 12x + 8$
	$(u - v)^3 = u^3 - 3u^2v + 3uv^2 - v^3$	$(2x - 1)^3 = (2x)^3 - 3(2x)^2(1) + 3(2x)(1)^2 - (1)^3$ $= 8x^3 - 12x^2 + 6x - 1$

EXAMPLE 3
Special Binomial Products

Find the products:

(a) $(x - 7)(5x + 2)$ (b) $(6x - \sqrt{5})^2$

(c) $(5x + 9)(5x - 9)$ (d) $(3x + 2)^3$

Solution:

(a) The product is of the form (binomial)(binomial). We can handle this product mentally according to the following scheme commonly called the **FOIL Method.**

F: Multiply **F**irst Terms

O: Multiply **O**utside Terms

I: Multiply **I**nside Terms

L: Multiply **L**ast Terms

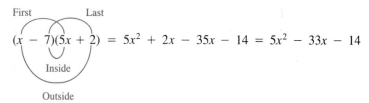

$$(x - 7)(5x + 2) = 5x^2 + 2x - 35x - 14 = 5x^2 - 33x - 14$$

(b) The square of a binomial has the form

$$(u + v)^2 = u^2 + 2uv + v^2$$

Thus, we have

$$(6x - \sqrt{5})^2 = (6x)^2 + 2(6x)(-\sqrt{5}) + (-\sqrt{5})^2$$
$$= 36x^2 - 12\sqrt{5}x + 5$$

(c) The product of a sum and a difference of the *same* two terms has no middle term, and it takes the form

$$(u + v)(u - v) = u^2 - v^2$$

Thus,

$$(5x + 9)(5x - 9) = (5x)^2 - (9)^2$$
$$= 25x^2 - 81$$

(d) The cube of a binomial has the form

$$(u + v)^3 = u^3 + 3u^2v + 3uv^2 + v^3$$

Note the *decrease* of powers on the first term and the *increase* on the second term. Thus, we write

$$(3x + 2)^3 = (3x)^3 + 3(3x)^2(2) + 3(3x)(2)^2 + (2)^3$$
$$= 27x^3 + 54x^2 + 36x + 8$$

These special products can be generalized so that the u and v terms are replaced by quantities enclosed in parentheses. For instance,

$$[(a + 2) + (b - 1)]^2 = (a + 2)^2 + 2(a + 2)(b - 1) + (b - 1)^2$$

$$[u + v]^2 \qquad = \qquad u^2 + 2uv + v^2$$

where each of the resulting terms can be expanded according to the same rules.

EXAMPLE 4
Generalizing Special Products

Multiply $(x + y + 3)(x + y - 2)$.

Solution:

If we group x and y, then the given trinomials have binomial forms and we multiply (**FOIL** method) as follows:

$$
\begin{aligned}
(x + y + 3)&(x + y - 2) \\
&= [(x + y) + 3][(x + y) - 2] && \textit{Associative Law} \\
&= (x + y)^2 + (-2)(x + y) + 3(x + y) - 6 && \textit{\textbf{FOIL}} \\
&= (x + y)^2 + (x + y) - 6 && \textit{Combine like terms} \\
&= x^2 + 2xy + y^2 + x + y - 6 && \textit{Square of binomial}
\end{aligned}
$$

Remark: In Section 1.8 you will learn to factor trinomial forms like $(x + y)^2 + (x + y) - 6$, so keep this example in mind for later use.

Evaluating Algebraic Expressions

We frequently need to **evaluate** an algebraic expression. This means that a specific number is assigned to each variable in the expression. When evaluating, we need to be aware of what numbers are *permissible* to use in place of the variables. (In this context we consider permissible numbers as those that yield *real* valued answers.) For instance, consider the following two expressions.

Expression	Assigned Number	Permissible?
$\sqrt{x - 2}$	$x = 3$ $x = 0$ all $x \geq 2$	**yes,** $\sqrt{1} = 1$ **no,** $\sqrt{-2}$ is not real **yes,** $\sqrt{\text{nonnegative real}}$ is real
$\dfrac{3}{x^2 - 4}$	$x = 0$ $x = 2$ $x = -2$ all $x \neq \pm 2$	**yes,** $\dfrac{3}{-4}$ is real **no,** $\dfrac{3}{0}$ not defined **no,** $\dfrac{3}{0}$ not defined **yes,** real values obtained

The set of permissible values that can be assigned to the variable in an algebraic expression is commonly referred to as the **domain** of the variable. Thus, the domain of x in the expression $\sqrt{x - 2}$ is all $x \geq 2$. Similarly, the domain of x in the expression $3/(x^2 - 4)$ is all $x \neq \pm 2$.

EXAMPLE 5
Evaluating Algebraic Expressions

Evaluate the following expression for $x = -1$, $x = 0$, and $x = 2/3$.

$$2x^3 - 5x^2 + 3x$$

Solution:
For $x = -1$,

$$2x^3 - 5x^2 + 3x = 2(-1)^3 - 5(-1)^2 + 3(-1) \qquad \textit{Replace x with } -1$$
$$= 2(-1) - 5(1) - 3 \qquad\qquad \textit{Raise to powers}$$
$$= -2 - 5 - 3 = -10 \qquad\qquad \textit{Simplify}$$

For $x = 0$,

$$2x^3 - 5x^2 + 3x = 2(0)^3 - 5(0)^2 + 3(0) = 0$$

For $x = 2/3$,

$$2x^3 - 5x^2 + 3x = 2\left(\frac{2}{3}\right)^3 - 5\left(\frac{2}{3}\right)^2 + 3\left(\frac{2}{3}\right)$$

$$= \frac{16}{27} - \frac{20}{9} + 2$$

$$= \frac{16 - 60 + 54}{27} = \frac{10}{27}$$

Section Exercises 1.7

In Exercises 1–6, complete the table for the given polynomial.

Standard Form	Degree	Leading Coefficient	Terms	Coefficients
1. $2x^2 - x + 1$	_____	_____	_____	_____
2. $-3x^4 + 2x^2 - 5$	_____	_____	_____	_____
3. $x^5 - 1$	_____	_____	_____	_____
4. 3	_____	_____	_____	_____
5. $4x^5 + 6x^4 - 3x^3 + 10x^2 - x - 1$	_____	_____	_____	_____
6. $2x$	_____	_____	_____	_____

In Exercises 7–10, determine the degree of the given polynomial.

7. $2x^2y - 5xy^2 + 10x^2y^2$
8. $x^3 + 3x^2y + 3xy^2 + y^3$
9. $10x^2 - 7xyz + 4y^2 + 6z^4$
10. $5xyz$

In Exercises 11–30, perform the indicated operations and express each result as a polynomial.

11. Add: $-3x^2 - 13x + 7$ and $14x - 15$
12. Add: $3x^4 - 19x^2 + 1$ and $17x^3 + 4x$
13. Subtract: $-8x^3 - 14x^2 - 17$ from $15x^2 - 6$
14. Subtract: $13x^4 - 5x + 15$ from $15x^4 - 18x - 19$

15. $(9x^5 + 10x^3 + 8x) + (-7x^3 + 19x^2 - 7x - 9) +$
    $(5x^4 + 16x^3)$
16. $(-14x^4 - 5x^2 - 6) + (-2x^3 - 9x + 5) +$
    $(-8x^4 + 7x)$
17. $(18x^7 + 16x^3 - 7) - (-7x^7 + 19)$
18. $(-4x^2 - 2x - 8) - (9x^2 - 10x) + (-7x - 4)$
19. $(-17x^3 - 16xy - 9y^2) + (18x^3 + 4xy)$
20. $(-13x^2 - 19y^2 + 6xy) - (9xy + 2x^2 - 16y^2)$
21. Multiply: $(x^3 - 2x + 1)$ by $(x - 5)$
22. Multiply: $(x - y + 1)$ by $(x + y - 1)$
23. $(x^2 + 9)(x^2 - x - 4)$
24. $(x - 2)(x^2 + 2x + 4)$
25. $(x + 3)(x^2 - 3x + 9)$
26. $(2x^2 + 3y^3)(4x^4 - 6x^2y^3 + 9y^6)$
27. $(x^2 + 1)(x + 1)(x - 1)$
28. $(x^2 + x - 2)(x^2 - x + 2)$
29. $(x + \sqrt{5})(x - \sqrt{5})(x + 4)$
30. $(2x - y)(x + 3y) + 3(2x - y)$

In Exercises 31–55, use the special products given in this section to perform the indicated multiplication of binomials.

31. $(x + 3)(x + 4)$ 32. $(x - 5)(x + 10)$
33. $(3x - 5)(2x + 1)$ 34. $(7x - 2)(4x - 3)$

35. $(x + 6)^2$ 36. $(3x - 2)^2$
37. $(2x - 5y)^2$ 38. $[(x + 1) - y]^2$
39. $[(x - 3) + y]^2$ 40. $(x + 3)(x - 3)$
41. $(x + 2y)(x - 2y)$ 42. $(2x + 3y)(2x - 3y)$
43. $(2r^2 - 5)(2r^2 + 5)$ 44. $(3a^3 - 4b^2)(3a^3 + 4b^2)$
45. $(x + 1)^3$ 46. $(x - 2)^3$
47. $(2x - y)^3$ 48. $(3x + 2y)^3$
49. $[(x + 1) - y]^3$ 50. $[(z - 2) + y]^3$
51. $(\sqrt{x} + \sqrt{y})(\sqrt{x} - \sqrt{y})$ 52. $(5 + 2\sqrt{s})(5 - 2\sqrt{s})$
53. $(4r^3 - 3s^2)^2$
54. $(8y + \sqrt{z})^2$
55. $(10m - 3n)(4n + 3m)$
56. What is the degree of the product of two polynomials? (Assume one is of degree m and the other of degree n.)

Calculator Exercises

57. Use a calculator to evaluate $6x^3 - 4.2x^2 + 2.7$ when
    (a) $x = 5$ (b) $x = 1.5$
    (c) $x = -0.43$ (d) $x = \frac{2}{3}$
58. Use a calculator to evaluate $5/(x^2 + 2x)$ when
    (a) $x = 3.2$ (b) $x = -1.5$
    (c) $x = -2$ (d) $x = \frac{4}{3}$

Factoring   1.8

As you study this section you may get the feeling that you are learning algebra both forwards and backwards. In the last section we showed how to multiply polynomials to get new polynomials, whereas in this section we show how to find the factors whose product will yield a given polynomial. This process of writing a polynomial as a product is called **factoring.** Factoring is an important tool for reducing fractional expressions and for solving equations and inequalities. Unless noted otherwise, we will limit our discussion of factoring to polynomials whose factors have integer coefficients.

Factoring Out a Monomial

Our approach in this section is to work backwards with each of the basic products that we studied in the last section. We start with polynomials that can be written as the product of a monomial and another polynomial. The technique here is to use the Distributive Laws

$$a(b + c) = ab + ac \quad \text{and} \quad (a + b)c = ac + bc$$

from *right to left* rather than from left to right. We look for a monomial that is common to each term of the polynomial. For instance, the polynomial $9x^3 - 27x^4y + 18x^2$ has the monomial $9x^2$ as a factor in each term. Hence, we write

$$9x^3 - 27x^4y + 18x^2 = (9x^2)(x) - (9x^2)(3x^2y) + (9x^2)(2)$$
$$= 9x^2(x - 3x^2y + 2)$$

EXAMPLE 1

Factoring Out the Greatest Common Factor

Factor out the greatest common factor in each of the following.

(a) $6y^3z - 4yz^2 + 2y^2z^3$ (b) $(x + 2)(a + b) + (x + 2)(a - b)$

Solution:

(a) $2yz$ is common to all three terms, so we write

$$6y^3z - 4yz^2 + 2y^2z^3 = (2yz)(3y^2) - (2yz)(2z) + (2yz)(yz^2)$$
$$= 2yz(3y^2 - 2z + yz^2)$$

(b) In this case, the binomial factor $(x + 2)$ is common to both terms, so we write

$$(x + 2)(a + b) + (x + 2)(a - b)$$
$$= (x + 2)[(a + b) + (a - b)] \quad \textit{Distributive Law}$$
$$= (x + 2)[a + b + a - b] \quad \textit{Remove parentheses}$$
$$= (x + 2)(2a) \quad \textit{Combine terms}$$

Remark: You may want to skip the step that shows the factored form of each term.

Factoring polynomials is more complicated than finding products of polynomials. In fact, it is quite difficult to factor polynomials of degree greater than 2, unless the polynomial has the form of one of the special products listed in Section 1.7. The formulas listed next are the reverse forms of those special products.

FACTORING SPECIAL POLYNOMIAL FORMS

Polynomial	Factored Form	Example
Difference of Two Squares	$u^2 - v^2 = (u + v)(u - v)$	$9x^2 - 4 = (3x)^2 - (2)^2$ $= (3x + 2)(3x - 2)$
Perfect Square Trinomial	$u^2 + 2uv + v^2 = (u + v)^2$	$x^2 + 6x + 9 = x^2 + 2(x)(3) + (3)^2$ $= (x + 3)^2$
Trinomial with Binomial Factors	$mx^2 + nx + r = (ax + b)(cx + d)$ $[m = ac, r = bd, n = ad + bc]$	$3x^2 - 2x - 5 = (3x - 5)(x + 1)$
Sum and Difference of Two Cubes	$u^3 + v^3 = (u + v)(u^2 - uv + v^2)$	$x^3 + 8 = (x)^3 + (2)^3$ $= (x + 2)(x^2 - 2x + 4)$
	$u^3 - v^3 = (u - v)(u^2 + uv + v^2)$	$27x^3 - 1 = (3x)^3 - (1)^3$ $= (3x - 1)(9x^2 + 3x + 1)$

Difference of Two Squares

One of the easiest special polynomial forms to recognize and to factor is the difference of two squares. Think of this form as

$$u^2 \ominus v^2 = (u + v)(u - v)$$

$\swarrow$ difference   opposite signs

To recognize perfect square terms, look for terms whose coefficients are squares of integers and whose variables have *even* powers.

EXAMPLE 2
Factoring the Difference of Two Squares

Factor the following:

(a) $9 - 25x^2$ (b) $(x + 2)^2 - y^2$ (c) $16x^4 - y^4$

Solution:

(a) Using the difference of two squares formula with $u = 3$ and $v = 5x$, we get

$$9 - 25x^2 = (3)^2 - (5x)^2 = (3 + 5x)(3 - 5x)$$

(b) Using the difference of two squares formula with $u = (x + 2)$ and $v = y$, we get

$$(x + 2)^2 - y^2 = [(x + 2) + y][(x + 2) - y]$$
$$= (x + y + 2)(x - y + 2)$$

(c) Applying the formula twice, we get

$$16x^4 - y^4 = (4x^2)^2 - (y^2)^2 \qquad \text{\textit{1st application}}$$
$$= (4x^2 + y^2)(4x^2 - y^2)$$
$$= (4x^2 + y^2)[(2x)^2 - (y)^2]$$
$$= (4x^2 + y^2)(2x + y)(2x - y) \quad \text{\textit{2nd application}}$$

Perfect Square Trinomials

A perfect square trinomial is a trinomial that is the square of a binomial. It has the form

$$u^2 + 2uv + v^2 = (u + v)^2 \qquad \text{or} \qquad u^2 - 2uv + v^2 = (u - v)^2$$

same sign   same sign

Note the following characteristics of a perfect square trinomial:

1. The first term is a square, u^2.
2. The last term is a square, v^2.
3. The middle term is twice the product of the terms u and v.
4. The sign of the middle term determines the sign in the binomial.

EXAMPLE 3
Factoring Perfect
Square Trinomials

Factor

(a) $16x^2 + 8x + 1$ (b) $9x^2 - 6xy + y^2$ (c) $8x^4 + 24x^2 + 18$

Solution:

(a) In this case,

$$u^2 = 16x^2 = (4x)^2$$
$$v^2 = (1)^2$$
$$2uv = 2(4x)(1) = 8x$$

Thus,

$$16x^2 + 8x + 1 = (4x + 1)^2$$

(b) We have

$$u^2 = 9x^2 = (3x)^2$$
$$v^2 = (y)^2$$
$$-2uv = -2(3x)(y) = -6xy$$

Thus,

$$9x^2 - 6xy + y^2 = (3x - y)^2$$

(c) In this instance, neither the first nor the last term is a perfect square. However, if we first factor out the common factor 2, we get

$$8x^4 + 24x^2 + 18 = 2(4x^4 + 12x^2 + 9)$$

For the new trinomial, we have

$$u^2 = 4x^4 = (2x^2)^2$$
$$v^2 = (3)^2$$
$$2uv = 2(2x^2)(3) = 12x^2$$

Thus,

$$8x^4 + 24x^2 + 18 = 2(4x^4 + 12x^2 + 9)$$
$$= 2(2x^2 + 3)^2$$

Remark: When attempting to factor using one of the four special polynomial forms, you should first check for any common monomial factors and remove them before proceeding further.

Trinomials with Binomial Factors

Some trinomials that are not perfect squares can be factored into the product of two binomials according to the formula

$$mx^2 + nx + r = (ax + b)(cx + d)$$

This is simply the **FOIL** method in reverse. The goal is to find a combination of factors of m and r so that the outside and inside (**O** and **I**) products yield the middle term nx. This means that in the following scheme we want $nx = \mathbf{O} + \mathbf{I}$.

$$mx^2 + nx + r = (?x + ?)(?x + ?)$$

$$nx = \mathbf{O} + \mathbf{I} \qquad \mathbf{I} \qquad \mathbf{O}$$

Remark: It is impossible to factor some trinomials $mx^2 + nx + r$ into the product of two binomials with rational coefficients. Such trinomials are called **irreducible quadratic polynomials,** and more will be said about them in Sections 1.10 and 2.3.

EXAMPLE 4
Trinomials with Binomial Factors

Factor

(a) $x^2 - 7x + 12$ (b) $2x^2 + x - 6$ (c) $8x^2 + 22x + 9$

Solution:

(a) Consider

$$mx^2 + nx + r = x^2 - 7x + 12$$

Since $r = +12$, its factors have like signs, and since $n = -7$, both signs will be negative. Let's try $r = 12 = (-2)(-6)$.

$$x^2 - 7x + 12 \overset{?}{=} (x - 2)(x - 6) \qquad \textit{Test possible factors}$$
$$\mathbf{O} + \mathbf{I} = -8x \neq -7x \qquad \textit{Fails } \mathbf{O} + \mathbf{I} \textit{ Test}$$

A quick $\mathbf{O} + \mathbf{I}$ test (in your head) of the factors $(x - 1)$ and $(x - 12)$ will show that they don't work either, so we are left with the following (correct) factorization.

$$x^2 - 7x + 12 \overset{?}{=} (x - 3)(x - 4) \qquad \textit{Test possible factors}$$
$$\mathbf{O} + \mathbf{I} = -7x \qquad \textit{Satisfies } \mathbf{O} + \mathbf{I} \textit{ Test}$$

(b) Since $r = -6$, the factors of r must have *unlike* signs. We try $m = (2)(1)$ and $r = (-6)(1)$.

$$2x^2 + x - 6 \overset{?}{=} (2x - 6)(x + 1) \qquad \textit{Test possible factors}$$
$$\mathbf{O} + \mathbf{I} = -4x \neq x \qquad \textit{Fails } \mathbf{O} + \mathbf{I} \textit{ Test}$$

Try $r = (3)(-2)$, then

$$2x^2 + x - 6 \overset{?}{=} (2x + 3)(x - 2) \qquad \textit{Test possible factors}$$
$$\mathbf{O} + \mathbf{I} = -x \neq x \qquad \textit{Fails } \mathbf{O} + \mathbf{I} \textit{ Test}$$

Here we merely had a wrong sign, so we change signs for *both* factors of r, and we find the correct factorization to be

$$2x^2 + x - 6 \overset{?}{=} (2x - 3)(x + 2) \quad \textit{Test possible factors}$$

$$\mathbf{O} + \mathbf{I} = x \qquad\qquad\qquad \textit{Satisfies } \mathbf{O} + \mathbf{I} \textit{ Test}$$

(c) To factor $8x^2 + 22x + 9$, we choose combinations of factors from among

factors of 8	factors of 9
1, 2, 4, 8	1, 3, 9

The following tabular solution method is one of the most efficient ways to find and test *all* combinations of factors of m and r.

TABULAR METHOD FOR FACTORING TRINOMIALS

$$8x^2 + 22x + 9 = (?x + ?)(?x + ?)$$

Factors of $m = 8$	Factors of $r = 9$	Factor Combinations $(?x + ?)(?x + ?)$	$\mathbf{O} + \mathbf{I} \overset{?}{=} 22$
1, 2, 4, 8	1, 3, 9	$(x + 1)(8x + 9)$	$\mathbf{O} + \mathbf{I} = 17$
		$(x + 9)(8x + 9)$	$\mathbf{O} + \mathbf{I} = 73$
		$(x + 3)(8x + 3)$	$\mathbf{O} + \mathbf{I} = 27$
		$(2x + 1)(4x + 9)$	$\mathbf{O} + \mathbf{I} = 22$
		$(2x + 9)(4x + 1)$	$\mathbf{O} + \mathbf{I} = 38$
		$(2x + 3)(4x + 3)$	$\mathbf{O} + \mathbf{I} = 18$

Correct Factorization: $8x^2 + 22x + 9 = (2x + 1)(4x + 9)$

Remark: You will find that much of this work can be done in your head, and you can save even more time finding the correct factorization of trinomials. Note also that if $\mathbf{O} + \mathbf{I}$ yields the right number but the wrong sign, simply change the sign of *both* factors of r.

Sum or Difference of Cubes

The next two formulas show that sums and differences of cubes factor quite simply. Pay special attention to the signs of the terms.

$$u^3 + v^3 = (u + v)(u^2 - uv + v^2) \qquad u^3 - v^3 = (u - v)(u^2 + uv + v^2)$$

like signs   like signs

unlike signs   unlike signs

EXAMPLE 5
Sum or Difference of Cubes

Factor

(a) $y^3 - 27x^3$ (b) $(x - 2)^3 + 8$

Solution:

(a) Consider $u = y$ and $v = (3x)$, then

$$y^3 - 27x^3 = (y)^3 - (3x)^3$$
$$= (y - 3x)[y^2 + y(3x) + (3x)^2]$$
$$= (y - 3x)(y^2 + 3xy + 9x^2)$$

(b) Consider $u = (x - 2)$ and $v = 2$, then

$$(x - 2)^3 + 8 = (x - 2)^3 + (2)^3$$
$$= [(x - 2) + 2][(x - 2)^2 - (x - 2)(2) + (2)^2]$$
$$= (x)(x^2 - 4x + 4 - 2x + 4 + 4)$$
$$= x(x^2 - 6x + 12)$$

Factoring by Grouping

Sometimes polynomials with more than three terms can be factored by a method called **factoring by grouping.** It is not always obvious which terms to group, and sometimes several different groupings will work. The goal is to find groupings that lead to the special factorizations discussed in this section.

EXAMPLE 6
Factoring by Grouping

Factor completely

(a) $2xy - 8 + 8y - 2x$ (b) $4x^2 - 4x - y^2 + 1$

Solution:

(a) If we group as

$$2xy - 8 + 8y - 2x = (2xy - 8) + (8y - 2x) \quad \textit{Group terms}$$
$$= 2(xy - 4) + 2(4y - x) \quad \textit{Factor groups}$$

we get no common factors to continue our factoring. So we try

$$2xy - 8 + 8y - 2x = (2xy + 8y) - (8 + 2x) \quad \textit{Group terms}$$
$$= 2y(x + 4) - 2(4 + x) \quad \textit{Factor groups}$$
$$= (2y - 2)(x + 4) \quad \textit{Common factor}$$
$$= 2(y - 1)(x + 4) \quad \textit{Monomial factor}$$

(b) In this case, consider a grouping that leads to a special trinomial

$$4x^2 - 4x - y^2 + 1$$
$$= (4x^2 - 4x + 1) - y^2 \quad \textit{Perfect square trinomial}$$

$$= (2x - 1)^2 - y^2 \qquad \textit{Difference of squares}$$
$$= [(2x - 1) + y][(2x - 1) - y] \qquad \textit{Factor}$$
$$= (2x - 1 + y)(2x - 1 - y) \qquad \textit{Simplify}$$

A general guideline to follow in factoring polynomials is to factor out any common monomial factor, factor according to one of the special polynomial forms, or factor by grouping.

Section Exercises 1.8

In Exercises 1–8, factor out the common factor.

1. $3x + 6$
2. $5y - 30$
3. $xy - xz$
4. $-2cx - 10cy$
5. $9a^2b - 12ab^3$
6. $16x^2y^2 + 8xy^2 - 20xy$
7. $(x + y)z^2 - (x + y)$
8. $(x - 1)(y + z) + (x - 1)(y - z)$

In Exercises 9–16, factor each difference of squares.

9. $x^2 - 36$
10. $z^2 - 100$
11. $16y^2 - 9$
12. $49 - 9y^2$
13. $(x - 1)^2 - 4$
14. $25 - (z + 5)^2$
15. $81 - y^2$
16. $x^4 - 16$

In Exercises 17–24, factor each perfect square trinomial.

17. $x^2 - 4x + 4$
18. $x^2 + 10x + 25$
19. $4x^2 + 4x + 1$
20. $9x^2 - 12x + 4$
21. $x^2 - 4xy + 4y^2$
22. $25y^2 + 40yz + 16z^2$
23. $a^2b^2 - 2abc + c^2$
24. $x^4 - 2x^2z + z^2$

In Exercises 25–32, factor each trinomial with distinct binomial factors.

25. $x^2 + x - 2$
26. $x^2 + 5x + 6$
27. $x^2 - 5x + 6$
28. $2x^2 - x - 1$
29. $3x^2 - 5x + 2$
30. $x^2 - xy - 2y^2$
31. $2y^2 + 21yz - 36z^2$
32. $60x^2 - 61xy + 15y^2$

In Exercises 33–40, factor each sum or difference of cubes.

33. $x^3 - 8$
34. $x^3 - 27$
35. $y^3 + 64$
36. $z^3 + 125$
37. $x^3 - 27y^3$
38. $64x^3 + 125y^3$
39. $x^6 + 64$
40. $(x - a)^3 + b^3$

In Exercises 41–46, factor by grouping.

41. $xy - y + xz - z$
42. $xy - xz + 5y - 5z$
43. $5xy - 3y + 10x - 6$
44. $2ab + 6a - 7b - 21$
45. $7 - 10x - 7y + 10xy$
46. $ar - as - 8r + 8s$

In Exercises 47–60, completely factor each expression.

47. $2x^3 - 2x^2 - 4x$
48. $2ay^3 - 7ay^2 - 15ay$
49. $63rs^2 - 7r^3$
50. $80 - 5z^2$
51. $4x^3y - 4x^2y^2 + xy^3$
52. $150x + 120x^2 + 24x^3$
53. $6x^4 - 48xy^3$
54. $(x^2 + y^2)^2 - 4x^2y^2$
55. $(x^2 + 2y^2)^2 - 9x^2y^2$
56. $9x(3x - 9)^2 + (3x - 9)^3$
57. $4x^2(2x - 1) + 2x(2x - 1)^2$
58. $27x^2y - x^2y^4$
59. $2x^2z^2 + 2x^2z + 4xyz^2 + 4xyz$
60. $2r^2u + 10ru - 6r^2v - 30rv$
61. Use the formula

$$(x + a)^5$$
$$= x^5 + 5x^4a + 10x^3a^2 + 10x^2a^3 + 5xa^4 + a^5$$

to factor

$$x^5 - 10x^4 + 40x^3 - 80x^2 + 80x - 32$$

62. Use the formula

$$(x + a)^3 = x^3 + 3x^2a + 3xa^2 + a^3$$

to factor

$$27x^3 + 27x^2y + 9xy^2 + y^3$$

Rational Expressions   1.9

Quotients of algebraic expressions are called **fractional expressions** (or **algebraic fractions**). In particular, a quotient of two polynomials is called a **rational expression.** Some examples are

$$\frac{3x - 7}{5x^2 - 2x + 1}, \qquad \frac{4}{9xy^2} \qquad \text{and} \qquad \frac{x^3 - 4x}{x^2 + x - 2}$$

The rules for operating with algebraic fractions are much like those for numerical fractions (see p. 7). The key to success lies in your ability to *factor* polynomials. Note in the next example how factoring plays the key role in simplifying algebraic fractions.

EXAMPLE 1
Simplifying Algebraic Fractions

Reduce to lowest terms.

(a) $\dfrac{x^2 + 4x - 12}{3x - 6}$ (b) $\dfrac{x^3 - 4x}{x^2 + x - 2}$ (c) $\dfrac{12 + x - x^2}{2x^2 - 9x + 4}$

Solution:

(a) Factoring both numerator and denominator and then reducing, we have

$$\frac{x^2 + 4x - 12}{3x - 6} = \frac{(x + 6)(x - 2)}{3(x - 2)} \qquad \textit{Factor completely}$$

$$= \frac{x + 6}{3} \qquad \textit{Reduce}$$

Don't make the common mistake of trying to further reduce by canceling *terms:*

$$\frac{x + \overset{2}{\cancel{6}}}{\cancel{3}} \qquad \text{DOES NOT EQUAL} \qquad x + 2$$

Remember: *To reduce fractions, we cancel factors, not terms.*

(b) $\dfrac{x^3 - 4x}{x^2 + x - 2} = \dfrac{x(x^2 - 4)}{(x + 2)(x - 1)}$

$$= \frac{x(\cancel{x + 2})(x - 2)}{(\cancel{x + 2})(x - 1)} \qquad \textit{Factor completely}$$

$$= \frac{x(x - 2)}{x - 1} \qquad \textit{Cancel common factors}$$

(c) $\dfrac{12 + x - x^2}{2x^2 - 9x + 4} = \dfrac{(4 - x)(3 + x)}{(2x - 1)(x - 4)} \qquad \textit{Factor completely}$

$$= \frac{-(\cancel{x - 4})(3 + x)}{(2x - 1)(\cancel{x - 4})} \qquad (4 - x) = -1(x - 4)$$

$$= \frac{-(3 + x)}{2x - 1} \qquad \textit{Reduce}$$

$$= \frac{3 + x}{1 - 2x} \qquad \textit{Simplest form}$$

Remark: Reducing to lowest terms (or simplifying) involves the cancellation of *all* factors common to both the numerator and the denominator. It is okay, and often preferred, to leave answers in factored form, as in part (b). Sometimes changing signs of a fraction may make it look simpler, as seen in part (c).

EXAMPLE 2
Multiplying and Dividing
Algebraic Fractions

Perform the indicated operations and simplify.

(a) $\dfrac{2x^2 + x - 6}{x^2 + 4x - 5} \cdot \dfrac{x^3 - 3x^2 + 2x}{4x^2 - 6x}$

(b) $\dfrac{x^3 - 8}{x^2 - 4} \div \dfrac{x^2 + 2x + 4}{x^3 + 8}$

Solution:

(a) $\dfrac{2x^2 + x - 6}{x^2 + 4x - 5} \cdot \dfrac{x^3 - 3x^2 + 2x}{4x^2 - 6x}$

$$= \frac{(2x - 3)(x + 2)}{(x + 5)(x - 1)} \cdot \frac{x(x - 2)(x - 1)}{2x(2x - 3)} \qquad \textit{Factor completely}$$

$$= \frac{(2x - 3)(x + 2)(x)(x - 2)(x - 1)}{(x + 5)(x - 1)(2)(x)(2x - 3)} \qquad \textit{Multiply and reduce}$$

$$= \frac{(x + 2)(x - 2)}{2(x + 5)} = \frac{x^2 - 4}{2x + 10} \qquad \textit{Simplest form}$$

(b) $\dfrac{x^3 - 8}{x^2 - 4} \div \dfrac{x^2 + 2x + 4}{x^3 + 8}$

$$= \frac{(x - 2)(x^2 + 2x + 4)}{(x + 2)(x - 2)} \cdot \frac{(x + 2)(x^2 - 2x + 4)}{x^2 + 2x + 4}$$

$$= \frac{(x - 2)(x^2 + 2x + 4)(x + 2)(x^2 - 2x + 4)}{(x + 2)(x - 2)(x^2 + 2x + 4)}$$

$$= x^2 - 2x + 4$$

Combining Rational Expressions

To add or subtract algebraic expressions, we use the familiar LCD (lowest common denominator) method, or the basic definition

$$\frac{a}{b} \pm \frac{c}{d} = \frac{ad \pm bc}{bd}$$

EXAMPLE 3
Adding or Subtracting
Algebraic Fractions

Perform the given operations and simplify.

(a) $\dfrac{x}{x-3} - \dfrac{2}{3x+4}$

(b) $\dfrac{3}{x-1} - \dfrac{2}{x} + \dfrac{x+3}{x^2-1}$

(c) $\dfrac{x-2}{x^2+6x+9} - \dfrac{x+2}{2x^2-18}$

Solution:

(a) The basic definition is very efficient when you are working with just two fractions that have no common factors in their denominators. This method is often called **cross multiplication.**

$$\dfrac{x}{x-3} \times \dfrac{2}{3x+4} = \dfrac{x(3x+4)-2(x-3)}{(x-3)(3x+4)} \quad \textit{Cross multiply}$$

$$= \dfrac{3x^2+4x-2x+6}{(x-3)(3x+4)} \quad \textit{Remove parentheses}$$

$$= \dfrac{3x^2+2x+6}{(x-3)(3x+4)} \quad \textit{Combine like terms}$$

Missing Factor Model

(b) For three or more fractions, the LCD method works well. However, we can shortcut this LCD method by recognizing that the numerator of *each* fraction gets multiplied by the factors in the LCD that its denominator lacks. We refer to these as **missing factors** (or mf's). Using the missing factor model, we have

$$\dfrac{3}{x-1} - \dfrac{2}{x} + \dfrac{x+3}{(x+1)(x-1)}$$

$$= \dfrac{3(\mathrm{mf}_1) - 2(\mathrm{mf}_2) + (x+3)(\mathrm{mf}_3)}{\mathrm{LCD}}$$

$$= \dfrac{3(x)(x+1) - 2(x+1)(x-1) + (x+3)(x)}{x(x+1)(x-1)}$$

$$= \dfrac{3x^2+3x-2x^2+2+x^2+3x}{x(x+1)(x-1)}$$

$$= \dfrac{2x^2+6x+2}{x(x+1)(x-1)}$$

(c) Using our missing factor model, we have

Factored Denominators   **LCD**

$(x+3)^2,\ 2(x+3)(x-3)$ ⟹ $2(x+3)^2(x-3)$

$$\frac{x-2}{(x+3)^2} - \frac{x+2}{2(x+3)(x-3)}$$

$$= \frac{(x-2)(2)(x-3) - (x+2)(x+3)}{2(x+3)^2(x-3)}$$

$$= \frac{2(x^2 - 5x + 6) - (x^2 + 5x + 6)}{2(x+3)^2(x-3)}$$

$$= \frac{x^2 - 15x + 6}{2(x+3)^2(x-3)}$$

Remark: Sometimes the numerator of the answer has a factor in common with the denominator. In such cases the answer should be reduced.

Compound Fractions

Thus far in this section we have limited our operations to algebraic fractions that are rational expressions. Algebraic fractions that have separate fractions in the numerator, or denominator, or both, are called **compound fractions.** Some examples are

$$\frac{\dfrac{2}{x} - 3}{1 - \dfrac{1}{x-1}}, \qquad \frac{\dfrac{1}{x+h} - \dfrac{1}{x}}{h}, \qquad \text{and} \qquad \frac{\sqrt{4-x^2} + \dfrac{x^2}{\sqrt{4-x^2}}}{4-x^2}$$

Compound fractions can be simplified in either of two ways.

METHODS FOR SIMPLIFYING COMPOUND FRACTIONS

Combining Fractions	**Rationalizing Fractions**
1. Combine numerator's terms into one fraction.	1. Find LCD of all fractions within the compound fraction.
2. Combine denominator's terms into one fraction.	2. Multiply each term of compound fraction by LCD.
3. Invert denominator and multiply by numerator.	3. Simplify each product.
4. Reduce result to lowest terms.	4. Reduce result to lowest terms.

EXAMPLE 4
Simplifying Compound Fractions

Simplify

$$\frac{\dfrac{2}{x} - 3}{1 - \dfrac{1}{x-1}}$$

Solution:

Using the *combining method,* we have

$$\frac{\dfrac{2}{x} - 3}{1 - \dfrac{1}{x-1}} = \frac{\dfrac{2 - 3(x)}{x}}{\dfrac{1(x-1) - 1}{x-1}}$$ *Combine fractions*

$$= \frac{\dfrac{2 - 3x}{x}}{\dfrac{x-2}{x-1}}$$ *Simplify*

$$= \frac{2 - 3x}{x} \cdot \frac{x-1}{x-2}$$ *Invert denominator*

$$= \frac{-2 + 5x - 3x^2}{x^2 - 2x}$$ *Multiply and simplify*

Algebra of Calculus

The rationalizing method works well with radicals and negative exponents—situations frequently encountered in calculus. Our last example demonstrates the effectiveness of this method on some simplification problems taken from a calculus text.

EXAMPLE 5
Rationalizing Radicals
and Negative Exponents

Simplify

(a) $\dfrac{\sqrt{4 - x^2} + \dfrac{x^2}{\sqrt{4 - x^2}}}{4 - x^2}$ (b) $\dfrac{2x(6x - 4)^{-2/3} - \frac{1}{3}(6x - 4)^{1/3}}{x^2}$

Solution:

(a) $\dfrac{\sqrt{4 - x^2} + \dfrac{x^2}{\sqrt{4 - x^2}}}{4 - x^2}$

$$= \frac{\left(\sqrt{4 - x^2} + \dfrac{x^2}{\sqrt{4 - x^2}}\right)\sqrt{4 - x^2}}{(4 - x^2)\sqrt{4 - x^2}}$$ *Multiply by LCD*

$$= \frac{(4 - x^2) + x^2}{(4 - x^2)^{3/2}}$$ *Distribute and reduce*

$$= \frac{4}{\sqrt{(4 - x^2)^3}}$$ *Simplify*

(b) Note in this case that

$$(6x - 4)^{-2/3} = \frac{1}{(6x - 4)^{2/3}}$$

Hence, the LCD for the given numerator is $3(6x - 4)^{2/3}$, and we get

$$\frac{2x(6x - 4)^{-2/3} - \tfrac{1}{3}(6x - 4)^{1/3}}{x^2}$$

$$= \frac{[2x(6x - 4)^{-2/3} - \tfrac{1}{3}(6x - 4)^{1/3}](3)(6x - 4)^{2/3}}{x^2(3)(6x - 4)^{2/3}}$$

$$= \frac{6x(1) - (6x - 4)}{3x^2(6x - 4)^{2/3}}$$

$$= \frac{4}{3x^2 \sqrt[3]{(6x - 4)^2}}$$

Remark: In Section 1.10 we will investigate in more detail some special techniques of the algebra of calculus.

Section Exercises 1.9

In Exercises 1–8, reduce each fraction to lowest terms.

1. $\dfrac{3xy}{xy + x}$

2. $\dfrac{9x^2 + 9xy}{xy + y^2}$

3. $\dfrac{x^3 + 5x^2 + 6x}{x^2 - 4}$

4. $\dfrac{x^2 + 8x - 20}{x^2 + 11x + 10}$

5. $\dfrac{xy + yz + 2x + 2z}{xy + xz}$

6. $\dfrac{a^2b + a^2c - 2ab^2 - 2abc}{b^3 - bc^2}$

7. $\dfrac{a^3 - 8}{a^2 + 2a + 4}$

8. $\dfrac{c^3 - 2c^2 - 3c}{c^3 + 1}$

In Exercises 9–40, perform the indicated operations and simplify your answer.

9. $\dfrac{5}{x - 1} + \dfrac{x}{x - 1}$

10. $\dfrac{2x - 1}{x + 3} + \dfrac{1 - x}{x + 3}$

11. $\dfrac{4}{x} - \dfrac{3}{x^2}$

12. $\dfrac{5}{x - 1} + \dfrac{3}{x}$

13. $\dfrac{2}{x + 2} - \dfrac{1}{x - 2}$

14. $\dfrac{x}{x^2 + x - 2} - \dfrac{1}{x + 2}$

15. $\dfrac{1}{x^2 - x - 2} - \dfrac{x}{x^2 - 5x + 6}$

16. $\dfrac{x - 1}{x^2 + 5x + 4} + \dfrac{2}{x^2 - x - 2} + \dfrac{10}{x^2 + 2x - 8}$

17. $-\dfrac{1}{x} + \dfrac{2}{x^2 + 1}$

18. $\dfrac{2}{x + 1} + \dfrac{1 - x}{x^2 - 2x + 3}$

19. $\dfrac{5}{x - 1} \cdot \dfrac{x - 1}{25(x - 2)}$

20. $\dfrac{(x + 5)(x - 3)}{x + 2} \cdot \dfrac{1}{(x + 5)(x + 2)}$

21. $\dfrac{(x - 9)(x + 7)}{x + 1} \cdot \dfrac{x}{9 - x}$

22. $\dfrac{x + 13}{x^3(3 - x)} \cdot \dfrac{x(x - 3)}{5}$

23. $\dfrac{r}{r - 1} \cdot \dfrac{r^2 - 1}{r^2}$

24. $\dfrac{4y - 16}{5y + 15} \cdot \dfrac{2y + 6}{4 - y}$

25. $\dfrac{t^2 - t - 6}{t^2 + 6t + 9} \cdot \dfrac{t + 3}{t^2 - 4}$

26. $\dfrac{y^3 - 8}{2y^3} \cdot \dfrac{4y}{y^2 - 5y + 6}$

27. $\dfrac{x^2 + xy - 2y^2}{x^3 + x^2y} \cdot \dfrac{x}{x^2 + 3xy + 2y^2}$

28. $\dfrac{x^3 - y^3}{x + y} \cdot \dfrac{x^2 + y^2}{x^2 - y^2}$

29. $\dfrac{3(x + y)}{4} \div \dfrac{x + y}{2}$

30. $\dfrac{x + 2}{5(x - 3)} \div \dfrac{x - 2}{5(x - 3)}$

31. $\dfrac{(xy)^2}{(x + y)^2} \div \dfrac{xy}{(x + y)^3}$

32. $\dfrac{\dfrac{x^2 - y^2}{xy}}{(x - y)^2}$

33. $\dfrac{\dfrac{x-4}{5x}}{\dfrac{x^3-64}{x^3}}$

34. $\dfrac{\dfrac{x^2-8x-240}{x^3+4x^2}}{\dfrac{x^2-144}{x^2}}$

35. $\dfrac{1}{(x-y)} \cdot \left(\dfrac{x}{y}-\dfrac{y}{x}\right)$

36. $\left(\dfrac{1}{x^2+y^2}\right)\left(\dfrac{x}{y}+\dfrac{y}{x}\right)$

37. $(x^2-1)\left(\dfrac{1}{x+1}-\dfrac{1}{x-1}\right)$

38. $\left(\dfrac{2x}{x^2+1}-\dfrac{1}{x}\right) \cdot \dfrac{x}{x+1}$

39. $\dfrac{x^2+1}{x+1} \cdot \left(\dfrac{1}{x}+\dfrac{1-x}{x^2+1}\right)$

40. $\dfrac{1}{x^4} \cdot \left[x+3+\dfrac{6}{x-1}+\dfrac{4}{(x-1)^2}+\dfrac{1}{(x-1)^3}\right]$

In Exercises 41–52, simplify the complex fractions by (a) the combining method, and (b) the rationalizing method.

41. $\dfrac{\dfrac{x}{y}-1}{x-y}$

42. $\dfrac{x-y}{\dfrac{x}{y}-\dfrac{y}{x}}$

43. $\dfrac{\left(\dfrac{x+3}{x-3}\right)^2}{\dfrac{1}{x+3}+\dfrac{1}{x-3}}$

44. $\dfrac{\dfrac{1}{x}-\dfrac{1}{x+1}}{\dfrac{1}{x+1}}$

45. $\dfrac{\dfrac{5}{y}-\dfrac{6}{2y+1}}{\dfrac{5}{y}+4}$

46. $\dfrac{\dfrac{5}{x^2-x-2}-\dfrac{6}{x^2+x-6}}{\dfrac{4}{x^2+4x+3}}$

47. $\dfrac{\dfrac{1}{(x+h)^2}-\dfrac{1}{x^2}}{h}$

48. $\dfrac{\dfrac{x+h}{x+h+1}-\dfrac{x}{x+1}}{h}$

49. $\dfrac{x^2+x^{-1}y^2}{x^{-1}y^2-y+x}$

50. $\dfrac{a^2b^{-1}+a^{-1}b^2}{a^{-1}+b^{-1}}$

51. $\dfrac{x^4-2^{-4}}{4x^2-1}$

52. $\dfrac{-2x^2(x+1)^{-3}+2x(x+1)^{-2}}{(x+1)^{-1}-(x+1)^{-2}}$

In Exercises 53–60, rationalize the radicals and simplify.

53. $\dfrac{\sqrt{x}-\dfrac{1}{2\sqrt{x}}}{\sqrt{x}}$

54. $\dfrac{\sqrt{1-x^2}+\dfrac{x^2}{\sqrt{1-x^2}}}{1-x^2}$

55. $\dfrac{\dfrac{t^2}{\sqrt{t^2+1}}-\sqrt{t^2+1}}{t^2}$

56. $\dfrac{(2x+1)^{1/3}-\dfrac{4x}{3(2x+1)^{2/3}}}{(2x+1)^{2/3}}$

57. $\dfrac{(x^2+1)^{1/4}-x^2(1+x^2)^{-3/4}}{2(x^2+1)^{1/2}}$

58. $\dfrac{x^2(1+x^2)^{1/2}-x^2(1+x^2)^{-1/2}}{1+x^2}$

59. $\dfrac{-x^3(1-x^2)^{-1/2}-2x(1-x^2)^{1/2}}{x^4}$

60. $\dfrac{x^{1/3}-(1/3)(x^{-2/3})}{x^{-2/3}}$

Algebraic Errors and the Algebra of Calculus

1.10

Before we wrap up our review of the fundamental concepts of algebra, we want to look at some common algebraic errors. Many of these errors are made because they are the *easiest* things to do. This is a strong temptation, especially when the error makes the remainder of the problem much simpler (a big temptation in calculus). Regardless of *why* they are made, we feel that it is helpful to review a list of errors to avoid.

ALGEBRAIC ERRORS TO AVOID

	Error	**Correct Form**	**Comment**
1. Parentheses	$a - (x - b) \neq a - x - b$	$a - (x - b) = a - x + b$	Change all signs when distributing negative through parentheses.
	$(a + b)^2 \neq a^2 + b^2$	$(a + b)^2 = a^2 + 2ab + b^2$	Don't forget middle term when squaring binomials.
	$\left(\dfrac{1}{2}a\right)\left(\dfrac{1}{2}b\right) \neq \dfrac{1}{2}(ab)$	$\left(\dfrac{1}{2}a\right)\left(\dfrac{1}{2}b\right) = \dfrac{1}{4}(ab) = \dfrac{ab}{4}$	1/2 occurs twice as a factor.
2. Fractions	$\dfrac{a}{x + b} \neq \dfrac{a}{x} + \dfrac{a}{b}$	Leave as $\dfrac{a}{x + b}$	Don't add denominators when adding fractions.
	$\dfrac{\left(\dfrac{x}{a}\right)}{b} \neq \dfrac{bx}{a}$	$\dfrac{\left(\dfrac{x}{a}\right)}{b} = \left(\dfrac{x}{a}\right)\left(\dfrac{1}{b}\right) = \dfrac{x}{ab}$	Multiply by reciprocal when dividing.
	$\dfrac{1}{a} + \dfrac{1}{b} \neq \dfrac{1}{a + b}$	$\dfrac{1}{a} + \dfrac{1}{b} = \dfrac{a + b}{ab}$	Use definition for adding fractions.
	$\dfrac{1}{3x} \neq \dfrac{1}{3}x$	$\dfrac{1}{3x} = \dfrac{1}{3} \cdot \dfrac{1}{x}$	Use definition for multiplying fractions.
	$\dfrac{1}{3}x \neq \dfrac{1}{3x}$	$\dfrac{1}{3}x = \dfrac{1}{3} \cdot x = \dfrac{x}{3}$	Be careful when using a slash to denote division.
	$\dfrac{1}{x} + 2 \neq \dfrac{1}{(x + 2)}$	$\dfrac{1}{x} + 2 = \dfrac{1}{x} + 2$ $= \dfrac{1 + 2x}{x}$	Be careful when using a slash to denote division.
3. Exponents and Radicals	$(x^2)^3 \neq x^5$	$(x^2)^3 = x^{2 \cdot 3} = x^6$	Multiply exponents when an exponential form is raised to a power.
	$x^2 \cdot x^3 \neq x^6$	$x^2 \cdot x^3 = x^{2+3} = x^5$	Add exponents when multiplying exponentials with like bases.
	$2x^3 \neq (2x)^3$	$2x^3 = 2(x^3)$	Exponents have priority over coefficients.
	$\dfrac{1}{x^{1/2} - x^{1/3}} \neq x^{-1/2} - x^{-1/3}$	Leave as $\dfrac{1}{x^{1/2} - x^{1/3}}$	Don't move term-by-term from denominator to numerator.

$\sqrt{5x} \neq 5\sqrt{x}$	$\sqrt{5x} = \sqrt{5}\sqrt{x}$	Radicals apply to every factor inside radical.
$\sqrt{x^2 + a^2} \neq x + a$	Leave as $\sqrt{x^2 + a^2}$	Don't apply radicals term-by-term.
$\sqrt{-x^2 + a^2} \neq -\sqrt{x^2 - a^2}$	Leave as $\sqrt{-x^2 + a^2}$ or write as $\sqrt{a^2 - x^2}$	Don't factor negatives out of square roots.

4. Cancellations

$\dfrac{a + bx}{a} \neq 1 + bx$	$\dfrac{a + bx}{a} = \dfrac{a}{a} + \dfrac{bx}{a}$ $= 1 + \dfrac{b}{a}x$	Cancel common factors, *not* common terms.
$\dfrac{a + ax}{a} \neq a + x$	$\dfrac{a + ax}{a} = \dfrac{a(1 + x)}{a}$ $= 1 + x$	Factor *before* canceling.
$1 + \dfrac{x}{2x} \neq 1 + \dfrac{1}{x}$	$1 + \dfrac{x}{2x} = 1 + \dfrac{1}{2} = \dfrac{3}{2}$	When canceling factors with like bases, subtract exponents, *not* coefficients.

The Algebra of Calculus

In our review of algebra to this point we have required that most answers be expressed in simplest form. In calculus you will often have to reverse this procedure. At times we need *unsimplified* algebraic forms in order to perform the operations of calculus. Consequently, it is often necessary to take a simplified algebraic expression and **unsimplify** it. The following list, taken from a standard calculus text, shows some of the "backwards" algebra needed in calculus.

SOME ALGEBRA OF CALCULUS

	Required Simplest Algebraic Form	Desired Unsimplified Form for Calculus	Comment
1. Unusual Factoring	$\dfrac{5x^4}{8}$	$\dfrac{5}{8}(x^4)$	Factor out fractional coefficient.
	$\dfrac{x^2 + 3x}{-6}$	$-\dfrac{1}{6}(x^2 + 3x)$	Factor out fractional coefficient.
	$2x^2 - x - 3$	$2\left(x^2 - \dfrac{x}{2} - \dfrac{3}{2}\right)$	Factor out leading coefficient.

	$\dfrac{x}{2}(x+1)^{-1/2} + (x+1)^{1/2}$	$\dfrac{(x+1)^{-1/2}}{2}[x+2(x+1)]$	Remove factor with negative exponent.
2. Inserting Required Factors	$(2x-1)^3$	$\dfrac{1}{2}(2x-1)^3(2)$	Multiply and divide by desired factor.
	$7x^2(4x^3-5)^{1/2}$	$\dfrac{7}{12}(4x^3-5)^{1/2}(12x^2)$	Multiply and divide by desired factor.
	$\dfrac{4x^2}{9} - 4y^2 = 1$	$\dfrac{x^2}{9/4} - \dfrac{y^2}{1/4} = 1$	Invert and *divide*.
3. Rewriting with Negative Exponents	$\dfrac{9}{-5x^3}$	$\dfrac{9}{-5}(x^{-3})$	Move factor to numerator and change sign of exponent.
	$\dfrac{7}{\sqrt{2x-3}}$	$7(2x-3)^{-1/2}$	Move factor to numerator and change sign of exponent.
4. Writing a Fraction as a Sum of Terms	$\dfrac{x+2x^2+1}{\sqrt{x}}$	$x^{1/2} + 2x^{3/2} + x^{-1/2}$	Divide each term by $x^{1/2}$.
	$\dfrac{1+x}{x^2+1}$	$\dfrac{1}{x^2+1} + \dfrac{x}{x^2+1}$	Rewrite fraction as sum of fractions.
	$\dfrac{2x}{x^2+2x+1}$	$\dfrac{2x+2-2}{x^2+2x+1} =$	Add and subtract terms in numerator.
		$\dfrac{2x+2}{x^2+2x+1} - \dfrac{2}{(x+1)^2}$	Rewrite fraction as difference of fractions.
	$\dfrac{x^2-2}{x+1}$	$x-1-\dfrac{1}{x+1}$	Long division.
	$\dfrac{x+7}{x^2-x-6}$	$\dfrac{2}{x-3} - \dfrac{1}{x+2}$	Use method of *partial fractions*.

Remark: (a) Notice that factorization is *not* limited to integer factors. Non-integer factoring simply boils down to *multiplying and dividing* by the desired factor.

(b) When we multiply like factors we add exponents. When we factor we are undoing multiplication and so we *subtract* exponents. Hence

$$x(x+1)^{-1/2} + (x+1)^{1/2}$$

factors as

$$(x+1)^{-1/2}[x+(x+1)]$$

(c) *Long division* is needed to express

$$\frac{x^2 + 2}{x + 1} \quad \text{as} \quad x - 1 - \frac{1}{x + 1}$$

and we will cover that in Section 4.3.

(d) *Partial fractions* are needed to express

$$\frac{x + 7}{x^2 - x - 6} \quad \text{as} \quad \frac{2}{x - 3} - \frac{1}{x + 2}$$

and we will cover that in Section 5.3.

The next two examples fill in some details for many of the steps given in the table.

EXAMPLE 1
Inserting Factors and Rewriting
with Negative Exponents

(a) Explain the following

$$\frac{4x^2}{9} - 4y^2 = \frac{x^2}{9/4} - \frac{y^2}{1/4}$$

(b) Rewrite the expression so that the denominator is free of terms involving x.

$$\frac{2}{x^3} - \frac{1}{\sqrt{x}} + \frac{3}{16x^2}$$

Solution:

(a) To get rid of fractions in a denominator we invert and multiply. Hence to *put* fractions in the denominator we do the opposite: Invert and *divide*. Thus, we have

$$\frac{4x^2}{9} - 4y^2 = \frac{4}{9}(x^2) - 4(y^2) = \frac{x^2}{9/4} - \frac{y^2}{1/4}$$

(b) $\dfrac{2}{x^3} - \dfrac{1}{\sqrt{x}} + \dfrac{3}{16x^2} = \dfrac{2}{x^3} - \dfrac{1}{x^{1/2}} + \dfrac{3}{(4x)^2}$

$$= 2x^{-3} - x^{-1/2} + 3(4x)^{-2}$$

EXAMPLE 2
Writing a Fraction
as a Sum of Terms

Rewrite each fraction as the sum of two or more terms.

(a) $\dfrac{x + 2x^2 + 1}{\sqrt{x}}$
(b) $\dfrac{1 + x}{x^2 + 1}$
(c) $\dfrac{2x}{x^2 + 2x + 1}$

Solution:

(a) If the denominator is a monomial, we make one fraction for each term in the numerator, then reduce each. Thus, we have

$$\frac{x + 2x^2 + 1}{\sqrt{x}} = \frac{x}{x^{1/2}} + \frac{2x^2}{x^{1/2}} + \frac{1}{x^{1/2}} \qquad \text{Separate into 3 fractions}$$

$$= x^{1/2} + 2x^{3/2} + x^{-1/2} \qquad \text{Subtract exponents}$$

(b) Again, we can make a fraction for each term in the numerator, but reductions are not possible. We have

$$\frac{1 + x}{x^2 + 1} = \frac{1}{x^2 + 1} + \frac{x}{x^2 + 1} \qquad \text{Separate into 2 fractions}$$

(c) We can always add and subtract the same terms in an expression. So we can write

$$\frac{2x}{x^2 + 2x + 1} = \frac{2x + 2 - 2}{x^2 + 2x + 1} \qquad \text{Add and subtract 2}$$

$$= \frac{2x + 2}{x^2 + 2x + 1} - \frac{2}{x^2 + 2x + 1} \qquad \text{Separate into 2 fractions}$$

$$= \frac{2x + 2}{x^2 + 2x + 1} - \frac{2}{(x + 1)^2}$$

Section Exercises 1.10

In Exercises 1–24, find and correct any errors.

1. $2x - (3y + 4) = 2x - 3y + 4$

2. $\dfrac{4}{16x - (2x + 1)} = \dfrac{4}{14x + 1}$

3. $5z + 3(x - 2) = 5z + 3x - 2$

4. $x(yz) = (xy)(xz)$

5. $-\dfrac{x - 3}{x - 1} = \dfrac{3 - x}{1 - x}$

6. $\dfrac{(x - 1)(x + 3)}{(5 - x)(-x)} = \dfrac{(1 - x)(x + 3)}{x(5 - x)}$

7. $a\left(\dfrac{x}{y}\right) = \dfrac{ax}{ay}$

8. $(5z)(6z) = 30z$

9. $(4x)^2 = 4x^2$

10. $\left(\dfrac{x}{y}\right)^3 = \dfrac{x^3}{y}$

11. $\sqrt{x + 9} = \sqrt{x} + 3$

12. $\dfrac{x + 5}{y + 5} = \dfrac{x}{y} + 1$

13. $\dfrac{6x + y}{6x - y} = \dfrac{x + y}{x - y}$

14. $(-2)^6 = -2^6$

15. $\dfrac{1}{x + y^{-1}} = \dfrac{y}{x + 1}$

16. $\dfrac{1}{a^{-1} + b^{-1}} = \left(\dfrac{1}{a + b}\right)^{-1}$

17. $\dfrac{4 + x}{xy^{-1}} = \dfrac{4 + xy}{x}$

18. $x(x + 5)^{1/2} = (x^2 + 5x)^{1/2}$

19. $\sqrt[3]{x^3 + 7x^2} = x^2\sqrt[3]{x} + 7$

20. $\dfrac{1}{x^{1/2} + y^{1/2}} = (x + y)^{-1/2}$

21. $\dfrac{3}{x} + \dfrac{4}{y} = \dfrac{7}{x + y}$

22. $\dfrac{7x - 5(x + 3)}{x(x + 3)} = 12$

23. $\dfrac{1}{2y} = \dfrac{1}{2}y$

24. $4y + \dfrac{1}{2y} = 4.5y$

In Exercises 25–39, factor each expression so that at least one factor is a polynomial with integer coefficients.

25. $\frac{2}{3}x^2 + \frac{1}{4}x + 5$

26. $\frac{3}{4}x + \frac{1}{2}$

27. $\sqrt{x} + (\sqrt{x})^3$

28. $x^{1/3} - 5x^{4/3}$

29. $(2x)x^{1/2} + \frac{1}{2}x^{-1/2}(x^2 + 1)$

30. $2x^{1/3} + \frac{1}{3}x^{-2/3}(2x - 1)$

31. $-\frac{2}{3}x(1 - 2x)^{-2/3} + (1 - 2x)^{1/3}$

32. $\dfrac{\frac{1}{2}x^{-1/2}(x + 2)^3 - 3x^{1/2}(x + 2)^2}{(x + 2)^6}$

33. $\dfrac{\dfrac{x^2}{\sqrt{x^2 + 1}} - \sqrt{x^2 + 1}}{x^2}$

34. $\dfrac{1}{2\sqrt{x}} + 5x^{3/2} - 10x^{5/2}$

35. $\frac{1}{10}(2x + 1)^{5/2} - \frac{1}{6}(2x + 1)^{3/2}$

36. $\frac{4}{3}(x + 3)^{3/2} - 7(x + 3)^{1/2}$

37. $\frac{3}{7}(t + 1)^{7/3} - \frac{3}{4}(t + 1)^{4/3}$

38. $\frac{2}{3}(1 - x)^{3/2} - \frac{4}{5}(1 - x)^{5/2} + \frac{2}{7}(1 - x)^{7/2}$

39. $\frac{1}{20}(2x - 1)^{5/2} + \frac{1}{6}(2x - 1)^{3/2} - \frac{3}{4}(2x - 1)^{1/2}$

In Exercises 40–49, insert the required factor in the parentheses.

40. $x^2(x^3 - 1)^4 = ($ $)(x^3 - 1)^4(3x^2)$

41. $x(1 - 2x^2)^3 = ($ $)(1 - 2x^2)^3(-4x)$

42. $5x\sqrt[3]{1 + x^2} = ($ $)\sqrt[3]{1 + x^2}(2x)$

43. $\dfrac{1}{\sqrt{x}(1 + \sqrt{x})^2} = ($ $)\dfrac{1}{(1 + \sqrt{x})^2}\left(\dfrac{1}{2\sqrt{x}}\right)$

44. $\dfrac{4x + 6}{(x^2 + 3x + 7)^3} = ($ $)\dfrac{1}{(x^2 + 3x + 7)^3}(2x + 3)$

45. $\dfrac{x + 1}{(x^2 + 2x - 3)^2} = ($ $)\dfrac{1}{(x^2 + 2x - 3)^2}(2x + 2)$

46. $\dfrac{1}{(x - 1)\sqrt{(x - 1)^4 - 4}} = \dfrac{(\quad)}{(x - 1)^2\sqrt{(x - 1)^4 - 4}}$

47. $\dfrac{9x^2}{25} + \dfrac{16y^2}{49} = \dfrac{x^2}{(\quad)} + \dfrac{y^2}{(\quad)}$

48. $\dfrac{36(x - 1)^2}{169} + (y + 5)^2 = \dfrac{(x - 1)^2}{(\quad)} + (y + 5)^2$

49. $\dfrac{3}{x} + \dfrac{5}{2x^2} - \dfrac{3}{2}x = ($ $)(6x + 5 - 3x^3)$

In Exercises 50–55, write each fraction as a sum of terms, as in Example 2.

50. $\dfrac{x^3 - 5x^2 + 4}{x^2}$

51. $\dfrac{16 - 5x - x^2}{x}$

52. $\dfrac{2x^5 - 3x^3 + 5x - 1}{x^{3/2}}$

53. $\dfrac{4x^3 - 7x^2 + 1}{x^{1/3}}$

54. $\dfrac{3x^2 - 5}{x^3 + 1}$

55. $\dfrac{x^2 + 4x + 8}{x^4 + 1}$

Review Exercises / Chapter 1

In Exercises 1–24, describe the *error* and then make the necessary correction.

1. $\frac{7}{16} + \frac{3}{16} = \frac{10}{32}$

2. $\frac{15}{32} - \frac{21}{32} = \frac{-6}{0}$

3. $10(4 \cdot 7) = 40 \cdot 70$

4. $(\frac{1}{4}x)(\frac{1}{3}y) = \frac{1}{4}xy$

5. $4(\frac{3}{7}) = \frac{12}{28}$

6. $\frac{2}{9} \times \frac{4}{9} = \frac{8}{9}$

7. $\frac{15}{16} \div \frac{2}{3} = \frac{5}{8}$

8. $15 \div 2 + 3 = 15 \div 5 = 3$

9. $12 + 8 \times 6 = 20 \times 6 = 120$

10. $\frac{-3}{4} = -\frac{3}{4}$

11. $2[5 - (3 - 2)] = 2[5 - 3 - 2]$

12. $-3(-x + y) = 3x + 3y$

13. $(2x)^4 = 2x^4$

14. $\left(\dfrac{y}{8}\right)^2 = \dfrac{y^2}{8}$

15. $(5 + 8)^2 = 5^2 + 8^2$

16. $(-x)^6 = -x^6$

17. $(3^4)^4 = 3^8$

18. $6^{-2} = -6^2$

19. $\sqrt{3^2 + 4^2} = 3 + 4$

20. $\sqrt{10x} = 10\sqrt{x}$

21. $\sqrt{7x}\sqrt[3]{2} = \sqrt{14x}$

22. $\sqrt[4]{\sqrt[4]{2}} = \sqrt[8]{2}$

23. Since $-5 < -3$, it follows that $-2(-5) < -2(-3)$

24. Since $4 < 7$, it follows that $\frac{1}{4} < \frac{1}{7}$

In Exercises 25–40, perform the indicated operations.

25. $-10(7 - 5)$

26. $-10(7)(-5)$

27. $-|16 - 5|$

28. $|5 - 16|$

29. $|-3| + 4(-2) - 6$

30. $16 - 8 \div 4$

31. $\sqrt{5}\sqrt{125}$

32. $\dfrac{\sqrt{72}}{\sqrt{2}}$

33. $6[4 - 2(6 + 8)]$

34. $-4[16 - 3(7 - 10)]$

35. $\left(\dfrac{3^2}{5^2}\right)^{-3}$

36. $6^{-4}(-3)^5$

37. $2(-5)^2$

38. $\left(\dfrac{25}{16}\right)^{-1/2}$

39. $(3 \times 10^4)^2$

40. $\dfrac{1}{(4 \times 10^{-2})^3}$

In Exercises 41–44, use absolute value notation to describe each distance.

41. The distance between x and 7 is at least 4.

42. The distance between x and 25 is no more than 10.

43. The distance between y and -30 is less than 5.

44. The distance between y and $1/2$ is more than 2.

In Exercises 45–48, graph each interval on the real number line.

45. $|x - 2| < 1$

46. $|x| \leq 4$

47. $|x - \frac{3}{2}| \geq \frac{3}{2}$

48. $|x + 3| > 4$

In Exercises 49–60, perform the indicated operations and write the result in standard form.

49. $(7 + 5i) + (-4 + 2i)$

50. $-(6 - 2i) + (-8 + 3i)$

51. $\left(\dfrac{\sqrt{2}}{2} - \dfrac{\sqrt{2}}{2}i\right) - \left(\dfrac{\sqrt{2}}{2} + \dfrac{\sqrt{2}}{2}i\right)$

52. $(13 - 8i) - 5i$

53. $5i(13 - 8i)$

54. $(1 + 6i)(5 - 2i)$

55. $(10 - 8i)(2 - 3i)$

56. $i(6 + i)(3 - 2i)$

57. $\dfrac{6 + i}{i}$

58. $\dfrac{3 + 2i}{5 + i}$

59. $\dfrac{4}{(-3i)}$

60. $\dfrac{1}{(2 + i)^4}$

In Exercises 61–84, perform the indicated operations and simplify.

61. $(x^2 - 2x + 1)(x^3 - 1)$

62. $(x^3 - 3x)(2x^2 + 3x + 5)$

63. $\left(x^2 - \dfrac{1}{x}\right)(x^2 + 1)$

64. $(t^5 - 3t)\left(\dfrac{1}{t^2} + t\right)$

65. $(y^2 - y)(y^2 + 1)(y^2 + y + 1)$

66. $(3z^3 + 4z)(z - 5)(z + 1)$

67. $\dfrac{x}{x^3 - 1} \cdot \dfrac{x - 1}{x^3}$

68. $\dfrac{4x^2 - 1}{(2x)(x^2 + 2x + 1)} \cdot \dfrac{x + 1}{4x^2 + 4x + 1}$

69. $\dfrac{x^2}{x^4 - 2x^2 - 8} \cdot \dfrac{x^2 + 4x + 2}{x^2 + 2}$

70. $\dfrac{x^2 - 1}{x^3 + x} \cdot \dfrac{x^4 - 1}{(x + 1)^2}$

71. $\dfrac{1}{x - 1} - \dfrac{1}{x + 2}$

72. $\dfrac{2}{x} - \dfrac{3}{x - 1} + \dfrac{4}{x + 1}$

73. $x - 1 + \dfrac{1}{x + 2} + \dfrac{1}{x - 1}$

74. $2x + \dfrac{3}{2(x - 4)} - \dfrac{1}{2(x + 2)}$

75. $x + 3 + \dfrac{6}{x - 1} + \dfrac{4}{(x - 1)^2} + \dfrac{1}{(x - 1)^3}$

76. $\dfrac{1}{x - 1} + \dfrac{1 - x}{x^2 + x + 1}$

77. $\dfrac{1}{x} - \dfrac{x - 1}{x^2 + 1}$

78. $\dfrac{1}{6(x - 2)} - \dfrac{1}{6(x + 2)} + \dfrac{1}{3(x^2 + 2)}$

79. $\dfrac{1}{x - 2} + \dfrac{1}{(x - 2)^2} + \dfrac{1}{x + 2} - \dfrac{1}{(x + 2)^2}$

80. $\dfrac{1}{L}\left(\dfrac{1}{y} - \dfrac{1}{L - y}\right)$ where L is a constant

81. $\dfrac{x^3}{x^3 - 1} \div \dfrac{x^2}{x^2 - 1}$

82. $\dfrac{4x - 6}{(x - 1)^2} \div \dfrac{2x^2 - 3x}{x^2 + 2x - 3}$

83. $\dfrac{\dfrac{x^2(5x - 6)}{2x + 3}}{\dfrac{5x}{2x + 3}}$

84. $\dfrac{\dfrac{x^3 + y^3}{x^2 + y^2}}{x^2 - xy + y^2}$

In Exercises 85–90, simplify each expression.

85. $\dfrac{\dfrac{1}{x} - \dfrac{1}{y}}{x^2 - y^2}$

86. $\dfrac{\dfrac{1}{x} - \dfrac{1}{y}}{\dfrac{1}{x} + \dfrac{1}{y}}$

87. $\dfrac{\left(\dfrac{3a}{a^2} - 1\right)}{\dfrac{a}{x} - 1}$

88. $\dfrac{\dfrac{1}{2x - 3} - \dfrac{1}{2x + 3}}{\dfrac{1}{2x} - \dfrac{1}{2x + 3}}$

89. $\dfrac{\dfrac{3}{2(x - 4)} - \dfrac{1}{2x}}{x^2 - 3x - 10}$

90. $\dfrac{1 - \dfrac{1}{1 + 1/a}}{\dfrac{1}{a + 1} - \dfrac{1}{a - 1}}$

In Exercises 91–100, insert the missing factor or factors.

91. $x^3 - 1 = (x - 1)(\quad)$

92. $x^6 - y^6 = (x - y)(x + y)(\quad)(\quad)$

93. $x^4 - 2x^2y^2 + y^4 = (x + y)^2(\quad)^2$

94. $a^6 + 2a^3b^3 + b^6 = (a + b)^2(\quad)^2$

95. $\dfrac{3}{4}x^2 - \dfrac{5}{6}x + 4 = \dfrac{1}{12}(\quad)$

96. $\dfrac{2}{3}x^4 - \dfrac{3}{8}x^3 + \dfrac{5}{6}x^2 = \dfrac{x^2}{24}(\quad)$

97. $\dfrac{(x + 1)^{1/2} - \dfrac{x}{2(x + 1)^{1/2}}}{x + 1} = \left[\dfrac{1}{(\quad)}\right](x + 2)$

98. $\dfrac{x^3}{(x^2 - 1)^{1/2}} + 2x(x^2 - 1)^{1/2} = \left[\dfrac{x}{(x^2 - 1)^{1/2}}\right](\quad)$

99. $z^4 - 5z^3 + 8z - 40 = (z - 5)(\quad)(\quad)$

100. $x^2 - 7xy + 10y^2 + x^3 - 2x^2y = (x - 2y)(\quad)$

Calculator Exercises

101. Calculate 15^4 in two ways. First use the exponential key $\boxed{y^x}$. Second, enter 15 and then press the square key $\boxed{x^2}$ twice. Why do these two methods give the same result?

102. Calculate $\sqrt[5]{107} \ \sqrt[5]{1145}$ in two ways. First use the key-stroke sequence

$$107 \ \boxed{y^x} \ .2 \ \boxed{\times} \ 1145 \ \boxed{y^x} \ .2 \ \boxed{=}$$

Second, use the sequence

$$107 \ \boxed{\times} \ 1145 \ \boxed{=} \ \boxed{y^x} \ .2 \ \boxed{=}$$

Why do these two methods give the same result?

103. Enter any number between 0 and 1 in a calculator and press the square key $\boxed{x^2}$ repeatedly. What number does the calculator display seem to be approaching?

104. Enter any positive number other than 1 in a calculator. What number is approached as the square root key $\boxed{\sqrt{x}}$ is pressed repeatedly?

105. Complete the following table.

n	1	10	10^2	10^4	10^6	10^{10}
$\dfrac{5}{n}$						

What number is $5/n$ approaching as n increases without bound? (See Exercise 50 of Section 1.2.)

CHAPTER

2 | ALGEBRAIC EQUATIONS AND INEQUALITIES

Linear Equations   2.1

We have at this point completed our review of the *fundamentals* of algebra. We now want to *use* these fundamentals to solve problems that can be expressed in the form of equations, inequalities, or systems of equations. Such problems are common in science, business, industry, government, and the marketplace.

Equations

An **equation** is a statement that two algebraic expressions are equal. Some examples of equations in one variable x are

$$3x - 5 = 7, \qquad x^2 - x - 6 = 0, \qquad x^2 - 9 = (x + 3)(x - 3)$$

$$\sqrt{2x} = 4, \qquad \frac{x}{x + 2} = \frac{2}{3}$$

To **solve** an equation means to find all values of the unknown (the variable) for which the equation is a true statement. Those values for which an equation is true are called **solutions** or **roots** of the equation. For instance, $x = 4$ is a solution of the equation $3x - 5 = 7$, because $3(4) - 5 = 7$ is a true statement. Similarly, $x = 8$ is a solution to $\sqrt{2x} = 4$, because $\sqrt{2(8)} = \sqrt{16} = 4$.

Equations that are true for *every* real number for which all terms of the equation are defined are called **identities.** For instance

$$x^2 - 9 = (x + 3)(x - 3)$$

is an identity because every real number is a solution to this equation.

71

Most equations have values in their domains that are not solutions. For example, $x = 1$ is not a solution to $x^2 - 4 = 0$. We call such equations **conditional equations.**

The algebraic process of *solving an equation* usually generates a chain of intermediate equations, each with the same solution(s) as the original. Such equations are called **equivalent equations.** The operations that yield equivalent equations come from the substitution property and cancellation laws of Section 1.2.

GENERATING EQUIVALENT EQUATIONS

A given equation is transformed into an *equivalent equation* by:

1. Adding or subtracting the same quantity from both sides.
2. Multiplying or dividing both sides by the same nonzero* quantity.

*When multiplying or dividing by a *variable,* check to see that the resulting equation has the same solutions as the given one.

EXAMPLE 1
Solving Equations

Solve the following for x:

(a) $6(x - 1) + 4 = 3(7x + 3)$

(b) $\dfrac{x}{x - 3} - 2 = \dfrac{4}{x - 3}$

Solution:

(a)
$$6(x - 1) + 4 = 3(7x + 3) \qquad \textit{Given}$$
$$6x - 6 + 4 = 21x + 9 \qquad \textit{Remove parentheses}$$
$$6x - 2 = 21x + 9 \qquad \textit{Combine like terms}$$
$$-15x = 11 \qquad \textit{Simplify}$$
$$\frac{-15x}{-15} = \frac{11}{-15} \qquad \textit{Divide both sides by } -15$$
$$x = -\frac{11}{15} \qquad \textit{Simplify}$$

These six equations are all equivalent, and we say they form the **steps** of the solution.

(b)
$$\frac{x}{x - 3} - 2 = \frac{4}{x - 3}$$
$$(x - 3)\frac{x}{x - 3} - (x - 3)(2) = (x - 3)\frac{4}{x - 3}$$
$$x - (2x - 6) = 4$$
$$-x + 6 = 4$$
$$x = 2$$

Checking a Solution: Since we multiplied by the variable quantity $(x - 3)$, we must check to see that $x = 2$ is actually a solution to the original equation. We leave that for you to try.

Our students sometimes tell us that a solution looks easy when we work it out in class, but that they don't see where to begin when they try it alone. Our response to such comments is usually this: In the first place, no one (not even great mathematicians) can expect to look at every mathematical problem and immediately know where to begin. Many problems involve some trial and error before a solution is found. In the second place, mathematics is a skill, and like all skills it requires lots of practice! To make algebra work for you, you must put in a great deal of time, you must expect to try solution methods that end up not working, and you must learn from both your successes and your failures.

Clues to Solving Equations

Let's look first at a familiar conversion problem to get some clues about solving equations. Temperature on the Celsius scale is converted to degrees Fahrenheit by the formula

$$F = \frac{9}{5}C + 32$$

The structure of this equation follows the standard *priority of operations.* First, C is multiplied by $\frac{9}{5}$. Then 32 is added to the result. To solve this equation for C in terms of F, we need to undo what has been done to C. That is, we do the *opposite operations in the reverse order.*

$$F = \frac{9}{5}C + 32 \qquad\qquad \textit{Given}$$

$$F - 32 = \frac{9}{5}C \qquad\qquad \textit{Subtract 32 to undo addition of 32}$$

$$\frac{F - 32}{9/5} = \frac{(9/5)C}{9/5} \qquad\qquad \textit{Divide by 9/5 to undo multiplication by 9/5}$$

$$\frac{5}{9}(F - 32) = C \qquad\qquad \textit{Simplify}$$

Linear Equations

Practice in identifying the kind and order of operations in a given equation will enhance your skill both in *solving* equations and in *building* equations (Section 2.2). For the remainder of this section, we will concentrate on solving equations which can be simplified to the form

$$ax + b = 0, \qquad a \neq 0$$

We call such equations **linear.**

Linear equations have exactly one solution. To see this, consider the following steps. (Remember that $a \neq 0$.)

$$ax + b = 0 \qquad \text{\textit{Given}}$$

$$ax + b - b = -b \qquad \text{\textit{Subtract b from both sides}}$$

$$ax = -b \qquad \text{\textit{Simplify}}$$

$$\frac{ax}{a} = \frac{-b}{a} \qquad \text{\textit{Divide both sides by a}}$$

$$x = -\frac{b}{a} \qquad \text{\textit{Simplify}}$$

SOLUTION OF A LINEAR EQUATION

The linear equation

$$ax + b = 0, \qquad a \neq 0$$

has exactly one solution

$$x = -\frac{b}{a}$$

We believe the following guidelines will enhance your success in solving linear equations.

GUIDELINES FOR SOLVING LINEAR EQUATIONS

1. Remove all symbols of grouping. (For equations with fractions, multiply both sides by the LCD.)
2. Collect and combine like terms.
3. Identify the operations and their order of priority. (Do this mentally.)
4. Solve for the unknown by doing the opposite operations in reverse order.
5. Check answer. (This is a must if you multiplied or divided by a variable quantity.)

EXAMPLE 2
Solving Linear Equations

Solve for x

(a) $2(3a - 4x) - 3b = 5x + 2b$

(b) $\dfrac{4}{x + 2} = 1 - \dfrac{x}{x - 2}$

Solution:

(a) $2(3a - 4x) - 3b = 5x + 2b$ *Given*

       $6a - 8x - 3b = 5x + 2b$ *Remove parentheses*

       $6a - 3b - 2b = 5x + 8x$ *Collect like terms*

       $6a - 5b = 13x$ *Combine like terms*

       $\dfrac{6a - 5b}{13} = x$ *Divide by 13*

(b) In this case, the LCD is $(x + 2)(x - 2)$. Thus we have

$$\frac{4}{x + 2} = 1 - \frac{x}{x - 2}$$

$$(x + 2)(x - 2)\frac{4}{x + 2} = (x + 2)(x - 2)\left(1 - \frac{x}{x - 2}\right)$$

$$4(x - 2) = (x + 2)(x - 2) - (x + 2)x$$

$$4x - 8 = x^2 - 4 - x^2 - 2x$$

$$6x = 4$$

$$x = \frac{4}{6} = \frac{2}{3}$$

Check the solution by substituting $x = \frac{2}{3}$ back into the original equation.

Remark: Multiplying a fractional *equation* by its LCD very neatly eliminates all fractions in the equation. But be sure you see that this only works with *equations,* not with *expressions.*

The next example demonstrates the importance of a check when you have multiplied or divided by a variable.

EXAMPLE 3
An Equation with No Solutions

Solve for y

$$\frac{5y + 2}{y - 3} + 2 = \frac{4y + 5}{y - 3}$$

Solution:
The LCD is $y - 3$, and we have

$$\frac{5y + 2}{y - 3} + 2 = \frac{4y + 5}{y - 3} \qquad \begin{array}{l}\textit{Original equation}\\ \textit{(y = 3 is not a solution.)}\end{array}$$

$$(y - 3)\left(\frac{5y + 2}{y - 3} + 2\right) = (y - 3)\left(\frac{4y + 5}{y - 3}\right)$$

$$5y + 2 + 2(y - 3) = 4y + 5 \qquad \begin{array}{l}\textit{Nonequivalent equation}\\ \textit{(y = 3 is a solution.)}\end{array}$$

$$5y + 2y - 4y = 5 + 6 - 2$$
$$3y = 9$$
$$y = 3$$

In this case $y = 3$ is *not* a solution, since it yields a denominator of zero in the original equation. This means that the given equation has no solutions.

Extraneous Solutions

In Example 3, our solution steps produced an **extraneous solution**—a y-value that is not a solution to the original equation. Extraneous solutions are obtained when one of the solution steps results in an equation that is not equivalent to the original equation.

EXAMPLE 4
Cross Multiplying to Solve an Equation

Solve for y

$$\frac{3y - 2}{2y + 1} = \frac{6y - 9}{4y + 3}$$

Solution:

An equation with a single fraction on each side is readily cleared of denominators by *cross multiplying*.

$$\frac{3y - 2}{2y + 1} = \frac{6y - 9}{4y + 3} \qquad \textit{Given}$$

$$(3y - 2)(4y + 3) = (6y - 9)(2y + 1) \quad \textit{Cross multiply}$$

$$12y^2 + y - 6 = 12y^2 - 12y - 9 \qquad \textbf{FOIL}$$

$$13y = -3 \qquad \textit{Combine like terms}$$

$$y = -\frac{3}{13} \qquad \textit{Divide by 13}$$

You should verify that this solution does not yield any zero denominators.

Equations for Calculators

Up to this point, we have been carefully choosing our equations so that the calculations are not very messy to perform. This is rather artificial, since real-world problems frequently involve numbers that are not simple integers or fractions. In such cases a calculator is useful.

EXAMPLE 5
Solving Equations with the Help of a Calculator

Solve for x in

$$\frac{1}{9.38} - \frac{3}{x} = \frac{5}{0.3714}$$

Solution:
Round-off error will be minimized if we solve for x before doing any calculations. In this case the LCD is $(9.38)(0.3714)(x)$ and we have

$$\frac{1}{9.38} - \frac{3}{x} = \frac{5}{0.3714}$$

$$(9.38)(0.3714)(x)\left(\frac{1}{9.38} - \frac{3}{x}\right) = (9.38)(0.3714)(x)\left(\frac{5}{0.3714}\right)$$

$$0.3714x - 3(9.38)(0.3714) = (9.38)(5)(x)$$

$$0.3714x - 5(9.38)x = 3(9.38)(0.3714)$$

$$[0.3714 - 5(9.38)]x = 3(9.38)(0.3714)$$

$$x = \frac{3(9.38)(0.3714)}{0.3714 - 5(9.38)}$$

$$x \approx -0.225 \qquad \text{(Rounded to 3 places)}$$

Remark: Because of round-off procedures, a check of a decimal solution may not yield exactly the same values for both sides of the original equation. The difference, however, will usually be quite small.

Section Exercises 2.1

In Exercises 1–10, determine if the given value of the unknown is a solution of the equation.

1. $5x - 3 = 3x + 5$
   (a) $x = 0$ (b) $x = -5$ (c) $x = 4$ (d) $x = 10$
2. $7 - 3x = 5x - 17$
   (a) $x = -3$ (b) $x = 0$ (c) $x = 8$ (d) $x = 3$
3. $3x^2 + 2x - 5 = 2x^2 - 2$
   (a) $x = -3$ (b) $x = 1$ (c) $x = 4$ (d) $x = -5$
4. $5x^3 + 2x - 3 = 4x^3 + 2x - 11$
   (a) $x = 2$ (b) $x = -2$ (c) $x = 0$ (d) $x = 10$
5. $\frac{5}{2x} - \frac{4}{x} = 3$
   (a) $x = -\frac{1}{2}$ (b) $x = 4$ (c) $x = 0$ (d) $x = \frac{1}{4}$
6. $3 + \frac{1}{x + 2} = 4$
   (a) $x = -1$ (b) $x = -2$ (c) $x = 0$ (d) $x = 5$
7. $(x + 5)(x - 3) = 20$
   (a) $x = 3$ (b) $x = -5$ (c) $x = 5$ (d) $x = -7$
8. $\sqrt[3]{x - 8} = 3$
   (a) $x = 2$ (b) $x = -2$ (c) $x = 35$ (d) $x = 8$
9. $x + \frac{1}{2}\sqrt{x} - 3 = 0$
   (a) $x = -4$ (b) $x = 4$ (c) $x = \frac{1}{2}$ (d) $x = \frac{9}{4}$
10. $x^4 - 3x^3 = 4x^2 + 12x$
    (a) $x = 0$ (b) $x = -2$ (c) $x = 2$ (d) $x = 3$

In Exercises 11–20, determine if the equation is conditional or an identity.

11. $3x - 10 = 4x$
12. $x^2 + 2(3x - 2) = x^2 + 8(x + 2) - 2x - 20$
13. $x^2 - 8x + 5 = (x - 4)^2 - 11$
14. $4[(x + \frac{1}{2})^2 - 6] = 4x^2 - 4x - 23$
15. $\frac{x}{3} + \frac{x}{4} = 1$
16. $\frac{5}{x} + \frac{3}{x} = 24$
17. $3 + \frac{1}{x + 1} = \frac{4x}{x + 1}$
18. $2x(x^2 - 7x + 12) = 2x(x^2 - 4x) - 6(x^2 - 4x)$
19. $\frac{1}{t}(t + 2)(t - 1) = \frac{t(t + 2) - t - 2}{t}$
20. $(x + 2)^2(x + 1)^2 = 0$

In Exercises 21–50, solve the given equation and check your answer.

21. $8x - 5 = 3x + 10$
22. $7x + 3 = 3x - 13$
23. $\frac{x}{2} - 6 = 4x - \frac{3}{4}$
24. $\frac{x}{5} - \frac{x}{2} = 3$
25. $2(x + 5) - 7 = 3(x - 2)$

26. $4(13t - 15) + 3(2t - 38) = 0$
27. $6[x - (2x + 3)] = 8 - 5x$
28. $8(x + 2) - 3(2x + 1) = 2(x + 5)$
29. $\frac{3}{2}(z + 5) - \frac{1}{4}(z + 24) = 0$
30. $0.6z + 1.1 = 0.3z - 4$
31. $0.25x + 0.75(10 - x) = 3$
32. $0.60x + 0.40(100 - x) = 50$
33. $\dfrac{100 - 4u}{3} = \dfrac{5u + 6}{4} + 6$
34. $\dfrac{16 + y}{y} + \dfrac{32 + y}{y} = 100$
35. $\dfrac{5x - 4}{5x + 4} = \dfrac{2}{3}$ 36. $\dfrac{10x + 3}{5x + 6} = \dfrac{1}{2}$
37. $10 - \dfrac{13}{x} = 4 + \dfrac{5}{x}$ 38. $\dfrac{15}{x} - 4 = \dfrac{6}{x} + 3$
39. $\dfrac{7}{2x + 1} - \dfrac{8x}{2x - 1} = -4$
40. $\dfrac{4}{u - 1} + \dfrac{6}{3u + 1} = \dfrac{15}{3u + 1}$
41. $\dfrac{1}{x - 3} + \dfrac{1}{x + 3} = \dfrac{10}{x^2 - 9}$
42. $\dfrac{1}{x - 2} + \dfrac{3}{x + 3} = \dfrac{4}{x^2 + x - 6}$
43. $(x + 2)^2 + 5 = (x + 3)^2$
44. $(x + 1)^2 + 2(x - 2) = (x + 1)(x - 2)$
45. $(x + 2)^2 - x^2 = 4(x + 1)$
46. $4(x + 1) - 3x = x + 5$
47. $(2x + 1)^2 = 4(x^2 + x + 1)$
48. $6x + ax = 2x + 5$
49. $4 - 2(x - 2b) = ax + 3$
50. $5 + ax = 12 - bx$

Calculator Exercises

In Exercises 51–55, solve each equation for x and then use a calculator to write the solution in decimal form. (Round your answer to three decimal places.)

51. $0.275x + 0.725(500 - x) = 300$
52. $2.763 - 4.5(2.1x - 5.1432) = 6.32x + 5$
53. $\dfrac{x}{0.6321} + \dfrac{x}{0.0692} = 1000$
54. $\dfrac{2}{7.398} - \dfrac{4.405}{x} = \dfrac{1}{x}$
55. $(x + 5.62)^2 + 10.83 = (x + 7)^2$

It is important to remember that rounding calculations more than once in a given problem can result in substantial round-off error. One way to avoid this is to store intermediate results in your calculator's memory, rather than writing them down and then re-entering them. In Exercises 56–61, perform each calculation two ways: (a) calculate entirely on your calculator, and then round the answer to two decimal places, and (b) round both the numerator and the denominator to two decimal places before dividing, and then round the final answer to two decimal places. Determine if the second method introduced an additional round-off error.

56. $\dfrac{1 + 0.86603}{1 - 0.86603}$ 57. $\dfrac{1 + 0.73205}{1 - 0.73205}$
58. $\dfrac{2 + 0.57735}{1 + 0.57735}$ 59. $\dfrac{2 - 1.63254}{(2.58)(0.135)}$
60. $\dfrac{1.73205 - 1.19195}{3 - (1.73205)(1.19195)}$ 61. $\dfrac{3.33 + (1.98/0.74)}{4 + (6.25/3.15)}$

Formulas and Applications   2.2

Real-world applications of algebra do not generally come to us expressed in mathematical language. A truly challenging part of algebra is to learn to translate oral or verbal descriptions of problems into mathematical statements. Our goal in this section is to provide suggestions and to demonstrate procedures that will help you increase your skill in this translation process.

Formula Applications

Certain categories of problems can be put into concise mathematical packages using formulas. A **formula** can be thought of as an algorithm (or recipe) for doing a calculation. For instance, to find the perimeter P of a rectangle of length $L = 15$ inches and width $W = 7$ inches, we can use the formula $P = 2L + 2W$ and get $P = 2(15) + 2(7) = 30 + 14 = 44$ inches.

This same formula can be used to find the width W, if the perimeter and length are known. We can algebraically *manipulate* this formula and solve for W in terms of P and L. We have

$$P = 2L + 2W \qquad \text{\textit{Given}}$$

$$P - 2L = 2W \qquad \text{\textit{Subtract 2L from both sides}}$$

$$\frac{P - 2L}{2} = W \qquad \text{\textit{Divide both sides by 2}}$$

Formulas you should know include the following.

COMMON FORMULAS

	Square	**Rectangle**	**Circle**	**Triangle**
Area:	$A = s^2$	$A = LW$	$A = \pi r^2$	$A = \frac{1}{2}bh$
Perimeter:	$P = 4s$	$P = 2L + 2W$	$C = 2\pi r$	

	Cube	**Rectangular Solid**	**Cylinder**	**Sphere**
Volume:	$V = s^3$	$V = LWH$	$V = \pi r^2 h$	$V = \frac{4}{3}\pi r^3$

Simple Interest	**Balance (Savings)**	**Distance**
$I = Prt$	$A = P(1 + r)^t$	$D = rt$

If you know ready-made formulas like these, then verbal descriptions of problems involving area, volume, interest, distance, and so on can be readily translated into mathematical language.

EXAMPLE 1
Manipulating Formulas

(a) Solve for r in $I = Prt$. \qquad (b) Solve for P in $B = P + Prt$.

(c) Solve for R in $\dfrac{1}{R} = \dfrac{1}{R_1} + \dfrac{1}{R_2}$

Solution:

(a) This formula describes the interest I on P dollars at rate r and time t. To solve for r, we have

$$I = Prt \qquad \text{\textit{Given}}$$

$$I = (Pt)r \qquad \text{\textit{Commutative Law (do this mentally)}}$$

$$\frac{I}{Pt} = r \qquad \text{\textit{Divide by Pt}}$$

(b) This formula describes the balance B, due on a loan of P dollars. To solve for P, we have

$$B = P + Prt \qquad \textit{Given}$$

$$B = P(1 + rt) \qquad \textit{Distributive Law}$$

$$\frac{B}{1 + rt} = P \qquad \textit{Divide by } 1 + rt$$

(c) This formula involves resistance in an electrical circuit. In solving for R, there is a great *temptation* simply to invert each term, to get $R/1 = R_1/1 + R_2/1$. This is *absolutely illegal!* Instead, we can proceed as follows:

$$\frac{1}{R} = \frac{1}{R_1} + \frac{1}{R_2} \qquad \textit{Given}$$

$$\frac{1}{R} = \frac{R_2 + R_1}{R_1 R_2} \qquad \textit{Combine fractions}$$

$$R = \frac{R_1 R_2}{R_2 + R_1} \qquad \textit{Invert both sides}$$

EXAMPLE 2
Formula Applications

(a) A soft drink company wants its cylindrical can to have a volume of 300 cubic centimeters and a radius of 3 centimeters (cm). How tall will the can be?

(b) A deposit of \$1250 was placed into a savings account that compounds interest once a year at a rate of 12%. What is the amount in this account at the end of the third year?

Solution:

(a) This is a *volume of a cylinder* problem, so we use the formula $V = \pi r^2 h$. Since the height of the can is unknown, we can solve for h to get

$$\frac{V}{\pi r^2} = h$$

Thus, for $V = 300 \text{ cm}^3$ and $r = 3$ cm, we have

$$h = \frac{300 \text{ cm}^3}{\pi (3 \text{ cm})^2} = \frac{300 \text{ cm}^3}{9\pi \text{ cm}^2}$$

$$h \approx \frac{300}{9(3.14)} \text{ cm} \approx 10.6 \text{ cm}$$

(b) Since this is a *savings account problem,* we choose the formula $A = P(1 + r)^t$ where, in this case, $P = 1250$, $r = 12\% = 0.12$, and $t = 3$. Therefore,

$$A = P(1 + r)^t$$

$$= 1250(1 + 0.12)^3 \qquad \textit{Use } \boxed{y^x} \textbf{key}$$

$$\approx 1250(1.405)$$
$$A \approx \$1756.16$$

There is no single rule to show how to set up applied problems if we do not have an associated formula. However, there are some useful general guidelines.

GUIDELINES FOR SOLVING WORD PROBLEMS

1. Read (and reread) the problem carefully.
2. Write an informal equation or model tying together stated or implied *operations* of algebra. Draw a picture, when appropriate.
3. Label the known and unknown quantities.
4. Write the formal algebraic equation.
5. Solve the equation and check the result.

Remark: For many students, the two main hangups in solving word problems seem to be the lack of recognition of the vocabulary of algebra and the frustrating mistake of trying to solve the problem all at one time. Most problems are solved in stages, so don't expect quick one-step solutions.

Building Equations

Since an equation is built by putting together a sequence of operations, it is vitally important that you recognize the vocabulary of algebraic operations. We have compiled some key words and phrases that help identify the operations used to build equations.

TRANSLATING ENGLISH INTO ALGEBRA

Key Words or Phrases	Algebraic Operation	Examples
equals, equal to, is, are, was, will be, represents, same as	=	
sum, plus, greater, increased by, added to, together, exceeds, total of	+	If Eric's salary of \$22,400 is *increased by* 9%, what is his new salary? *Informal equation:* (salary) + 9%(salary) = (new salary)

difference, minus, less, decreased by, subtracted from, reduced by, the remainder	$-$	All dresses in a store are *reduced by* 20%. Find the original price of a dress selling for $36.76. *Informal equation:* $$\left(\dfrac{\text{original}}{\text{price}}\right) - 20\% \left(\dfrac{\text{original}}{\text{price}}\right) = \$36.76$$
product, times, multiplied by, twice, percent of	$\times$ or $\cdot$	If five hundred dollars represents 25 *percent of* Lori's monthly salary, find her monthly salary. *Informal equation:* $$500 = 25\% \times (\text{monthly salary})$$
quotient, divided by, ratio, per	$\div$ or $/$	One number is 16 more than twice the other, and their *ratio* is 6. Find the two numbers. *Informal equation:* $$\left(\dfrac{\text{larger}}{\text{number}}\right) \div \left(\dfrac{\text{smaller}}{\text{number}}\right) = 6$$

EXAMPLE 3
Building Linear Equations

(a) One number is one-third of another number. If their difference is 28, find the two numbers.

(b) A rectangle is twice as long as it is wide, and its perimeter is 132 inches. Find its dimensions.

Solution:

(a) After reading the problem carefully, we see the relationship.

Informal equation: (larger #) $-$ (smaller #) = 28

Labels: x = larger number, $\dfrac{1}{3}x$ = smaller number

Algebraic equation: $x - \dfrac{x}{3} = 28$

Solution: $3x - x = 28(3)$ *Multiply by LCD*

$2x = 84$

$x = 42$ *(larger number)*

$\dfrac{x}{3} = 14$ *(smaller number)*

(b) In this case, a picture (see Figure 2.1) is appropriate.

Informal equation: 2(length) + 2(width) = 132 inches

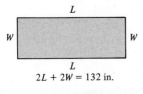

L

W W

L
$2L + 2W = 132$ in.

FIGURE 2.1

Labels: $W =$ width, $L =$ length $= 2W$

Equation: $2(2W) + 2W = 132$

Solution: $4W + 2W = 132$

$6W = 132$

$W = 22$ *(width)*

$L = 2W = 44$ *(length)*

Several important categories of applied problems involve percentages and discounts, distance, simple interest, coins, mixtures, and work. You will notice in each of the following examples that the key to the solution is an understanding of the basic **model** (informal equation) that ties together the algebraic operations.

EXAMPLE 4
Percentages and Discounts

(a) Eighty-five dollars is what percent of $250?
(b) Two hundred is 110% of what number?
(c) A fire sale has all items reduced by 45%. What was the original price of a clock radio now priced at $41.25?

Solution:

(a)   *Model:* $85 =$ (percentage) $\times$ (250)

Label: $p =$ percentage   *Written in decimal form*

Equation: $85 = (p)(250)$

Solution: $p = \dfrac{85}{250}$

$p = 0.34$ *(decimal form)*

$p = 34\%$ *(percentage form)*

(b)   *Model:* $200 = (110\%) \times$ (what number)

Labels: $x =$ number, $110\% = 1.1$

Equation: $200 = 1.1 \times x$

Solution: $200 = 1.1x$

$x = \dfrac{200}{1.1} \approx 181.82$

Remark: Note that the actual calculation in part (b) is done with the *decimal form* of the percentage p.

(c)   *Model:*   (original price) $-$ (amount of discount) $=$ sale price

Labels: $x =$ original price, $0.45x =$ discount

Equation: $x - 0.45x = 41.25$

Solution: $0.55x = 41.25$

$$x = \frac{41.25}{0.55} = \$75 \quad \textit{(original price)}$$

EXAMPLE 5
Distance Problems

(a) Chad and Pam are running in a 10 kilometer race. Chad runs at 12 kilometers per hour and Pam at 10 kilometers per hour. How long will it be before they are $\frac{2}{3}$ kilometer apart?

(b) Greg traveled 20 miles downstream by boat in the same time it took him to travel 12 miles upstream. If the speed of the boat in still water is 10 miles per hour, what was the speed of the stream?

Solution:

The key to the solution of a distance problem involves the manipulation of the basic formula

distance = (rate) × (time)

$$d = rt$$

($r = d/t$ and $t = d/r$ are alternative forms of the same equation.)

(a) After careful reading, we can see that the appropriate model is

Model: (Chad's distance) − (Pam's distance) = constant

Labels: *for Chad:* $d_1 = $ (rate)(time) $= (12)(t)$

for Pam: $d_2 = $ (rate)(time) $= (10)(t)$

Equation: $12t - 10t = \frac{2}{3}$

Solution: $2t = \frac{2}{3}$

$t = \frac{1}{3}$ hour = 20 minutes

(b) *Model:* time down = time up

Labels: *rates:* $x = $ rate of stream

$10 + x = $ rate of boat downstream

$10 - x = $ rate of boat upstream

distance: 20 = distance downstream

12 = distance upstream

Equation: $t = \dfrac{d}{r}$

$$\frac{20}{10 + x} = \frac{12}{10 - x}$$

Solution: $20(10 - x) = 12(10 + x)$

$200 - 20x = 120 + 12x$

$80 = 32x$

$x = 2.5$ mph   *(rate of stream)*

EXAMPLE 6
Coin and Simple Interest Models

(a) Kim has saved $21.20 in dimes and quarters. If there are 119 coins in all, how many of each coin has she saved?

(b) Eric invests $10,000 received from a trust fund in two ways. Some is at $9\frac{1}{2}\%$ and the rest is at 11%. How much is invested at each rate if he receives $1038.50 simple interest per year?

Solution:

(a) Coin problems generally fit the model

$$(\text{coin value})\begin{pmatrix}\text{number} \\ \text{of} \\ \text{coins}\end{pmatrix} + (\text{coin value})\begin{pmatrix}\text{number} \\ \text{of} \\ \text{coins}\end{pmatrix} = \text{total value}$$

In this case, we have

Model: $\left(\dfrac{\text{dime}}{\text{value}}\right)(\text{number}) + \left(\dfrac{\text{quarter}}{\text{value}}\right)(\text{number}) = \text{total value}$

Labels: *dimes:*   value = 0.10,   number = x
         quarters: value = 0.25,   number = $119 - x$

Equation: $0.10(x) + 0.25(119 - x) = 21.20$

Solution: $0.1x - 0.25x + 29.75 = 21.20$
$$-0.15x = -8.55$$
$$x = 57 \text{ dimes}$$
$$119 - x = 62 \text{ quarters}$$

(b) Simple interest problems are based on the model

    $(\text{rate})(\text{principal}) = \text{total interest}$

In this case, we have

Model: $(r_1)(P_1) + (r_2)(P_2) = I$

Labels: $r_1 = 9\frac{1}{2}\% = 0.095,$ $P_1 = x$
         $r_2 = 11\% = 0.11,$ $P_2 = 10{,}000 - x$
         $I = \text{interest} = 1038.50$

Equation: $0.095(x) + 0.11(10{,}000 - x) = 1038.50$

Solution: $0.095x - 0.11x + 1100 = 1038.50$
$$-0.015x = -61.50$$
$$x = \$4100 \text{ at } 9\tfrac{1}{2}\%$$
$$10{,}000 - x = \$5900 \text{ at } 11\%$$

EXAMPLE 7
Mixture Model

(a) A pharmacist needs to strengthen a 15% alcohol solution to one of 32% alcohol. How much pure alcohol should be added to 200 milliliters of the 15% solution?

(b) A department store has \$30,000 of inventory in 12-inch and 19-inch color TVs. The profit on a 12-inch set is 22%, while the profit on a 19-inch set is 40%. If the profit on the entire stock is 35%, how much was invested in each type of TV?

Solution:

Mixture problems often fit the model

$$(\text{percent})(\text{amount}) + (\text{percent})(\text{amount}) = (\text{percent})(\text{total amount})_,$$

(a) In this case, we have

$$\textit{Model:}\quad (P_1)\binom{\text{original}}{\text{solution}} + (P_2)\binom{\text{pure}}{\text{alcohol}} = (P_3)\binom{\text{final}}{\text{solution}}$$

Labels: *original solution:* $P_1 = 0.15$,   amount $= 200$

pure alcohol: $P_2 = 1.00$,   amount $= x$

final solution: $P_3 = 0.32$,   amount $= 200 + x$

Equation: $0.15(200) + 1.00(x) = 0.32(200 + x)$

Solution: $30 + x = 64 + 0.32x$

$$0.68x = 34$$

$$x = 50 \text{ milliliters of pure alcohol}$$

(b) In this case, we have

$$\textit{Model:}\,(P_1)\binom{\text{inventory}}{\text{value}} + (P_2)\binom{\text{inventory}}{\text{value}} = (P_3)\binom{\text{total}}{\text{inventory}}$$

Labels: *12-inch TV:* $P_1 = 0.22$,   value $= x$

19-inch TV: $P_2 = 0.40$,   value $= 30,000 - x$

total inventory: $P_3 = 0.35$,   value $= 30,000$

Equation: $0.22(x) + 0.40(30,000 - x) = 0.35(30,000)$

Solution: $0.22x - 0.4x + 12,000 = 10,500$

$$-0.18x = -1,500 \qquad \textit{(12-inch models)}$$

$$x = \$8333.33$$

$$30,000 - x = \$21,666.67 \qquad \textit{(19-inch models)}$$

EXAMPLE 8
Work Model

Using a push type mower, Steve can mow the lawn in $6\frac{1}{2}$ hours. After mowing for 2 hours, Steve's neighbor joins the mowing with his tractor mower. Together they finish the job in 1 hour and 15 minutes. How long would it take to mow the entire lawn using the tractor?

Solution:

Work problems fit the model

$$(\text{rate})(\text{time}) + (\text{rate})(\text{time}) = 1$$

where rate $= \dfrac{1}{(\text{total time to complete job})}$.

In this case, we have

$$\textit{Model:} \quad \begin{pmatrix} \text{small} \\ \text{mower} \\ \text{rate} \end{pmatrix}\begin{pmatrix} \text{small} \\ \text{mower} \\ \text{time} \end{pmatrix} + \begin{pmatrix} \text{large} \\ \text{mower} \\ \text{rate} \end{pmatrix}\begin{pmatrix} \text{large} \\ \text{mower} \\ \text{time} \end{pmatrix} = 1$$

$$\textit{Labels:} \quad \text{small mower:} \quad \text{rate} = \dfrac{1}{(13/2)} = \dfrac{2}{13},$$

$$\text{time} = 2 + 1\dfrac{1}{4} = \dfrac{13}{4}$$

$$\text{large mower:} \quad \text{rate} = \dfrac{1}{x},$$

$$\text{time} = \dfrac{5}{4}$$

$$\textit{Equation:} \quad \left(\dfrac{2}{13}\right)\left(\dfrac{13}{4}\right) + \left(\dfrac{1}{x}\right)\left(\dfrac{5}{4}\right) = 1$$

$$\textit{Solution:} \qquad\qquad 2x + 5 = 4x$$

$$5 = 2x$$

$$x = 2\dfrac{1}{2} \text{ hours for tractor}$$

Section Exercises 2.2

In Exercises 1–12, solve for the indicated variable.

1. Area of Triangle: $\qquad\qquad\qquad A = \frac{1}{2}bh$ $\qquad\qquad$ (Solve for h)
2. Perimeter of a Rectangle: $\qquad\quad P = 2L + 2W$ $\qquad$ (Solve for L)
3. Volume of a Rectangular Solid: $\quad V = LWH$ $\qquad\quad$ (Solve for L)
4. Volume of Right Circular Cylinder: $V = \pi r^2 h$ $\qquad\quad$ (Solve for h)
5. Markup: $\qquad\qquad\qquad\qquad\quad S = C + RC$ $\qquad$ (Solve for C)
6. Discount: $\qquad\qquad\qquad\qquad\; S = L - RL$ $\qquad$ (Solve for L)
7. Investment at Simple Interest: $\quad A = P + PRT$ $\qquad$ (Solve for R)
8. Investment at Compound Interest: $A = P\left(1 + \dfrac{R}{n}\right)^{nT}$ $\qquad$ (Solve for P)
9. Area of a Trapezoid: $\qquad\qquad\; A = \frac{1}{2}(a + b)h$ $\qquad$ (Solve for b)
10. Area of a Sector of a Circle: $\qquad A = \dfrac{\pi r^2 \theta}{360}$ $\qquad\qquad$ (Solve for θ)
11. Volume of a Spherical Segment: $\;\; V = \frac{1}{3}\pi h^2(3r - h)$ $\qquad$ (Solve for r)
12. Volume of an Oblate Spheroid: $\quad V = \frac{4}{3}\pi a^2 b$ $\qquad\quad$ (Solve for b)

13. The sum of two consecutive natural numbers is 525. Find the two numbers.

14. The sum of three consecutive natural numbers is 804. Find the three numbers.

15. One number is 5 times another number. Find the numbers if their difference is 148.

16. Tom's weekly paycheck is $45 greater than Bill's. Find each paycheck amount if together they total $627.

17. A rectangle is 1.5 times as long as it is wide, and its perimeter is 75 inches. Find its dimensions.

18. Three linear feet of wood is to be cut to form a picture frame. To be aesthetically pleasing, the frame must have a width of 0.62 times the length. Find the dimensions of the frame.

19. To get an A in a course a student must have an average of at least 90. What score must a student have on the final to get an A if her test scores are 87, 92, and 84?

20. Repeat Exercise 19, assuming that the final exam counts twice as much as each of the other scores.

21. What is 30% of 45?

22. What is 175% of 360?

23. What is 0.045% of 2,650,000?

24. 432 is what percent of 1600?

25. 459 is what percent of 340?

26. 12 is $\frac{1}{2}$% of what number?

27. 70 is 40% of what number?

28. $825 is 250% of what number?

29. A family has annual loan payments that equal 58.6% of their income. If they made payments of $13,077.75 during the year, what was their income?

30. The sale price of a swimming pool is $849. Find the list price if the sale price represents a 16.5% discount.

31. Two families meet at a designated park for a picnic. At the end of the day one family travels east at an average speed of 42 miles per hour and the other travels west at an average speed of 50 miles per hour. Both families have approximately 160 miles to travel. Find
    (a) The time it takes each family to get home.
    (b) The time that will have elapsed when they are 100 miles apart.
    (c) The distance the eastbound family has to travel after the westbound family has arrived home.

32. Students are going to a football game 135 miles away in two cars. The first car leaves on time and travels at an average speed of 45 miles per hour. The second car starts $\frac{1}{2}$ hour later and travels at an average speed of 55 miles per hour. How long will it take the second carload of students to catch up, and will they catch up prior to their arrival at the game?

33. On a certain day a corporate executive flew in the corporate jet to a meeting in a city 1500 miles away. On the return flight that evening he noted that he still had 300 miles to go after traveling as long as the total trip that morning. If the air speed of the plane is 600 miles per hour, how fast was the wind blowing at the altitude he was flying? (Assume that the wind direction was parallel to his flight path and constant during the day.)

34. Find the time in minutes required for light to travel from the sun to the earth if light travels at a speed of 3.0×10^8 meters per second and the distance is 1.5×10^{11} meters.

35. Radio waves travel at the same speed as light, 3.0×10^8 meters per second. Find the time required for a radio wave to travel from mission control in Houston to NASA astronauts on the surface of the moon 3.86×10^8 meters away.

36. Find the distance to a star that is 50 light years (distance traveled by light in one year) away if light travels at 186,000 miles per hour.

37. John has $10.75 in quarters and 50-cent pieces. How many of each coin does John have if there are 35 coins in all?

38. Nancy has $175 in $5 and $10 bills. If there are a total of 23 bills, how many of each denomination does she have?

39. Denise invests $12,000 in two funds paying $10\frac{1}{2}$% and 13% simple interest, respectively. How much is invested in each if the yearly interest is $1447.50?

40. Robert invests $25,000 in two funds paying 11% and $12\frac{1}{2}$% simple interest, respectively. How much is invested in each if the yearly interest is $2975.00?

41. Jack invests $12,000 in a fund paying $9\frac{1}{2}$% simple interest, and $8,000 in a fund where the interest rate is variable. At the end of the year Jack received notification that his total interest for both funds was $2,054.40. Find the equivalent simple interest rate on the variable rate fund.

42. Mary has $10,000 on deposit earning simple interest with the interest rate linked to the *prime rate*. Because of a fall in the prime rate, the rate on her investment dropped by $1\frac{1}{2}$% for the last quarter of the year. If her annual earnings on the fund were $1,112.50, find the interest rate for the first three quarters of the year and the last quarter.

43. A company has fixed costs of $10,000 per month, and variable costs of $8.50 per unit manufactured. How many units can they manufacture if in a given month they have $85,000 available to cover costs? (*Fixed costs* are those that occur regardless of the level of production. *Variable costs* depend on the level of production.)

44. Jean was 30 years old when her daughter Ruth was born. How old will Ruth be when
    (a) Her age is one-third that of her mother?
    (b) Their combined ages total 100?

45. Using the values from the table below, determine the amounts of Solutions 1 and 2, respectively, needed to obtain the desired amount and concentration of the final mixture.

	Concentration of Solution 1	Concentration of Solution 2	Amount of Final Solution	Concentration of Final Solution
(a)	10%	30%	100 gallons	25%
(b)	25%	50%	5 liters	30%
(c)	15%	45%	10 quarts	30%
(d)	70%	90%	25 gallons	75%

46. A 55-gallon barrel contains a mixture with a concentration of 40%. How much of this mixture must be withdrawn and replaced by 100% concentrate to bring the mixture up to 75% concentration?

47. A farmer mixed gasoline and oil to have 2 gallons of mixture for his two-cycle chain saw engine. This mixture was 32 parts gasoline to 1 part two-cycle oil. How much gasoline must be added to bring the mixture to 40 parts gasoline and 1 part oil?

48. A grocer mixes two kinds of nuts, costing $2.49 per pound and $3.89 per pound respectively, to make 100 pounds of a mixture costing $3.19 per pound. How much of each kind of nut was put into the mixture?

49. A farmer can till a 200-acre field in 4 hours with one of his tractors. With another tractor the job takes 5 hours. How long would it take the farmer and his son if they tilled the field using both tractors?

50. Referring to Exercise 49, how long will each person till if the son drives the slower tractor and he starts (upon his return from school) $\frac{1}{2}$ hour after his father?

51. There are two pumps connected to a large gasoline storage tank. One pump will fill the tank in 5 hours, and the second will fill the tank in 8 hours. How long will it take to fill the tank if both pumps are used?

52. Referring to Exercise 51, how long will each pump run to fill the tank if the faster one is started 90 minutes after the slower pump?

53. Referring to Exercise 51, assume that the tank is half full and the pumps are working against each other. For example, suppose the slower pump is being used to transfer gasoline from the tank to an awaiting train of tank cars, and the faster pump is being used to pump gasoline into the storage tank. How long will it take to fill the tank?

54. Find two consecutive natural numbers such that the difference of their reciprocals equals one-fourth the reciprocal of the smaller number.

In Exercises 55–66, solve for the indicated variable.

55. Capacitance in Series Circuits:
$$C = \frac{1}{(1/C_1) + (1/C_2)}$$
(Solve for C_1)

56. Inductance in Parallel Circuits:
$$L = \frac{1}{\dfrac{1}{L_1} + \dfrac{1}{L_2} + \dfrac{1}{L_3}}$$
(Solve for L_3)

57. Thermal Expansion:
$$L = L_0[1 + \alpha(\Delta t)]$$
(Solve for α)

58. Freely Falling Body:
$$h = v_0 t + \frac{1}{2} at^2$$
(Solve for a)

59. Newton's Law of Universal Gravitation:
$$F = \alpha \frac{m_1 m_2}{r^2}$$
(Solve for m_2)

60. Heat Flow:
$$H = \frac{KA(t_2 - t_1)}{L}$$
(Solve for t_1)

61. Lensmaker's Equation: $\dfrac{1}{f} = (n - 1)\left(\dfrac{1}{R_1} - \dfrac{1}{R_2}\right)$ (Solve for R_1)

62. Thermometer Construction: $L = \dfrac{v_0}{A_0}(B_m - B_\gamma)t$ (Solve for B_γ)

63. Arithmetic Progression: $L = a + (n - 1)d$ (Solve for n)

64. Arithmetic Progression: $S = \dfrac{n}{2}[2a + (n - 1)d]$ (Solve for a)

65. Geometric Progression: $S = \dfrac{rL - a}{r - 1}$ (Solve for r)

66. Prismoidal Formula: $V = \dfrac{1}{6}H(S_0 + 4S_1 + S_2)$ (Solve for S_1)

Quadratic Equations   2.3

So far in this chapter we have carefully chosen our problems so that the equations obtained would reduce to linear form. We now focus our attention on equations that can be expressed in the **standard quadratic form**

$$ax^2 + bx + c = 0, \qquad a \neq 0$$

where a, b, and c are real numbers. We will discuss two methods for solving quadratic equations: *factoring* and *completing the square*.

Solution by Factoring

Solution by factoring is based upon a property of zero discussed in Section 1.2:

$$a \cdot b = 0 \qquad \Longrightarrow \qquad a = 0 \text{ or } b = 0$$

This means that if we can factor a quadratic equation (in standard form) into two linear factors, then we can find its solutions by setting each factor equal to zero. For instance, since

$$x^2 - 3x - 10 \text{ factors as } (x - 5)(x + 2)$$

the roots of the equation $x^2 - 3x - 10 = 0$ can be found as follows:

$$x^2 - 3x - 10 = 0$$
$$(x - 5)(x + 2) = 0 \qquad\qquad \textit{Factored form}$$
$$x - 5 = 0 \qquad x + 2 = 0 \qquad \textit{Set each factor equal to zero}$$
$$x = 5 \qquad\qquad x = -2 \qquad \textit{Solutions}$$

One word of caution: *Don't* use the above factoring procedure on equations that are not in standard form. For instance, the following is *incorrect algebra:*

$$x^2 - 3x - 10 = 7 \qquad \textit{Not in standard form}$$
$$(x - 5)(x + 2) = 7 \qquad \textit{Product} \neq 0$$
$$x - 5 = 7 \qquad x + 2 = 7 \qquad \textit{Invalid}$$
$$x = 12 \qquad\qquad x = 5$$

EXAMPLE 1
Solving Quadratic Equations by Factoring

Solve each of the following by factoring.

(a) $2x^2 + 9x + 7 = 3$ (b) $6x^2 - 3x = 0$ (c) $9x^2 - 6x + 1 = 0$

Solution:

(a)
$$2x^2 + 9x + 4 = 0 \qquad \textit{Standard form}$$
$$(2x + 1)(x + 4) = 0 \qquad \textit{Factored form}$$
$$2x + 1 = 0 \qquad x + 4 = 0 \qquad \textit{Set factors equal to zero}$$
$$x = -\frac{1}{2} \qquad x = -4 \qquad \textit{Solutions}$$

(b) Note the lack of a constant term.
$$6x^2 - 3x = 0 \qquad \textit{Given}$$
$$3x(2x - 1) = 0 \qquad \textit{Factor out common monomial}$$
$$3x = 0 \qquad 2x - 1 = 0 \qquad \textit{Set factors equal to zero}$$
$$x = 0 \qquad x = \frac{1}{2} \qquad \textit{Solutions}$$

(c)
$$9x^2 - 6x + 1 = 0 \qquad \textit{Given}$$
$$(3x - 1)^2 = 0 \qquad \textit{Perfect square trinomial}$$
$$3x - 1 = 0 \qquad 3x - 1 = 0 \qquad \textit{Set factors equal to zero}$$
$$x = \frac{1}{3} \qquad x = \frac{1}{3} \qquad \textit{Solutions}$$

The number $x = \frac{1}{3}$ is referred to as a **double** or **repeated root** of $9x^2 - 6x + 1 = 0$.

The special quadratic form

$$x^2 = d, \qquad d \geq 0$$

has two simple methods of solution: by factoring,

$$x^2 - d = 0 \quad \Longrightarrow \quad (x + \sqrt{d})(x - \sqrt{d}) = 0$$

or by *taking the square root of both sides,* as indicated next.

SQUARE ROOT PRINCIPLE

If $x^2 = d$ and $d \geq 0$, then $x = \pm\sqrt{d}$.

Our next example shows how effective this square root principle can be.

EXAMPLE 2
Solving the Quadratic
Form $x^2 = d$

Solve each of the following:

(a) $4x^2 = 12$ (b) $(x - 3)^2 = 7$

Solution:

(a) In this case, we have

$$4x^2 = 12$$
$$x^2 = \frac{12}{4} = 3$$
$$x = \pm\sqrt{3}$$

Note that $x^2 - 3 = 0$ factors as $(x + \sqrt{3})(x - \sqrt{3}) = 0$, which gives the same solutions.

(b) In this case, an extra step is needed after taking square roots.

$$(x - 3)^2 = 7$$
$$x - 3 = \pm\sqrt{7}$$
$$x = 3 \pm \sqrt{7}$$

Thus,

$$x = 3 + \sqrt{7} \approx 5.646 \qquad \text{or} \qquad x = 3 - \sqrt{7} \approx 0.354$$

Note that $(x - 3)^2 - 7 = 0$ factors as

$$[(x - 3) + \sqrt{7}][(x - 3) - \sqrt{7}] = 0$$

which gives the same solutions.

Completing the Square

The equation in part (b) of Example 2 can lead us into an interesting situation. Suppose we expand that equation to obtain

$$(x - 3)^2 = 7$$
$$x^2 - 6x + 9 = 7$$
$$x^2 - 6x + 2 = 0$$

We know this latter equation is equivalent to the original, and thus has the same two solutions, $x = 3 \pm \sqrt{7}$. However, the latter equation is not factorable, and we could not find its solutions unless we could somehow *reverse* the steps shown above. A procedure for reversing these steps is called **completing the square.** Observe how it works on our equation above.

$$x^2 - 6x + 2 = 0 \qquad \qquad \textit{Given}$$
$$x^2 - 6x = -2 \qquad \qquad \textit{Move constant to right side}$$

$$x^2 - (6)x + (3)^2 = -2 + (3)^2$$

Take half of 6, square it, and add to both sides

$$\underset{(\text{half})^2}{\underbrace{\qquad\uparrow\qquad}}$$

$$x^2 - 6x + 9 = 7 \qquad\qquad \textit{Simplify}$$
$$(x - 3)^2 = 7 \qquad\qquad \textit{Perfect square trinomial}$$

Now the solution can be obtained by using the square root principle.

EXAMPLE 3
Solution by Completing the Square

Solve

$$3x^2 - 4x - 5 = 0$$

Solution:
Since solution by factoring is easier than completing the square, always check quickly to see if a given quadratic is factorable. Since the leading coefficient is not one ($a \neq 1$), we first divide both sides by a.

$$3x^2 - 4x - 5 = 0$$
$$3x^2 - 4x = 5$$
$$x^2 - \frac{4}{3}x = \frac{5}{3}$$
$$x^2 - \frac{4}{3}x + \left(\frac{2}{3}\right)^2 = \frac{5}{3} + \left(\frac{2}{3}\right)^2$$

$$\underset{(\text{half})^2}{\underbrace{\qquad\uparrow\qquad}}$$

$$\left(x - \frac{2}{3}\right)^2 = \frac{19}{9}$$
$$x - \frac{2}{3} = \pm\sqrt{19/9} = \pm\frac{\sqrt{19}}{3}$$
$$x = \frac{2}{3} \pm \frac{\sqrt{19}}{3} = \frac{1}{3}(2 \pm \sqrt{19})$$

In the next section we will see how completing the square can be used to develop a general formula for solving quadratic equations. It is also useful in working with conics (Section 5.4) and in rewriting algebraic expressions to simplify calculus operations. The next example shows how to complete the square on an algebraic *expression* rather than an equation. There are subtle differences.

EXAMPLE 4
Completing the Square for Algebraic Expressions

For each of the following expressions (taken from a calculus text) rewrite the denominator so that it involves the sum or difference of two squares.

(a) $\dfrac{1}{2x^2 - x - 3}$

(b) $\dfrac{1}{\sqrt{3x - x^2}}$

Solution:

(a) Since we need to write equivalent expressions, we must *add and subtract* the same constant value. In this case, we first factor out the coefficient of x^2.

$$2x^2 - x - 3$$

$$= 2\left(x^2 - \frac{1}{2}x - \frac{3}{2}\right) \qquad \textit{Factor out 2}$$

$$= 2\left[x^2 - \frac{1}{2}x + \left(\frac{1}{4}\right)^2 - \frac{3}{2} - \left(\frac{1}{4}\right)^2\right] \qquad \begin{array}{l}\textit{Add and subtract}\\(1/4)^2\end{array}$$

$$= 2\left[\left(x - \frac{1}{4}\right)^2 - \frac{25}{16}\right] \qquad \textit{Factor trinomial}$$

Therefore,

$$\frac{1}{2x^2 - x - 3} = \frac{1}{2\left[\left(x - \frac{1}{4}\right)^2 - \left(\frac{5}{4}\right)^2\right]}$$

Although $2x^2 - x - 3$ is factorable, some operations in calculus are simpler with the completed square form than with the factored form.

(b) The negative coefficient on x^2 complicates things a bit. Note how we deal with it.

$$3x - x^2 = -(x^2 - 3x) \qquad \begin{array}{l}\textit{Factor out negative}\\\textit{coefficient of } x^2\end{array}$$

$$= -\left[x^2 - 3x + \left(\frac{3}{2}\right)^2 - \left(\frac{3}{2}\right)^2\right] \qquad \begin{array}{l}\textit{Add and subtract}\\(3/2)^2\end{array}$$

$$= -\left[\left(x - \frac{3}{2}\right)^2 - \frac{9}{4}\right] \qquad \textit{Factor trinomial}$$

$$= \frac{9}{4} - \left(x - \frac{3}{2}\right)^2 \qquad \textit{Remove brackets}$$

Therefore,

$$\frac{1}{\sqrt{3x - x^2}} = \frac{1}{\sqrt{\left(\frac{3}{2}\right)^2 - \left(x - \frac{3}{2}\right)^2}}$$

Section Exercises 2.3

In Exercises 1–20, solve the quadratic equations by factoring.

1. $x^2 - 2x - 8 = 0$
2. $x^2 - 10x + 9 = 0$
3. $x^2 + 10x + 25 = 0$
4. $x^2 - 14x + 49 = 0$
5. $4x^2 - 12x + 9 = 0$
6. $16x^2 + 56x + 49 = 0$
7. $6x^2 + 3x = 0$
8. $11x^2 + 33x = 0$
9. $16x^2 - 9 = 0$
10. $9x^2 - 1 = 0$

11. $3 + 5x - 2x^2 = 0$

12. $144 - 73x + 4x^2 = 0$

13. $50 + 5x = 3x^2$

14. $2x^2 = 19x + 33$

15. $12x = x^2 + 27$

16. $26x = 8x^2 + 15$

17. $x^2 + 2ax + a^2 = 0$

18. $(x + a)^2 - b^2 = 0$

19. $(x + 3)^2 - 4 = 0$

20. $a^2x^2 - b^2 = 0$

In Exercises 21–35, solve the equations by taking the square root of both sides.

21. $x^2 = 16$

22. $x^2 = 144$

23. $x^2 = 7$

24. $x^2 = 27$

25. $x^2 = 64$

26. $9x^2 = 25$

27. $3x^2 = 36$

28. $7x^2 = 32$

29. $(x + 3)^2 = 81$

30. $(x - 5)^2 = 8$

31. $(x - 12)^2 = 18$

32. $(x + 13)^2 = 21$

33. $(x - 7)^2 = (x + 3)^2$

34. $(x + 5)^2 = (x + 4)^2$

35. $(x + 1)^2 = x^2$

In Exercises 36–50, solve the quadratic equations by completing the square.

36. $x^2 - 2x + 2 = 0$

37. $x^2 + 6x + 13 = 0$

38. $x^2 - 2x - 3 = 0$

39. $x^2 - 2x + 5 = 0$

40. $-x^2 - 4x = 0$

41. $x^2 - 2x = 0$

42. $4x - x^2 = 0$

43. $9x^2 - 18x + 5 = 0$

44. $2x^2 + 4x + 8 = 0$

45. $5 + 4x - x^2 = 0$

46. $4x^2 - 4x - 99 = 0$

47. $50x^2 - 60x - 7 = 0$

48. $9x^2 + 12x + 3 = 0$

49. $4x = 4x^2 - 3$

50. $80 + 6x = 9x^2$

In Exercises 51–60, complete the square on the quadratic portion of the algebraic expressions.

51. $\dfrac{1}{x^2 + 2x + 5}$

52. $\dfrac{1}{x^2 - 4x + 13}$

53. $\dfrac{1}{x^2 - 4x - 12}$

54. $\dfrac{4}{4x^2 + 4x - 3}$

55. $\dfrac{1}{\sqrt{6x - x^2}}$

56. $\dfrac{4}{\sqrt{15 - 8x - 16x^2}}$

57. $\dfrac{2}{\sqrt{4x^2 - 4x + 5}}$

58. $\dfrac{1}{25x^2 + 20x + 7}$

59. $\dfrac{-1}{x^2 - 10x - 16}$

60. $\dfrac{1}{\sqrt{16 - 6x - x^2}}$

The Quadratic Formula and Applications

2.4

In Section 2.3 we introduced the technique of completing the square to solve irreducible quadratic equations. In that setting, we were required to complete the square on each equation separately. In this section we will use the completing the square procedure *once* in a general setting to obtain the **quadratic formula,** a short-cut for solving quadratic equations.

As you continue in mathematics, you will encounter several techniques in which the long way of doing the problem is presented first. Afterwards, the longer technique will be used to develop shorter techniques that are more efficient. The long way stresses understanding and the short-cut stresses efficiency. In mathematics, we value both of these goals.

Derivation of the Quadratic Formula

$$ax^2 + bx + c = 0 \qquad \text{\textit{Given, } } a \neq 0$$

$$ax^2 + bx = -c \qquad \text{\textit{Move constant}}$$

$$x^2 + \frac{b}{a}x = -\frac{c}{a} \qquad \text{\textit{Divide by coefficient of } } x^2$$

$$x^2 + \frac{b}{a}x + \left(\frac{b}{2a}\right)^2 = -\frac{c}{a} + \left(\frac{b}{2a}\right)^2 \qquad \textit{Add } \left(\frac{1}{2} \cdot \frac{b}{a}\right)^2$$

$$\underbrace{\qquad}_{(\text{half})^2}$$

$$\left(x + \frac{b}{2a}\right)^2 = \frac{b^2 - 4ac}{4a^2} \qquad \textit{Factor left side, simplify right side}$$

$$x + \frac{b}{2a} = \pm\sqrt{\frac{b^2 - 4ac}{4a^2}} \qquad \textit{Take square roots}$$

$$x = -\frac{b}{2a} \pm \frac{\sqrt{b^2 - 4ac}}{2a} \qquad \textit{Solve for } x$$

You should memorize this important formula.

THE QUADRATIC FORMULA

The solutions of a quadratic equation in standard form

$$ax^2 + bx + c = 0, \qquad a \neq 0$$

are obtained from the formula

$$x = \frac{-b \pm \sqrt{b^2 - 4ac}}{2a}$$

Since the square root of a negative number is not a real number, some concern should exist about the quantity under the radical sign, $b^2 - 4ac$. This quantity is called the **discriminant,** and it determines the nature of the roots of a quadratic equation.

ROOTS OF A QUADRATIC EQUATION

Given $ax^2 + bx^2 + c = 0 \qquad (a \neq 0)$:

1. If $b^2 - 4ac > 0$, the equation has **two distinct real roots**

$$x = \frac{-b \pm \sqrt{b^2 - 4ac}}{2a}$$

2. If $b^2 - 4ac = 0$, the equation has a **double root**

$$x = -\frac{b}{2a}$$

3. If $b^2 - 4ac < 0$, the equation has **no real roots.**

Remark: When the discriminant is negative ($b^2 - 4ac < 0$), its square root is **imaginary** and the quadratic formula yields two **complex roots.**

EXAMPLE 1
Using the Discriminant

Without solving the equations, determine how many real roots each of the following has.

(a) $4x^2 - 20x + 25 = 0$ (b) $13x^2 + 7x + 1 = 0$ (c) $5x^2 = 8x$

Solution:

(a) We have $a = 4$, $b = -20$, and $c = 25$, with a discriminant of

$$b^2 - 4ac = 400 - 4(4)(25) = 400 - 400 = 0$$

Since the discriminant is zero, there is one double root.

(b) In this case, $a = 13$, $b = 7$, and $c = 1$, with

$$b^2 - 4ac = 49 - 4(13)(1) = 49 - 52 = -3 < 0$$

Since the discriminant is negative, there are *no* real roots.

(c) In standard form the equation is

$$5x^2 - 8x = 0$$

with $a = 5$, $b = -8$, and $c = 0$. Thus,

$$b^2 - 4ac = 64 - 4(5)(0) = 64 > 0$$

Since the discriminant is positive, there are *two* real roots.

EXAMPLE 2
Using the Quadratic Formula

Solve by using the quadratic formula. If the discriminant is negative, list the complex roots in $a + bi$ form.

(a) $x^2 + 3x = 9$ (b) $6x^2 - 2x + 5 = 0$ (c) $8x^2 - 24x + 18 = 0$

Solution:

(a) In standard form, the equation is

$$x^2 + 3x - 9 = 0$$

where $a = 1$, $b = 3$, and $c = -9$. By the quadratic formula, we have

$$x = \frac{-b \pm \sqrt{b^2 - 4ac}}{2a} = \frac{-3 \pm \sqrt{(3)^2 - 4(1)(-9)}}{2(1)}$$

$$= \frac{-3 \pm \sqrt{45}}{2} = \frac{-3 \pm 3\sqrt{5}}{2}$$

Thus,

$$x = \frac{-3 + 3\sqrt{5}}{2} \quad \text{or} \quad x = \frac{-3 - 3\sqrt{5}}{2}$$

(b) We have $a = 6$, $b = -2$, $c = 5$, and

$$x = \frac{-b \pm \sqrt{b^2 - 4ac}}{2a} = \frac{-(-2) \pm \sqrt{4 - 4(6)(5)}}{2(6)}$$

$$= \frac{2 \pm \sqrt{-116}}{12} = \frac{2 \pm 2\sqrt{-29}}{12}$$

In $a + bi$ form, the complex roots are

$$x = \frac{1}{6} + \frac{\sqrt{29}}{6}\, i \quad \text{and} \quad x = \frac{1}{6} - \frac{\sqrt{29}}{6}\, i$$

(c) To simplify our calculations, we begin by factoring 2 out of the quadratic.

$8x^2 - 24x + 18 = 0$	*Given*
$2(4x^2 - 12x + 9) = 0$	*Remove common factor*
$4x^2 - 12x + 9 = 0$	*Divide both sides by 2*

Thus, $a = 4$, $b = -12$, and $c = 9$, and by the quadratic formula we have

$$x = \frac{-b \pm \sqrt{b^2 - 4ac}}{2a} = \frac{-(-12) \pm \sqrt{(-12)^2 - 4(4)(9)}}{2(4)}$$

$$= \frac{12 \pm \sqrt{0}}{8} = \frac{3}{2}$$

Note that when the discriminant turns out to be a perfect square (zero in this case), it means we could have factored the quadratic. This particular quadratic factors as $4x^2 - 12x + 9 = (2x - 3)^2 = 0$.

Remark: When solving a quadratic equation, first try to factor. If that doesn't work, then use the quadratic formula—it always works!

When you use a calculator to evaluate the quadratic formula, use the memory key to store the square root of the discriminant.

SOLVING QUADRATIC EQUATIONS WITH A CALCULATOR

A good calculator approach with the quadratic formula is
1. Compute $\sqrt{b^2 - 4ac}$ and store the result.
2. Add result to $-b$ and divide by $2a$ to get *first root*.
3. Subtract result from $-b$ and divide by $2a$ to get *second root*.

EXAMPLE 3
The Quadratic Formula on a Calculator

Solve

$$16.3x^2 - 197.6x + 7.042 = 0$$

Solution:

In this case, $a = 16.3$, $b = -197.6$, and $c = 7.042$.

Calculator Steps	Display
1. 4 $\boxed{\times}$ 16.3 $\boxed{\times}$ 7.042 $\boxed{=}$	459.13840
$\boxed{+/-}$ $\boxed{+}$ 197.6 $\boxed{x^2}$ $\boxed{=}$ $\boxed{\sqrt{}}$ $\boxed{\text{STO}}$	196.43478
2. $\boxed{+}$ 197.6 $\boxed{=}$ $\boxed{\div}$ 2 $\boxed{\div}$ 16.3 $\boxed{=}$	12.086956
3. 197.6 $\boxed{-}$ $\boxed{\text{RCL}}$ $\boxed{=}$ $\boxed{\div}$ 2 $\boxed{\div}$ 16.3 $\boxed{=}$	0.0357430

Thus, $x \approx 12.087$ or $x \approx 0.036$.

Keep the Guidelines for Solving Word Problems (Section 2.2) in mind as you study the next few examples and work the exercises for this section.

EXAMPLE 4
Falling Objects

(a) The formula $s = -16t^2 + v_0t + s_0$ is called a **position equation,** and in this particular case it gives the height above ground of a falling object.

s = height above ground for any time, t

s_0 = initial (original) height when $t = 0$

v_0 = initial velocity when $t = 0$

Solve for t in

$$s = -16t^2 + v_0t + s_0$$

(b) From the top of a 300-foot river gorge, how long will it take a rock to hit the water below if (i) it is dropped, (ii) it is thrown vertically downward at 50 feet per second?

Solution:

(a) To solve for t in $s = -16t^2 + v_0t + s_0$, we can use the quadratic formula as follows:

$$-16t^2 + v_0t + (s_0 - s) = 0 \qquad \textit{Write in standard form}$$

where $a = -16$, $b = v_0$, and $c = s_0 - s$.

$$t = \frac{-v_0 \pm \sqrt{(v_0)^2 - 4(-16)(s_0 - s)}}{2(-16)} \qquad \textit{Substitute into formula}$$

$$t = \frac{-v_0 \pm \sqrt{v_0{}^2 + 64(s_0 - s)}}{32} \qquad \textit{Simplify}$$

(b) Figure 2.2 shows the path of the rock as it falls. The initial height is $s_0 = 300$. Since we want to know the time when the rock hits the water ($s = 0$), we need to solve the equation

$$0 = -16t^2 + v_0t + s_0 = -16t^2 + v_0t + 300$$

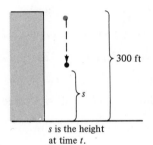

300 ft

s

s is the height at time t.

FIGURE 2.2

(i) Since the object is dropped, we know that the initial velocity is $v_0 = 0$, and we have

$$0 = -16t^2 + 300$$

$$16t^2 = 300$$

$$t^2 = \frac{300}{16} = \frac{75}{4}$$

$$t = \pm\frac{\sqrt{75}}{\sqrt{4}} = \pm\frac{5\sqrt{3}}{2}$$

Now, choosing the positive time value, we have

$$t = \frac{5\sqrt{3}}{2} \approx 4.33 \text{ seconds}$$

(ii) In this case, the rock is thrown downward at 50 feet per second and the initial velocity is $v_0 = -50$. (Negative velocity denotes a downward direction and positive velocity denotes an upward direction.) To find the time it hits the water, we set $s = 0$ as follows:

$$-16t^2 - 50t + 300 = 0$$

where $a = -16$, $b = -50$, and $c = 300$.

$$t = \frac{50 \pm \sqrt{(-50)^2 - 4(-16)(300)}}{32} = \frac{50 \pm \sqrt{21700}}{32}$$

Choosing the positive time value, we have

$$t \approx 3.04 \text{ seconds}$$

Remark: When you solve applied problems, try to get in the habit of asking yourself if your answer seems reasonable. For instance, in Example 4 it is reasonable that the rock thrown downward would hit the water sooner than the one that was dropped.

EXAMPLE 5
A Quadratic Application

A picture is 3 inches longer than it is wide and has an area of 120 square inches. It is to be enclosed in a frame of uniform width of 2 inches. What are the outer dimensions of the frame?

Solution:
In this case, our model includes the picture shown in Figure 2.3.

Model: area of picture = (width)(length) = 120

Labels: W = picture width, L = picture length
                $W + 4$ = total width, $L + 4$ = total length

Since the length of the picture is 3 inches more than the width, it follows that $L = W + 3$.

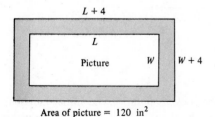

Area of picture = 120 in^2

FIGURE 2.3

Equation: $W(W + 3) = 120$

$$W^2 + 3W - 120 = 0$$

$$W = \frac{-3 \pm \sqrt{9 + 480}}{2}$$

$$= \frac{-3 \pm \sqrt{489}}{2}$$

Choosing the positive root, we have

$$W = \frac{-3 + \sqrt{489}}{2} \approx 9.56 \text{ inches}$$

Finally, the frame has outer dimensions

width $= W + 4 \approx 13.56$ inches

length $= L + 4 = W + 7 \approx 16.56$ inches

EXAMPLE 6
A Quadratic Application

The walkway from building A to building B on a college campus is L-shaped, with the total distance being 400 meters. By cutting diagonally across the grass, students shorten the walking distance to 300 meters. What are the lengths of the two legs of the existing walkway?

Solution:
Our model includes the picture shown in Figure 2.4.

Model: $a^2 + b^2 = c^2$ *(Pythagorean Theorem)*

Labels: $x =$ length of one leg $= a$

$400 - x =$ length of the other leg $= b$

$300 =$ length of diagonal $= c$

Equation: Using the Pythagorean Theorem, we have

$$x^2 + (400 - x)^2 = (300)^2$$
$$2x^2 - 800x + 160{,}000 = 90{,}000$$
$$2x^2 - 800x + 70{,}000 = 0$$
$$x^2 - 400x + 35{,}000 = 0$$

Now, by the quadratic formula, we have

$$x = \frac{400 \pm \sqrt{(400)^2 - 4(1)(35000)}}{2(1)}$$

$$= \frac{400 \pm \sqrt{20{,}000}}{2} = \frac{400 \pm 100\sqrt{2}}{2}$$

$$= 200 \pm 50\sqrt{2}$$

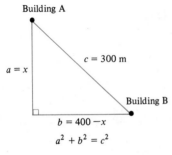

Building A

$c = 300$ m

$a = x$

Building B

$b = 400 - x$

$a^2 + b^2 = c^2$

FIGURE 2.4

Both solutions are positive, and it doesn't matter which one we choose. If we let

$$x = 200 + 50\sqrt{2} \approx 270.7 \text{ meters}$$

then the length of the other leg is

$$400 - x = 200 - 50\sqrt{2} \approx 129.3 \text{ meters}$$

Try choosing the other value of x to see that the same two lengths result.

EXAMPLE 7
Reduced Rates

A ski club chartered a bus for a ski trip at a cost of $480. In an attempt to lessen the bus fare per skier, the club invited nonmembers to go along. After 5 nonmembers joined the trip the fare per skier decreased by $4.80. How many club members are going on the bus?

Solution:

Model: $480 = (\text{cost per skier})(\text{number of skiers})$

Labels: *originally:* $x = \text{number of members}$

$$\frac{480}{x} = \text{cost per skier}$$

with more skiers: $x + 5 = \text{number of skiers}$

$$\frac{480}{x} - 4.80 = \text{cost per skier}$$

Equation:

$$480 = \left(\frac{480}{x} - 4.80\right)(x + 5)$$

$$480x = (480 - 4.8x)(x + 5)$$

$$480x = 480x - 4.8x^2 - 24x + 2400$$

$$4.8x^2 + 24x - 2400 = 0$$

$$x^2 + 5x - 500 = 0$$

$$(x + 25)(x - 20) = 0$$

$$x = 20 \text{ or } -25$$

Choosing the positive value of x, we have

$$x = 20 \text{ ski club members}$$

Section Exercises 2.4

In Exercises 1–8, use the discriminant to determine how many real roots each of the quadratic equations has.

1. $4x^2 - 4x + 1 = 0$ 2. $2x^2 - x - 1 = 0$

3. $3x^2 + 4x + 1 = 0$ 4. $x^2 + 2x + 4 = 0$
5. $2x^2 - 5x + 5 = 0$ 6. $3x^2 - 6x + 3 = 0$
7. $\frac{1}{5}x^2 + \frac{6}{5}x - 8 = 0$ 8. $\frac{1}{3}x^2 - 5x + 25 = 0$

In Exercises 9–30, use the quadratic formula to solve the equations. List any complex roots in $a + bi$ form.

9. $16x^2 + 8x - 3 = 0$
10. $25x^2 - 20x + 3 = 0$
11. $x^2 - 2x - 2 = 0$
12. $x^2 - 10x + 22 = 0$
13. $x^2 + 14x + 44 = 0$
14. $x^2 + 6x - 4 = 0$
15. $x^2 + 8x - 4 = 0$
16. $4x^2 - 4x - 4 = 0$
17. $9x^2 - 12x - 3 = 0$
18. $16x^2 - 40x + 22 = 0$
19. $36x^2 + 24x - 7 = 0$
20. $x^2 + 3x - 1 = 0$
21. $4x^2 + 4x - 7 = 0$
22. $16x^2 - 40x + 5 = 0$
23. $49x^2 - 28x + 4 = 0$
24. $9x^2 + 24x + 16 = 0$
25. $25h^2 + 80h + 61 = 0$
26. $5s^2 - 6s + 3 = 0$
27. $16t^2 + 4t + 3 = 0$
28. $8t = 5 + 2t^2$
29. $(y - 5)^2 = 2y$
30. $(z + 6)^2 = -2z$

Calculator Exercises

In Exercises 31–35, use a calculator to solve the quadratic equations. Round your answers to three decimal places.

31. $5.1x^2 - 1.7x - 3.2 = 0$
32. $10.4x^2 + 8.6x + 1.2 = 0$
33. $7.06x^2 - 4.85x + 3.92 = 0$
34. $-0.052x^2 + 0.101x - 0.193 = 0$
35. $422x^2 - 506x - 347 = 0$

Applications

36. A manufacturer has determined that the total weekly cost C of operating a certain facility is given by

$$C = 0.5x^2 + 15x + 5000$$

where x is the number of units produced. If, for a certain week, the total costs were \$11,500, how many units were produced?

37. A company estimates that the daily cost of producing x units of a certain product is given by

$$C = 800 + 0.04x + 0.0002x^2$$

If, for a certain day, the costs were \$1680, how many units were produced?

38. A manufacturer of lighting fixtures has daily production costs of

$$C = 800 - 10x + \frac{x^2}{4}$$

where x is the number of fixtures produced. If, for a certain day, the costs were \$896, how many fixtures were produced?

39. An open box (see Figure 2.5) having a square base is to be constructed from 108 square inches of material. What

should be the dimensions of the base if the height of the box is to be 3 inches? (Hint: The surface area is given by $S = x^2 + 4xh$.)

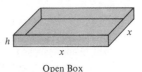

Open Box

FIGURE 2.5

40. A rancher has 200 feet of fencing to enclose two adjacent rectangular corrals (see Figure 2.6). Find the dimensions so that the enclosed area will be 1400 square feet.

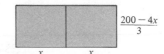

FIGURE 2.6

41. Find two positive numbers whose sum is 100 and product is 2500.

42. Find two consecutive positive numbers whose product is 72.

43. Find a number such that the sum of the number and its reciprocal is $\frac{26}{5}$.

44. Find two consecutive even positive integers whose product is 440.

45. Find two consecutive positive integers, the sum of whose squares is 113.

46. Use the equation for falling objects from Example 4 to find the time when an object strikes the ground given the following conditions:
    (a) The object is dropped from a balloon at a height of 1600 feet.
    (b) The object is thrown vertically upward from ground level with an initial velocity of 60 feet per second.

47. Use the equation for falling objects from Example 4 to find the time when an object strikes the ground given the following conditions:
    (a) The object is dropped from a height of 550 feet (approximate height of the Washington Monument).
    (b) The object is dropped at a height of 64 feet from a balloon rising vertically at the rate of 16 feet per second. (The balloon's velocity becomes the object's initial velocity.)

48. A rectangular classroom seats 72 students. If 3 more seats were added to each row, the number of rows could be reduced by 2. Find the original number of seats in each row.

49. Two brothers must mow a rectangular lawn that is 100 feet by 200 feet, and each is to do one-half. The first starts by

mowing around the outside of the lawn. How wide a strip must he mow on each of the four sides? Approximately how many times must he go around the lawn if the mower has a 24-inch cut?

50. Repeat Exercise 49, assuming that the first brother agrees to mow three-fourths of the lawn.

51. A windlass is used to tow a boat to the dock (see Figure 2.7). The rope is attached to the boat at a point 15 feet below the level of the windlass. Use the Pythagorean Theorem to find the distance from the boat to the dock when there is 75 feet of rope out.

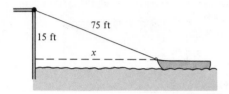

FIGURE 2.7

52. Each day a pilot for a small commuter airline flies to three cities whose locations approximate the vertices of an isosceles right triangle. Each day she returns to the city from which she departed in the morning. If the hypotenuse is 600 miles, use the Pythagorean Theorem to find the other two sides of the triangle.

53. Two planes leave simultaneously from the same airport, one flying due east and the other due south. The eastbound plane is flying 50 miles per hour faster than the southbound. If after 3 hours the planes are 2440 miles apart, find the ground speed of each plane.

54. A college charters a bus for $1700 to take students to the World's Fair. When 6 more students join the trip, the cost per student drops by $7.50. How many students were in the original group?

55. Together two machines can do a job in 8 hours. Alone, it takes machine A 2 hours more than machine B to do the job. Find the time required for each machine to do the job.

56. A certain business is now selling 2000 units per month at a price of $10.00 each. If it can sell 250 more units per month for each $0.25 reduction in price, what price per unit will yield a monthly revenue of $36,000?

57. An open box is to be made from a square piece of material by cutting 2-inch squares from each corner and turning up the sides (see Figure 2.8). If the box is to contain 200 cubic inches, find the size of the original piece of material.

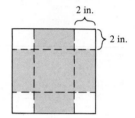

FIGURE 2.8

58. Repeat Exercise 57, assuming that 4-inch squares are cut from each corner of the material.

59. A man invests $10,000 for 2 years at $r\%$ compounded annually. If at the end of the 2-year period the investment has increased to $12,321, find r.

60. Repeat Exercise 59, assuming that the investment increased to $11,990.25.

Other Types of Equations   2.5

In this section we will extend our solution techniques to include special categories of nonlinear and nonquadratic equations. We begin with equations involving polynomials with recognizable factors. (You may want to review the special polynomial forms listed in Section 1.8.)

EXAMPLE 1
Equations Involving Polynomials with Recognizable Factors

Solve each of the following by factoring completely.

(a) $3x^6 - 48x^2 = 0$
(b) $x^3 - 3x^2 - 3x + 9 = 0$

Solution:

(a)

$$3x^6 - 48x^2 = 0 \qquad \textit{Given}$$

$$3(x^2)(x^4 - 16) = 0 \qquad \textit{Common monomial factor}$$

$$3(x^2)(x^2 + 4)(x^2 - 4) = 0 \qquad \textit{Difference of squares}$$

$$3(x^2)(x^2 + 4)(x + 2)(x - 2) = 0 \qquad \textit{Difference of squares}$$

Setting each variable factor equal to zero, we have

$$x^2 = 0 \qquad x^2 + 4 = 0 \qquad x + 2 = 0 \qquad x - 2 = 0$$
$$\text{(no real}$$
$$x = 0 \qquad \text{solutions)} \qquad x = -2 \qquad x = 2$$

Thus, the equation has three real roots: 0, -2, and 2.

(b) Grouping terms works well in this case.

$$x^3 - 3x^2 - 3x + 9 = 0 \qquad \textit{Given}$$

$$x^2(x - 3) - 3(x - 3) = 0 \qquad \textit{Group terms}$$

$$(x - 3)(x^2 - 3) = 0 \qquad \textit{Factor out } (x - 3)$$

$$x - 3 = 0 \qquad x^2 - 3 = 0 \qquad \textit{Set factors to zero}$$

$$x = 3 \qquad x = \pm\sqrt{3} \qquad \textit{Solve for x}$$

We conclude that there are three roots: 3, $\sqrt{3}$, and $-\sqrt{3}$.

Remark: A common mistake in a problem such as part (a) of Example 1 is to divide out the variable factor (x^2) before attempting to solve the equation. This would lead to the loss of one of the solutions. Be sure to factor completely and *then* set all variable factors equal to zero.

Until now our basic strategy for solving an equation has been:

In Section 2.1 we saw that multiplying both sides of an equation by a variable expression is a valuable solution technique, even though it may produce an equation that is not equivalent to the original. Thus, with this technique we must check the resulting solutions back in the original equation. This situation is illustrated in the following diagram.

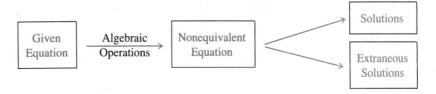

When we solve equations involving radicals or fractional powers, it is often useful to raise both sides of the equation to the same power. The potential then exists for introducing extraneous solutions, and we must validate each solution with a check in the original equation.

EXAMPLE 2
Radical Equations

Find all real solutions to

(a) $\sqrt{2x + 7} - x = 2$ (b) $\sqrt{1 + 5y} + \sqrt{1 - y} = 6$

Solution:

(a) It is best to isolate the radical before squaring both sides.

$$\sqrt{2x + 7} - x = 2 \qquad \textit{Given}$$
$$\sqrt{2x + 7} = x + 2 \qquad \textit{Isolate radical}$$
$$2x + 7 = x^2 + 4x + 4 \qquad \textit{Square both sides}$$
$$0 = x^2 + 2x - 3 \qquad \textit{Standard form}$$
$$0 = (x + 3)(x - 1) \qquad \textit{Factor}$$

$$x + 3 = 0 \qquad x - 1 = 0 \qquad \textit{Set factors to zero}$$
$$x = -3 \qquad x = 1 \qquad \textit{Possible solutions}$$

Solution check:

$$\sqrt{2(-3) + 7} - (-3) \overset{?}{=} 2 \qquad \textit{Check: } x = -3 \textit{ is not a}$$
$$\sqrt{1} + 3 \neq 2 \qquad \textit{solution}$$

$$\sqrt{2(1) + 7} - 1 \overset{?}{=} 2 \qquad \textit{Check: } x = 1 \textit{ is a solution}$$
$$\sqrt{9} - 1 = 2$$

(b) Again, it is a good idea to isolate one radical before squaring.

$$\sqrt{1 + 5y} + \sqrt{1 - y} = 6$$
$$\sqrt{1 + 5y} = 6 - \sqrt{1 - y}$$
$$1 + 5y = 36 - 12\sqrt{1 - y} + (1 - y)$$
$$6y - 36 = -12\sqrt{1 - y}$$
$$6 - y = 2\sqrt{1 - y}$$
$$36 - 12y + y^2 = 4(1 - y)$$
$$y^2 - 8y + 32 = 0$$
$$y = \frac{8 \pm \sqrt{64 - 4(1)(32)}}{2(1)}$$
$$= \frac{8 \pm \sqrt{-64}}{2}$$

Since the discriminant is negative, there are *no* real solutions.

EXAMPLE 3
Equations with
Fractional Exponents

Find all real solutions to

(a) $(x^2 - x - 4)^{3/4} = 8$
(b) $3x^2(2x - 1)^{1/2} + 2x(2x - 1)^{3/2} = 0$

Solution:

(a) In this case, we raise both sides to the *reciprocal* power, $4/3$.

$$(x^2 - x - 4)^{3/4} = 8$$
$$(x^2 - x - 4) = (8)^{4/3} = (\sqrt[3]{8})^4$$
$$x^2 - x - 4 = 16$$
$$x^2 - x - 20 = 0$$
$$(x - 5)(x + 4) = 0$$

$$x - 5 = 0 \qquad x + 4 = 0$$
$$x = 5 \qquad\qquad x = -4$$

Solution check:

$$(25 - 5 - 4)^{3/4} \overset{?}{=} 8 \qquad\qquad \textit{Check: } x = 5 \textit{ is a solution}$$
$$(16)^{3/4} \overset{?}{=} 8$$
$$(\sqrt[4]{16})^3 \overset{?}{=} 8$$
$$(2)^3 = 8$$
$$(16 + 4 - 4)^{3/4} \overset{?}{=} 8 \qquad\qquad \textit{Check: } x = -4 \textit{ is a solution}$$
$$(16)^{3/4} = 8$$

(b) This equation comes from a standard calculus text. We proceed by factoring out the common factors.

$3x^2(2x - 1)^{1/2} + 2x(2x - 1)^{3/2} = 0$	*Given*
$x(2x - 1)^{1/2}[3x + 2(2x - 1)^1] = 0$	*Factor out $x(2x - 1)^{1/2}$*
$x\sqrt{2x - 1}(7x - 2) = 0$	*Simplify*

$$x = 0 \qquad \sqrt{2x - 1} = 0 \qquad 7x - 2 = 0 \qquad \textit{Set factors to zero}$$
$$x = 0 \qquad\qquad x = \frac{1}{2} \qquad\qquad x = \frac{2}{7} \qquad \textit{Possible solutions}$$

Solution check:
Note that $x = 0$ and $x = \frac{2}{7}$ are not solutions because

$$\sqrt{2(0) - 1} = \sqrt{-1} \qquad \text{and} \qquad \sqrt{2\left(\frac{2}{7}\right) - 1} = \sqrt{-\frac{3}{7}}$$

are not real values. Hence the only real solution is $x = \frac{1}{2}$.

An equation is said to be of **quadratic type** if it can be written in the form

$$au^2 + bu + c = 0, \qquad a \neq 0$$

where u is an expression in some variable.

EQUATIONS OF QUADRATIC TYPE

Given Form	Quadratic Form	u
$x^4 - 5x^2 + 6 = 0$	$(x^2)^2 - 5(x^2) + 6 = 0$	$u = x^2$
$x^6 + 2x^3 + 1 = 0$	$(x^3)^2 + 2(x^3) + 1 = 0$	$u = x^3$
$2x^{2/3} + 3x^{1/3} - 9 = 0$	$2(x^{1/3})^2 + 3(x^{1/3}) - 9 = 0$	$u = x^{1/3}$
$\sqrt{x} - 3 + x = 0$	$(\sqrt{x})^2 + (\sqrt{x}) - 3 = 0$	$u = \sqrt{x}$
		$= x^{1/2}$

Remark: Notice that the variable factor in one term is the *square* of that factor in another term. Initially, you may want to write the given quadratic types in terms of the variable u. As you gain experience, we encourage you to do the u-substitution in your head.

EXAMPLE 4
Polynomial Equations
of Quadratic Type

Find all real solutions to

(a) $x^6 + 2x^3 + 1 = 0$
(b) $x^4 - 5x^2 - 2 = 0$

Solution:

(a) This equation is of quadratic type, where $u = x^3$. We have

$$x^6 + 2x^3 + 1 = (x^3)^2 + 2(x^3) + 1 = 0$$
$$u^2 + 2u + 1 = 0$$
$$(u + 1)^2 = 0$$
$$u + 1 = 0$$

Back substituting x^3 for u, we have

$$x^3 + 1 = 0$$
$$(x + 1)(x^2 - x + 1) = 0$$
$$x + 1 = 0 \quad \Longrightarrow \quad x = -1$$
$$x^2 - x + 1 = 0 \quad \Longrightarrow \quad x = \frac{1 \pm \sqrt{1 - 4}}{2}$$

Since $\sqrt{1 - 4} = \sqrt{-3}$ is not real, we conclude that the *only* possible real solution is $x = -1$. A quick check validates this solution.

(b) In this case, let $u = x^2$, then

$$x^4 - 5x^2 - 2 = (x^2)^2 - 5(x^2) - 2 = 0$$
$$u^2 - 5u - 2 = 0$$

By the quadratic formula we have

$$x^2 = u = \frac{5 \pm \sqrt{25 + 8}}{2} = \frac{5 \pm \sqrt{33}}{2}$$

Since $(5 - \sqrt{33})/2$ is negative, the only possible roots are

$$x = \pm \sqrt{\frac{5 + \sqrt{33}}{2}}$$

A check will show that both of these roots satisfy the original **equation.**

EXAMPLE 5
Radical Equations
of Quadratic Type

Find all real solutions to

$$2x^{2/3} + 3x^{1/3} - 9 = 0$$

Solution:
Let $u = x^{1/3}$, then

$$2x^{2/3} + 3x^{1/3} - 9 = 2(x^{1/3})^2 + 3(x^{1/3}) - 9 = 0$$
$$2u^2 + 3u - 9 = 0$$
$$(2u - 3)(u + 3) = 0$$

Setting these factors equal to zero, we have

$$2u - 3 = 0 \qquad u + 3 = 0$$
$$u = \frac{3}{2} \qquad u = -3$$
$$x^{1/3} = \frac{3}{2} \qquad x^{1/3} = -3$$
$$x = \frac{27}{8} \qquad x = -27$$

A check will show that both solutions are valid.

Section Exercises 2.5 ────────────────────────────

In Exercises 1–14, solve each of the equations by factoring completely.

1. $4x^4 - 18x^2 = 0$
2. $20x^3 - 125x = 0$
3. $x^3 - 2x^2 - 3x = 0$
4. $2x^4 - 15x^3 + 18x^2 = 0$

5. $x^4 - 81 = 0$
6. $x^6 - 64 = 0$
7. $5x^3 + 30x^2 + 45x = 0$
8. $9x^4 - 24x^3 + 16x^2 = 0$
9. $x^3 - 3x^2 - x + 3 = 0$
10. $x^3 + 2x^2 + 3x + 6 = 0$

11. $x^4 - x^3 + x - 1 = 0$
12. $x^4 + 2x^3 - 8x - 16 = 0$
13. $ax + bx - 5a - 5b = 0$
14. $ax^2 - bx^2 - 3a + 3b = 0$

In Exercises 15–30, solve the radical equations and check your solutions.

15. $\sqrt{x - 10} - 4 = 0$
16. $\sqrt{5 - x} - 3 = 0$
17. $\sqrt[3]{2x + 5} + 3 = 0$
18. $\sqrt[3]{3x + 1} - 5 = 0$
19. $\sqrt{3x + 2} = 3x$
20. $\sqrt{3 - 10x} = 5x$
21. $x = \sqrt{11x - 30}$
22. $2x - \sqrt{15 - 4x} = 0$
23. $-\sqrt{26 - 11x} + 4 = x$
24. $x + \sqrt{31 - 9x} = 5$
25. $\sqrt{x + 1} - 3x = 1$
26. $\sqrt{x + 5} = \sqrt{x - 5}$
27. $\sqrt{x + 5} + \sqrt{x - 5} = 10$
28. $2\sqrt{x + 1} - \sqrt{2x + 3} = 1$
29. $\sqrt{x + 9} - \sqrt{2x + 1} = 2$
30. $2\sqrt{2x - 11} + 8\sqrt{11 - x} = 14$

In Exercises 31–40, solve the equations with fractional exponents and check your solutions.

31. $(x^2 - 5)^{2/3} = 16$
32. $(x^2 - x - 22)^{4/3} = 16$
33. $\frac{3}{2}x(x - 1)^{1/2} + (x - 1)^{3/2} = 0$
34. $\frac{4}{3}x^2(x - 1)^{1/3} + 2x(x - 1)^{4/3} = 0$
35. $\frac{8x^2}{3}(x^2 - 1)^{1/3} + (x^2 - 1)^{4/3} = 0$
36. $-5x^3(1 - 2x)^{3/2} + 3x^2(1 - 2x)^{5/2} = 0$
37. $\frac{2x\sqrt{2 + 3x}}{3} - \frac{4}{27}(2 + 3x)^{3/2} = 0$
38. $2x(x^2 - 1)^{1/2} - x^3(x^2 - 1)^{-1/2} = 0$
39. $2x(2x - 1)^{1/2} - x^2(2x - 1)^{-1/2} = 0$
40. $(x - 1)^{1/3} - \frac{x}{3}(x - 1)^{-2/3} = 0$

In Exercises 41–56, find the real roots of the given quadratic type equation and check your solutions.

41. $x^4 + 5x^2 - 36 = 0$
42. $x^4 - 4x^2 + 3 = 0$
43. $\frac{1}{t^2} + \frac{8}{t} + 15 = 0$
44. $6\left(\frac{s}{s + 1}\right)^2 + 5\left(\frac{s}{s + 1}\right) - 6 = 0$
45. $4x^4 - 65x^2 + 16 = 0$
46. $9t^4 - 12t^2 - 3 = 0$
47. $36t^4 + 29t^2 - 7 = 0$
48. $5t^4 - 6t^2 + 3 = 0$
49. $x^6 + 7x^3 - 8 = 0$
50. $x^6 + 3x^3 + 2 = 0$
51. $x^6 + 10x^3 + 25 = 0$
52. $4h^6 + 4h^3 - 7 = 0$
53. $2x + 9\sqrt{x} - 5 = 0$
54. $6x - 7\sqrt{x} - 3 = 0$
55. $5 - 3x^{1/3} - 2x^{2/3} = 0$
56. $9t^{2/3} + 24t^{1/3} + 16 = 0$

Calculator Exercises

In Exercises 57–60, use a calculator to solve each equation. Round your answers to three decimal places.

57. $3.2x^4 - 1.5x^2 - 2.1 = 0$
58. $7.08x^6 + 4.15x^3 - 9.6 = 0$
59. $1.8x - 6\sqrt{x} - 5.6 = 0$
60. $4x^{2/3} + 8x^{1/3} + 3.6 = 0$

Applications

61. An airline offers daily flights between Chicago and Denver. The total monthly cost for these flights is given by

$$C = \sqrt{0.2x + 1}$$

where C is measured in millions of dollars and x is measured in thousands of passengers. Find the number of passengers if the monthly cost was 2.5 million dollars.

62. The demand equation for a certain product is given by

$$p = 40 - \sqrt{x - 1}$$

where x is the number of units demanded per day and p is the price per unit. Find the demand if the price is set at $34.70.

63. A power station is on one side of a river that is one-half mile wide. A factory is 6 miles downstream on the other side of the river. It costs $18 per foot to run power lines overland, and $24 per foot to run them underwater. If the total cost for the project is $616,877.27, find x as shown in Figure 2.9.

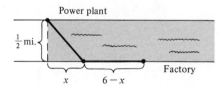

Power plant

$\frac{1}{2}$ mi.

Factory

x $6 - x$

FIGURE 2.9

64. Find the dimensions of a rectangle if the diagonal is 2 inches longer than the length, which in turn is 4 inches longer than the width.

65. The surface area of a cone is given by

$$S = \pi r\sqrt{r^2 + h^2}$$

Solve this equation for h.

66. An equation describing a circuit containing inductance and capacitance is

$$i = \pm\sqrt{\frac{1}{LC}}\sqrt{Q^2 - q}$$ Solve this equation for Q.

Solving Inequalities   2.6

Inequalities were introduced in Section 1.3 in the context of *order* on the real number line. Here we look at inequalities involving a variable. To *solve* such inequalities means to find the set of all real numbers for which the statement is true. In most cases, these solution sets consist of intervals on the real line and we call them **solution intervals** of the inequality. You may want to review the properties of inequalities given in Section 1.3.

EXAMPLE 1
Solving Linear Inequalities

Solve the following inequalities:

(a) $5x - 7 < 3x + 9$ (b) $1 - \dfrac{3x}{2} \ge x - 4$ (c) $-3 \le 6x - 1 < 3$

Solution:

(a)
$5x - 7 < 3x + 9$	*Given*
$5x - 3x < 9 + 7$	*Subtract 3x and add 7*
$2x < 16$	*Combine terms*
$x < 8$	*Divide by 2*

Thus, the solution interval is $x < 8$.

(b)
$1 - \dfrac{3x}{2} \ge x - 4$	*Given*
$2 - 3x \ge 2x - 8$	*Multiply by LCD*
$-5x \ge -10$	*Subtract 2x, subtract 2*
$x \le 2$	*Divide by -5 and reverse the inequality*

Thus, the solution interval is $x \le 2$.

(c) In this case, we have two inequalities which we solve simultaneously.

$-3 \le 6x - 1 < 3$	*Given*
$-2 \le 6x < 4$	*Add 1*
$-\dfrac{2}{6} \le x < \dfrac{4}{6}$	*Divide by 6*
$-\dfrac{1}{3} \le x < \dfrac{2}{3}$	*Reduce fractions*

Many important uses of inequalities involve absolute values. In Section 1.3 we used absolute value to denote the distance between two points on the number line. That is,

$$d(a, b) = |a - b| = \text{distance between points } a \text{ and } b$$

This means that

$$d(x, 0) = |x| = \text{distance between } x \text{ and } 0$$

As a consequence, we give the following interpretation of absolute value inequalities.

ABSOLUTE VALUE INEQUALITIES

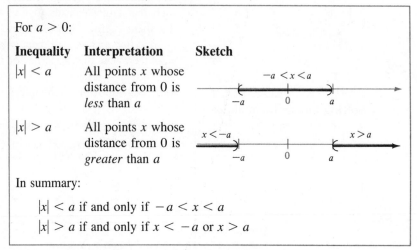

For $a > 0$:

Inequality	Interpretation	Sketch
$|x| < a$	All points x whose distance from 0 is *less* than a	
$|x| > a$	All points x whose distance from 0 is *greater* than a	

In summary:

$|x| < a$ if and only if $-a < x < a$

$|x| > a$ if and only if $x < -a$ or $x > a$

Remark: Informally, it may help to think of *less than* as denoting ''insideness'' and *greater than* as denoting ''outsideness.'' That is,

$|x| < a$ has solutions *inside* the interval $(-a, a)$

$|x| > a$ has solutions *outside* the interval $(-a, a)$

EXAMPLE 2
Solving Absolute Value Inequalities

Solve the inequalities and sketch the solution intervals.

(a) $|x - 5| < 2$ (b) $|x + 3| \geq 7$

Solution:

(a) We seek all points x whose distance from 5 is less than 2, as shown in Figure 2.10.

$$|x - 5| < 2 \qquad \textit{Given}$$
$$-2 < x - 5 < 2 \qquad \textit{Interpret absolute value}$$
$$3 < x < 7 \qquad \textit{Add 5}$$

$$|x - 5| < 2$$

FIGURE 2.10

(b) We seek all points x whose distance from -3 is greater than or equal to 7, as shown in Figure 2.11.

$$|x + 3| \geq 7 \qquad\qquad\qquad\qquad \textit{Given}$$

$$x + 3 \leq -7 \quad \text{or} \quad x + 3 \geq 7 \quad \textit{Interpret absolute value}$$

$$x \leq -10 \qquad\qquad\quad x \geq 4 \quad \textit{Solve separately}$$

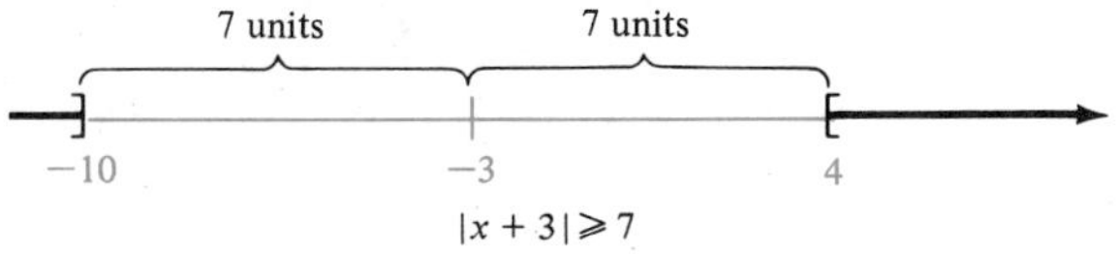

7 units   7 units

$$-10 \qquad\qquad -3 \qquad\qquad 4$$

$$|x + 3| \geq 7$$

FIGURE 2.11

Solving Inequalities by Factoring

Many nonlinear inequalities lend themselves to two solution methods used for quadratic equations—*factoring* and *completing the square*. With the factoring method, we use the principle that a polynomial can change signs *only* at its zeros (the values that make the polynomial zero). Between two consecutive zeros a polynomial must be entirely positive or entirely negative. This means that when the real zeros of a polynomial are put in order, they divide the real line into intervals in which the polynomial has no sign changes. For example, the polynomial

$$x^2 - x + 6 = (x - 3)(x + 2)$$

can change signs only at $x = -2$ and $x = 3$. In the context of polynomial inequalities, we call these values the **critical numbers** and the resulting intervals on the real line the **test intervals** of the inequality. We need to test only *one value* from each interval to solve a polynomial inequality.

EXAMPLE 3
Solving Inequalities by Factoring

Find the solution intervals for each of the following:

(a) $x^2 < x + 6$ (b) $2x^3 + 5x^2 \geq 12x$

Solution:

(a) $\qquad\qquad\qquad x^2 < x + 6 \qquad\qquad \textit{Given}$

$$x^2 - x - 6 < 0 \qquad\qquad \textit{Standard form}$$

$$(x - 3)(x + 2) < 0 \qquad\qquad \textit{Factor}$$

Critical numbers:

$$x = -2, \quad x = 3$$

Test intervals:

$$x < -2, \quad -2 < x < 3, \quad x > 3$$

    To test an interval, we choose a representative number in the interval and compute the sign of each factor. For example, for any $x < -2$, both of the factors $(x - 3)$ and $(x + 2)$ are negative. Consequently, the product (of two negatives) is positive and the inequality is *not* satisfied in the interval $x < -2$. We suggest the testing format shown in Figure 2.12.

Test: Is $(x - 3)(x + 2) < 0$?

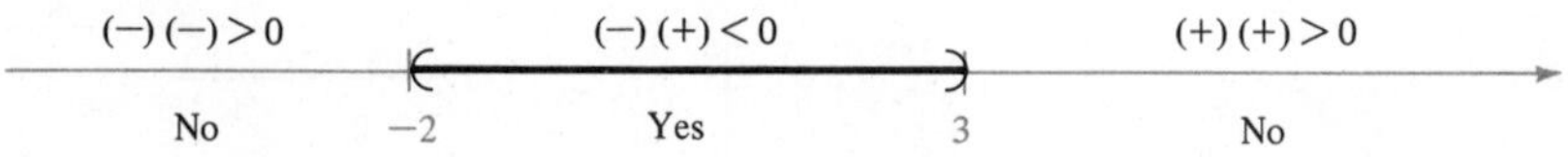

$(-)(-)>0$		$(-)(+)<0$		$(+)(+)>0$
No	-2	Yes	3	No

FIGURE 2.12

Since the inequality is satisfied only by the center test interval, we conclude that the solution interval is

$$-2 < x < 3$$

(b) $2x^3 + 5x^2 \geq 12x$

$2x^3 + 5x^2 - 12x \geq 0$

$x(2x - 3)(x + 4) \geq 0$

Critical numbers:

$$x = -4, \qquad x = 0, \qquad x = \tfrac{3}{2}$$

Test intervals:

$$x < -4, \qquad -4 < x < 0, \qquad 0 < x < \tfrac{3}{2}, \qquad x > \tfrac{3}{2}$$

Test: Is $x(2x - 3)(x + 4) \geq 0$?

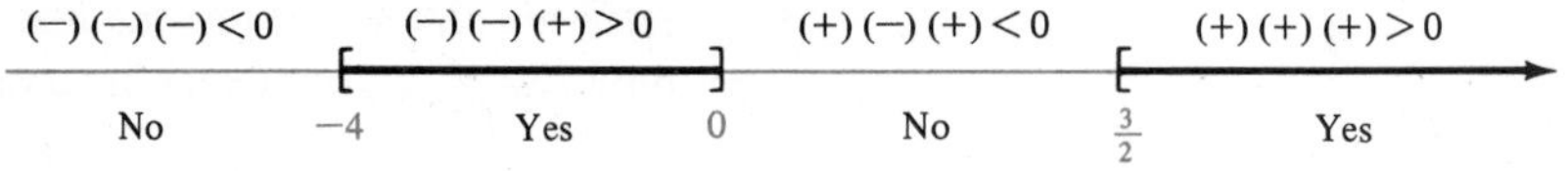

$(-)(-)(-)<0$		$(-)(-)(+)>0$		$(+)(-)(+)<0$		$(+)(+)(+)>0$
No	-4	Yes	0	No	$\tfrac{3}{2}$	Yes

FIGURE 2.13

Solution intervals:

$$-4 \leq x \leq 0 \qquad \text{or} \qquad x \geq \tfrac{3}{2}$$

The concept of critical numbers can be extended to inequalities involving fractional expressions. Specifically, an expression that is the ratio of two polynomials can change signs only at its zeros (the values that make the numerator zero) *and* at its undefined values (the values that make the denominator zero). For example, the expression

$$\frac{x - 1}{(x - 2)(x + 3)}$$

can change signs only at $x = 1$, $x = 2$, and $x = -3$. Thus, these three values are the critical numbers of the inequality

$$\frac{x - 1}{(x - 2)(x + 3)} < 0$$

EXAMPLE 4
Inequality Involving
a Fractional Expression

Solve

$$\frac{2x - 7}{x - 5} \le 3$$

Solution:
We should rewrite the inequality with 0 alone on the right.

$$\frac{2x - 7}{x - 5} \le 3 \qquad\qquad \textit{Given}$$

$$\frac{2x - 7}{x - 5} - 3 \le 0 \qquad\qquad \textit{Subtract 3}$$

$$\frac{2x - 7 - 3x + 15}{x - 5} \le 0 \qquad\qquad \textit{Combine terms}$$

$$\frac{-x + 8}{x - 5} \le 0$$

Critical numbers:

$$x = 5, \qquad x = 8$$

Test intervals:

$$x < 5, \qquad 5 < x < 8, \qquad x > 8$$

Test: Is $(8 - x)/(x - 5) \le 0$?

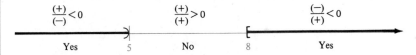

FIGURE 2.14

Solution intervals:

$$x < 5 \qquad \text{or} \qquad x \ge 8$$

Note that we allow the critical number 8 in the solution intervals, but we do not allow $x = 5$ since it yields a zero denominator in the original expression.

Before looking at our next example, note these two properties of absolute value:

$$|ab| = |a| \cdot |b| \qquad \text{and} \qquad \left|\frac{a}{b}\right| = \frac{|a|}{|b|}, \qquad b \ne 0$$

In Exercises 69 and 70 you are asked to prove these properties.

EXAMPLE 5
An Inequality Involving Absolute Value

Solve

$$|x - 2| \leq |4x + 1|$$

Solution:

$|x - 2| \leq |4x + 1|$	*Given*
$\dfrac{|x - 2|}{|4x + 1|} \leq 1$	*Divide by $|4x + 1|$*
$\left|\dfrac{x - 2}{4x + 1}\right| \leq 1$	*Property of absolute value*
$-1 \leq \dfrac{x - 2}{4x + 1} \leq 1$	*Interpret absolute value*

$$\frac{x - 2}{4x + 1} \geq -1 \qquad\qquad \frac{x - 2}{4x + 1} \leq 1$$

$$\frac{x - 2}{4x + 1} + 1 \geq 0 \qquad\qquad \frac{x - 2}{4x + 1} - 1 \leq 0$$

$$\frac{5x - 1}{4x + 1} \geq 0 \qquad\qquad \frac{-3x - 3}{4x + 1} \leq 0$$

Critical numbers: *Critical numbers:*

$$x = -\tfrac{1}{4}, \quad x = \tfrac{1}{5} \qquad\qquad x = -1, \quad x = -\tfrac{1}{4}$$

Now, combining these two sets of critical numbers, we have the following test intervals:

Test intervals:

$$x < -1, \quad -1 < x < -\tfrac{1}{4}, \quad -\tfrac{1}{4} < x < \tfrac{1}{5}, \quad x > \tfrac{1}{5}$$

To test these intervals it is convenient to choose x-values and substitute back into the original equation.

Interval	x-value	$|x - 2|$	$|4x + 1|$	$|x - 2| \leq |4x + 1|$?
$x < -1$	-2	4	7	yes
$-1 < x < -\tfrac{1}{4}$	$-\tfrac{1}{2}$	$\tfrac{5}{2}$	1	no
$-\tfrac{1}{4} < x < \tfrac{1}{5}$	0	2	1	no
$x > \tfrac{1}{5}$	1	1	5	yes

Since $x = -1$ and $x = \tfrac{1}{5}$ both satisfy the original inequality, the solution intervals are as follows.

Solution intervals: $x \leq -1$ or $x \geq -\tfrac{1}{5}$

For inequalities involving an irreducible quadratic, the factoring method is inappropriate, and we use a different method for finding the solution intervals. This method involves *completing the square* and is based on the following relationship:

PERFECT SQUARE INEQUALITIES

For $a > 0$:

$$x^2 < a \text{ if and only if } -\sqrt{a} < x < \sqrt{a}$$
$$x^2 > a \text{ if and only if } x < -\sqrt{a} \text{ or } x > \sqrt{a}$$

Remark: This property is quite obvious if we use our alternative definition of absolute value (Section 1.4), $|x| = \sqrt{x^2}$.

EXAMPLE 6
Solving Inequalities by Completing the Square

Find the solution interval(s) for

$$3x^2 - 6x \le 8$$

Solution:

$3x^2 - 6x \le 8$	*Given*
$3x^2 - 6x - 8 \le 0$	*Standard form*

Since this quadratic is not factorable, we complete the square.

$x^2 - 2x \le \frac{8}{3}$	*Divide by coefficient of x^2*
$x^2 - 2x + 1 \le \frac{8}{3} + 1$	*Add $(\frac{1}{2} \cdot 2)^2$*
$(x - 1)^2 \le \frac{11}{3}$	*Perfect square inequality*
$|x - 1| < \sqrt{\frac{11}{3}}$	*Absolute value form*
$-\sqrt{\frac{11}{3}} < x - 1 < \sqrt{\frac{11}{3}}$	*Interpret absolute value*
$1 - \sqrt{\frac{11}{3}} < x < 1 + \sqrt{\frac{11}{3}}$	*Solution interval*
$-0.91 < x < 2.9$	*Decimal approximation*

EXAMPLE 7
Applications of Inequalities

(a) A projectile is fired straight upward from ground level with an initial velocity of 384 feet per second. During what time period will its height exceed 2048 feet?

(b) A subcompact car can be rented from Company A for $180 per week with no charge for mileage. A similar car can be rented from Company B for $100 per week, plus 20 cents per mile driven. For what weekly mileage m does it cost less to rent from Company A?

Solution:

(a) Recall that the position of an object moving in a vertical path is given by $s = -16t^2 + v_0 t + s_0$. In this case, $s_0 = 0$ and $v_0 = 384$. Thus, we seek the solution interval for the inequality

$$-16t^2 + 384t > 2048$$

$$t^2 - 24t < -128 \qquad \textit{Divide by } -16 \textit{ and reverse inequality}$$

$$t^2 - 24t + 128 < 0 \qquad \textit{Standard form}$$

$$(t - 8)(t - 16) < 0 \qquad \textit{Factored form}$$

Critical numbers: $t = 8, \qquad t = 16$

Test intervals: $t < 8, \qquad 8 < t < 16, \qquad t > 16$

Solution interval: $8 < t < 16$

(b) We need to solve the inequality

(weekly cost from B) > (weekly cost from A)

$$100 + 0.20m > 180$$

$$0.20m > 80$$

Solution interval: $m > 400$ miles

Section Exercises 2.6

In Exercises 1–4, determine whether or not the given value of x satisfies the inequality.

1. $5x - 12 > 0$
    (a) $x = 3$ (b) $x = -3$ (c) $x = \frac{5}{2}$ (d) $x = \frac{3}{2}$

2. $x + 1 < \dfrac{2x}{3}$
    (a) $x = 0$ (b) $x = 4$ (c) $x = -4$ (d) $x = -3$

3. $0 < \dfrac{x - 2}{4} < 2$
    (a) $x = 4$ (b) $x = 10$ (c) $x = 0$ (d) $x = \frac{7}{2}$

4. $-1 < \dfrac{3 - x}{2} \leq 1$
    (a) $x = 0$ (b) $x = \sqrt{5}$ (c) $x = 1$ (d) $x = 5$

In Exercises 5–40, solve the inequality and graph the solution on the real number line.

5. $x - 5 \geq 7$

6. $2x > 3$

7. $4x + 1 < 2x$

8. $2x + 7 < 3$

9. $2x - 1 \geq 0$

10. $3x + 1 \geq 2$

11. $4 - 2x < 3$

12. $x - 4 \leq 2$

13. $-4 < \dfrac{2x - 3}{3} < 4$

14. $0 \leq \dfrac{x + 3}{2} < 5$

15. $\frac{3}{4} > x + 1 > \frac{1}{4}$

16. $-1 < -\dfrac{x}{3} < 1$

17. $|x| < 5$

18. $|2x| < 6$

19. $\left|\dfrac{x}{2}\right| > 3$

20. $|5x| > 10$

21. $|x + 2| < 5$

22. $|3x + 1| \geq 4$

23. $\left|\dfrac{x - 3}{2}\right| \geq 5$

24. $|2x + 1| < 5$

25. $|9 - 2x| < 1$

26. $\left|1 - \dfrac{2x}{3}\right| < 1$

27. $x^2 \leq 9$

28. $x^2 < 5$

29. $x^2 > 4$

30. $(x - 3)^2 \geq 1$

31. $(x + 2)^2 < 25$

32. $(x + 6)^2 \leq 8$

33. $x^2 + 4x + 4 \geq \frac{9}{4}$

34. $x^2 - 6x + 9 < \frac{16}{9}$

35. $x^2 + x - 6 < 0$

36. $4x^3 - 12x^2 > 0$

37. $3(x - 1)(x + 1) > 0$

38. $6(x + 2)(x - 1) < 0$

39. $4x^3 - 6x^2 < 0$

40. $x^2 + 2x - 3 > 0$

In Exercises 41–48, find the interval(s) in which the given expression is defined.

41. $\sqrt{x^2 - 7x + 12}$

42. $\sqrt{x^2 - 4}$

43. $\sqrt[4]{4 - x^2}$

44. $\sqrt{144 - 9x^2}$

45. $\sqrt{12 - x - x^2}$

46. $\sqrt{x^2 + 4}$

47. $\sqrt{x^2 - 3x + 3}$

48. $\sqrt[4]{-x^2 + 2x - 2}$

49. Use absolute values to define each interval (or pair of intervals) on the real line.

(a)

(b)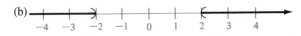

50. Use absolute values to define each interval (or pair of intervals) on the real line.

(a) All real numbers within 10 units of 12.

(b)

In Exercises 51–60, solve each inequality.

51. $\dfrac{x + 6}{x + 1} < 2$

52. $\dfrac{x + 12}{x + 2} \geq 3$

53. $\dfrac{3x - 5}{x - 5} > 4$

54. $\dfrac{5 - 2x}{1 + 3x} < 6$

55. $\dfrac{4}{x + 5} > \dfrac{1}{2x + 3}$

56. $\dfrac{5}{x - 6} > \dfrac{3}{x + 2}$

57. $\dfrac{4}{5x - 6} \leq \dfrac{6}{7x + 9}$

58. $|x + 6| > |2x - 5|$

59. $|1 - x| \leq |x + 2|$

60. $|2x + 3| < |3x + 4|$

61. A projectile is fired straight upward from ground level with an initial velocity of 160 feet per second.
    (a) At what instant will it be back at ground level?
    (b) During what time period will its height exceed 384 feet?

62. A rectangle with a perimeter of 100 meters is to have an area of at least 500 square meters. Within what bounds must the length of the rectangle lie?

63. P dollars invested at simple interest rate r for t years grows to an amount

$$A = P + Prt$$

If an investment of $1000 is to grow to an amount greater than $1250 in two years, then the interest rate must be greater than what percentage?

64. A family establishes a business selling mini-donuts at a shopping mall. The cost of making a dozen donuts is $1.45, and the donuts sell for $2.95 per dozen. In addition to the cost of the ingredients, the business must pay $25 per day for rent and utilities. If the daily profit varies between $50 and $200, between what levels (in dozens) do the daily sales vary?

65. In the manufacture and sale of a certain product, the revenue for selling x units is

$$R = 115.95x$$

and the cost of producing x units is

$$C = 95x + 750$$

In order for a profit to be realized, it is necessary that R be greater than C. For what values of x will this product return a profit?

66. A utility company has a fleet of vans for which the annual operating cost per van is estimated to be

$$C = 0.32m + 2300$$

where C is measured in dollars and m is measured in miles. If the company wants the annual operating cost to be less than $10,000, then m must be less than what value?

67. The heights, h, of two-thirds of the members of a certain population satisfy the inequality

$$\left| \frac{h - 68.5}{2.7} \right| \leq 1$$

where h is measured in inches. Determine the interval on the real line in which these heights lie.

68. The estimated daily production, p, at a refinery is given by

$$|p - 2{,}250{,}000| < 125{,}000$$

where p is measured in barrels of oil. Determine the high and low production levels.

69. Given any two real numbers a and b, prove that

$$|ab| = |a| \cdot |b|$$

70. Given any two real numbers a and b, prove that

$$\left| \frac{a}{b} \right| = \frac{|a|}{|b|}$$

Review Exercises / Chapter 2

Solve the equations in Exercises 1–30.

1. $5x^4 - 12x^3 = 0$

2. $4x^3 - 6x = 0$

3. $6x = 3x^2$

4. $2x = \dfrac{2}{x^3}$

5. $3\left(1 - \dfrac{1}{5t}\right) = 0$

6. $3x^2 + 1 = 0$

7. $2 - x^{-2} = 0$

8. $2x + 4x^{-2} = 0$

9. $\dfrac{8}{x^3} = 1$

10. $-t^2 + 6t = 4$

11. $4t^3 - 12t^2 + 8t = 0$

12. $12t^3 - 84t^2 + 120t = 0$

13. $(x - 1)(2x - 3) + (x^2 - 3x + 2) = 0$

14. $\dfrac{(t^2 + 2t + 2) - (t + 1)(2t + 2)}{(t^2 + 2t + 2)^2} = 0$

15. $\dfrac{(x - 1)(2x) - x^2}{(x - 1)^2} = 0$

16. $\dfrac{1}{(t + 1)^2} = 1$

17. $x^2 - 6x + 10 = 0$

18. $\sqrt{x - 2} - 8 = 0$

19. $(x - 1)^{2/3} - 25 = 0$

20. $\sqrt{x^2 + 1} = \sqrt{2x}$

21. $2(x - 2)\sqrt{x} + \dfrac{(x - 2)^2}{2\sqrt{x}} = 0$

22. $3\sqrt{3x} + \dfrac{3(x + 2)}{2\sqrt{3x}} = 0$

23. $\dfrac{4}{(x - 4)^2} = 1$

24. $\dfrac{1}{x - 2} = 3$

25. $\dfrac{(x^2 + 3)^2 - 4x^2(x^2 + 3)}{(x^2 + 3)^2} = 0$

26. $(x + 4)^{1/2} + 5x(x + 4)^{3/2} = 0$

27. $\frac{8}{3}x^2(x^2 - 4)^{1/3} + (x^2 - 4)^{4/3} = 0$

28. $2x(1 - x)^{1/3} + \dfrac{x^2}{3}(1 - x)^{-2/3} = 0$

29. $\sqrt{2x + 3} + \sqrt{x - 2} = 2$

30. $5\sqrt{x} - \sqrt{x - 1} = 6$

Solve the inequalities in Exercises 31–40.

31. $\dfrac{1}{2}(3 - x) > \dfrac{1}{3}(2 - 3x)$

32. $\dfrac{x}{5} - 6 \le -\dfrac{x}{2} + 6$

33. $x^2 - 4 \le 0$

34. $x^2 - 2x \ge 3$

35. $\dfrac{x - 5}{3 - x} < 0$

36. $|4 - x| < 2$

37. $|2 + 3x| > 5$

38. $x^3 + 2x^2 + x > 0$

39. $\dfrac{2}{x + 1} \le \dfrac{3}{x - 1}$

40. $|x - 5| \le |x + 2|$

In Exercises 41–46, solve each equation for the indicated unknown.

41. $V = \frac{1}{3}\pi r^2 h$ (Solve for r)

42. $A = 2\pi r^2 + 2\pi rh$ (Solve for r)

43. $S = V_0 t - 16t^2$ (Solve for t)

44. $Z = \sqrt{R^2 - X^2}$ (Solve for X)

45. $L = \dfrac{k}{3\pi r^2 p}$ (Solve for p)

46. $E = k(2w)\left(\dfrac{v}{2}\right)^2$ (Solve for v)

47. If the distance from a spacecraft to the horizon is 1000 miles, find x, the altitude of the craft. Assume the radius of the earth is 4000 miles. (See Figure 2.15.)

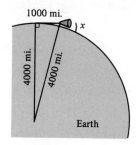

1000 mi.

4000 mi.   4000 mi.

x

Earth

FIGURE 2.15

48. Each week a salesman must make a 180-mile trip to pick up supplies. If he were to increase his average speed by 5 miles per hour, the trip would take 24 minutes less than usual. Find his usual average speed.

49. A group of farmers agrees to share equally in the cost of a $48,000 piece of machinery. If they could find two more farmers to join the group, each person's share of the cost would decrease by $4000. Find the present number in the group.

50. Daily demand for a product is 90 units when the selling price is $90 per unit. If for each $0.25 reduction in selling price average demand will increase by 10 units, find the selling price that will yield daily revenues of $11,570.

CHAPTER

3 | FUNCTIONS AND GRAPHS

The Cartesian Plane and the Distance Formula

3.1

Just as we can represent the real numbers geometrically by points on the real line, we can represent ordered pairs of real numbers by points in a plane. An **ordered pair** (x, y) of real numbers has x as its *first* member and y as its *second* member. The model for representing ordered pairs is called the **rectangular coordinate system,** or the **Cartesian plane.** It is developed by considering two real lines intersecting at right angles (Figure 3.1).

The horizontal real line is traditionally called the **x-axis,** and the vertical real line is called the **y-axis.** Their point of intersection is called the **origin,** and the lines divide the plane into four parts called **quadrants** (Figure 3.2).

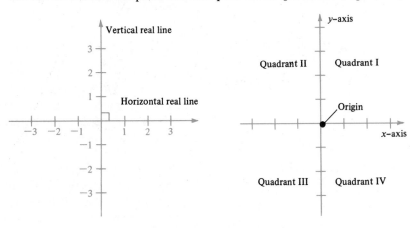

FIGURE 3.1

FIGURE 3.2   The Cartesian Plane

121

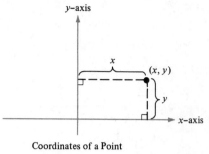

Coordinates of a Point

FIGURE 3.3

We identify each point in the plane by an ordered pair (x, y) of real numbers x and y, called the **coordinates** of the point. The number x represents the directed distance from the y-axis to the point, and y represents the directed distance from the x-axis to the point (Figure 3.3). For the point (x, y), the first coordinate is referred to as the x-coordinate or **abscissa,** and the second or y-coordinate is referred to as the **ordinate.**

Remark: We use the notation (x, y) to denote both a point in the plane and an open interval on the real line. Generally this should cause no confusion because the nature of a specific problem will show which we are talking about.

EXAMPLE 1
Plotting Points
in the Cartesian Plane

Locate the points $(-1, 2)$, $(3, 4)$, $(0, 0)$, $(3, 0)$, and $(-2, -3)$ in the Cartesian plane.

Solution:
The solution is shown in Figure 3.4.

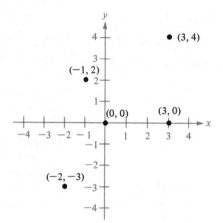

FIGURE 3.4

Development of the Distance Formula

In Chapter 1 we determined the distance between two points x_1 and x_2 on the real line. We will now find the distance between two points in the plane. Recall from the Pythagorean Theorem that, for a right triangle with hypotenuse c and sides a and b, we have the relationship $a^2 + b^2 = c^2$. Conversely, if $a^2 + b^2 = c^2$, then the triangle is a right triangle (Figure 3.5).

Suppose we wish to determine the distance d between two points (x_1, y_1) and (x_2, y_2) in the plane. Using these two points, a right triangle can be formed, as shown in Figure 3.6. We see that the length of the vertical side of the triangle is $|y_2 - y_1|$. Similarly, the length of the horizontal side of the triangle is $|x_2 - x_1|$. By the Pythagorean Theorem, we then have

$$d^2 = |x_2 - x_1|^2 + |y_2 - y_1|^2 \quad \text{or} \quad d = \sqrt{|x_2 - x_1|^2 + |y_2 - y_1|^2}$$

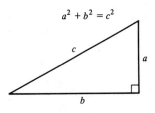

FIGURE 3.5   Pythagorean Theorem

Replacing $|x_2 - x_1|^2$ and $|y_2 - y_1|^2$ by the equivalent expressions $(x_2 - x_1)^2$ and $(y_2 - y_1)^2$, we can write

$$d = \sqrt{(x_2 - x_1)^2 + (y_2 - y_1)^2}$$

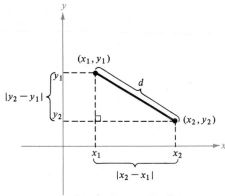

FIGURE 3.6   Distance Between Two Points

We choose the positive square root for d because the distance between two points is not a directed distance. We have therefore established the following rule:

DISTANCE FORMULA

The distance d between two points (x_1, y_1) and (x_2, y_2) in the plane is given by

$$d = \sqrt{(x_2 - x_1)^2 + (y_2 - y_1)^2}$$

EXAMPLE 2
Finding the Distance
Between Two Points

Find the distance between the points $(-2, 1)$ and $(3, 4)$.

Solution:
Applying the Distance Formula, we have

$$d = \sqrt{[3 - (-2)]^2 + (4 - 1)^2}$$
$$= \sqrt{(5)^2 + (3)^2} = \sqrt{25 + 9}$$
$$= \sqrt{34} \approx 5.83$$

EXAMPLE 3
An Application of the
Distance Formula

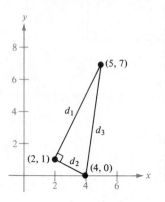

FIGURE 3.7

Use the Distance Formula to show that the points (2, 1), (4, 0), and (5, 7) are the vertices of a right triangle.

Solution:
Refer to Figure 3.7. The three sides have lengths

$$d_1 = \sqrt{(5-2)^2 + (7-1)^2} = \sqrt{9+36} = \sqrt{45}$$
$$d_2 = \sqrt{(4-2)^2 + (0-1)^2} = \sqrt{4+1} = \sqrt{5}$$
$$d_3 = \sqrt{(5-4)^2 + (7-0)^2} = \sqrt{1+49} = \sqrt{50}$$

Since

$$d_1{}^2 + d_2{}^2 = 45 + 5 = 50 = d_3{}^2$$

we can apply the Pythagorean Theorem to conclude that the triangle must be a right triangle.

Remark: In Example 3, the figure provided was not really essential to the solution of the problem. *Nevertheless,* we strongly recommend that you get in the habit of including sketches with your problem solutions even if they are not specifically required.

Next we introduce a rule for finding the coordinates of the midpoint of the line segment joining two points in the plane.

MIDPOINT RULE

The midpoint of the line segment joining points (x_1, y_1) and (x_2, y_2) is

$$\left(\frac{x_1 + x_2}{2}, \frac{y_1 + y_2}{2} \right)$$

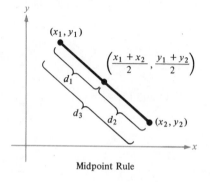

Midpoint Rule

FIGURE 3.8

Proof:
To find the midpoint of a line segment, we merely find the "average" values of the respective coordinates of the two endpoints. To prove this we refer to Figure 3.8, and show that

$$d_1 = d_2 \qquad \text{and} \qquad d_1 + d_2 = d_3$$

Using the Distance Formula, we obtain

$$d_1 = \sqrt{\left(\frac{x_1 + x_2}{2} - x_1\right)^2 + \left(\frac{y_1 + y_2}{2} - y_1\right)^2}$$
$$= \frac{1}{2}\sqrt{(x_2 - x_1)^2 + (y_2 - y_1)^2}$$

$$d_2 = \sqrt{\left(x_2 - \frac{x_1 + x_2}{2}\right)^2 + \left(y_2 - \frac{y_1 + y_2}{2}\right)^2}$$

$$= \frac{1}{2}\sqrt{(x_2 - x_1)^2 + (y_2 - y_1)^2}$$

$$d_3 = \sqrt{(x_2 - x_1)^2 + (y_2 - y_1)^2}$$

Thus, it follows that

$$d_1 = d_2 \quad \text{and} \quad d_1 + d_2 = d_3$$

EXAMPLE 4
Finding the Midpoint
of a Line Segment

Find the midpoint of the line segment joining the points $(-3, -5)$ and $(3, 9)$.

Solution:
Figure 3.9 shows the two given points together with their midpoint. By the Midpoint Rule, we have

$$\left(\frac{-3 + 3}{2}, \frac{-5 + 9}{2}\right) = (0, 2)$$

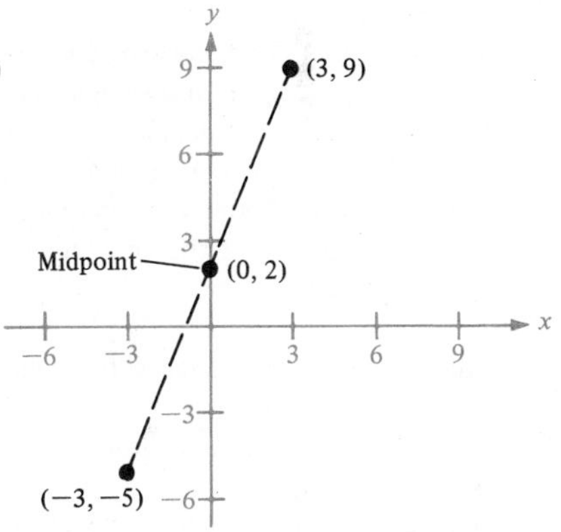

FIGURE 3.9

EXAMPLE 5
Finding Points at a Specified
Distance from a Given Point

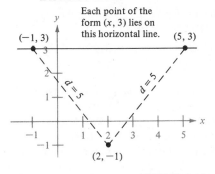

FIGURE 3.10

Find x so that the distance between $(x, 3)$ and $(2, -1)$ is 5.

Solution:
As we begin this problem we don't know how many values of x satisfy the given requirements. Even so, we can use the Distance Formula to write the following equation:

$$d = 5 = \sqrt{(x - 2)^2 + (3 + 1)^2}$$

$$25 = (x^2 - 4x + 4) + 16$$

$$0 = x^2 - 4x - 5$$

$$0 = (x - 5)(x + 1)$$

$$x = 5 \text{ or } -1$$

Now we see that there are two solutions and conclude that the points $(5, 3)$ and $(-1, 3)$ each lie 5 units from the point $(2, -1)$. (See Figure 3.10.)

Section Exercises 3.1

In Exercises 1–8, plot the points, find the distance between the points, and find the midpoint of the line segment joining the points.

1. $(2, 1)$, $(4, 5)$
2. $(-3, 2)$, $(3, -2)$
3. $(\frac{1}{2}, 1)$, $(-\frac{3}{2}, -5)$
4. $(\frac{2}{3}, -\frac{1}{3})$, $(\frac{5}{6}, 1)$
5. $(2, 2)$, $(4, 14)$
6. $(-3, 7)$, $(1, -1)$
7. $(1, \sqrt{3})$, $(-1, 1)$
8. $(-2, 0)$, $(0, \sqrt{2})$

9. Show that the points $(4, 0)$, $(2, 1)$, $(-1, -5)$ are vertices of a right triangle.
10. Show that the points $(1, -3)$, $(3, 2)$, $(-2, 4)$ are vertices of an isosceles triangle.
11. Show that the points $(0, 0)$, $(1, 2)$, $(2, 1)$, $(3, 3)$ are vertices of a rhombus. (A rhombus is a four-sided figure whose sides are all of the same length.)
12. Show that the points $(0, 1)$, $(3, 7)$, $(4, 4)$, $(1, -2)$ are vertices of a parallelogram.
13. Use the Distance Formula to determine if the points $(0, -4)$, $(2, 0)$, and $(3, 2)$ lie on a straight line.
14. Use the Distance Formula to determine if the points $(0, 4)$, $(7, -6)$, and $(-5, 11)$ lie on a straight line.
15. Use the Distance Formula to determine if the points $(-2, 1)$, $(-1, 0)$, and $(2, -2)$ lie on a straight line.
16. Find y so that the distance from the origin to the point $(3, y)$ is 5.
17. Find x so that the distance from the origin to the point $(x, -4)$ is 5.
18. Find x so that the distance from $(2, -1)$ to the point $(x, 2)$ is 5.
19. Find the relationship between x and y so that the point (x, y) is equidistant from $(4, -1)$ and $(-2, 3)$.
20. Find the relationship between x and y so that the point (x, y) is equidistant from $(3, \frac{5}{2})$ and $(-7, -1)$.
21. Use the Midpoint Rule successively to find the three points that divide the line segment joining (x_1, y_1) and (x_2, y_2) into four equal parts.
22. Use the result of Exercise 21 to find the points that divide into four equal parts the line segment joining these points:
    (a) $(1, -2)$, $(4, -1)$ (b) $(-2, -3)$, $(0, 0)$
23. Prove that
$$\left(\frac{2x_1 + x_2}{3}, \frac{2y_1 + y_2}{3} \right)$$
is one of the points of trisection of the line segment joining (x_1, y_1) and (x_2, y_2). Also, find the midpoint of the line segment joining
$$\left(\frac{2x_1 + x_2}{3}, \frac{2y_1 + y_2}{3} \right) \quad \text{and} \quad (x_2, y_2)$$

to find the second point of trisection of the line segment joining (x_1, y_1) and (x_2, y_2).
24. Use the results of Exercise 23 to find the points of trisection of the line segment joining these points:
    (a) $(1, -2)$, $(4, 1)$ (b) $(-2, -3)$, $(0, 0)$

In Exercises 25 and 26, use Figure 3.11 showing average rates for home mortgages between January, 1980 and May, 1981.

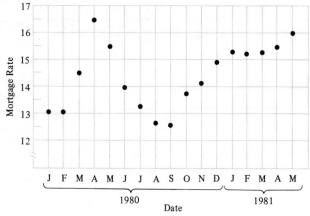

FIGURE 3.11

25. Approximate the average mortgage rate for
    (a) May, 1980   (b) December, 1980
    (c) July, 1980   (d) May, 1981
26. Approximate the *increase* in average mortgage rates from
    (a) July, 1980 to May, 1981
    (b) January, 1981 to May, 1981

In Exercises 27–30, use Figure 3.12 showing the Dow-Jones Industrial Average (DJIA) from 1929 to 1980.

27. Approximate the DJIA for
    (a) June, 1949   (b) January, 1970
    (c) December, 1953   (d) March, 1963
28. Approximate the *increase* (or *decrease*) in the DJIA from
    (a) the high of 1929 to the low of 1932
    (b) the low of 1974 to the high of 1976
29. In which years did the DJIA go above 1000?
30. Approximate the *percentage increase* (or *decrease*) in the DJIA from
    (a) January, 1940 to January, 1950
    (b) January, 1973 to January, 1975

FIGURE 3.12   Dow–Jones Average of Industrial Stock Prices

Calculator Exercises

In Exercises 31–36, use a calculator to (a) find the midpoint of the line segment joining the two points, and (b) find the distance between the points.

31. $(6.2, 5.4), (-3.7, 1.8)$

32. $(-16.8, 12.3), (5.6, 4.9)$
33. $(-36, -18), (48, -72)$
34. $(1.451, 3.051), (5.906, 11.360)$
35. $(0.721, -1.106), (-0.345, -0.093)$
36. $(-8.62, 18.25), (-8.62, 4.67)$

Graphs of Equations   3.2

News magazines frequently show graphs that compare the rate of inflation, the gross national product, wholesale prices, or the unemployment rate to the time of year. Industrial firms and businesses use graphs to report their monthly production and sales statistics. Such graphs provide a simple geometrical picture of the way one quantity changes with respect to another.

    Frequently, the relationship between two quantities is expressed in the form of an equation. In this section, we introduce the basic procedure for determining the geometric picture associated with an algebraic equation.

    Consider the equation

$$3x + y = 7$$

If $x = 2$ and $y = 1$, the equation is satisfied, and we call the point $(2, 1)$ a **solution point** of the equation. Of course, there are other solution points. To make up a table of solution points, we choose arbitrary values for x and

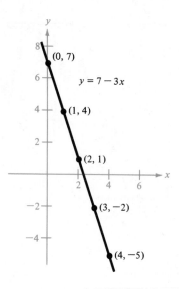

FIGURE 3.13

determine the corresponding values for y. To determine the values for y, it is convenient to replace the equation by the equivalent form

$$y = 7 - 3x$$

x	0	1	2	3	4
y	7	4	1	-2	-5

Thus, $(0, 7)$, $(1, 4)$, $(2, 1)$, $(3, -2)$, and $(4, -5)$ are all solution points of the equation $3x + y = 7$. We could continue this process indefinitely and obtain infinitely many solution points for the equation $3x + y = 7$. We call the collection of all such solution points the **graph** of the equation $3x + y = 7$, as shown in Figure 3.13.

DEFINITION OF THE GRAPH OF AN EQUATION

> The **graph of an equation** involving two variables x and y is the collection of all points in the plane that are solution points to the equation.

EXAMPLE 1
Sketching the Graph
of an Equation

Sketch the graph of the equation $y = x^2 - 2$.

Solution:
First, we make a table of values (solution points) by choosing several convenient values of x and calculating the corresponding values of y.

x	-2	-1	0	1	2	3
$y = x^2 - 2$	2	-1	-2	-1	2	7

Next, we plot these points in the plane, as in Figure 3.14. Finally, we connect the points by a smooth curve, as in Figure 3.15.

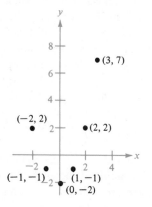

FIGURE 3.14

FIGURE 3.15   Graph of $y = x^2 - 2$

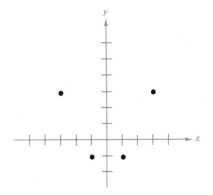

We call this method of sketching a graph the **point-plotting method.** It has three basic steps.

THE POINT-PLOTTING METHOD OF GRAPHING

1. Make up a table of several solution points of the equation.
2. Plot these points in the plane.
3. Connect the points with a smooth curve.

FIGURE 3.16

Steps 1 and 2 of the point-plotting method can usually be accomplished with ease. However, Step 3 can be the source of some major difficulties. For instance, how would you connect the four points in Figure 3.16? Without additional points or further information about the equation, any one of the three graphs in Figure 3.17 would be reasonable.

Obviously, with too few solution points, we could badly misrepresent the graph of a given equation. Just how many points should be plotted? For a straight-line graph, two points are sufficient. For more complicated graphs we need many more points. More sophisticated techniques will be discussed in later sections, but for now plot enough points so as to reveal the essential behavior of the graph. A programmable calculator is useful for determining the many solution points needed for an accurate graph.

We suggest that in choosing points to plot you start with those points that are easiest to calculate. Two such points are those having zero as either their x- or y-coordinate. These points are called **intercepts,** because they are points at which the graph intersects the x- or y-axis.

DEFINITION OF INTERCEPTS

The point $(a, 0)$ is called an ***x*-intercept** of the graph of an equation if it is a solution point of the equation. To find the x-intercepts, let y be zero and solve the equation for x.

The point $(0, b)$ is called a ***y*-intercept** of the graph of an equation if it is a solution point of the equation. To find the y-intercepts, let x be zero and solve the equation for y.

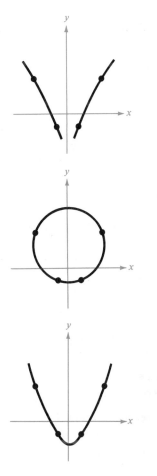

Remark: Some texts denote the x-intercept as the x-coordinate of the point $(a, 0)$ rather than the point itself. Unless it is necessary to make a distinction, we will use ''intercept'' to mean either the point or the coordinate.

Of course, it is possible that a particular graph will have no intercepts, or it may have several. For instance, consider the four graphs in Figure 3.18.

FIGURE 3.17

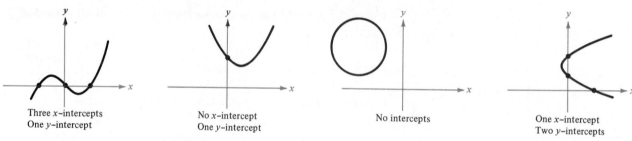

FIGURE 3.18

EXAMPLE 2
Finding *x*- and *y*-Intercepts

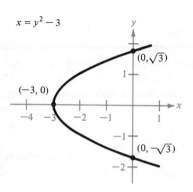

FIGURE 3.19

Find the *x*- and *y*-intercepts for the graph of

$$y^2 - 3 = x$$

Solution:

Let $y = 0$. Then $-3 = x$.

   x-intercept: $(-3, 0)$

Let $x = 0$. Then $y^2 - 3 = 0$ has solutions $y = \pm\sqrt{3}$.

   y-intercepts: $(0, \sqrt{3}), (0, -\sqrt{3})$

(See Figure 3.19.)

The graph shown in Figure 3.19 is said to be "symmetric" with respect to the *x*-axis. This means that if the Cartesian plane were folded along the *x*-axis, the portion of the graph above the *x*-axis would then coincide with the portion below the *x*-axis. Symmetry with respect to the *y*-axis can be described in a similar manner.

Knowing the symmetry of a graph *before* attempting to sketch it is beneficial, for then we need only half as many solution points as we otherwise would. We define three basic types of symmetry as follows (see Figure 3.20).

DEFINITION OF SYMMETRY

> A graph is said to be **symmetric with respect to the *y*-axis** if, whenever (x, y) is on the graph, $(-x, y)$ is also on the graph.
>
> A graph is said to be **symmetric with respect to the *x*-axis** if, whenever (x, y) is on the graph, $(x, -y)$ is also on the graph.
>
> A graph is said to be **symmetric with respect to the origin** if, whenever (x, y) is on the graph, $(-x, -y)$ is also on the graph.

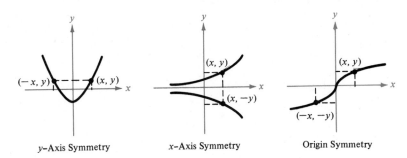

y–Axis Symmetry   *x*–Axis Symmetry   Origin Symmetry

FIGURE 3.20

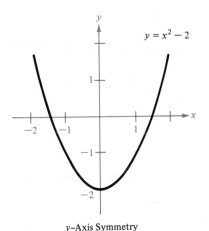

$y = x^2 - 2$

y–Axis Symmetry

FIGURE 3.21

Suppose we apply this definition of symmetry to the graph of the equation $y = x^2 - 2$. By replacing x by $-x$, we obtain

$$y = (-x)^2 - 2 \quad \text{or} \quad y = x^2 - 2$$

Since this substitution does not change the equation, it follows that if (x, y) is a solution point of the equation, then $(-x, y)$ must also be a solution point. Therefore, the graph of $y = x^2 - 2$ is symmetric with respect to the *y*-axis. (See Figure 3.21.)

A similar test can be made for symmetry with respect to the *x*-axis or to the origin. These three tests are summarized as follows.

TESTS FOR SYMMETRY

The graph of an equation is symmetric with respect to:

1. the *y*-axis if replacing x by $-x$ yields an equivalent equation
2. the *x*-axis if replacing y by $-y$ yields an equivalent equation
3. the origin if replacing x by $-x$ *and* y by $-y$ yields an equivalent equation

EXAMPLE 3
Testing for Symmetry

Show that the graph of $y = 2x^3 - x$ is symmetric with respect to the origin.

Solution:
By replacing x by $-x$ and y by $-y$, we have

$$-y = 2(-x)^3 - (-x)$$
$$-y = -2x^3 + x$$

Now by multiplying both sides of the equation by -1, we have

$$y = 2x^3 - x$$

which is the original equation. Therefore, the graph of $y = 2x^3 - x$ is symmetric with respect to the origin. See Figure 3.22.

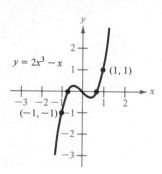

FIGURE 3.22   Origin Symmetry

EXAMPLE 4
Using Symmetry
as an Aid to Graphing

Sketch the graph of $x - y^2 = 1$.

Solution:
The graph is symmetric with respect to the x-axis, since replacing y by $-y$ yields

$$x - (-y)^2 = 1$$
$$x - y^2 = 1$$

This means that the graph below the x-axis is a mirror-image of the graph above the x-axis. Hence we can first sketch the portion above the x-axis and then reflect it to obtain the entire graph (Figure 3.23).

$x = y^2 + 1$	1	2	5
y	0	1	2

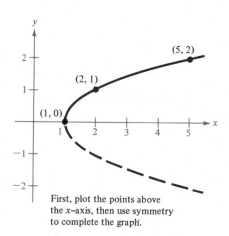

First, plot the points above the x-axis, then use symmetry to complete the graph.

FIGURE 3.23

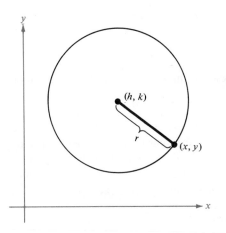

FIGURE 3.24

So far in this section we have studied the Point-Plotting Method of graphing and two additional concepts (intercepts and symmetry) that can be used to streamline this graphing procedure. A third graphing aid is that of *equation recognition*, the ability to recognize the general shape of a graph simply by looking at its equation. A **circle** is one type of graph whose equation is easily recognized.

Figure 3.24 shows a circle of radius r with center at (h, k). The point (x, y) is on this circle if and only if its distance from the center (h, k) is r. This means that a circle consists of the set of all points (x, y) that are at a given positive distance r from a fixed point (h, k). Expressing this relationship in terms of the Distance Formula, we have

$$\sqrt{(x - h)^2 + (y - k)^2} = r$$

By squaring both sides of this equation, we obtain the **standard form of the equation of a circle.**

STANDARD FORM OF THE EQUATION OF A CIRCLE

> The point (x, y) lies on the circle of radius r and center (h, k) if and only if
>
> $$(x - h)^2 + (y - k)^2 = r^2$$

As a special case, *the equation of a circle with its center at the origin* is simply

$$x^2 + y^2 = r^2$$

EXAMPLE 5

Finding an Equation for a Circle

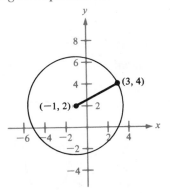

FIGURE 3.25

The point $(3, 4)$ lies on a circle whose center is at $(-1, 2)$ (Figure 3.25). Find an equation for the circle.

Solution:
The radius of the circle is the distance between $(-1, 2)$ and $(3, 4)$. Thus,

$$r = \sqrt{[3 - (-1)]^2 + (4 - 2)^2}$$
$$= \sqrt{16 + 4} = \sqrt{20}$$

And we conclude that the standard equation for this circle is

$$[x - (-1)]^2 + (y - 2)^2 = (\sqrt{20})^2$$
$$(x + 1)^2 + (y - 2)^2 = 20$$

If we remove parentheses in the standard equation in Example 5, we obtain

$$(x + 1)^2 + (y - 2)^2 = 20$$
$$x^2 + 2x + 1 + y^2 - 4y + 4 = 20$$
$$x^2 + y^2 + 2x - 4y - 15 = 0$$

where the latter equation is in the **general form of the equation of a circle:**

$$Ax^2 + Ay^2 + Dx + Ey + F = 0 \quad (A \neq 0)$$

The general form of the equation is less useful than the equivalent standard form. For instance, it is not immediately apparent from the general equation of the circle in Example 5 that the center is at $(-1, 2)$ and the radius is $\sqrt{20}$. To graph the equation of a circle, it is best to write the equation in standard form. This can be accomplished by **completing the square** (Section 2.3), as demonstrated in the following example.

EXAMPLE 6
Completing the Square

Sketch the graph of the circle whose general equation is

$$4x^2 + 4y^2 + 20x - 16y + 37 = 0$$

Solution:
To complete the square we will first divide by 4 so that the coefficients of x^2 and y^2 are both 1. Thus, we have

$$x^2 + y^2 + 5x - 4y + \frac{37}{4} = 0$$

Then we write

$$(x^2 + 5x + \quad) + (y^2 - 4y + \quad) = -\frac{37}{4}$$

$$\left(x^2 + 5x + \frac{25}{4}\right) + (y^2 - 4y + 4) = -\frac{37}{4} + \frac{25}{4} + 4$$

$$\underset{(\text{half})^2}{\underline{\qquad\uparrow}} \qquad\qquad \underset{(\text{half})^2}{\underline{\qquad\uparrow}}$$

$$\left(x + \frac{5}{2}\right)^2 + (y - 2)^2 = 1$$

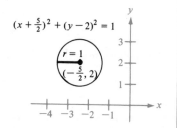

$(x + \frac{5}{2})^2 + (y - 2)^2 = 1$

FIGURE 3.26

Therefore, the circle is centered at $(-\frac{5}{2}, 2)$, and its radius is 1 (Figure 3.26).

Section Exercises 3.2

In Exercises 1–6, match the given equation with its graph. [The graphs are labeled (a), (b), (c), (d), (e), and (f).]

1. $y = x - 2$
2. $y = -\frac{1}{2}x + 2$
3. $y = x^2 + 2x$
4. $y = \sqrt{9 - x^2}$
5. $y = |x| - 2$
6. $y = x^3 - x$

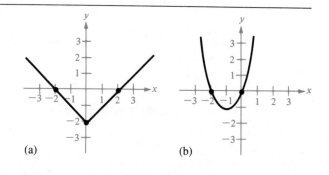

(a)　　　　　(b)

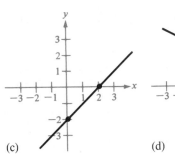

(c)

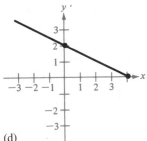

(d)

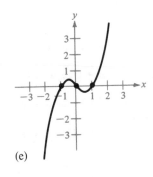

(e)

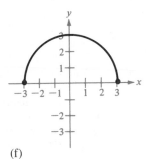

(f)

In Exercises 7–16, find the intercepts.

7. $y = 2x - 3$ 8. $y = (x - 1)(x - 3)$
9. $y = x^2 + x - 2$ 10. $y^2 = x^3 - 4x$
11. $y = x^2\sqrt{9 - x^2}$ 12. $xy = 4$
13. $y = \dfrac{x - 1}{x - 2}$ 14. $y = \dfrac{x^2 + 3x}{(3x + 1)^2}$
15. $x^2y - x^2 + 4y = 0$ 16. $y = 2x - \sqrt{x^2 + 1}$

In Exercises 17–26, check for symmetry about both axes and the origin.

17. $y = x^2 - 2$ 18. $y = x^4 - x^2 + 3$
19. $x^2y - x^2 + 4y = 0$ 20. $(xy)^2 - x^2 - 4y^2 = 0$
21. $y^2 = x^3 - 4x$ 22. $xy^2 = -10$
23. $y = x^3 + x$ 24. $xy = 1$
25. $y = \dfrac{x}{x^2 + 1}$ 26. $y = x^3 + x - 3$

In Exercises 27–45, use the methods of this section to sketch the graph of each equation. Identify the intercepts and test for symmetry.

27. $y = x$ 28. $y = x + 3$
29. $y = -3x + 2$ 30. $y = 2x - 3$
31. $y = 1 - x^2$ 32. $y = x^2 + 3$
33. $y = -2x^2 + x + 1$ 34. $y = x^3 - 3x$
35. $y = x^3 + 2$ 36. $y = x^3 - 1$
37. $x^2 + 4y^2 = 4$ 38. $x = y^2 - 4$

39. $y = (x + 2)^2$ 40. $y = \dfrac{1}{x^2 + 1}$
41. $y = \dfrac{1}{x}$ 42. $y = 2x^4$
43. $y = |x - 2|$ 44. $y = -|x - 2|$
45. $y = \sqrt{x - 3}$

46. (a) Sketch the graph of $y = 3x^4 - 4x^3$ by completing the accompanying table and plotting the resulting points.

x	-1	0	1	2
y				

(b) Find additional points satisfying $y = 3x^4 - 4x^3$ by completing the accompanying table. Now refine the graph of part (a).

x	-0.75	-0.50	-0.25	0.25	0.5	1.33
y						

In Exercises 47–52, write the general equation of the specified circle.

47. Center $(0, 0)$; radius $= 3$
48. Center $(0, 0)$; radius $= 5$
49. Center $(2, -1)$; radius $= 4$
50. Center $(-4, 3)$; radius $= \frac{5}{8}$
51. Center $(-1, 2)$; point on circle $(0, 0)$
52. Center $(3, -2)$; point on circle $(-1, 1)$

In Exercises 53–60, write the given equation (of a circle) in standard form and sketch its graph.

53. $x^2 + y^2 - 2x + 6y + 6 = 0$
54. $x^2 + y^2 - 2x + 6y - 15 = 0$
55. $x^2 + y^2 - 2x + 6y + 10 = 0$
56. $3x^2 + 3y^2 - 6y - 1 = 0$
57. $2x^2 + 2y^2 - 2x - 2y - 3 = 0$
58. $4x^2 + 4y^2 - 4x + 2y - 1 = 0$
59. $16x^2 + 16y^2 + 16x + 40y - 7 = 0$
60. $x^2 + y^2 - 4x + 2y + 3 = 0$

In Exercises 61–64, determine whether the given points lie on the graph of the given equation.

61. Equation: $2x - y - 3 = 0$
    (a) $(1, 2)$ (b) $(1, -1)$ (c) $(4, 5)$

62. Equation: $x^2 + y^2 = 4$
    (a) $(1, -\sqrt{3})$ (b) $(\frac{1}{2}, -1)$ (c) $(\frac{3}{2}, \frac{7}{2})$
63. Equation: $x^2y - x^2 + 4y = 0$
    (a) $(1, \frac{1}{5})$ (b) $(2, \frac{1}{2})$ (c) $(-1, -2)$
64. Equation: $x^2 - xy + 4y = 3$
    (a) $(0, 2)$ (b) $(-2, -\frac{1}{6})$ (c) $(3, -6)$
65. The Consumer Price Index for the 1970s is given in the table below. A mathematical model for the CPI during this 10-year period is

$$y = 0.55t^2 + 5.85t + 114.41$$

where y represents the CPI and t represents the year, with $t = 0$ corresponding to 1970.
    (a) Graphically compare the actual CPI during the 10-year period with that predicted by the model.
    (b) Use the model to predict the CPI for 1985.

Year	1970	1971	1972	1973	1974
CPI	116.3	121.3	125.3	133.1	147.7

Year	1975	1976	1977	1978	1979
CPI	161.2	170.5	181.5	195.3	211.1

66. From the model in Exercise 65, we obtain the model

$$V = \frac{100}{0.55t^2 + 5.85t + 114.41}$$

where V represents the purchasing power of the dollar (in terms of constant 1967 dollars) and t represents the year, with $t = 0$ corresponding to 1970. Use this model to complete the following table, and then graph your results.

t	0	2	4	6	8	10	12
V							

67. The farm population in the United States as a percentage of the total population is given in the following table:

Year	1950	1955	1960	1965	1970	1975	1979
%	15.3	11.6	8.7	6.4	4.8	4.2	3.4

A mathematical model for these data is given by

$$y = \frac{100}{4.90 + 0.79t}$$

where y represents the percentage and t represents the year, with $t = 0$ corresponding to 1950.
    (a) Graphically compare the actual percentage with that given by the model.
    (b) Use the model to predict the farm population as a percentage of the total population in 1990.
68. The average number of acres per farm in the United States is given in the following table:

Year	1950	1960	1965	1970	1975	1978
Number of Acres	213	297	340	374	391	401

A mathematical model for these data is given by

$$y = -0.13t^2 + 10.43t + 211.3$$

where y represents the acreage and t represents the year, with $t = 0$ corresponding to 1950.
    (a) Graphically compare the actual number of acres per farm with that given by the model.
    (b) Use the model to predict the average number of acres per farm in the United States in 1985.

Functions   3.3

Many common relationships involve two variables in such a way that the value of one of the variables depends on the value of the other. For example, the sales tax on an item depends on its selling price. The distance an object moves in a given time depends on its speed.

Consider the relationship between the area of a circle and its radius. This relationship can be expressed by the equation $A = \pi r^2$. We have within the set of positive numbers a free choice for the value of r. The value of A then depends on our choice of r. Thus, we refer to A as the **dependent variable** and r as the **independent variable.**

Of particular interest are relationships such that to every value of the independent variable there corresponds *one and only one* value of the dependent variable. We call this type of correspondence a **function.**

DEFINITION OF A FUNCTION

A **function** is a relationship between two variables such that to each value of the independent variable there corresponds exactly one value of the dependent variable.

The collection of all values assumed by the independent variable is called the **domain** of the function, and the collection of all values assumed by the dependent variable is called the **range** of the function.

Functions and Formulas

We often specify functions by formulas. In such cases we *exclude* from the domain of the function all real numbers for which the formula is undefined. Two common instances in which the domain of a function is restricted by its formula are:

1. *Even Roots of Negative Numbers.* For example,

Domain:

$$y = \sqrt{x + 2} \qquad\qquad x \geq -2$$

has negative values inside the square root if $x < -2$. Thus, the domain is restricted to all real numbers greater than or equal to -2.

2. *Division by Zero.* For example,

Domain:

$$y = \frac{4}{x^2 - 9} \qquad\qquad \text{all real } x \neq \pm 3$$

has a zero denominator when $x = \pm 3$. Hence the domain is restricted to all real numbers except $x = \pm 3$.

EXAMPLE 1
Determining Functional
Relationships from Equations

Which of the following equations define functional relationships between the variables x and y?

(a) $x + y = 1$

(b) $x^2 + y^2 = 1$

(c) $x^2 + y = 1$

(d) $x + y^2 = 1$

Solution:

To determine if an equation defines a functional relationship between its variables, we isolate the dependent variable on the left-hand side, as shown in the second and fourth columns of Table 3.1. Note that those equations that assign two values ($\pm$) to the dependent variable for each assigned value of the independent variable do not define functions.

TABLE 3.1

Original equation	*x as the dependent variable*	*Is x a function of y?*	*y as the dependent variable*	*Is y a function of x?*
(a) $x + y = 1$	$x = 1 - y$	yes	$y = 1 - x$	yes
(b) $x^2 + y^2 = 1$	$x = \pm\sqrt{1 - y^2}$	no (two values of x for some values of y)	$y = \pm\sqrt{1 - x^2}$	no (two values of y for some values of x)
(c) $x^2 + y = 1$	$x = \pm\sqrt{1 - y}$	no	$y = 1 - x^2$	yes
(d) $x + y^2 = 1$	$x = 1 - y^2$	yes	$y = \pm\sqrt{1 - x}$	no

EXAMPLE 2
Finding the Domain and Range of a Function

Determine the domain and range for the function of x defined by

$$y = \sqrt{x - 1}$$

Solution:

Since $\sqrt{x - 1}$ is not defined for $x - 1 < 0$ (that is, for $x < 1$), we must have $x \geq 1$. Therefore, the domain of the function is $x \geq 1$.

To find the range, we observe that $y = \sqrt{x - 1}$ is never negative. Moreover, as x takes on the various values in the domain, y takes on all nonnegative values, and we find the range to be $y \geq 0$.

The graph of the function lends further support to our conclusions (Figure 3.27).

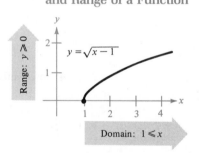

Range: $y \geqslant 0$

$y = \sqrt{x - 1}$

Domain: $1 \leqslant x$

FIGURE 3.27

EXAMPLE 3
A Function Defined by More Than One Equation

Determine the domain and range for the function of x given by

$$y = \begin{cases} \sqrt{x - 1}, & \text{if } x \geq 1 \\ 1 - x & \text{if } x < 1 \end{cases}$$

Solution:

Since $x \geq 1$ or $x < 1$, the domain of the function is the entire set of real numbers.

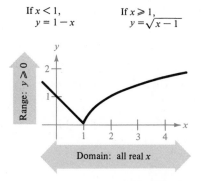

If $x < 1$,
$y = 1 - x$

If $x \geq 1$,
$y = \sqrt{x - 1}$

Range: $y \geq 0$

Domain: all real x

FIGURE 3.28

On the portion of the domain for which $x \geq 1$, the function behaves as in Example 2. For $x < 1$, $1 - x$ is positive, and therefore the range of the function is the interval $[0, \infty)$.

Again, a graph of the function helps to verify our conclusions (Figure 3.28).

Remark: Sometimes the domain is *implied,* as in Example 2 ($\sqrt{x - 1}$ is defined only if $x \geq 1$). On other occasions, the physical nature of the problem may restrict the domain to a certain subset of the real numbers. For instance, in the function for the area of a circle, $A = \pi r^2$, we list the domain as all $r > 0$.

Functional Notation

When using an equation to define a function, we generally isolate the dependent variable on the left side of the equation. For instance, writing the equation $x + 2y = 1$ in the form

$$y = \frac{1 - x}{2}$$

indicates that y is the dependent variable. In **functional notation** this equation has the form

$$f(x) = \frac{1 - x}{2}$$

This notation has the advantage of clearly identifying the dependent variable as $f(x)$ while at the same time providing a name "f" for the function. [The symbol $f(x)$ is read "f of x."]

To denote the value of the dependent variable when $x = 3$, we use the symbol $f(3)$ as follows:

$$f(3) = \frac{1 - (3)}{2} = \frac{-2}{2} = -1$$

Similarly,

$$f(0) = \frac{1 - (0)}{2} = \frac{1}{2}$$

$$f(-2) = \frac{1 - (-2)}{2} = \frac{3}{2}$$

The values $f(3)$, $f(0)$, and $f(-2)$ are called **functional values,** and they lie in the range of f. This means that the values $f(3)$, $f(0)$, and $f(-2)$ are y-values, and thus the points $(3, f(3))$, $(0, f(0))$, and $(-2, f(-2))$ lie on the graph of f. (See Figure 3.29.)

The role of the variable x, in an equation that defines a function, is simply that of a "placeholder." For instance, the function

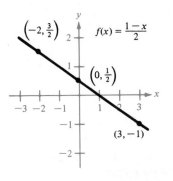

$\left(-2, \frac{3}{2}\right)$

$f(x) = \frac{1-x}{2}$

$\left(0, \frac{1}{2}\right)$

$(3, -1)$

FIGURE 3.29

$$f(x) = 2x^2 - 4x + 1$$

can be properly described by the form

$$f(\ ) = 2(\ )^2 - 4(\ ) + 1$$

where parentheses are used instead of x. Therefore, to evaluate $f(-2)$, we simply place -2 in each set of parentheses:

$$f(-2) = 2(-2)^2 - 4(-2) + 1 = 2(4) + 8 + 1 = 17$$

EXAMPLE 4
Evaluating a Function

For the function f defined by $f(x) = x^2 - 4x + 7$, evaluate

(a) $f(3x)$ (b) $f(x - 1)$

*(c) $\dfrac{f(x + \Delta x) - f(x)}{\Delta x}$

Solution:
We begin by writing the equation for f in the form

$$f(\ ) = (\ )^2 - 4(\ ) + 7$$

(a) $f(3x) = (3x)^2 - 4(3x) + 7$ *Replace x with 3x*

$\qquad\quad = 9x^2 - 12x + 7$ *Expand terms*

(b) $f(x - 1) = (x - 1)^2 - 4(x - 1) + 7$ *Replace x with (x − 1)*

$\qquad\qquad = x^2 - 2x + 1 - 4x + 4 + 7$ *Expand terms*

$\qquad\qquad = x^2 - 6x + 12$ *Collect like terms*

(c) $\dfrac{f(x + \Delta x) - f(x)}{\Delta x}$

$$= \frac{[(x + \Delta x)^2 - 4(x + \Delta x) + 7] - [x^2 - 4x + 7]}{\Delta x}$$

$$= \frac{x^2 + 2x\Delta x + (\Delta x)^2 - 4x - 4\Delta x + 7 - x^2 + 4x - 7}{\Delta x}$$

$$= \frac{2x\Delta x + (\Delta x)^2 - 4\Delta x}{\Delta x}$$

$$= \frac{\Delta x(2x + \Delta x - 4)}{\Delta x}$$

$$= 2x + \Delta x - 4$$

*Δ is the uppercase Greek letter "delta." The symbol Δx is read as "delta-x" and is commonly used in calculus to denote a small change in x.

A function whose range consists of a single value is called a **constant function.** For example, $f(x) = -4$ is a constant function and no matter what value we choose for x, the resulting functional value is -4. That is, $f(-2) = -4$, $f(0) = -4$, $f(10) = -4$, and so on.

Although we generally use f as a convenient function name and x as the independent variable, we can use other symbols. For instance, the following equations all define the same function:

$$f(x) = x^2 - 4x + 7$$
$$f(t) = t^2 - 4t + 7$$
$$g(s) = s^2 - 4s + 7$$

Functions and Calculators

The basic steps in evaluating a function by means of a programmable calculator are

1. Program the function into the calculator.

2. Evaluate the function at specified values.

For example, for the function

$$C(x) = \frac{5000 + 0.56x}{x}$$

the calculator steps for a programmable (Texas Instruments) calculator are

| LRN | | STO | 1 | × | .56 | + | 5000 | = | ÷ | RCL | 1 | = |
| R/S | RST | LRN | RST |

Once this program has been entered, the function can be evaluated at several points simply by entering various x-values as follows:

Calculator Steps		Display
10	R/S	500.56
100	R/S	50.56
1000	R/S	5.56
2000	R/S	3.06
5000	R/S	1.56
10000	R/S	1.06

We conclude this section with a glossary of terms related to the concept of a function.

GLOSSARY OF THE TERMINOLOGY OF FUNCTIONS

Function: A relationship between two variables such that to each value of the independent variable there corresponds exactly one value of the dependent variable.

Function Notation: $y = f(x)$

  f is the **name** of the function
  y is the **dependent variable**
  x is the **independent variable**
  $f(x)$ is the **value of the function at** x

Domain: The collection of all values of the independent variable for which the function is defined. If x is in the domain of f, we say that f is **defined** at x. If x is not in the domain of f, we say that f is **undefined** at x.

Range: The collection of all values assumed by the dependent variable.

Implied Domain: If f is defined by an equation and the domain is not specified, then we assume the domain to consist of all real numbers for which the equation is defined. [For example, the implied domain of the function $f(x) = \sqrt{x}$ is the collection of nonnegative real numbers.]

Constant Function: A function whose range consists of a single number.

Section Exercises 3.3

1. Given $f(x) = 2x - 3$, find
   (a) $f(1)$ (b) $f(0)$
   (c) $f(-3)$ (d) $f(b)$
   (e) $f(x - 1)$ (f) $f(\frac{1}{4})$
2. Given $f(x) = x^2 - 2x + 2$, find
   (a) $f(\frac{1}{2})$ (b) $f(3)$
   (c) $f(-1)$ (d) $f(c)$
   (e) $f(x + \Delta x)$ (f) $f(2)$
3. Given $f(x) = \sqrt{x + 3}$, find
   (a) $f(-3)$ (b) $f(-2)$
   (c) $f(0)$ (d) $f(6)$
   (e) $f(x + \Delta x)$ (f) $f(c)$
4. Given $f(x) = 1/\sqrt{x}$, find
   (a) $f(1)$ (b) $f(4)$
   (c) $f(2)$ (d) $f(\frac{1}{4})$
   (e) $f(x + \Delta x)$ (f) $f(x + \Delta x) - f(x)$
5. Given $f(x) = |x|/x$, find
   (a) $f(2)$ (b) $f(-2)$
   (c) $f(-100)$ (d) $f(100)$
   (e) $f(x^2)$ (f) $f(x - 1)$

6. Given $f(x) = |x| + 4$, find
   (a) $f(2)$ (b) $f(-2)$
   (c) $f(3)$ (d) $f(x^2)$
   (e) $f(x + \Delta x)$ (f) $f(x + \Delta x) - f(x)$
7. Given $f(x) = x^2 - x + 1$, find
   $$\frac{f(2 + \Delta x) - f(2)}{\Delta x}$$
8. Given $f(x) = 1/x$, find
   $$\frac{f(1 + \Delta x) - f(1)}{\Delta x}$$
9. Given $f(x) = x^3$, find
   $$\frac{f(x + \Delta x) - f(x)}{\Delta x}$$
10. Given $f(x) = 3x - 1$, find
   $$\frac{f(x) - f(1)}{x - 1}$$

11. Given $f(x) = 1/\sqrt{x - 1}$, find

$$\frac{f(x) - f(2)}{x - 2}$$

12. Given $f(x) = x^3 - x$, find

$$\frac{f(x) - f(1)}{x - 1}$$

13. Let $f(x) = x^2 - 1$. Find all real numbers x such that $f(x) = 8$.

14. Let $f(x) = x^3 - x$. Find all real numbers x such that $f(x) = 0$.

15. Let

$$f(x) = \frac{3}{x - 1} + \frac{4}{x - 2}$$

Find all real numbers x such that $f(x) = 0$.

16. Let

$$f(x) = a + \frac{b}{x}$$

Find all real numbers x such that $f(x) = 0$.

In Exercises 17–26, find the domain and range of the given function.

17. $f(x) = \sqrt{x - 1}$

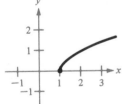

18. $f(x) = \sqrt{1 - x}$

19. $f(x) = x^2$

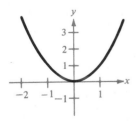

20. $f(x) = 4 - x^2$

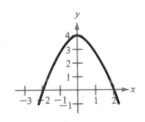

21. $f(x) = \sqrt{9 - x^2}$

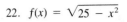
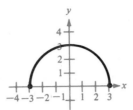

22. $f(x) = \sqrt{25 - x^2}$

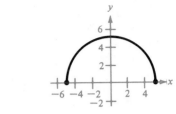

23. $f(x) = 1/|x|$

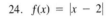
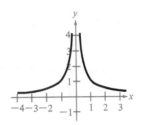

24. $f(x) = |x - 2|$

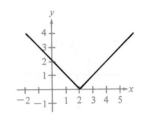

25. $f(x) = |x|/x$

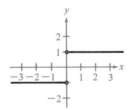

26. $f(x) = \sqrt{x^2 - 4}$

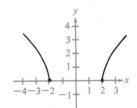

In Exercises 27–36, identify the equations that determine y as a function of x.

27. $x^2 + y^2 = 4$ 28. $x = y^2$

29. $x^2 + y = 4$ 30. $x + y^2 = 4$

31. $2x + 3y = 4$

32. $x^2 + y^2 - 2x - 4y + 1 = 0$

33. $y^2 = x^2 - 1$ 34. $y = \pm\sqrt{x}$

35. $x^2y - x^2 + 4y = 0$ 36. $xy - y - x - 2 = 0$

In Exercises 37 and 38, find a formula for the given function $V = f(x)$ and give its domain.

37. The value V of a farm having $500,000 worth of buildings, livestock, and equipment in terms of the number of acres on the farm. (Each acre is valued at $1750.)

38. The value V of wheat at $4.45 per bushel as a function of the number of bushels.

39. A company produces a product for which the variable cost is $12.30 per unit and the fixed costs are $98,000. The

product sells for $17.98. Let x be the number of units produced.
(a) Write the total cost C as a function of the number of units produced.
(b) Write the revenue R as a function of the number of units produced.
(c) Write the profit P as a function of the number of units produced. (Note: $P = R - C$.)

40. The inventor of a new game believes that the variable cost for producing the game is $0.95 per unit and the fixed costs are $6,000. He plans to wholesale the game for $1.69. Let x be the number of games sold.
(a) Write the total cost C as a function of the number of games sold.
(b) Write the average cost per unit $\overline{C} = C/x$ as a function of x.

41. The demand function for a particular commodity is given by

$$p = \frac{14.75}{1 + 0.01x}, \qquad 0 \le x$$

where p is the price per unit and x is the number of units sold.
(a) Find x as a function of p.
(b) Use the result of part (a) to find the number of units sold when the price is $10.00.

42. A power station is on one side of a river that is one-half mile wide. A factory lies downstream three miles on the other side of the river. It costs $10 per foot to run the power lines on land and $15 per foot to run them underwater. Write the cost C of running the line from the power station to the factory as a function of x. (See Figure 3.30.)

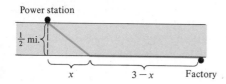

FIGURE 3.30

43. A radio manufacturer charges $90 per unit for units that cost $60 to produce. To encourage large orders from distributors, the manufacturer will reduce the price by $0.01 per unit for each unit in excess of 100. (For example, an order of 200 units would have a price of $89 per unit.) This price reduction is discontinued when the price per unit drops to $75.
(a) Write the price per unit p as a function of the order size x.

(b) Write the profit P as a function of the order size x. (Note: $P = R - C = px - 60x$.)
(c) Find the total profit for an order of 1000 units.

44. Assume that the amount of money deposited in a bank is proportional to the square of the interest rate the bank pays on the money. That is, $d = kr^2$, where d is the total deposit, r is the interest rate, and k is the proportionality constant. Assuming the bank can reinvest the money for a return of 18%, write the bank's profit P as a function of the interest rate r.

Calculator Exercises

45. The force (in tons) of water pressure against the face of a certain dam is a function of the depth y of the water and is given by

$$F(y) = 149.76\sqrt{10}y^{5/2}$$

Complete the following table by finding the force for various water levels.

y	5	10	20	30	40
$F(y)$					

46. The work W (in foot-pounds) required to fill a storage tank to a depth of y feet is given by

$$W(y) = 25\pi\left(4y^2 - \frac{y^3}{3}\right)$$

Complete the following table by finding the work done in filling the tank to various levels.

y	1	2	4	6	8	10
$W(y)$						

47. The number N of bacteria in a certain culture is a function of time t (in hours) and is approximated by

$$N(t) = 500\left(1 + \frac{4t}{50 + t^2}\right)$$

Complete the following table estimating the size of the population at the indicated times.

t	0	1	2	4	6	8	10
$N(t)$							

48. The temperature T of food placed in a refrigerator is a function of the time t (in hours) and is approximated by

$$T(t) = 10\left(\frac{4t^2 + 16t + 25}{t^2 + 4t + 10}\right)$$

Complete the following table estimating the temperature of the food at the indicated time.

t	0	2	6	8	10
$T(t)$					

Graphs of Functions   3.4

In Section 3.3 we discussed functions from an algebraic (analytic) point of view. Here we study functions from a geometric (graphic) perspective. In this setting it is convenient to view a function f as a correspondence between two sets of real numbers.

ALTERNATIVE DEFINITION OF A FUNCTION

> A **function** f from a set X to a set Y is a correspondence that assigns to each element x in X exactly one element y in Y. We call y the **image** of x under f and denote it by $f(x)$. The **domain** of f is the set X, while the **range** consists of all images of elements in X.

This section is divided into three parts, each related to one aspect of the graph of a function:

1. The **shape** of the graph.
2. The **position** of the graph relative to the x- and y-axes.
3. **Transformations** of the graph.

As you study this section, remember that the graph of the function $y = f(x)$ consists of all points $(x, f(x))$ where

x = the directed distance from the y-axis

$f(x)$ = the directed distance from the x-axis

as shown in Figure 3.31.

Since (by the definition of a function) there corresponds exactly one y-value for each x-value, it follows that a vertical line can intersect the graph of a function at most once. This observation provides us with a convenient visual test for functions.

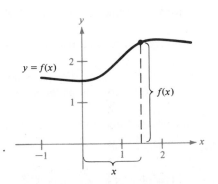

FIGURE 3.31

THE VERTICAL LINE TEST FOR FUNCTIONS

> An equation defines y as a function of x if no vertical line in the plane intersects the graph of the equation at more than one point.

EXAMPLE 1
Vertical Line Test for Functions

Which of the graphs in Figure 3.32 represent functions of x?

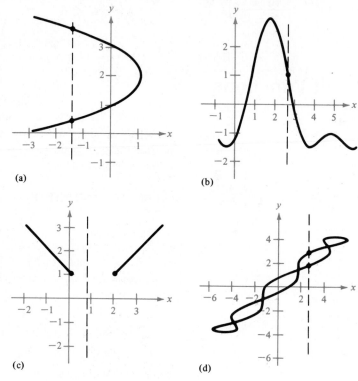

(a)

(b)

(c)

(d)

FIGURE 3.32

Solution:

(a) This is *not* the graph of a function, since we can find a vertical line that intersects the graph twice.

(b) This *is* the graph of a function, since every vertical line intersects this graph at most once.

(c) This *is* the graph of a function. (Note that if a vertical line does not intersect the graph, it simply means that the function is undefined for this particular value of x.)

(d) This is *not* the graph of a function, since we can find a vertical line that intersects the graph twice.

In Section 3.3 we mentioned that a function is called **constant** if its range consists of a single y-value. For example,

$$h(x) = 2 \quad \text{and} \quad g(x) = -1$$

are constant functions. Since the y-value of a constant function is the same for all x, it follows that the graph of a constant function consists of a horizontal line. (See Figure 3.33.)

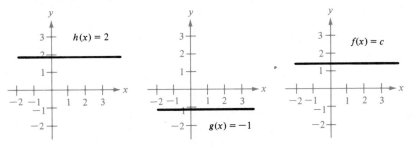

FIGURE 3.33   The graph of a constant function is a horizontal line.

GRAPH OF A CONSTANT FUNCTION

> The graph of $f(x) = c$ is a horizontal line with a y-intercept at $(0, c)$.

The more we know about the graph of a function, the more we know about the function itself. Consider the graph shown in Figure 3.34. As we move from left to right, this graph falls for negative values of x, is constant from $x = 0$ to $x = 2$, and then rises for x greater than 2. Correspondingly, we say that the function is:

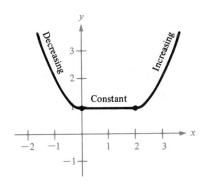

FIGURE 3.34

decreasing on the interval $(-\infty, 0)$,

constant on the interval $(0, 2)$, and

increasing on the interval $(2, \infty)$.

DEFINITION OF INCREASING AND DECREASING FUNCTIONS

> A function f is said to be **increasing** on an interval if, for any two numbers x_1 and x_2 in the interval,
>
> $\quad x_1 < x_2 \quad$ implies $\quad f(x_1) < f(x_2)$
>
> A function f is said to be **decreasing** on an interval if, for any two numbers x_1 and x_2 in the interval,
>
> $\quad x_1 < x_2 \quad$ implies $\quad f(x_1) > f(x_2)$
>
> A function f is said to be **constant** on an interval if, for any two numbers x_1 and x_2 in the interval,
>
> $\quad f(x_1) = f(x_2)$

EXAMPLE 2
Increasing and
Decreasing Functions

In Figure 3.35 determine the intervals over which each function is increasing, decreasing, or constant.

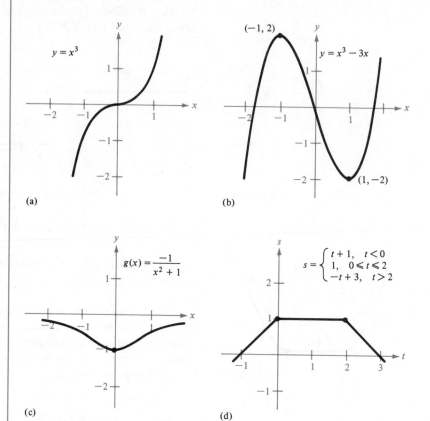

(a)

(b)

(c)

(d)

FIGURE 3.35

Solution:

(a) Although it might appear that there is an interval about zero over which this function is constant, we see that if $x_1 < x_2$, then $x_1^3 < x_2^3$, and we conclude that the function is increasing for all x.

(b) This function is

> increasing on the interval $(-\infty, -1)$,
>
> decreasing on the interval $(-1, 1)$, and
>
> increasing on the interval $(1, \infty)$.

(c) This function is

> decreasing on the interval $(-\infty, 0)$ and
>
> increasing on the interval $(0, \infty)$.

(d) This function is

      increasing on the interval $(-\infty, 0)$,

      constant on the interval $(0, 2)$, and

      decreasing on the interval $(2, \infty)$.

Remark: The points at which a function changes its increasing, decreasing, or constant behavior are especially important in producing an accurate graph of the function. These points often identify maximum or minimum values of the function. Techniques for finding the exact location of these special points are developed in calculus.

Position of a Function's Graph Relative to the x- and y-Axes

In Section 3.2 we defined an x-intercept to be a point $(x, 0)$ at which a graph crosses the x-axis. When the graph of a function crosses the x-axis, we say that the function has a **zero** at that point. For example, the function $f(x) = x - 4$ has a zero at $(4, f(4))$, since $f(4) = 4 - 4 = 0$.

    In Section 3.2, we also discussed different types of symmetry. In the terminology of functions, we say that a function is **even** if its graph is symmetric with respect to the y-axis; a function is **odd** if its graph is symmetric with respect to the origin. Thus, our symmetry tests in Section 3.2 yield the following test for even and odd functions.

TEST FOR EVEN AND ODD FUNCTIONS

> The function $y = f(x)$ is **even** if
>
> $$f(-x) = f(x)$$
>
> The function $y = f(x)$ is **odd** if
>
> $$f(-x) = -f(x)$$

EXAMPLE 3
Even and Odd Functions

Determine whether the following functions are even, odd, or neither.

(a) $g(x) = x^3 - x$ (b) $h(x) = x^2 + 1$ (c) $f(x) = x^3 - 1$

Solution:

(a) This function is odd, since

$$g(-x) = (-x)^3 - (-x) = -x^3 + x = -(x^3 - x) = -g(x)$$

(b) This function is even, since

$$h(-x) = (-x)^2 + 1 = x^2 + 1 = h(x)$$

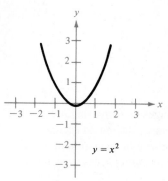

Original Graph

FIGURE 3.36

(c) By substituting $-x$ for x, we have

$$f(-x) = (-x)^3 - 1 = -x^3 - 1$$

Now, since

$$f(x) = x^3 - 1 \quad \text{and} \quad -f(x) = -x^3 + 1$$

we conclude that

$$f(-x) \neq f(x) \quad \text{and} \quad f(-x) \neq -f(x)$$

which implies that the function is neither even nor odd.

Transformations of a Function's Graph

Some families of graphs all have the same basic shape. Consider the graph of $y = x^2$, as shown in Figure 3.36. Now compare this graph to those shown in Figure 3.37.

Each of the graphs in Figure 3.37 is a **transformation** of the graph of $y = x^2$. The three basic types of transformations involved in these eight graphs are (1) horizontal shifts, (2) vertical shifts, and (3) reflections.

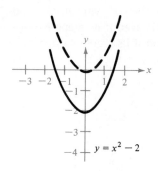

(a) Vertical shift
downward

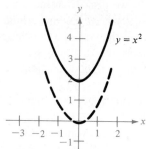

(b) Vertical shift
upward

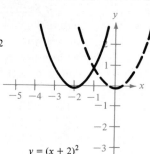

(c) Horizontal shift
to the left

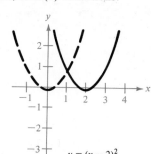

(d) Horizontal shift
to the right

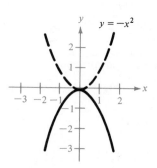

(e) Reflection

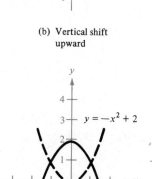

(f) Reflection and
vertical shift

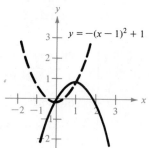

(g) Reflection, vertical
and horizontal shifts

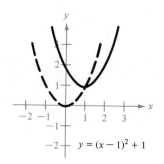

(h) Vertical and
horizontal shifts

FIGURE 3.37   *Transformations of the Graph of $y = x^2$*

BASIC TYPES OF TRANSFORMATIONS $(c > 0)$

Original Graph: $y = f(x)$
Horizontal Shift c units to the **right:** $y = f(x - c)$
Horizontal Shift c units to the **left:** $y = f(x + c)$
Vertical Shift c units **downward:** $y = f(x) - c$
Vertical Shift c units **upward:** $y = f(x) + c$
Reflection (about the x-axis): $y = -f(x)$
Reflection (about the y-axis): $y = f(-x)$

EXAMPLE 4
Transformations of the
Graph of a Function

Use the graph of $y = x^3$ shown in Figure 3.38 to sketch the graph of each of the following functions:

(a) $y = x^3 + 1$ (b) $y = (x - 1)^3$
(c) $y = -x^3$ (d) $y = (x + 2)^3 + 1$

Solution:
The graphs are shown in Figure 3.39.

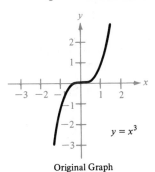

Original Graph

FIGURE 3.38

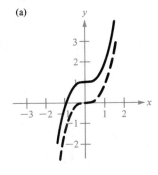

(a)

Vertical shift: 1 up
$y = f(x) + 1 = x^3 + 1$

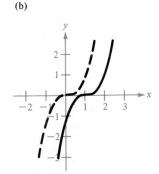

(b)

Horizontal shift: 1 right
$y = f(x - 1) = (x - 1)^3$

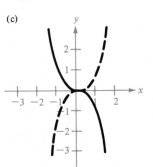

(c)

Reflection
$y = -f(x) = -x^3$

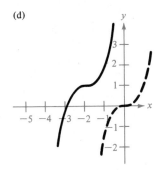

(d)

Vertical shift: 1 up
Horizontal shift: 2 left
$y = f(x + 2) + 1 = (x + 2)^3 + 1$

FIGURE 3.39

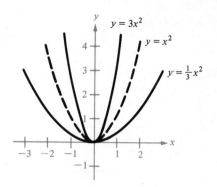

FIGURE 3.40

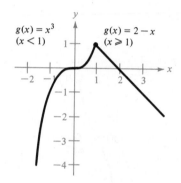

FIGURE 3.41

Horizontal shifts, vertical shifts, and reflections are called **rigid** transformations because the basic shape of the graph is unchanged. As illustrated in Figures 3.37 and 3.39, the only change caused by a rigid transformation is the *position* of the graph in the xy-plane.

Nonrigid transformations are those which cause a *distortion* of the original graph. For example, the graph of

$$y = cf(x)$$

is a nonrigid vertical transformation of the graph of $y = f(x)$. If $c > 1$, we say the transformation **stretches,** and if $0 < c < 1$ we say the transformation **shrinks** the graph of $y = f(x)$. Figure 3.40 shows the effect of the coefficients $c = 3$ and $c = \frac{1}{3}$ on the graph of $f(x) = x^2$.

On occasion we encounter a function that is defined differently on distinct subsets of its domain. Such is the case with

$$g(x) = \begin{cases} x^3, & \text{if } x < 1 \\ 2 - x, & \text{if } x \geq 1 \end{cases}$$

whose graph is shown in Figure 3.41. One commonly used function that fits into this category is the **greatest integer function**

$$f(x) = [x] = \text{the greatest integer less than or equal to } x$$

More explicitly, f is described by

$$f(x) = \begin{cases} \vdots & \vdots \\ -1, & \text{if } -1 \leq x < 0 \\ 0, & \text{if } \;\; 0 \leq x < 1 \\ 1, & \text{if } \;\; 1 \leq x < 2 \\ \vdots & \vdots \end{cases}$$

and its graph is shown in Figure 3.42.

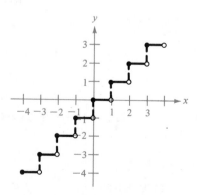

FIGURE 3.42

Greatest Integer Function

Section Exercises 3.4

In Exercises 1–10, use the vertical line test to determine if y is a function of x.

1. $y = x^2$

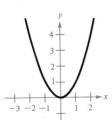

2. $y = x^3 - 1$

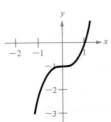

3. $x - y^2 = 0$

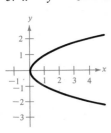

4. $x^2 + y^2 = 9$

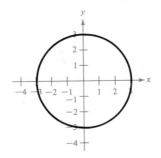

5. $\sqrt{x^2 - 4} - y = 0$

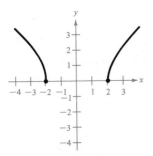

6. $x - xy + y + 1 = 0$

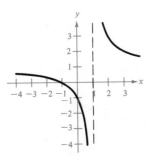

7. $x^2 = xy - 1$

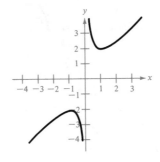

8. $x = |y|$

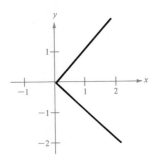

9. $x^2 - 4y^2 + 4 = 0$

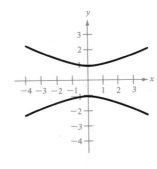

10. $y = |x - 3| + |x - 1|$

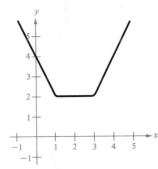

In Exercises 11–20, (a) determine the intervals over which the function is increasing, decreasing, or constant, and (b) determine if the function is even, odd, or neither.

11. $f(x) = 2x$

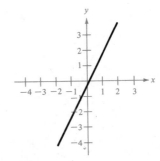

12. $f(x) = x^2 - 2x$

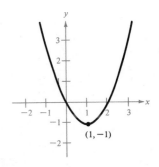

(1, −1)

13. $f(x) = x^3 - 3x^2$

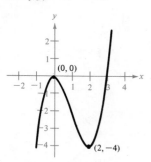

14. $f(x) = \sqrt{x^2 - 4}$

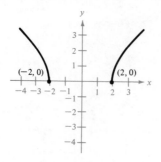

15. $f(x) = 3x^4 - 6x^2$

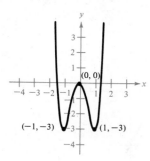

16. $f(x) = x^{2/3}$

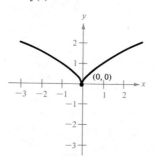

17. $f(x) = |x + 1| + |x - 1|$

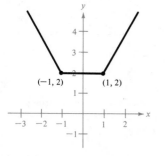

18. $f(x) = x^3 - 12x$

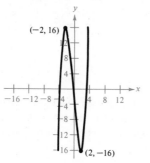

19. $f(x) = x\sqrt{x + 3}$

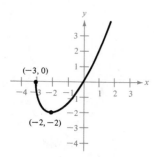

20. $f(x) = |2x - 3|$

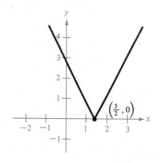

In Exercises 21–30, sketch the graph of the function.

21. $f(x) = 5 - 3x$

22. $f(x) = 2x - 3$

23. $f(x) = x^2 - 4$

24. $f(x) = -x^2 + 2x$

25. $g(t) = \sqrt[3]{t - 1}$

26. $h(t) = \dfrac{1}{t^2 + 1}$

27. $f(x) = |x + 2|$

28. $f(x) = |x| + |x + 2|$

29. $f(x) = \begin{cases} x + 3, & \text{if } x \le 0 \\ 3, & \text{if } 0 < x \le 2 \\ 2x - 1, & \text{if } x > 2 \end{cases}$

30. $f(x) = \begin{cases} 2x + 1, & \text{if } x \le -1 \\ x^2 - 2, & \text{if } x > -1 \end{cases}$

31. Use the accompanying graph of $f(x) = \sqrt{x}$ to sketch the graph of each of the following.
 (a) $y = f(x) + 2 = \sqrt{x} + 2$
 (b) $y = -f(x) = -\sqrt{x}$
 (c) $y = f(x - 2) = \sqrt{x - 2}$
 (d) $y = f(x + 3) = \sqrt{x + 3}$
 (e) $y = 2 - f(x - 4) = 2 - \sqrt{x - 4}$
 (f) $y = f(2x) = \sqrt{2x}$
 (g) $y = 2f(x) = 2\sqrt{x}$

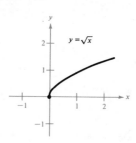

32. Use the accompanying graph of $f(x) = \sqrt[3]{x}$ to sketch the graph of each of the following.
 (a) $y = f(x) - 1 = \sqrt[3]{x} - 1$
 (b) $y = f(x + 1) = \sqrt[3]{x + 1}$
 (c) $y = f(x - 1) = \sqrt[3]{x - 1}$
 (d) $y = -f(x - 2) = -\sqrt[3]{x - 2}$
 (e) $y = f(x + 1) - 1 = \sqrt[3]{x + 1} - 1$
 (f) $y = f(x/2) = \sqrt[3]{x/2}$
 (g) $y = \tfrac{1}{2}f(x) = \tfrac{1}{2}\sqrt[3]{x}$

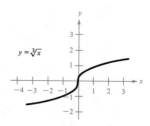

33. Use the accompanying graph of $f(x) = x\sqrt{x + 3}$ to write formulas for the functions whose graphs are shown in parts (a) through (d).

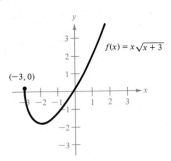

$f(x) = x\sqrt{x + 3}$

$(-3, 0)$

(a)

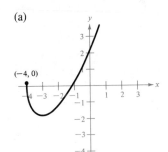

$(-4, 0)$

(b)

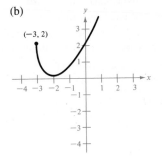

$(-3, 2)$

(c)

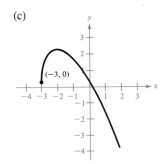

$(-3, 0)$

(d)

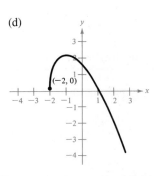

$(-2, 0)$

(a)

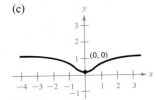

$(0, 0)$

(b)

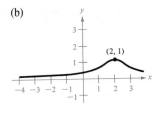

$(2, 1)$

(c)

$(0, 0)$

(d)

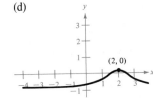

$(2, 0)$

35. Show that a function of the form

$$f(x) = a_{2n+1}x^{2n+1} + \cdots + a_3 x^3 + a_1 x$$

is odd.

36. Show that a function of the form

$$f(x) = a_{2n}x^{2n} + a_{2n-2}x^{2n-2} + \cdots + a_2 x^2 + a_0$$

is even.

37. Show that the product of two even functions is an even function.

38. Show that the product of two odd functions is an even function.

39. Show that the product of an odd function and an even function is odd.

34. Use the accompanying graph of $f(x) = 1/(x^2 + 1)$ to write formulas for the functions whose graphs are shown in parts (a) through (d).

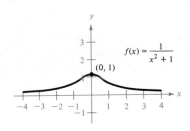

$f(x) = \dfrac{1}{x^2 + 1}$

$(0, 1)$

Linear Functions and Lines in the Plane

3.5

The simplest (and perhaps the most useful) type of function is a **linear function**

$$f(x) = ax + b, \qquad a \neq 0$$

As we shall see in this section, the graph of a linear function is a (nonvertical) line. In this text, we follow the convention of using the term **line** to mean a *straight* line.

The Slope of a Line

By the **slope** of a (nonvertical) line, we mean the number of units a line rises (or falls) vertically for each unit of horizontal change from left to right. For instance, consider the two points (x_1, y_1) and (x_2, y_2) on the line in Figure 3.43. As we move from left to right along this line, a change of $(y_2 - y_1)$ units in the vertical direction corresponds to a change of $(x_2 - x_1)$ units in the horizontal direction:

$$y_2 - y_1 = \text{the change in } y$$

and

$$x_2 - x_1 = \text{the change in } x$$

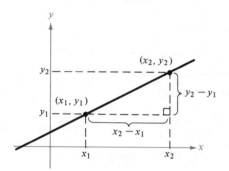

FIGURE 3.43

Thus, the slope of the line in Figure 3.43 is given by the ratio of these changes.

DEFINITION OF THE SLOPE OF A LINE

The **slope** of the line passing through the points (x_1, y_1) and (x_2, y_2) is

$$m = \frac{y_2 - y_1}{x_2 - x_1}$$

where $x_1 \neq x_2$.

Remark: Note that

$$\frac{y_2 - y_1}{x_2 - x_1} = \frac{-(y_1 - y_2)}{-(x_1 - x_2)} = \frac{y_1 - y_2}{x_1 - x_2}$$

Hence it does not matter which pair of coordinates we subtract to find the slope *as long as* we are consistent and both "subtracted coordinates" come from the same point.

In our first example (Figure 3.44), you will see that as we move from left to right:

1. A line with positive slope ($m > 0$) *rises*.
2. A line with negative slope ($m < 0$) *falls*.
3. A line with zero slope ($m = 0$) is *horizontal*.

EXAMPLE 1

Finding the Slope of a Line Passing Through Two Points

Find the slope of the line containing each of the following pairs of points:

(a) $(-2, 0)$ and $(3, 1)$ (b) $(-1, 2)$ and $(2, 2)$ (c) $(0, 4)$ and $(1, -1)$

Solution:

(a) The slope is

$$m = \frac{1 - 0}{3 - (-2)} = \frac{1}{3 + 2} = \frac{1}{5}$$

(b) The slope is

$$m = \frac{2 - 2}{2 - (-1)} = \frac{0}{3} = 0$$

(c) The slope is

$$m = \frac{-1 - 4}{1 - 0} = \frac{-5}{1} = -5$$

See Figure 3.44.

FIGURE 3.44

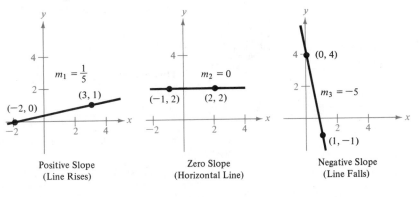

| Positive Slope (Line Rises) | Zero Slope (Horizontal Line) | Negative Slope (Line Falls) |

Remark: In Example 1, note that we did not consider the slope of a vertical line. This was not an oversight! The definition of slope does not apply to vertical lines. Informally, we say that the slope of a vertical line is *undefined*. Consider the points $(3, 4)$ and $(3, 1)$ on the line shown in Figure 3.45. In attempting to find the slope of this line, we obtain

$$m = \frac{4 - 1}{3 - 3}$$ *Undefined division by zero*

Since division by zero is undefined, it follows that the slope of a vertical line must also be undefined.

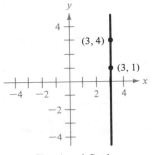

Slope is undefined.
(vertical line)

FIGURE 3.45

It is important to realize that the slope of a line is independent of the particular points used to calculate the slope. *Any* two points on the line can be used. This can be verified from the similar triangles shown in Figure 3.46. (Recall that the ratios of corresponding sides of similar triangles are equal.)

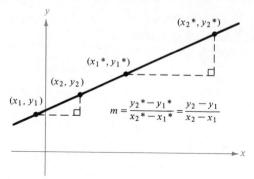

$$m = \frac{y_2{}^* - y_1{}^*}{x_2{}^* - x_1{}^*} = \frac{y_2 - y_1}{x_2 - x_1}$$

Any two points on a line can be used to determine its slope.

FIGURE 3.46

Equations of Lines

If we know the slope of a line and one point on the line, how can we determine the equation of the line? Figure 3.47 leads us to the answer to this question. For, if (x_1, y_1) is a point lying on a line of slope m, and (x, y) is any *other* point on the line, then

$$\frac{y - y_1}{x - x_1} = m$$

This equation, involving the two variables x and y, can be rewritten in the form

$$y - y_1 = m(x - x_1)$$

which is called the **point-slope equation** of a line.

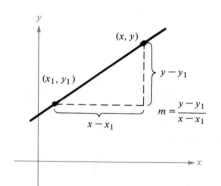

FIGURE 3.47

POINT-SLOPE EQUATION OF A LINE

The equation of the line with slope m passing through the point (x_1, y_1) is given by

$$y - y_1 = m(x - x_1)$$

EXAMPLE 2
The Point-Slope Equation of a Line

Find the equation of the line with slope of 3 and passing through the point $(1, -2)$.

Solution:
Using the point-slope form,

$$y - y_1 = m(x - x_1)$$

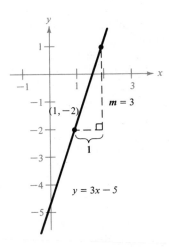

FIGURE 3.48

we have

$$y - (-2) = 3(x - 1)$$
$$y + 2 = 3x - 3$$

or

$$y = 3x - 5$$

See Figure 3.48.

We can combine the definition of the slope of a line with the point-slope equation to obtain the following **two-point equation** of a line.

TWO-POINT EQUATION OF A LINE

The equation of the line passing through the points (x_1, y_1) and (x_2, y_2) is given by

$$y - y_1 = \frac{y_2 - y_1}{x_2 - x_1}(x - x_1)$$

where $x_1 \neq x_2$.

EXAMPLE 3
The Two-Point Equation of a Line

The total U.S. sales (including inventories) during the first two quarters of 1978 were 539.9 and 560.2 billion dollars, respectively. Assuming a *linear growth pattern*, predict the total sales during the fourth quarter of 1978.

Solution:
In Figure 3.49 we let x represent the quarter and y represent the sales in billions of dollars, and we let $(1, 539.9)$ and $(2, 560.2)$ be two points on the line representing the total U.S. sales. Using the two-point equation of a line, we have

$$y - y_1 = \frac{y_2 - y_1}{x_2 - x_1}(x - x_1)$$

$$y - 539.9 = \frac{560.2 - 539.9}{2 - 1}(x - 1)$$

$$y = 20.3(x - 1) + 539.9$$

$$y = 20.3x + 519.6$$

Now, we estimate the fourth quarter sales ($x = 4$) to be

$$y = (20.3)(4) + 519.6 = 600.8 \text{ billion dollars}$$

(In this particular case, the estimate proves to be quite good. The actual fourth quarter sales in 1978 were 600.5 billion dollars.)

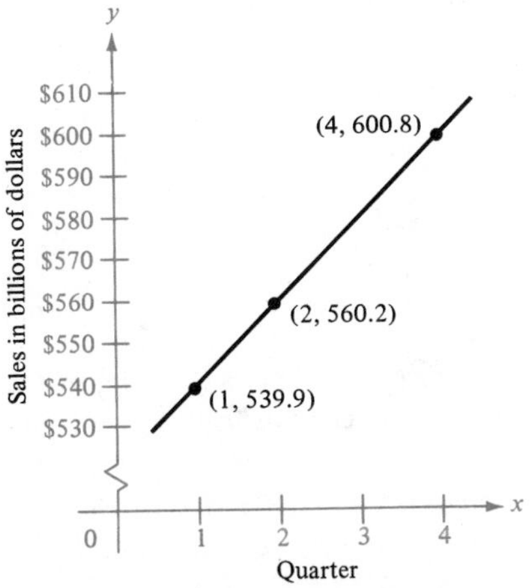

Total U.S. Sales in 1978

FIGURE 3.49

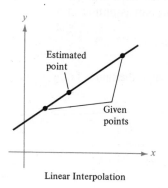

Linear Extrapolation

Linear Interpolation

FIGURE 3.50

The prediction method illustrated in Example 3 is called **linear extrapolation.** Note that the extrapolated point does not lie between the given points. (See Figure 3.50.) When the estimated point lies *between* two given points, we call the procedure **linear interpolation.**

Graphing Lines in the Plane

Many problems in analytic geometry can be classified in two basic categories:

1. Given a graph, what is its equation?
2. Given an equation, what is its graph?

We can use the point-slope and two-point equations for the first category—to find an equation of a line given its geometric description. However, these two forms *are not* particularly useful for the second category. The form that is best suited to graphing lines is called the **slope-intercept** form for the equation of a line.

THE SLOPE-INTERCEPT FORM FOR THE EQUATION OF A LINE

The graph of the equation

$$y = mx + b$$

is a line having a *slope* of m and a *y-intercept* at $(0, b)$.

Remark: Note that the linear function

$$f(x) = ax + b$$

is in slope-intercept form. Thus, the graph of a linear function is a line having a slope of a and a y-intercept at $(0, b)$.

EXAMPLE 4
Graphing Linear Equations

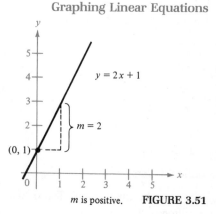

m is positive.   **FIGURE 3.51**

Sketch the graphs of the following linear equations:

(a) $y = 2x + 1$

(b) $y = 2$

(c) $3y + x - 6 = 0$

Solution:

(a) The y-intercept occurs at $b = 1$, and since the slope is 2 we know that this line rises 2 units for each unit the line moves to the right. (See Figure 3.51.)

(b) The y-intercept occurs at $b = 2$, and since the slope is zero we know that the line is horizontal. That is, it doesn't rise or fall. (See Figure 3.52.)

(c) We begin by writing the equation in slope-intercept form:

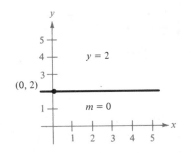

m is zero.

FIGURE 3.52

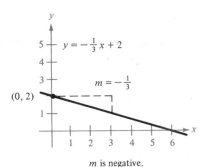

m is negative.

FIGURE 3.53

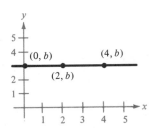

Horizontal Line: $y = b$

FIGURE 3.54

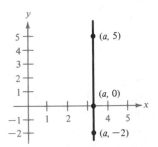

Vertical Line: $x = a$

FIGURE 3.55

$$3y + x - 6 = 0$$
$$3y = -x + 6$$
$$y = -\frac{1}{3}x + 2$$

Now we see that the *y*-intercept occurs at $b = 2$ and the slope is $m = -\frac{1}{3}$. Thus, this line falls 1 unit for every 3 units the line moves to the right. (See Figure 3.53.)

From the slope-intercept equation of a line, we can see that a horizontal line ($m = 0$) has an equation of the form

$$y = (0)x + b \quad \text{or} \quad y = b \qquad \textit{Horizontal line}$$

This is consistent with the fact that each point on a horizontal line through $(0, b)$ has a *y*-coordinate of *b*. (See Figure 3.54.)

Similarly, each point on a vertical line through $(a, 0)$ has an *x*-coordinate of *a*. (See Figure 3.55.) Hence a vertical line has an equation of the form

$$x = a \qquad \textit{Vertical line}$$

This equation cannot be written in the slope-intercept form, since the slope of a vertical line is undefined. However, *every* line has an equation that can be written in the **general form**

$$Ax + By + C = 0 \qquad \textit{General form}$$

where *A* and *B* are not *both* zero. If $A = 0$ (and $B \neq 0$), the equation can be reduced to the form $y = b$, which represents a horizontal line. If $B = 0$ (and $A \neq 0$), the general equation can be reduced to the form $x = a$, which represents a vertical line.

We now have identified the following six forms of equations of lines.

EQUATIONS OF LINES

General Equation:	$Ax + By + C = 0$
Vertical Line:	$x = a$
Horizontal Line:	$y = b$
Slope-Intercept Equation:	$y = mx + b$
Point-Slope Equation:	$y - y_1 = m(x - x_1)$
Two-Point Equation:	$y - y_1 = \dfrac{y_2 - y_1}{x_2 - x_1}(x - x_1)$

Parallel and Perpendicular Lines

The slope of a line is a convenient tool for determining when two lines are parallel or perpendicular. This is seen in the following two rules.

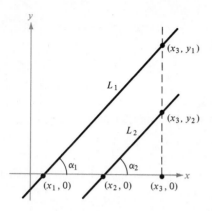

Parallel lines have
equal slopes.

FIGURE 3.56

PARALLEL LINES

> Two distinct nonvertical lines are parallel if and only if their slopes are
> equal.

Proof: Recall that the phrase "if and only if" is a way of stating two rules
in one. One rule says, "If two nonvertical lines are parallel, then they must
have equal slopes." The other rule is the converse, which says, "If two
distinct lines have equal slopes, they must be parallel." We will prove the
first of these two rules and leave the proof of the converse as an exercise
(see Exercise 59).

Assume that we have two parallel lines L_1 and L_2 with slopes m_1 and
m_2. If these lines are both horizontal, then $m_1 = m_2 = 0$, and the rule is
established. If L_1 and L_2 are not horizontal, then they must intersect the
x-axis at points $(x_1, 0)$ and $(x_2, 0)$, as shown in Figure 3.56. Since L_1 and
L_2 are parallel, their intersection with the x-axis must produce equal angles
α_1 and α_2. (α is the Greek lowercase letter alpha.) Therefore, the two right
triangles with vertices

$$(x_1, 0),\ (x_3, 0),\ (x_3, y_1) \qquad \text{and} \qquad (x_2, 0)\ (x_3, 0),\ (x_3, y_2)$$

must be similar. From this we conclude that the ratios of their corresponding
sides must be equal, and thus

$$m_1 = \frac{y_1 - 0}{x_3 - x_1} = \frac{y_2 - 0}{x_3 - x_2} = m_2$$

Hence the lines L_1 and L_2 must have equal slopes.

PERPENDICULAR LINES

> Two nonvertical lines are perpendicular if and only if their slopes are
> related by the equation
>
> $$m_1 = -\frac{1}{m_2}$$

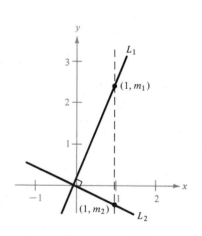

The slopes of perpendicular lines are negative
reciprocals of each other.

FIGURE 3.57

Proof: As in the previous rule, we will prove only one direction of the rule
and leave the other as an exercise (see Exercise 60). Let us assume that we
are given two nonvertical perpendicular lines L_1 and L_2 with slopes m_1 and
m_2. For simplicity's sake let these two lines intersect at the origin, as shown
in Figure 3.57. The vertical line $x = 1$ will intersect L_1 and L_2 at the
respective points $(1, m_1)$ and $(1, m_2)$. Since the triangle formed by these two
points and the origin is a right triangle, we can apply the Pythagorean Theorem and conclude that

$$\left(\begin{array}{c}\text{distance between}\\(0,0)\text{ and }(1,m_1)\end{array}\right)^2 + \left(\begin{array}{c}\text{distance between}\\(0,0)\text{ and }(1,m_2)\end{array}\right)^2 = \left(\begin{array}{c}\text{distance between}\\(1,m_1)\text{ and }(1,m_2)\end{array}\right)^2$$

Using the Distance Formula, we have

$$(\sqrt{1+m_1^2})^2 + (\sqrt{1+m_2^2})^2 = (\sqrt{0^2 + (m_1 - m_2)^2})^2$$

$$1 + m_1^2 + 1 + m_2^2 = (m_1 - m_2)^2$$

$$2 + m_1^2 + m_2^2 = m_1^2 - 2m_1 m_2 + m_2^2$$

$$2 = -2m_1 m_2$$

$$-1 = m_1 m_2$$

or

$$-\frac{1}{m_2} = m_1$$

EXAMPLE 5
Finding Parallel and
Perpendicular Lines

Find the equation of the line that passes through the point $(2, -1)$ and is

(a) parallel to the line $2x - 3y = 5$
(b) perpendicular to the line $2x - 3y = 5$

Solution:
Writing the equation $2x - 3y = 5$ in slope-intercept form, we have

$$3y = 2x - 5$$

or

$$y = \frac{2}{3}x - \frac{5}{3}$$

Therefore, the given line has a slope of $m = \frac{2}{3}$.

(a) Any line parallel to the given line must also have a slope of $\frac{2}{3}$. Thus, the line through $(2, -1)$ that is parallel to the given line has an equation of the form

$$y - (-1) = \frac{2}{3}(x - 2)$$

$$3(y + 1) = 2(x - 2)$$

$$3y + 3 = 2x - 4$$

$$-2x + 3y = -7$$

or

$$2x - 3y = 7$$

(Note the similarity to the original equation $2x - 3y = 5$.)

(b) Any line perpendicular to the given line must have a slope of $-\frac{3}{2}$. Therefore, the line through $(2, -1)$ that is perpendicular to the given line has the equation

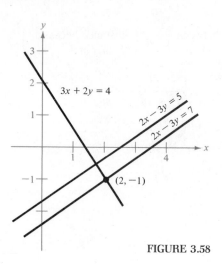

$$y - (-1) = -\frac{3}{2}(x - 2)$$

$$2(y + 1) = -3(x - 2)$$

$$3x + 2y = 4$$

See Figure 3.58.

FIGURE 3.58

Section Exercises 3.5

In Exercises 1–6, estimate the slope of the given line from its graph.

1.

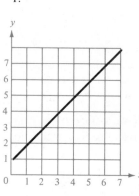

2.

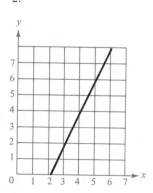

3.

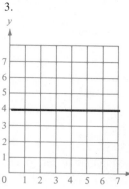

4.

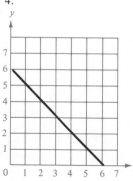

5.

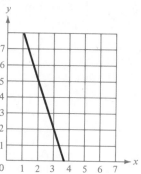

6.

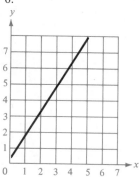

In Exercises 7–14, plot the points and find the slope of the line passing through each pair of points.

7. $(3, -4)$, $(5, 2)$
8. $(-2, 1)$, $(4, -3)$
9. $(\frac{1}{2}, 2)$, $(6, 2)$
10. $(-\frac{3}{2}, -5)$, $(2, -1)$
11. $(-6, -1)$, $(-6, 4)$
12. $(2, 1)$, $(2, 5)$
13. $(1, 2)$, $(-2, -2)$
14. $(\frac{7}{8}, \frac{3}{4})$, $(\frac{5}{4}, -\frac{1}{4})$

In Exercises 15–33, find an equation for the indicated line and sketch its graph.

15. through $(2, 1)$ and $(0, -3)$
16. through $(-3, -4)$ and $(1, 4)$
17. through $(0, 0)$ and $(-1, 3)$
18. through $(-3, 6)$ and $(1, 2)$
19. through $(2, 3)$ and $(2, -2)$

20. through $(6, 1)$ and $(10, 1)$
21. through $(1, -2)$ and $(3, -2)$
22. through $(\frac{7}{8}, \frac{3}{4})$ and $(\frac{5}{4}, -\frac{1}{4})$
23. through $(0, 3)$: $m = \frac{3}{4}$
24. through $(-1, 2)$: m is undefined
25. through $(0, 0)$: $m = \frac{2}{3}$
26. through $(-1, -4)$: $m = \frac{1}{4}$
27. through $(0, 5)$: $m = -2$
28. through $(-2, 4)$: $m = -\frac{3}{5}$
29. y-intercept at 2: $m = 4$
30. y-intercept at $-\frac{2}{3}$: $m = \frac{1}{6}$
31. y-intercept at $\frac{2}{3}$: $m = \frac{3}{4}$
32. y-intercept at 4: $m = 0$
33. vertical line with x-intercept at 3
34. Show that the line with intercepts $(a, 0)$ and $(0, b)$ has the following equation:

$$\frac{x}{a} + \frac{y}{b} = 1, \qquad a \neq 0, b \neq 0$$

In Exercises 35–42, use the result of Exercise 34 to write an equation of the indicated line.

35. x-intercept $(2, 0)$; y-intercept $(0, 3)$
36. x-intercept $(-3, 0)$; y-intercept $(0, 4)$
37. x-intercept $(-\frac{1}{6}, 0)$; y-intercept $(0, -\frac{2}{3})$
38. x-intercept $(-\frac{2}{3}, 0)$; y-intercept $(0, -2)$

39. Point on line $(1, 2)$
    x-intercept $(a, 0)$
    y-intercept $(0, a)$
    $a \neq 0$

40. Point on line $(-3, 4)$
    x-intercept $(a, 0)$
    y-intercept $(0, a)$
    $a \neq 0$

41. Point on line $(\frac{3}{2}, \frac{1}{2})$
    x-intercept $(2a, 0)$
    y-intercept $(0, a)$
    $a \neq 0$

42. Point on line $(-3, 1)$
    x-intercept $(a, 0)$
    y-intercept $(0, -a)$
    $a \neq 0$

In Exercises 43–48, write an equation of the line through the given point (a) parallel to the given line, and (b) perpendicular to the given line.

43. $(2, 1)$, $4x - 2y = 3$
44. $(-3, 2)$, $x + y = 7$
45. $(\frac{7}{8}, \frac{3}{4})$, $5x + 3y = 0$
46. $(-6, 4)$, $3x + 4y = 7$
47. $(2, 5)$, $x = 4$
48. $(-1, 0)$, $y = -3$

49. Find the equation of the line giving the relationship between the temperature in degrees Celsius, C, and degrees Fahrenheit, F. Use the fact that water freezes at $0°$ Celsius ($32°$ Fahrenheit) and boils at $100°$ Celsius ($212°$ Fahrenheit).

50. Use the result of Exercise 49 to complete the table.

C		$-10°$	$10°$			$177°$
F	$0°$			$68°$	$90°$	

51. A manufacturer pays its assembly line workers \$4.50 per hour, *plus* an additional piecework rate of \$0.75 per unit produced. Write a linear equation for the hourly wages W in terms of the number of units produced per hour x.

52. A small business purchases a piece of equipment for \$875. After five years the equipment will be outdated and have no value. Write a linear equation giving the value V of the equipment during the five years it will be used.

53. A company constructs a warehouse for \$825,000. It has an estimated useful life of 25 years, after which its value is expected to be \$75,000. If straight-line depreciation is used, write a linear equation giving the value V of the warehouse during its 25 years of useful life.

54. A real estate office handles an apartment complex with 50 units. When the rent is \$280 per month, all 50 units are occupied. However, when the rent is \$325 per month, the average number of occupied units drops to 47. Assume that the relationship between the monthly rent p and the demand x is linear.
    (a) Write the equation of the line giving the demand x in terms of the rent p.
    (b) (Linear Extrapolation) Use this equation to predict the number of units occupied if the rent is raised to \$355.
    (c) (Linear Interpolation) Predict the number of units occupied if the rent is lowered to \$295.

55. The amount (in billions of dollars) spent by the United States for energy imports between 1975 and 1978 is given in the following table:

Year	1975	1976	1977	1978
t	0	1	2	3
Imports: y	\$ 96	\$121	\$148	\$172

(a) Assuming an approximately linear relation between y and t, write an equation for the line passing through $(0, 96)$ and $(3, 172)$.
(b) (Linear Interpolation) Use this equation to estimate the amount spent in 1976 and 1977. Compare the estimate with the actual amount.
(c) (Linear Extrapolation) Predict the amount spent on energy imports in 1980.
(d) What information is given by the slope of the line in part (a)?

56. A particular brand of woodstove sells for \$739.40, and a cord of wood sells for \$105.00.
    (a) Write an equation giving the total cost, C, in terms of the number, x, of cords of wood purchased.

(b) Find the total cost of burning 6 cords of wood in this stove.

57. A contractor purchases a piece of equipment for $26,500. The equipment's operator is paid $9.50 per hour, and it uses an average of $5.25 per hour for fuel and maintenance.
    (a) Write a linear equation giving the total cost, C, of operating this equipment t hours.
    (b) If customers are charged $25 per hour of machine use, write an equation for the revenue, R, derived from t hours of use.
    (c) Use the formula for profit ($P = R - C$) to write an equation for the profit derived from t hours of use.
    (d) (Break-Even Point) Use the result of part (c) to find the number of hours this equipment must be used to yield a profit of 0 dollars.

58. A sales representative uses her own car as she travels for her company. The cost to the company is $75 per day for lodging and meals, plus $0.22 per mile driven. Write a linear equation giving the daily cost, C, to the company in terms of x, the number of miles driven.

59. Complete the proof of the theorem concerning the slopes of parallel lines by proving the following: If two distinct lines have equal slopes, they must be parallel.

60. Complete the proof of the theorem concerning the slopes of perpendicular lines by proving the following: If two nonvertical lines have slopes that are negative reciprocals, they must be perpendicular.

Composite and Inverse Functions   3.6

Two functions can be combined in various ways to create new functions. For example, if

$$f(x) = 2x - 3 \quad \text{and} \quad g(x) = x^2 - 1$$

we can form the functions

$$f(x) + g(x) = (2x - 3) + (x^2 - 1) = x^2 + 2x - 4 \qquad \textit{(Sum)}$$
$$f(x) - g(x) = (2x - 3) - (x^2 - 1) = -x^2 + 2x - 2 \qquad \textit{(Difference)}$$
$$f(x)g(x) = (2x - 3)(x^2 - 1) = 2x^3 - 3x^2 - 2x + 3 \qquad \textit{(Product)}$$
$$\frac{f(x)}{g(x)} = \frac{2x - 3}{x^2 - 1}, \quad x \neq \pm 1 \qquad \textit{(Quotient)}$$

These *combination* functions may have domains that differ from those of the original functions. For instance, the domain of $f(x)/g(x)$ has the restriction $x \neq \pm 1$, which does not occur in the original functions.

We can combine two functions in yet another way, called the **composition** of two functions.

DEFINITION OF COMPOSITE FUNCTION

Let f and g be functions such that the range of g is in the domain of f. Then the function whose values are given by $f(g(x))$ is called the **composite** of f with g.

Remark: Some texts use the notation $(f \circ g)(x)$ to denote the composite function $f(g(x))$.

It is important to realize that the composite of f with g may not be equal to the composite of g with f. This is illustrated in the following example.

EXAMPLE 1
Forming Composite Functions

Given $f(x) = 2x - 3$ and $g(x) = x^2 - 1$, find

(a) $f(g(x))$ (b) $g(f(x))$

Solution:

(a) Since $f(x) = 2(x) - 3$

   we have

$$f(g(x)) = 2(g(x)) - 3$$
$$= 2(x^2 - 1) - 3 = 2x^2 - 5$$

(b) Since $g(x) = (x)^2 - 1$

   we have

$$g(f(x)) = (f(x))^2 - 1 = (2x - 3)^2 - 1$$
$$= 4x^2 - 12x + 8$$

Although the compositions formed in Example 1 look fairly straightforward, this procedure cannot be performed haphazardly. That is, you must take care to see that the domains and ranges of f and g have a proper fit with each other.

TABLE 3.2
Forming Composite Functions

	$g(x) = -x^2$	$f(x) = \sqrt{x + 1}$	$f(g(x)) = \sqrt{-x^2 + 1}$
Domain	$[-1, 1]$	$[-1, \infty)$	$[-1, 1]$
Range	$[-1, 0]$	$[0, \infty)$	$[0, 1]$

Remark: The domain of g is restricted in order to make the range of g lie within the domain of f.

	$g(x) = \dfrac{1}{x^2 + 1}$	$f(x) = -\sqrt{x}$	$f(g(x)) = -\sqrt{\dfrac{1}{x^2 + 1}}$
Domain	$(-\infty, \infty)$	$[0, \infty)$	$(-\infty, \infty)$
Range	$(0, 1]$	$(-\infty, 0]$	$[-1, 0)$

Remark: The range of g lies within the domain of f.

Note in Example 1 that $f(g(x)) \neq g(f(x))$ and, in general, these two composite functions are *not* equal. An important case in which they are equal occurs when

$$f(g(x)) = g(f(x)) = x$$

We call such functions **inverses** of each other, as stated in the following definition.

DEFINITION OF INVERSE FUNCTIONS

Two functions f and g are **inverses** of each other if

$\quad\quad f(g(x)) = x \quad\quad$ for each x in the domain of g

and

$\quad\quad g(f(x)) = x \quad\quad$ for each x in the domain of f

We denote g by f^{-1} (read "f inverse").

Remark: For inverse functions f and g, the range of g must be equal to the domain of f, and vice versa. Also note that whenever we write $f^{-1}(x)$, we will *always* be referring to the inverse of the function f and *not* to the reciprocal of f.

EXAMPLE 2
Demonstrating Inverse Functions

Show that the following functions are inverses of each other:

$$f(x) = 2x^3 - 1 \quad\quad \text{and} \quad\quad g(x) = \sqrt[3]{\frac{x + 1}{2}}$$

Solution:
First, note that both composite functions exist, since the domain and range of both f and g consist of the set of all real numbers. The composite of f with g is

$$f(g(x)) = 2\left(\sqrt[3]{\frac{x + 1}{2}}\right)^3 - 1$$

$$= 2\left(\frac{x + 1}{2}\right) - 1 = x + 1 - 1 = x$$

The composite of g with f is

$$g(f(x)) = \sqrt[3]{\frac{(2x^3 - 1) + 1}{2}}$$

$$= \sqrt[3]{\frac{2x^3}{2}} = \sqrt[3]{x^3} = x$$

Since $f(g(x)) = g(f(x)) = x$, f and g are inverses of each other.

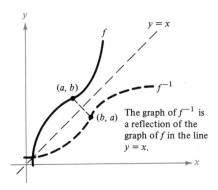

FIGURE 3.59

The following rule suggests a geometrical interpretation of inverse functions.

GRAPHS OF INVERSE FUNCTIONS

The graph of f contains the point (a, b) if and only if the graph of f^{-1} contains the point (b, a).

The rule can be interpreted geometrically to mean that the graph of f^{-1} can be obtained by reflecting the graph of f in the line $y = x$ (Figure 3.59). This geometric property suggests an *algebraic* procedure for finding the inverse of a function. Since (a, b) lies on the graph of f if and only if (b, a) lies on the graph of f^{-1}, we can find the inverse of a function by algebraically interchanging the roles of x and y.

EXAMPLE 3

Finding the Inverse of a Function

Find the inverse of the function given by $f(x) = \sqrt{2x - 3}$.

Solution:
Substituting y for $f(x)$, we have $y = \sqrt{2x - 3}$. Now to find the inverse function, we simply solve for x in terms of y. Since y is nonnegative, squaring both sides gives an equivalent equation:

$$\sqrt{2x - 3} = y \qquad \text{Given } y \text{ as a function of } x$$

$$2x - 3 = y^2$$

$$2x = y^2 + 3$$

$$x = \frac{y^2 + 3}{2} \qquad \text{Solve for } x \text{ as a function of } y$$

$$f^{-1}(y) = \frac{(y)^2 + 3}{2} \qquad \text{Write in functional notation}$$

$$f^{-1}(x) = \frac{x^2 + 3}{2} \qquad \text{Use } x \text{ as independent variable}$$

Now that we have a formula for f^{-1}, we find its domain by finding the range of f.

Range of f: $[0, \infty)$ ⟹   Domain of f^{-1}: $[0, \infty)$

Finally, we conclude that the inverse function is

$$f^{-1}(x) = \frac{x^2 + 3}{2}, \qquad x \geq 0$$

Note the reflective property of these two graphs in Figure 3.60.

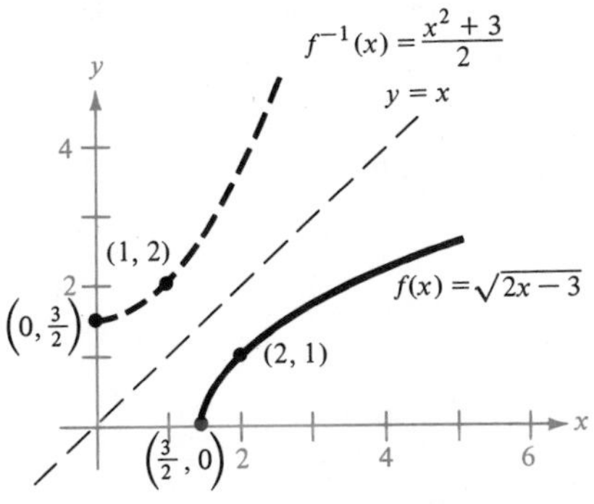

FIGURE 3.60

Remark: To avoid being confused by the interplay of variables in Example 3, remember that the independent variable is really a "dummy variable" which serves merely as a placeholder. Thus,

$$f^{-1}(y) = \frac{y^2 + 3}{2}, \quad f^{-1}(a) = \frac{a^2 + 3}{2}, \quad \text{and} \quad f^{-1}(x) = \frac{x^2 + 3}{2}$$

all represent the same function.

The following summary should help you remember the basic steps for finding the inverse of a function.

FINDING THE INVERSE OF A FUNCTION

To find the inverse of f:

1. Write the function in the form $\qquad\qquad$ $y = f(x)$
2. If possible, solve for x in terms of y: $\qquad$ $x = g(y)$
3. Interchange x and y: $\qquad\qquad\qquad$ $y = g(x)$
4. Check to see that $\qquad\qquad$ Domain of f = Range of g
   and $\qquad\qquad\qquad\qquad\qquad$ Range of f = Domain of g
5. The inverse of f is $\qquad\qquad\qquad$ $f^{-1}(x) = g(x)$

Not all functions possess an inverse. *For a function f to have an inverse it is necessary that f be one-to-one.* A function is **one-to-one** if no two elements in its domain correspond to the same element in its range.

Horizontal Line Test

Graphically, we can see that a function $y = f(x)$ is not one-to-one if a horizontal line intersects the graph of the function more than once. For instance, in Figure 3.61, the line $y = 3$ intersects the graph of the function twice, and hence the function is not one-to-one.

EXAMPLE 4
A Function That Has No Inverse

Find the inverse (if it exists) of the function given by

$$f(x) = x^2 - 1$$

(Assume the domain of f is the set of all real numbers.)

Solution:
We note that

$$f(2) = (2)^2 - 1 = 3$$

and $\qquad f(-2) = (-2)^2 - 1 = 3$

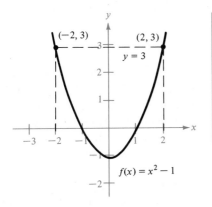

f is not one-to-one
and has no inverse.

FIGURE 3.61

Thus, f is not one-to-one, and it has no inverse. This same conclusion can be obtained by substituting y for $f(x)$ and solving for x as follows:

$$x^2 - 1 = y$$
$$x^2 = y + 1$$
$$x = \pm\sqrt{y + 1}$$

This last equation does not define x as a function of y, and thus f has no inverse. (See Figure 3.61.)

As you work through the exercises for this section, develop the habit of checking the range and domain of each function. The problem of finding an inverse function can be difficult (or even impossible) for two reasons. First, given $y = f(x)$, it may be algebraically difficult to solve for x in terms of y. Second, if f is not one-to-one, then f^{-1} does not exist.

In later chapters we will study two important classes of inverse functions: logarithmic functions and inverse trigonometric functions.

Section Exercises 3.6

In Exercises 1–8, find (a) $f(x) + g(x)$, (b) $f(x) - g(x)$, (c) $f(x) \cdot g(x)$, (d) $f(x)/g(x)$, (e) $f(g(x))$, if defined, and (f) $g(f(x))$, if defined.

1. $f(x) = x + 1$, $\quad g(x) = x - 1$
2. $f(x) = 2x - 5$, $\quad g(x) = 1 - x$
3. $f(x) = x^2$, $\quad g(x) = 1 - x$
4. $f(x) = 2x - 5$, $\quad g(x) = 5$
5. $f(x) = x^2 + 5$, $\quad g(x) = \sqrt{1 - x}$
6. $f(x) = \sqrt{x^2 - 4}$, $\quad g(x) = x^2/(x^2 + 1)$
7. $f(x) = 1/x$, $\quad g(x) = 1/x^2$
8. $f(x) = x/(x + 1)$, $\quad g(x) = x^3$

In Exercises 9–16, (a) show that f and g are inverse functions by showing that $f(g(x)) = x$ and $g(f(x)) = x$, and (b) graph f and g on the same set of coordinate axes.

9. $f(x) = x^3$, $\qquad\qquad g(x) = \sqrt[3]{x}$
10. $f(x) = 1/x$, $\qquad\qquad g(x) = 1/x$
11. $f(x) = 5x + 1$, $\qquad\quad g(x) = (x - 1)/5$
12. $f(x) = 3 - 4x$, $\qquad\quad g(x) = (3 - x)/4$
13. $f(x) = \sqrt{x - 4}$, $\qquad\; g(x) = x^2 + 4, x \geq 0$
14. $f(x) = 9 - x^2, x \geq 0$, $\quad g(x) = \sqrt{9 - x}, x \leq 9$
15. $f(x) = 1 - x^3$, $\qquad\quad g(x) = \sqrt[3]{1 - x}$

16. $f(x) = 1/(1 + x^2)$, $x \geq 0$, $\quad g(x) = \sqrt{(1 - x)/x}$, $0 < x \leq 1$

In Exercises 17–28, find the inverse of f. Then graph both f and f^{-1}.

17. $f(x) = 2x - 3$ $\qquad\qquad$ 18. $f(x) = 3x$
19. $f(x) = x^5$ $\qquad\qquad\quad$ 20. $f(x) = x^3 + 1$
21. $f(x) = \sqrt{x}$ $\qquad\qquad\;$ 22. $f(x) = x^2, x \geq 0$
23. $f(x) = \sqrt{4 - x^2}, 0 \leq x \leq 2$
24. $f(x) = \sqrt{x^2 - 4}, x \geq 2$
25. $f(x) = \sqrt[3]{x - 1}$ $\qquad\quad$ 26. $f(x) = 3\sqrt[5]{2x - 1}$
27. $f(x) = x^{2/3}, x \geq 0$ $\qquad$ 28. $f(x) = x^{3/5}$

In Exercises 29–40, determine if the given function is one-to-one and if so find its inverse.

29. $f(x) = ax + b, a \neq 0$ $\qquad$ 30. $f(x) = \sqrt{x - 2}$
31. $f(x) = x^2$ $\qquad\qquad\qquad$ 32. $f(x) = x^4$
33. $f(x) = x/\sqrt{x^2 + 5}$ $\qquad$ 34. $f(x) = |x - 2|$
35. $f(x) = 3$ $\qquad\qquad\qquad$ 36. $f(x) = x\sqrt{2 - x}$
37. $f(x) = 1/x$ $\qquad\qquad\quad$ 38. $f(x) = (3x + 4)/5$
39. $f(x) = \sqrt{2x + 3}$ $\qquad\;$ 40. $f(x) = x^2/(x^2 + 1)$

Variation and Mathematical Models

3.7

One of the goals of applied mathematics is to find equations that describe real-world phenomena. We call such equations **mathematical models,** and they arise in two basic ways: experimentally and theoretically. For models that arise experimentally, we usually have a number of *x*- and *y*-values and then we try to *fit a curve* (or line) to these points. We have already encountered a simple form of curve fitting in Section 3.5, when we looked at a method of finding the equation of a line passing through two points. We will encounter another example of curve fitting in Chapter 10.

Models can also arise through a theoretical (or verbal) description. We have already done some work with setting up "word problems." In this section we will study models related to the concept of **variation.**

DEFINITION OF DIRECT VARIATION

> The following statements are equivalent:
>
> 1. *y* **varies directly** as *x*.
> 2. *y* is **directly proportional** to *x*.
> 3. $y = kx$ for some constant k.
>
> We call k the **constant of variation** or the **constant of proportionality.**

In setting up a mathematical model described by some type of variation, we typically must be *given* specific values of *x* and *y* which then enable us to solve for the constant *k*.

EXAMPLE 1
Direct Variation

Hooke's Law for a spring states that the distance a spring stretches (or compresses) varies directly as the force on the spring. A force of 20 pounds stretches the spring 4 inches.

(a) Write an equation relating the distance stretched to the force applied.
(b) How far will a force of 30 pounds stretch the spring?

Solution:

(a) We begin by letting

d = distance spring is stretched (in inches)
F = force (in pounds)

Since we are given that d varies directly as F, we have the model

$$d = kF$$

Since $d = 4$ when $F = 20$, we have

$$4 = k(20) \qquad \Longrightarrow \qquad \frac{1}{5} = k$$

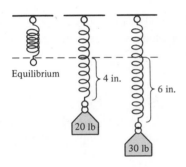

Equilibrium

4 in.

6 in.

20 lb

30 lb

FIGURE 3.62

Thus, the equation relating distance and force is

$$d = \frac{1}{5}F$$

(b) When $F = 30$,

$$d = \frac{1}{5}F = \frac{1}{5}(30) = 6 \text{ inches}$$

See Figure 3.62.

A second type of direct variation relates one variable to a *power* of another variable.

DEFINITION OF DIRECT VARIATION AS nTH POWER

The following statements are equivalent:

1. y **varies directly as the nth power** of x.
2. y is **directly proportional to the nth power** of x.
3. $y = kx^n$ for some constant k.

EXAMPLE 2
Direct Variation as a Power

A ball begins to roll down an inclined plane. The distance the ball rolls is directly proportional to the square of the time. During the first second the ball rolls 8 feet.

(a) Write an equation relating the distance traveled to the time.
(b) How far will the ball roll during the first 3 seconds?

Solution:

(a) If we let d be the distance rolled by the ball (in feet) and t be the time (in seconds), then we have the model

$$d = kt^2$$

Now, since $d = 8$ when $t = 1$, we can see that $k = 8$. Thus, the equation relating distance to time is

$$d = 8t^2$$

(b) When $t = 3$,

$$d = 8(3^2) = 8(9) = 72 \text{ feet}$$

See Figure 3.63.

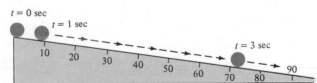

$t = 0$ sec
$t = 1$ sec
$t = 3$ sec

10 20 30 40 50 60 70 80 90

FIGURE 3.63

In Examples 1 and 2 we looked at models involving *direct* variation, in which an increase in one variable corresponds to an increase in the other variable. For example, in the model

$$d = \frac{1}{5}F, \qquad F > 0$$

an increase in F results in an increase in d. However, you should not assume that this always occurs with direct variation. For example, in the model

$$y = -3x$$

an increase in x results in a decrease in y, and yet we still say that y is directly proportional to x.

A third type of variation is called **inverse variation.**

DEFINITION OF INVERSE VARIATION

The following statements are equivalent:

1. y **varies inversely** as x.
2. y is **inversely proportional** to x.
3. $y = k/x$ for some constant k.

Remark: If x and y are related by an equation of the form

$$y = \frac{k}{x^n}$$

then we say that y varies inversely as the nth power of x (or y is inversely proportional to the nth power of x).

EXAMPLE 3
Inverse Variation

A gas law states that the volume of an enclosed gas varies directly as the temperature and inversely as the pressure. The pressure of a particular gas is 25 pounds per square inch when the temperature is $70°K$ and the volume is 500 cubic inches.

(a) Write an equation relating the pressure, temperature, and volume of this gas.
(b) Find the pressure when the temperature is $80°K$ and the volume is 400 cubic inches.

Solution:

(a) Let

$$V = \text{volume (in cubic inches)}$$
$$P = \text{pressure (in pounds per square inch)}$$
$$T = \text{temperature (in degrees Kelvin)}$$

Since we are given that V is directly proportional to T and inversely proportional to P, we have the model

$$V = \frac{kT}{P}$$

Note that the same constant of proportionality can be used for the direct proportion of T and the inverse proportion of P. Now, since $P = 25$ when $T = 70$ and $V = 500$,

$$500 = \frac{k(70)}{25}$$

$$\frac{500(25)}{70} = k$$

$$\frac{1250}{7} = k$$

Thus, the equation relating pressure, temperature, and volume is

$$V = \frac{1250}{7}\left(\frac{T}{P}\right)$$

(b) When $T = 80$ and $V = 400$,

$$P = \frac{1250}{7}\left(\frac{80}{400}\right) = \frac{250}{7} \approx 35.71 \text{ pounds per square inch}$$

Remark: Note that both direct and inverse variations are being described in the statement of Example 3: "Volume varies directly as the temperature *and* inversely as the pressure." In describing a direct and inverse variation in the same statement, we couple them with the word "and." Such statements require careful grammatical construction in order to diminish the possibility of confusion.

When we want to describe two different types of *direct* variation in the same statement, we use the word **jointly.**

DEFINITION OF JOINT VARIATION

The following statements are equivalent:
1. z **varies jointly** as x and y.
2. z is **jointly proportional** to x and y.
3. $z = kxy$ for some constant k.

Remark: If x, y, and z are related by an equation of the form

$$z = kx^n y^m$$

then we say that z varies jointly as the nth power of x and the mth power of y.

EXAMPLE 4
Joint Variation

The *simple* interest for a certain savings account is jointly proportional to the time and the principal. After one quarter (three months) the interest on a principal of $5,000 is $106.25.

(a) Write an equation relating the interest, principal, and time.
(b) Find the interest after three quarters.

Solution:

(a) Let I = interest (in dollars), P = principal (in dollars), and t = time (in years). Then, since we are given that I is jointly proportional to P and t, we have the model

$$I = kPt$$

Now, since $I = 106.25$ when $P = 5000$ and $t = 1/4$, we have

$$106.25 = k(5000)\left(\frac{1}{4}\right)$$

$$k = \frac{4(106.25)}{5000} = 0.085$$

Thus, the equation relating interest, principal, and time is

$$I = 0.085Pt$$

Note that this is the familiar equation for simple interest, and the constant of proportionality represents the annual percentage rate (which is 8.5% in this case).

(b) When $P = \$5,000$ and $t = 3/4$,

$$I = (0.085)(5000)\left(\frac{3}{4}\right) = \$318.75$$

Section Exercises 3.7

In Exercises 1–14, give the equation relating the variables and find the constant of proportionality.

1. y varies directly as x, and $y = 25$ when $x = 10$.
2. y is directly proportional to x, and $y = 8$ when $x = 24$.
3. s is directly proportional to the square of t, and $s = 64$ when $t = 2$.
4. A varies directly as the square of r, and $A = 9\pi$ when $r = 3$.

5. y varies inversely as x, and $y = 3$ when $x = 25$.
6. y is inversely proportional to x, and $y = 7$ when $x = 4$.
7. h is inversely proportional to the third power of t, and $h = \frac{3}{16}$ when $t = 4$.
8. R varies inversely as the square of s, and $R = 80$ when $s = \frac{1}{5}$.
9. z varies jointly as x and y, and $z = 64$ when $x = 4$ and $y = 8$.

10. z is jointly proportional to x and y, and $z = 32$ when $x = 10$ and $y = 16$.

11. F is jointly proportional to r and the third power of s, and $F = 4158$ when $r = 11$ and $s = 3$.

12. P varies directly as x and inversely as the square of y, and $P = \frac{28}{3}$ when $x = 42$ and $y = 9$.

13. z varies directly as the square of x and inversely as y, and $z = 6$ when $x = 6$ and $y = 4$.

14. v varies jointly as p and q and inversely as the square of s, and $v = 1.5$ when $p = 4.1$, $q = 6.3$, and $s = 1.2$.

15. Hooke's Law for a spring states that the distance a spring stretches (or compresses) varies directly as the force on the spring. A force of 50 pounds stretches a spring 5 inches.
    (a) How far will a force of 20 pounds stretch the spring?
    (b) What force is required to stretch the spring 1.5 inches?

16. Repeat Exercise 15 assuming that a force of 50 pounds stretches the spring 3 inches.

17. The maximum size (diameter) of a particle that can be moved by a stream varies as the square of the stream's velocity. Suppose that a stream with a velocity of 0.25 miles per hour can move sand particles with diameters up to 0.02 inches. What must the velocity be to carry particles with a diameter of 0.12 inches?

18. A stream of velocity v can move particles of diameter d or less. By what factor does d increase if the velocity is doubled? (See Exercise 17.)

19. The distance s that an object falls varies directly as the square of the time t that it has been falling. If an object falls a distance of 144 feet in 3 seconds, how far will it have fallen after 5 seconds?

20. A company has found that the demand for its product varies inversely as the price of the product. If there is a demand for 500 units when the price is \$3.75, approximate the demand at a price of \$4.25.

21. The amount of illumination from a light source varies inversely as the square of the distance from the source. If the distance from a light source is doubled, what happens to the amount of illumination?

22. The resistance of a wire carrying electrical current is directly proportional to its length and inversely proportional to its cross-sectional area. It is known that number 28 (diameter of 0.0126 inch) copper wire has a resistance of 66.17 ohms per thousand feet. If a resistance of 33.5 ohms is needed, what length of number 28 copper wire is required?

23. Use the information of Exercise 22 to determine the diameter of copper wire if a circuit with a length of 14 feet cannot have a resistance in excess of 0.05 ohm.

24. A horizontal beam supported at both ends can safely support an evenly distributed load. The weight of the load varies jointly as the width of the beam and the square of its depth, and inversely as the distance between the end supports. Determine what happens to the safe load if
    (a) the width and length of the beam are doubled
    (b) the width and the depth of the beam are doubled
    (c) all three of the dimensions are doubled
    (d) the depth of the beam is halved

Review Exercises / Chapter 3 ————————

In Exercises 1–10, find (a) the distance between the two points, (b) the coordinates of the midpoint of the line segment between the two points, (c) an equation of the line through the two points, and (d) an equation of the circle whose diameter is the line segment between the two points.

1. $(0, 0)$, $(6, 0)$
2. $(0, 0)$, $(0, 10)$
3. $(-2, -1)$, $(2, 2)$
4. $(-1, 4)$, $(2, 0)$
5. $(2, 1)$, $(14, 6)$
6. $(-2, 2)$, $(3, -10)$
7. $(5, -2)$, $(5, 6)$
8. $(-3, -4)$, $(-3, 5)$
9. $(-1, 0)$, $(6, 2)$
10. $(1, 6)$, $(4, 2)$

11. Determine the value of t so that the points $(-2, 5)$, $(0, t)$, and $(1, 1)$ are on the same line.

12. Determine the value of t so that the points $(-6, 1)$, $(1, t)$, and $(10, 5)$ are on the same line.

13. Show that the points $(1, 1)$, $(8, 2)$, $(9, 5)$, and $(2, 4)$ are vertices of a parallelogram.

14. Show that the points $(4, 5)$, $(-2, 4)$, and $(3, -1)$ are the vertices of an isosceles triangle.

15. (a) Determine the center and radius of the circle

$$4x^2 + 4y^2 - 4x - 40y + 92 = 0$$

    (b) Sketch the graph of the circle.

16. (a) Determine the center and radius of the circle

$$x^2 + y^2 - 20x - 10y + 100 = 0$$

    (b) Sketch the graph of the circle.

In Exercises 17–30, sketch the graph of the equation.

17. $y - 2x - 3 = 0$
18. $3x + 2y + 6 = 0$

19. $x - 5 = 0$

20. $y + 4 = 0$

21. $y = |x - 3|$

22. $y = 8 - |x|$

23. $y = \sqrt{5 - x}$

24. $y = x^2 - 4x$

25. $y = x^3 - x$

26. $y + 2x^2 = 0$

27. $y^2 - x + 3 = 0$

28. $x^2 + y^2 = 0$

29. $y = \sqrt{25 - x^2}$

30. $y = \sqrt{x + 2}$

31. Determine the domain and range of the function $f(x) = \sqrt{25 - x^2}$.

32. Determine the domain and range of the function $h(t) = |t + 1|$.

33. If $f(x) = x^2 + 1$, find
    (a) $f(2)$ (b) $f(-4)$ (c) $f(t^2)$
    (d) $-f(x)$ (e) $f(x + \Delta x)$

34. Given the functions $f(x) = 1/x$ and $g(x) = x^2 + 1$, find
    (a) $f[g(x)]$ (b) $g[f(x)]$

In Exercises 35–40, (a) find the inverse, f^{-1}, of the given func-

tion, (b) sketch the graphs of $f(x)$ and $f^{-1}(x)$ on the same axes, and (c) verify that $f^{-1}[f(x)] = f[f^{-1}(x)] = x$.

35. $f(x) = \frac{1}{2}x - 3$

36. $f(x) = 5x - 7$

37. $f(x) = \sqrt{x + 1}$

38. $f(x) = x^3 + 2$

39. $f(x) = x^2 - 5, x \geq 0$

40. $f(x) = \sqrt[3]{x + 1}$

41. The function $f(x) = 2(x - 4)^2$ does not have an inverse. Give a restriction on the domain of f so that the restricted function has an inverse, and then find the inverse.

42. Repeat Exercise 41 for the function $f(x) = |x - 2|$.

In Exercises 43 and 44, give the equation relating the variables and find the constant of proportionality.

43. z varies directly as the square of x and inversely as y, and $z = 16$ when $x = 5$ and $y = 2$.

44. w varies jointly as x and y and inversely as the third power of z, and $w = \frac{44}{9}$ when $x = 12$, $y = 11$, and $z = 6$.

4 POLYNOMIAL FUNCTIONS: GRAPHS AND ZEROS

Quadratic Functions and Their Graphs

4.1

In the first two sections of this chapter we will be studying polynomial functions. They are the most widely used functions in algebra, and you would be wise to become familiar with their equations and graphs.

DEFINITION OF A POLYNOMIAL FUNCTION

The function

$$f(x) = a_n x^n + a_{n-1} x^{n-1} + \cdots + a_2 x^2 + a_1 x + a_0$$

is called a **polynomial function of degree n.** The numbers a_i are called **coefficients,** with a_n the **leading coefficient** and a_0 the **constant term** of the polynomial function (n is a nonnegative integer).

In Section 3.5 we studied first degree polynomial functions of the form $f(x) = ax + b$ (linear functions), and we saw that their graphs are straight lines. In this section we will look at second degree polynomial functions of the form $f(x) = ax^2 + bx + c$.

DEFINITION OF A QUADRATIC FUNCTION

The function

$$f(x) = ax^2 + bx + c, \qquad a \neq 0$$

is called a **quadratic function,** and its graph is called a **parabola.**

Remark: It is common practice to use the subscript notation for coefficients of general polynomial functions, but for specific polynomial functions (of low degree) we generally use the simpler a, b, c, etc., form for the coefficients.

Oth Degree: $f(x) = a$ *Constant function*

1st Degree: $f(x) = ax + b$ *Linear function*

2nd Degree: $f(x) = ax^2 + bx + c$ *Quadratic function*

3rd Degree: $f(x) = ax^3 + bx^2 + cx + d$ *Cubic function*

The simplest type of quadratic function is $f(x) = ax^2$, in which b and c are both zero. We call this a **monomial** quadratic function because it has only one term. The graph of $f(x) = ax^2$ is a parabola that opens upward if $a > 0$ and downward if $a < 0$ (see Figure 4.1).

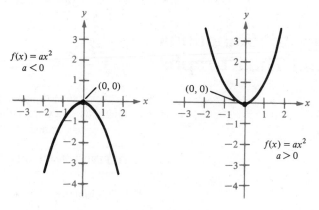

FIGURE 4.1

On the graph of $f(x) = ax^2$, we call the point $(0, 0)$ the **vertex** of the parabola. If $a > 0$, the vertex is a minimum point. If $a < 0$, the vertex represents a maximum point on the parabola.

EXAMPLE 1
Graphing Monomial
Quadratic Functions

Sketch the graph of

(a) $f(x) = x^2$ (b) $f(x) = \frac{1}{2}x^2$ (c) $f(x) = 2x^2$

Solution:

(a) We begin by plotting a few points on the graph.

x	-2	-1	0	1	2
$y = x^2$	4	1	0	1	4

Then we connect these points by a smooth curve, as shown in Figure 4.2.

(b) See Figure 4.3.
(c) See Figure 4.4.

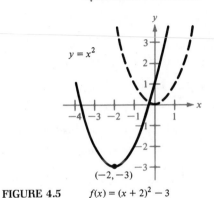

FIGURE 4.2

$f(x) = x^2$

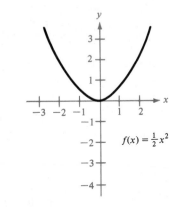

FIGURE 4.3

$f(x) = \frac{1}{2}x^2$

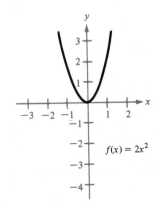

FIGURE 4.4

$f(x) = 2x^2$

Remark: Note that in Example 1 the graph of a monomial quadratic function ($y = ax^2$) has *symmetry with respect to the y-axis*. Also note that the coefficient a determines how widely the parabola opens. If $|a|$ is small, the parabola opens more widely than if $|a|$ is large.

We can use our knowledge of the basic shape of the graph of $f(x) = ax^2$ to graph any quadratic function by using the techniques developed in Section 3.4. You may wish to review Example 4 in Section 3.4.

EXAMPLE 2
Graphing General
Quadratic Functions

Sketch the graph of

$$f(x) = (x + 2)^2 - 3$$

Solution:
The basic graph we use to sketch this quadratic function is $y = x^2$. The graph of $f(x) = (x + 2)^2 - 3$ represents a horizontal shift of 2 units to the left and a vertical shift of 3 units down, as shown in Figure 4.5.

Remark: Note that the graph in Example 2 is symmetric with respect to the vertical line passing through the vertex of the parabola. We call this line the **axis** of the parabola.

The equation in Example 2 is written in **standard form** $f(x) = a(x - h)^2 + k$, which is especially convenient for sketching.

FIGURE 4.5 $f(x) = (x + 2)^2 - 3$

$y = x^2$

$(-2, -3)$

STANDARD FORM OF THE EQUATION OF A PARABOLA
(Vertical Axis)

The quadratic function

$$f(x) = a(x - h)^2 + k, \qquad a \neq 0$$

is said to be in **standard form.** The axis is given by the vertical line $x = h$, and the vertex occurs at the point (h, k). If $a > 0$, the parabola opens upward, and if $a < 0$, the parabola opens downward.

To write a quadratic function in standard form we use the procedure called *completing the square,* discussed in Section 2.3.

EXAMPLE 3
Writing a Quadratic Function
in Standard Form

Sketch the graph of the following quadratics by first writing each one in standard form.

(a) $f(x) = 2x^2 + 8x + 7$ (b) $f(x) = -x^2 + 6x - 8$

Solution:

(a) Since the coefficient of x^2 is different from 1, it is convenient to group the x terms and factor a 2 out of each x term before completing the square.

$$
\begin{aligned}
f(x) &= 2x^2 + 8x + 7 && \textit{Given} \\
&= 2(x^2 + 4x) + 7 && \textit{Factor 2 out of x terms} \\
&= 2(x^2 + 4x + 4 - 4) + 7 && \textit{Add and subtract 4} \\
& \qquad\qquad\quad \underset{(2)^2}{\underline{\qquad}} \\
&= 2(x^2 + 4x + 4) - 2(4) + 7 && \textit{Regroup terms} \\
&= 2(x^2 + 4x + 4) - 8 + 7 && \textit{Simplify} \\
&= 2(x + 2)^2 - 1 && \textit{Standard form}
\end{aligned}
$$

Now we see that this graph can be obtained from the graph of $y = 2x^2$ by shifting 2 units to the left and 1 unit down (Figure 4.6).

(b) In this case, we factor -1 out of the x terms.

$$
\begin{aligned}
f(x) &= -x^2 + 6x - 8 && \textit{Given} \\
&= -(x^2 - 6x) - 8 && \textit{Factor } -1 \textit{ out of x terms} \\
&= -(x^2 - 6x + 9 - 9) - 8 && \textit{Add and subtract 9} \\
& \qquad\qquad\quad \underset{(3)^2}{\underline{\qquad}}
\end{aligned}
$$

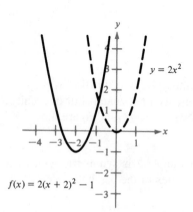

$y = 2x^2$

$f(x) = 2(x + 2)^2 - 1$

FIGURE 4.6

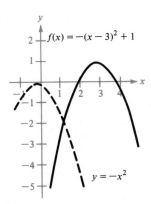

FIGURE 4.7

$$= -(x^2 - 6x + 9) - (-9) - 8 \quad \textit{Regroup terms}$$
$$= -(x^2 - 6x + 9) + 9 - 8 \quad \textit{Simplify}$$
$$= -(x - 3)^2 + 1 \quad\quad\quad \textit{Standard form}$$

Therefore, the graph can be obtained from the graph of $y = -x^2$ by shifting 3 units to the right and 1 unit up, as shown in Figure 4.7.

In addition to finding the vertex of the graph of a quadratic function, it is often useful to find the x- and y-intercepts. The graph of

$$y = ax^2 + bx + c, \quad a \neq 0$$

has a y-intercept at $(0, c)$. The x-intercepts can be determined by solving the quadratic equation $ax^2 + bx + c = 0$.

EXAMPLE 4
Finding Intercepts
of a Quadratic Function

Sketch the graphs of

(a) $f(x) = x^2 - 2x - 2$ (b) $f(x) = x^2 - 6x + 9$
(c) $f(x) = x^2 + 2x + 2$

and label the x- and y-intercepts.

Solution:

(a) The y-intercept occurs when $x = 0$:

$$y = f(0) = 0^2 - 2(0) - 2 = -2$$

The x-intercepts (if any) occur when $y = 0$:

$$y = x^2 - 2x - 2 = 0$$

By the quadratic formula, we have

$$x = \frac{-(-2) \pm \sqrt{(-2)^2 - 4(1)(-2)}}{2(1)}$$

$$= \frac{2 \pm \sqrt{12}}{2} = \frac{2 \pm 2\sqrt{3}}{2}$$

$$= 1 \pm \sqrt{3}$$

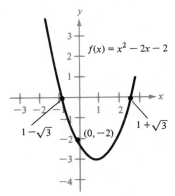

FIGURE 4.8

See Figure 4.8.

(b) The y-intercept occurs at $(0, 9)$ and the x-intercept occurs when

$$x^2 - 6x + 9 = 0$$

or

$$(x - 3)^2 = 0$$

which implies that there is only one x-intercept at $(3, 0)$, as shown in Figure 4.9.

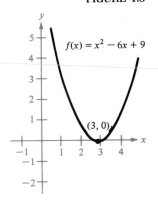

FIGURE 4.9

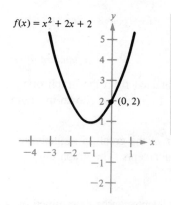

$f(x) = x^2 + 2x + 2$

FIGURE 4.10

(c) The y-intercept occurs at $(0, 2)$. When we try to find the x-intercepts of this function, we obtain

$$x^2 + 2x + 2 = 0$$

and since the discriminant

$$b^2 - 4ac = 2^2 - 4(1)(2) = 4 - 8 = -4$$

we see that there are no x-intercepts, as shown in Figure 4.10.

The graphs in Example 4 illustrate the three possible cases for the occurrence of x-intercepts for quadratic functions.

INTERCEPTS OF A PARABOLA

The graph of $f(x) = ax^2 + bx + c$ has
1. *two* x-intercepts if $b^2 - 4ac > 0$
2. *one* x-intercept if $b^2 - 4ac = 0$
3. *no* x-intercept if $b^2 - 4ac < 0$
The y-intercept is the point $(0, c)$.

Vertex and Axis of a Quadratic Function

If a parabola has two x-intercepts, then its axis is a vertical line halfway between its x-intercepts. Thus, by averaging the x-intercepts

$$x = \frac{-b + \sqrt{b^2 - 4ac}}{2a} \qquad \text{and} \qquad x = \frac{-b - \sqrt{b^2 - 4ac}}{2a}$$

we can obtain the midpoint

$$\frac{\dfrac{-b + \sqrt{b^2 - 4ac}}{2a} + \dfrac{-b - \sqrt{b^2 - 4ac}}{2a}}{2} = \frac{-(2b/2a)}{2} = -\frac{b}{2a}$$

Consequently, the graph of $f(x) = ax^2 + bx + c$ has a vertical axis at

$$x = -\frac{b}{2a}$$

and a vertex at

$$\left(-\frac{b}{2a}, f\left(-\frac{b}{2a} \right) \right)$$

This provides us with a way to find the vertex of the graph of $f(x) = ax^2 + bx + c$ without writing the equation in standard form. The formula is also valid for one or no x-intercept.

EXAMPLE 5
Finding the Vertex and Extreme
Values for a Quadratic Function

Find the vertex for the graph of

$$f(x) = 20 - 8x + 3x^2$$

and determine the minimum (or maximum) value of $f(x)$.

Solution:

By rewriting the function as

$$f(x) = 3x^2 - 8x + 20$$

we see that $a = 3$, $b = -8$, and $c = 20$. Thus, the x-coordinate of the vertex is

$$x = -\frac{b}{2a} = \frac{-(-8)}{2(3)} = \frac{4}{3}$$

and the y-coordinate is

$$f\left(\frac{4}{3}\right) = 3\left(\frac{16}{9}\right) - 8\left(\frac{4}{3}\right) + 20 = \frac{16}{3} - \frac{32}{3} + \frac{60}{3} = \frac{44}{3}$$

Therefore, the vertex is the point

$$\left(\frac{4}{3}, \frac{44}{3}\right)$$

Since $a = 3$ ($a > 0$), the graph opens upward and it follows that

$$f\left(\frac{4}{3}\right) = \frac{44}{3}$$

is the *minimum* value of $f(x)$.

EXAMPLE 6
An Application: Finding the
Maximum Value and Intercept

An object is propelled straight up into the air with an initial velocity of 32 feet per second (beginning at a height of 6 feet). The height at any time t is given by

$$s(t) = -16t^2 + 32t + 6$$

where $s(t)$ is measured in feet and t is measured in seconds.

(a) Find the maximum height attained by the object before it begins falling back to the ground.
(b) When does the object hit the ground?

Solution:

(a) The maximum height of the object occurs at the vertex of the graph of s. We note that $a = -16$, $b = 32$, and $c = 6$. Thus, the value of t at the vertex is

$$t = -\frac{b}{2a} = \frac{-32}{2(-16)} = 1$$

and the value of $s(t)$ is

$$s(1) = -16 + 32 + 6 = 22$$

Now since the vertex occurs at the point $(1, 22)$, we conclude that the maximum height is 22 feet.

(b) Note in part (a) that the object reached its maximum height after one second. To find the time that it hits the ground, we set $s(t) = 0$ and solve for t:

$$s(t) = -16t^2 + 32t + 6 = 0$$

Using the quadratic formula, we have

$$t = \frac{-(32) \pm \sqrt{(32)^2 - 4(-16)(6)}}{2(-16)}$$

$$= \frac{-32 \pm \sqrt{1408}}{-32}$$

$$= \frac{-32 \pm 8\sqrt{22}}{-32} = \frac{4 \pm \sqrt{22}}{4}$$

Finally, choosing the positive value of t, we find that the object hits the ground when

$$t = \frac{4 + \sqrt{22}}{4} \approx 2.17 \text{ seconds}$$

Section Exercises 4.1

In Exercises 1–10, use the basic graphs of $y = ax^2$, as given in Figure 4.1, and the concept of horizontal and vertical translations to match the quadratic function with the correct graph.

1. $f(x) = (x - 3)^2$
2. $f(x) = (x + 5)^2$
3. $f(x) = x^2 - 4$
4. $f(x) = x^2 + 1$
5. $f(x) = 5 - x^2$
6. $f(x) = 4 - (x - 1)^2$
7. $f(x) = (x + 2)^2 - 2$
8. $f(x) = 3x^2 - 5$
9. $f(x) = 9 - 2x^2$
10. $f(x) = 2(x - 1)^2 + 1$

(c)
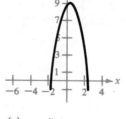

(d)
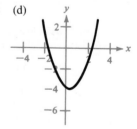

(a)

(b)
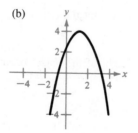

(e)
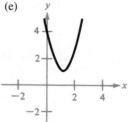

(f)
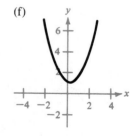

(g)
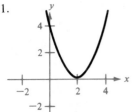

(h)
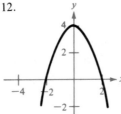

(i)
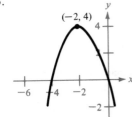

(j)
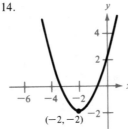

In Exercises 11–16, use the given graph to write (a) the standard form, and (b) the general form of the equation of the parabola.

11.
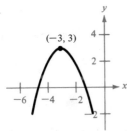

12.
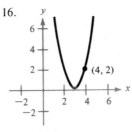

13.
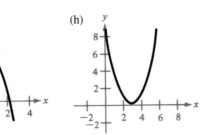
(−2, 4)

14.
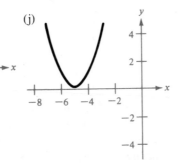
(−2, −2)

15.
(−3, 3)

16.
(4, 2)

In Exercises 17–36, sketch the graph of the quadratic function and identify the vertex and the intercepts.

17. $f(x) = x^2 - 5$
18. $f(x) = x^2 + 4$
19. $f(x) = 3x^2 + 1$
20. $f(x) = \frac{1}{2}x^2 - 4$
21. $f(x) = (x + 5)^2 - 6$
22. $f(x) = (x - 6)^2 + 3$
23. $f(x) = 16 - x^2$
24. $f(x) = 25 - x^2$
25. $f(x) = x^2 - 8x + 16$
26. $f(x) = x^2 + 2x + 1$
27. $f(x) = x^2 + 2x - 3$
28. $f(x) = x^2 - 4x - 1$
29. $f(x) = x^2 - 16x + 54$
30. $f(x) = x^2 + 8x + 11$
31. $f(x) = x^2 - x + \frac{5}{4}$
32. $f(x) = x^2 + 3x + \frac{1}{4}$
33. $f(x) = -x^2 + 2x + 5$
34. $f(x) = -x^2 - 4x + 1$
35. $f(x) = 4x^2 - 4x + 21$
36. $f(x) = 2x^2 - x + 1$

37. A rectangle with a perimeter of 100 inches has a width and length of x and $50 - x$ inches, respectively. Find x so that the area of the rectangle is maximum $[A = x(50 - x)]$.

38. Find the number of units x that produce a maximum revenue R if $R = 900x - 0.1x^2$.

39. A manufacturer of lighting fixtures has daily production costs of $C = 800 - 10x + 0.25x^2$. How many fixtures x should be produced each day to minimize costs?

40. Let x be the amount (in hundreds of dollars) a company spends on advertising, and let P be the profit. If $P = 230 + 20x - 0.5x^2$, what amount of advertising gives the maximum profit?

41. A manufacturer of radios charges $90 per unit when the average production cost per unit is $60. However, to encourage large orders from distributors, the manufacturer will reduce the charge by $0.10 per unit for each unit ordered in excess of 100 (for example, there would be a charge of $88 per radio for an order of 120 radios). Find the largest order size the manufacturer should allow so as to realize maximum profit.

42. Find the quadratic function that has a maximum at (3, 4) and passes through the point (1, 2).

43. Find the quadratic function that has a minimum at (5, 12) and passes through the point (7, 15).

44. Find two quadratic functions (one opening upward and the other downward) that have x-intercepts $(-1, 0)$ and $(3, 0)$.

45. Find two quadratic functions (one opening upward and the other downward) that have x-intercepts $(-2.5, 0)$ and $(2, 0)$.

Graphs of Polynomial Functions of Higher Degree

4.2

At this point you should be able to sketch an accurate graph of polynomial functions of degree 0, 1, or 2.

Degree	Function	Graph
0th Degree	$f(x) = a$	Horizontal line
1st Degree	$f(x) = ax + b$	Line of slope a
2nd Degree	$f(x) = ax^2 + bx + c$	Parabola

The graphs of polynomial functions of degree higher than 2 are more difficult to classify and to sketch. However, in this section you will begin to recognize some of the basic characteristics of these polynomial graphs. We continue to use the standard graphing techniques of point-plotting, finding x- and y-intercepts, and testing for symmetry about the y-axis or the origin.

Continuity

A polynomial function is continuous.* This means that the graph of a polynomial function has no breaks or gaps. (See Figure 4.11.)

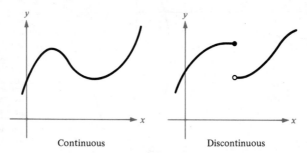

Continuous   Discontinuous

FIGURE 4.11

Smooth Curve

The graph of a polynomial function has only smooth rounded turns, as indicated in Figure 4.12.

*Continuous functions are dealt with much more completely in calculus.

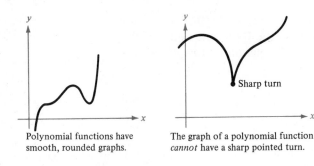

FIGURE 4.12

Polynomial functions have smooth, rounded graphs.

The graph of a polynomial function *cannot* have a sharp pointed turn.

The Leading Coefficient Test

Although the graph of a polynomial function can have several turns, eventually the graph will rise or fall without bound as x moves to the right or left. Symbolically, we write

$$f(x) \to \infty \qquad \text{as} \qquad x \to \infty$$

to mean that $f(x)$ increases without bound as x moves to the right without bound. (∞ is the mathematical symbol for infinity.) Whether the graph of

$$f(x) = a_n x^n + a_{n-1} x^{n-1} + \cdots + a_2 x^2 + a_1 x + a_0$$

eventually rises or falls can be determined by the function's degree (odd or even) and by the leading coefficient, as indicated in Figures 4.13 and 4.14.

Note that the dashed portions of the graphs in Figures 4.13 and 4.14 indicate that the leading coefficient test determines *only* the right and left behavior of the graph.

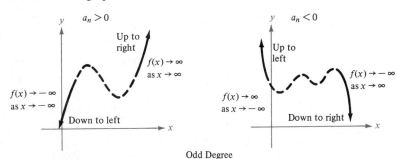

FIGURE 4.13

Odd Degree

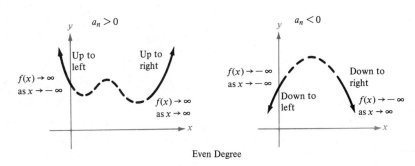

FIGURE 4.14

Even Degree

EXAMPLE 1
Determining the Right and Left Behavior of a Polynomial Graph

Use the Leading Coefficient Test to determine the right and left behavior of the graphs of the following polynomial functions:

(a) $f(x) = 3x^3 - x^2 + 4$ (b) $f(x) = -x^4 + 2x - 1$
(c) $f(x) = -2x^5 + 4x^4 + 2$

Solution:

(a) Since the degree is odd and the leading coefficient is positive, the graph moves up to the right and down to the left, as shown in Figure 4.15.

(b) Since the degree is even and the leading coefficient is negative, the graph moves down to the right and left, as shown in Figure 4.16.

(c) Since the degree is odd and the leading coefficient is negative, the graph moves down to the right and up to the left, as shown in Figure 4.17.

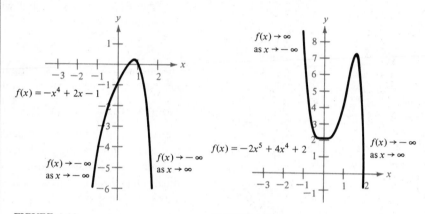

FIGURE 4.15   **FIGURE 4.16**   **FIGURE 4.17**

One of the most important problems in algebra is determining the *zeros* of polynomial functions. There is a strong interplay between graphical and algebraic approaches to this problem. Sometimes we use information about the graph of a function to help us find its zeros, and in other problems we use information about the zeros of a function to help us sketch its graph.

REAL ZEROS OF POLYNOMIAL FUNCTIONS

If f is a polynomial function and a is a real number, then the following statements are equivalent:

1. $x = a$ is a *zero* of the function.
2. $x = a$ is a *root* of the polynomial equation $f(x) = 0$.
3. $(x - a)$ is a *factor* of the polynomial $f(x)$.
4. $(a, 0)$ is an *x-intercept* of the graph of the function.

Graphing a polynomial function is often easier if we are able to factor the polynomial.

EXAMPLE 2

Finding Zeros of Polynomial Functions by Factoring

Find all real zeros of the following polynomial functions:

(a) $f(x) = x^3 - x^2 - 2x$ (b) $f(x) = -2x^4 + 2x^2$

(c) $f(x) = (-x^5 + 3x^3 + 4x)/4$

Solution:

(a) $f(x) = x^3 - x^2 - 2x = x(x^2 - x - 2)$

$$= x(x - 2)(x + 1)$$

Thus, the real zeros are $x = 0$, $x = 2$, and $x = -1$. From Figure 4.18, we see that $(0, 0)$, $(2, 0)$, and $(-1, 0)$ are the x-intercepts of the graph of this function.

(b) $f(x) = -2x^4 + 2x^2 = -2x^2(x^2 - 1)$

$$= -2x^2(x - 1)(x + 1)$$

Thus, the real zeros are $x = 0$, $x = 1$, and $x = -1$. From Figure 4.19, we see that $(0, 0)$, $(1, 0)$, and $(-1, 0)$ are the x-intercepts of the graph of this function.

(c) $f(x) = \frac{1}{4}(-x^5 + 3x^3 + 4x) = -\frac{1}{4}x(x^4 - 3x^2 - 4)$

$$= -\frac{1}{4}x(x^2 - 4)(x^2 + 1)$$

$$= -\frac{1}{4}x(x - 2)(x + 2)(x^2 + 1)$$

Thus, the real zeros are $x = 0$, $x = 2$, and $x = -2$. From Figure 4.20, we see that $(0, 0)$, $(2, 0)$, and $(-2, 0)$ are the x-intercepts of the graph of this function. Note that the quadratic factor $(x^2 + 1)$ has no real roots and consequently produces no real zeros of the function.

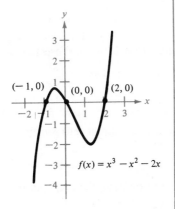

FIGURE 4.18

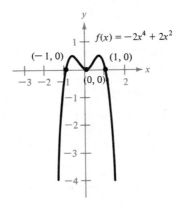

FIGURE 4.19

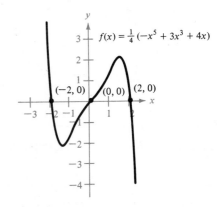

FIGURE 4.20

Remark: Note that in Example 2 each of the graphs happened to have three *x*-intercepts. An important result in algebra states that *the number of real zeros of a polynomial function cannot exceed its degree.* This result is related to the Fundamental Theorem of Algebra, to be discussed in Section 4.5. For now, we will use this important result and postpone its proof.

NUMBER OF REAL ZEROS OF A POLYNOMIAL FUNCTION

> A polynomial function of degree n has at most n real zeros.

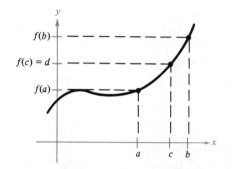

If d lies between $f(a)$ and $f(b)$ then there exists c between a and b such that $f(c) = d$.

FIGURE 4.21
Intermediate Value Theorem

Note that this theorem tells us nothing about how we might *find* the zeros, or even if any real zeros exist. In fact, many polynomial functions have no real zeros. For example, the function $f(x) = x^2 + 1$ has no real zeros.

The next theorem gives us information about the existence of real zeros of a polynomial function. It is called the Intermediate Value Theorem for Polynomial Functions, and it follows from the fact that polynomial functions are continuous. The theorem indicates that if $(a, f(a))$ and $(b, f(b))$ are two points on the graph of a polynomial such that $f(a) \neq f(b)$, then for any number d between $f(a)$ and $f(b)$ there must be a number c between a and b such that $f(c) = d$. See Figure 4.21.

INTERMEDIATE VALUE THEOREM

> If f is a polynomial function such that $a < b$ and $f(a) \neq f(b)$, then f takes on every value between $f(a)$ and $f(b)$ in the interval $[a, b]$.

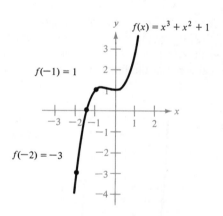

f has a zero between -2 and -1

FIGURE 4.22

This theorem helps us locate the real zeros of polynomial functions. Specifically, if we can find a value $x = a$ where a polynomial function is positive, and another value $x = b$ where it is negative, then we can conclude that the function has at least one real zero between these two values. For example, the function

$$f(x) = x^3 + x^2 + 1$$

is negative when $x = -2$ and positive when $x = -1$. It follows from the Intermediate Value Theorem that f must have a real zero somewhere between -2 and -1, as shown in Figure 4.22.

EXAMPLE 3
Graphically Estimating Zeros of Polynomial Functions

Use the Intermediate Value Theorem to estimate the real zero of

$$f(x) = x^3 - x^2 + 1$$

to the nearest tenth.

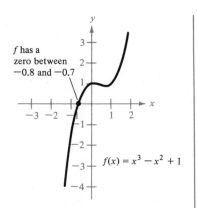

f has a
zero between
−0.8 and −0.7

$f(x) = x^3 - x^2 + 1$

FIGURE 4.23

Solution:
We begin by computing a few functional values as follows:

x	-2	-1	0	1
$f(x)$	-11	-1	1	1

Now since $f(-1)$ is negative and $f(0)$ is positive, we may conclude that the function has a zero between -1 and 0. To pinpoint this zero even further, we divide the interval $[-1, 0]$ into tenths and evaluate the function at each subdivision. Now we see that one real zero lies between -0.8 and -0.7. This is a tedious procedure, but we could continue it to estimate this zero to any desired accuracy. The graph of this function is shown in Figure 4.23.

x	-1	-0.9	-0.8	-0.7	-0.6	-0.5	-0.4	-0.3	-0.2	-0.1	0
$f(x)$	-1	-0.539	-0.152	0.167	0.424	0.625	0.776	0.883	0.952	0.989	1

In the last two examples in this section, we will demonstrate the use of the various sketching aids we have developed up to this point.

EXAMPLE 4
Sketching the Graph of a Polynomial Function

Sketch the graph of $f(x) = x^4 + 1$.

Solution:
Since the leading coefficient is positive and the degree is even, we know that the graph moves up to the right and left. Also, by setting $f(x)$ equal to zero we see that there are no x-intercepts, since $x^4 = -1$ has no real solutions. By plotting the y-intercept $(0, 1)$, we have the rough sketch shown in Figure 4.24.

To complete the sketch, we observe that the graph has symmetry with respect to the y-axis, and plot additional points, as shown in Figure 4.25.

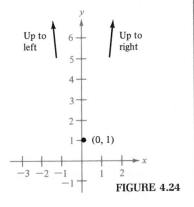

Up to left   Up to right

$(0, 1)$

FIGURE 4.24

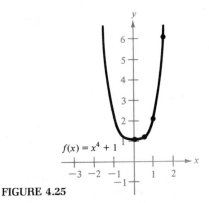

$f(x) = x^4 + 1$

FIGURE 4.25

x	0.5	1	1.5
$f(x)$	1.0625	2	6.0625

EXAMPLE 5
Sketching the Graph
of a Polynomial Function

Sketch the graph of

$$f(x) = 3x^4 - 4x^3$$

Solution:

Since the leading coefficient is positive and the degree is even, we know that the graph moves up to the right and left. Also, by setting $f(x)$ equal to zero, we have

$$f(x) = 3x^4 - 4x^3 = 0$$
$$x^3(3x - 4) = 0$$

Thus, the x-intercepts occur at $(0, 0)$ and $(\frac{4}{3}, 0)$, as shown in Figure 4.26.

To complete the sketch, we plot a few additional points, as shown in Figure 4.27.*

x	-1	0.5	1	1.5
$f(x)$	7	-0.3125	-1	1.6875

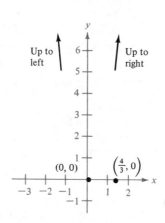

FIGURE 4.26

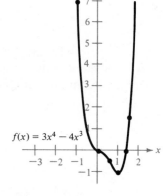

$f(x) = 3x^4 - 4x^3$

FIGURE 4.27

Section Exercises 4.2

In Exercises 1–10, use the Leading Coefficient Test or the number of real zeros of a polynomial function to match the polynomial functions with the correct graph.

1. $f(x) = -3x + 5$
2. $f(x) = 2x - 3$
3. $f(x) = x^2 - 2x$
4. $f(x) = -2x^2 - 8x - 9$
5. $f(x) = 2x^3 - 3x^2 - 12x + 8$
6. $f(x) = -\frac{1}{3}x^3 + x - \frac{2}{3}$
7. $f(x) = -\frac{1}{4}x^4 + 2x^2$
8. $f(x) = 3x^4 + 4x^3$
9. $f(x) = x^5 - 5x^3 + 4x$
10. $f(x) = x^3 - 1$

*In calculus you will learn how to find and then verify that a point such as $(1, -1)$ is a minimum point.

(a)
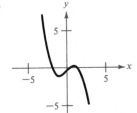

(b)
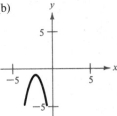

(c)
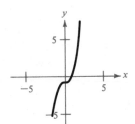

(d)
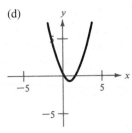

(e)
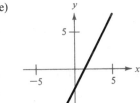

(f)
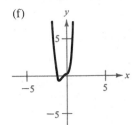

(g)
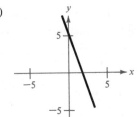

(h)
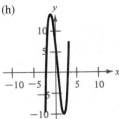

(i)
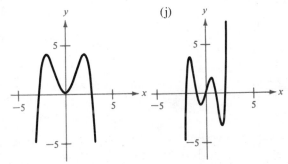

(j)

In Exercises 11–20, use the Leading Coefficient Test to determine the right- and left-hand behavior of the graph of the polynomial function.

11. $f(x) = 2x^2 - 3x + 1$
12. $f(x) = \frac{1}{3}x^3 + 5x$
13. $f(x) = 5 - \frac{7}{2}x - 3x^2$
14. $f(x) = -2.1x^5 + 4x^3 - 2$
15. $f(x) = 2x^5 - 5x + 7.5$
16. $f(x) = 1 - x^6$
17. $f(x) = 6 - 2x + 4x^2 - 5x^3$
18. $f(x) = \dfrac{3x^4 - 2x + 5}{4}$
19. $f(x) = -\frac{2}{3}(x^2 - 5x + 3)$
20. $f(x) = -\frac{7}{8}(x^3 + 5x^2 - 7x + 1)$

In Exercises 21–35, find all the real zeros of the polynomial function.

21. $f(x) = x^2 - 25$ 22. $f(x) = 49 - x^2$
23. $f(x) = x^2 - 6x + 9$ 24. $f(x) = x^2 + 10x + 25$
25. $f(x) = x^2 + x - 2$ 26. $f(x) = \frac{1}{2}(x^2 + 5x - 3)$
27. $f(x) = 3(x^2 - 4x + 1)$ 28. $f(x) = 5(x^2 - 2x - 1)$
29. $f(x) = x^3 - 4x^2 + 4x$ 30. $f(x) = x^4 - x^3 - 20x^2$
31. $f(x) = \frac{1}{2}(x^4 - 1)$ 32. $f(x) = x^5 + x^3 - 6x$
33. $f(x) = 2(x^4 - x^2 - 20)$ 34. $f(x) = x^5 - 6x^3 + 9x$
35. $f(x) = 5(x^4 + 3x^2 + 2)$

In Exercises 36–44, find a polynomial function that has the given zeros.

36. $0, -3$ 37. $0, 10$ 38. $-4, 5$
39. $2, -6$ 40. $0, 2, 5$ 41. $0, -2, -3$
42. $1, -\sqrt{2}, \sqrt{2}$ 43. $4, -3, 3, 0$ 44. $-2, -1, 0, 1, 2$

45. (a) Sketch the graph of $f(x) = x^3$.
    (b) Using the graph of part (a) as a model, sketch the graph of
       (i) $f(x) = (x - 2)^3$ (ii) $f(x) = x^3 - 2$
       (iii) $f(x) = (x - 2)^3 - 2$ (iv) $f(x) = -\frac{1}{2}x^3$

46. (a) Sketch the graph of $f(x) = x^4$.
    (b) Using the graph of part (a) as a model, sketch the graph of
       (i) $f(x) = (x + 3)^4$ (ii) $f(x) = x^4 - 3$
       (iii) $f(x) = 4 - x^4$ (iv) $f(x) = \frac{1}{4}(x - 1)^4$

In Exercises 47–56, sketch the graph of the polynomial function using the sketching aids of this section.

47. $f(x) = x^3 - 3x^2$ 48. $f(x) = 1 - x^5$
49. $f(x) = x^3 - 4x$ 50. $f(x) = -(x^4 - 2x^3)$
51. $f(x) = x^4 - 2x^3 + x^2$ 52. $f(x) = \frac{1}{4}x^4 - 2x^2$
53. $f(x) = 12x - x^3$ 54. $f(x) = (x + 1)^5 - 2$
55. $f(x) = x^7 - 2$ 56. $f(x) = 1 - x^6$

In Exercises 57–60, follow the procedure given in Example 3 to estimate the zero of $f(x)$ in the given interval $[a, b]$. (List your approximation to the nearest tenth.)

57. $f(x) = x^3 + x - 1,\ [0, 1]$

58. $f(x) = x^5 + x + 1,\ [-1, 0]$
59. $f(x) = x^4 - 10x^2 - 11,\ [3, 4]$
60. $f(x) = -x^3 + 3x^2 + 9x - 2,\ [4, 5]$

Polynomial Division and Synthetic Division

4.3

Up to this point in our study we have added, subtracted, and multiplied polynomials. In this section, we look at a procedure for dividing polynomials. This procedure has many important applications and is especially valuable in factoring and finding the zeros of polynomial functions.

Suppose that you sketched the graph of

$$f(x) = 6x^3 - 19x^2 + 16x - 4$$

As shown in Figure 4.28, a zero occurs near $x = 2$. You could verify this by evaluating the function at $x = 2$.

$$f(2) = 6(2^3) - 19(2^2) + 16(2) - 4$$
$$= 48 - 76 + 32 - 4 = 0$$

Since $x = 2$ is a zero of the polynomial function, we know that $(x - 2)$ is a factor of $f(x)$. This means that there exists a second degree polynomial $q(x)$ such that

$$f(x) = (x - 2) \cdot q(x)$$

To find $q(x)$, we can make use of a process called **long division of polynomials,** which is patterned after the long division for real numbers. As with real numbers, the algorithm makes repeated use of the pattern

partial quotient $\Rightarrow$ multiply $\Rightarrow$ subtract

Study the following example to see how this pattern is used.

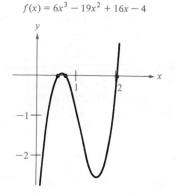

$f(x) = 6x^3 - 19x^2 + 16x - 4$

FIGURE 4.28

EXAMPLE 1
Long Division of Polynomials

Divide the polynomial $f(x) = 6x^3 - 19x^2 + 16x - 4$ by $(x - 2)$, and use the result to factor $f(x)$ completely.

Solution:

Partial quotients

$$
\require{enclose}
\begin{array}{r}
6x^2 - 7x + 2 \\
x - 2 \enclose{longdiv}{6x^3 - 19x^2 + 16x - 4} \\
\underline{6x^3 - 12x^2} \\
-\ 7x^2 + 16x \\
\underline{-\ 7x^2 + 14x} \\
2x - 4 \\
\underline{2x - 4} \\
0
\end{array}
$$

Multiply $6x^2(x - 2)$
Subtract
Multiply $-7x(x - 2)$
Subtract
Multiply $2(x - 2)$
Subtract

Now we see that

$$6x^3 - 19x^2 + 16x - 4 = (x - 2)(6x^2 - 7x + 2)$$

and by factoring the quadratic, we have

$$6x^3 - 19x^2 + 16x - 4 = (x - 2)(2x - 1)(3x - 2)$$

Note that this factorization agrees with the graph of f (Figure 4.28) in that the three x-intercepts occur at

$$x = 2, \qquad x = \frac{1}{2}, \qquad \text{and} \qquad x = \frac{2}{3}$$

The process of long division is (theoretically) summarized in a well known mathematical theorem called the **Division Algorithm.**

THE DIVISION ALGORITHM

If $f(x)$ and $d(x)$ are polynomials such that $d(x) \neq 0$, and the degree of $d(x)$ is less than (or equal to) the degree of $f(x)$, then there exist unique polynomials $q(x)$ and $r(x)$ such that

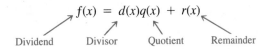

$$f(x) = d(x)q(x) + r(x)$$

Dividend   Divisor   Quotient   Remainder

where either $r(x) = 0$ or the degree of $r(x)$ is less than the degree of $d(x)$.

Remark: We can write a polynomial division in either

| *Division Tableau Form* | or | *Fractional Form* |

$$d(x)\overline{)f(x)}^{\,q(x)} \qquad\qquad \frac{f(x)}{d(x)} = q(x) + \frac{r(x)}{d(x)}$$

$$\begin{array}{c} \cdot \\ \cdot \cdot \\ \cdot \\ \overline{r(x)} \end{array}$$

where the rational expression $f(x)/d(x)$ is an **improper fraction** if the degree of the numerator is greater than (or equal to) the degree of the denominator. Similarly, we say that the rational expression $r(x)/d(x)$ is **proper** if the degree of the numerator is less than the degree of the denominator.

EXAMPLE 2
Long Division of Polynomials

Perform the indicated divisions.

(a) $\dfrac{3x - 1}{x + 2}$ \qquad\qquad\qquad (b) $(x^3 - 1) \div (x - 1)$

(c) $x^2 + 2x - 3\overline{)2x^4 + 4x^3 - 5x^2 + 2x - 2}$

Solution:

(a) This fraction is improper because the numerator and denominator are of the same degree. Since

$$
\begin{array}{r}
3 \\
x + 2\overline{)3x - 1} \\
\underline{3x + 6} \\
-7
\end{array}
$$

we have

$$
\frac{3x - 1}{x + 2} = 3 - \frac{7}{x + 2}
$$

(b) Because there is no x^2-term or x-term in the dividend, we line up the subtraction by using zero coefficients (or leaving a space) for these missing terms.

$$
\begin{array}{r}
x^2 + x + 1 \\
x - 1\overline{)x^3 + 0x^2 + 0x - 1} \\
\underline{x^3 - x^2} \\
x^2 \\
\underline{x^2 - x} \\
x - 1 \\
\underline{x - 1} \\
0
\end{array}
$$

(c)
$$
\begin{array}{r}
2x^2 + 1 \\
x^2 + 2x - 3\overline{)2x^4 + 4x^3 - 5x^2 + 2x - 2} \\
\underline{2x^4 + 4x^3 - 6x^2} \\
x^2 + 2x - 2 \\
\underline{x^2 + 2x - 3} \\
1
\end{array}
$$

Note that the first subtraction eliminated two terms from the dividend. When this happens, the quotient will contain a missing term. Thus, we write the quotient as

$$
\frac{2x^4 + 4x^3 - 5x^2 + 2x - 2}{x^2 + 2x - 3} = 2x^2 + 1 + \frac{1}{x^2 + 2x - 3}
$$

Synthetic Division

If the divisor is of the form $(x - a)$, there is a nice shortcut for this special case of long division by linear factors. The shortcut is called **synthetic division.** To see how synthetic division works, we take another look at Example 1.

$$
\begin{array}{r}
6x^2 - 7x + 2 \\
x - 2\overline{)6x^3 - 19x^2 + 16x - 4} \\
\underline{6x^3 - 12x^2} \\
-7x^2 + 16x \\
\underline{-7x^2 + 14x} \\
2x - 4 \\
\underline{2x - 4} \\
0
\end{array}
$$

We can retain the essential steps in this method by using only the coefficients, as follows:

$$
\begin{array}{r}
6 \quad -7 \quad\; 2 \\
-2\overline{)6 \quad -19 \quad 16 \quad -4} \\
\underline{6 \quad -12} \\
-7 \quad 16 \\
\underline{-7 \quad 14} \\
2 \quad -4 \\
\underline{2 \quad -4} \\
0
\end{array}
$$

Since the coefficients shown in color are duplicates of those in the quotient or the dividend, we can omit them and then condense vertically to the form

$$
\begin{array}{r}
6 \quad -7 \quad\; 2 \\
-2\overline{)6 \quad -19 \quad 16 \quad -4} \\
\underline{-12 \quad 14 \quad -4} \\
0
\end{array}
$$

Moving the quotient down to the bottom row, we have the alternative form

$$
\begin{array}{r}
-2\overline{)6 \quad -19 \quad 16 \quad -4} \\
\underline{-12 \quad 14 \quad -4} \\
6 \quad -7 \quad\; 2 \quad\; 0
\end{array}
$$

Now we can change from subtraction to addition (and reduce the likelihood of errors) by changing the sign of the divisor and of row two. Thus, we obtain the following synthetic division array for Example 1:

$$
\begin{array}{r|rrrr}
2 & 6 & -19 & 16 & -4 \\
 &   & 12 & -14 & 4 \\
\hline
 & 6 & -7 & 2 & 0
\end{array}
$$

Don't worry if you have a hard time following the equivalence of the tableaus for long division and synthetic division. The important thing is that you be able to *use* the synthetic division algorithm. We list the pattern for a cubic polynomial. The pattern for higher degree polynomials is similar.

SYNTHETIC DIVISION FOR A CUBIC POLYNOMIAL

To divide $ax^3 + bx^2 + cx + d$ by $(x - k)$, we use the following pattern:

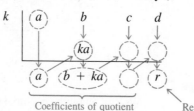

Coefficients of quotient   Remainder

Vertical Pattern: Add terms.
Diagonal Pattern: Multiply by k.

EXAMPLE 3
Using Synthetic Division

Use synthetic division to perform the following division:

$$x + 3 \overline{)x^4 \quad - 10x^2 - 2x + 4}$$

Solution:
We can set up the array as follows:

$$
\begin{array}{c|ccccc}
& x^4 & & -10x^2 & -2x & +4 \\
\hline
-3 & 1 & 0 & -10 & -2 & 4 \\
& & -3 & 9 & 3 & -3 \\
\hline
& 1 & -3 & -1 & 1 & 1 \longleftarrow \text{Remainder} \\
& x^3 & -3x^2 & -x & +1 &
\end{array}
$$

Dividend; Divisor $(x + 3)$; Quotient

Thus, we have

$$\frac{x^4 - 10x^2 - 2x + 4}{x + 3} = x^3 - 3x^2 - x + 1 + \frac{1}{x + 3}$$

The remainder obtained in the synthetic division process has an important interpretation which is described by a theorem called the Remainder Theorem.

THE REMAINDER THEOREM

If a polynomial $f(x)$ is divided by $(x - k)$, then the remainder is

$$r = f(k)$$

Proof:
From the Division Algorithm, we have

$$f(x) = (x - k)q(x) + r(x)$$

and since the degree of $r(x)$ must be less than the degree of $(x - k)$, we know that $r(x)$ must be a constant

$$r(x) = r$$

Now, by evaluating $f(x)$ at $x = k$, we have

$$f(k) = (k - k)q(k) + r$$
$$= (0)q(k) + r$$
$$= r$$

EXAMPLE 4
Evaluating a Polynomial
by the Remainder Theorem

Use the Remainder Theorem to evaluate the following function at $x = -2$:

$$f(x) = 3x^3 + 8x^2 + 5x - 7$$

Solution:
Using synthetic division, we have

$$
\begin{array}{r|rrrr}
-2 & 3 & 8 & 5 & -7 \\
   &   & -6 & -4 & -2 \\
\hline
   & 3 & 2 & 1 & -9
\end{array}
$$

Since the remainder is $r = -9$, we conclude that

$$f(-2) = -9$$

To check this result, we can use the standard procedure for evaluating functions.

$$f(-2) = 3(-2)^3 + 8(-2)^2 + 5(-2) - 7$$
$$= -24 + 32 - 10 - 7 = -9$$

We have already made considerable use of the equivalence of factors of polynomials and zeros of polynomial functions. Using the Remainder Theorem we can now prove this important result. It is known as the Factor Theorem.

FACTOR THEOREM

A polynomial $f(x)$ has a factor $(x - k)$ if and only if $f(k) = 0$.

Proof:
Using the division algorithm with the factor $(x - k)$, we have

$$f(x) = (x - k)q(x) + r(x)$$

By the Remainder Theorem, $r(x) = r = f(k)$, and we have

$$f(x) = (x - k)q(x) + f(k)$$

where $q(x)$ is a polynomial of lesser degree than $f(x)$. Now if $f(k) = 0$, then

$$f(x) = (x - k)q(x)$$

and we see that $(x - k)$ is a factor of $f(x)$. Conversely, if $(x - k)$ is a factor of $f(x)$, then division of $f(x)$ by $(x - k)$ yields a remainder of 0. Hence, by the Remainder Theorem, we have

$$f(k) = 0$$

EXAMPLE 5
Using Synthetic Division to Find Factors of a Polynomial

Show that $(x - 2)$ and $(x + 3)$ are factors of the polynomial

$$f(x) = 2x^4 + 7x^3 - 4x^2 - 27x - 18$$

Then find the remaining factors of $f(x)$.

Solution:
Using synthetic division with 2 and -3 successively, we get

$$
\begin{array}{r|rrrrr}
2 & 2 & 7 & -4 & -27 & -18 \\
  &   & 4 & 22 & 36 & 18 \\
\hline
  & 2 & 11 & 18 & 9 & 0 \\
\end{array}
$$
  ⟹ *0 remainder*
$(x - 2)$ is a factor

$$
\begin{array}{r|rrrr}
-3 & 2 & 11 & 18 & 9 \\
   &   & -6 & -15 & -9 \\
\hline
   & 2 & 5 & 3 & 0 \\
\end{array}
$$
  ⟹ *0 remainder*
$(x + 3)$ is a factor

The resulting quadratic factors as

$$2x^2 + 5x + 3 = (2x + 3)(x + 1)$$

Thus, the complete factorization of $f(x)$ is

$$\begin{aligned}
f(x) &= 2x^4 + 7x^3 - 4x^2 - 27x - 18 \\
&= (x - 2)(x + 3)(2x + 3)(x + 1)
\end{aligned}$$

Horner's Method

Synthetic division provides us with a method for evaluating polynomials that is especially useful on a calculator. Let's reconsider the polynomial function given in Example 4,

$$f(x) = 3x^3 + 8x^2 + 5x - 7$$

To evaluate $f(k)$, synthetic division yields

$$
\begin{array}{r|cccc}
k & 3 & 8 & 5 & -7 \\
  &   & 3k & (3k + 8)k & [(3k + 8)k + 5]k \\
\hline
  & 3 & 3k + 8 & (3k + 8)k + 5 & [(3k + 8)k + 5]k - 7 \\
\end{array}
$$

Now by the Remainder Theorem we know that

$$f(k) = [(3k + 8)k + 5]k - 7$$

In terms of x, we then have

$$3x^3 + 8x^2 + 5x - 7 = [(3x + 8)x + 5]x - 7$$

We call this **Horner's Method** of writing a polynomial. It can be applied to any polynomial by successively factoring out x from each non-constant term, as demonstrated in the following example.

EXAMPLE 6
Horner's Method

Use Horner's Method to rewrite the polynomial function

$$f(x) = 5x^4 - 3x^3 + x^2 - 8x + 7$$

Solution:

$$\begin{aligned}
f(x) &= 5x^4 - 3x^3 + x^2 - 8x + 7 \\
&= (5x^3 - 3x^2 + x - 8)x + 7 && \text{\textit{Factor x from first 4 terms}} \\
&= ([5x^2 - 3x + 1]x - 8)x + 7 && \text{\textit{Factor x from first 3 terms}} \\
&= ([(5x - 3)x + 1]x - 8)x + 7 && \text{\textit{Factor x from first 2 terms}}
\end{aligned}$$

Now, notice how easily we can evaluate $f(k)$ for the polynomial function in Example 6 by entering the following calculator steps:

$$5 \boxed{\times} k \boxed{-} 3 \boxed{=} \boxed{\times} k \boxed{+} 1$$
$$\boxed{=} \boxed{\times} k \boxed{-} 8 \boxed{=} \boxed{\times} k \boxed{+} 7$$

If k is a large number or a decimal value, we could save time by storing k and then using the $\boxed{\text{RCL}}$ key each time k is needed.

EVALUATING A POLYNOMIAL WITH A CALCULATOR

To evaluate the polynomial

$$f(x) = a_n x^n + a_{n-1} x^{n-1} + \cdots + a_1 x + a_0$$

on a calculator with algebraic logic, use the key sequence

$$x \boxed{\text{STO}} \; a_n \boxed{\times} \boxed{\text{RCL}} \boxed{+} a_{n-1} \boxed{=}$$
$$\boxed{\times} \boxed{\text{RCL}} \boxed{+} a_{n-2} \boxed{=}$$
$$\vdots$$
$$\boxed{\times} \boxed{\text{RCL}} \boxed{+} a_1 \boxed{=}$$
$$\boxed{\times} \boxed{\text{RCL}} \boxed{+} a_0 \boxed{=}$$

If $a_i = 0$, then omit $\boxed{+}$ a_i, and if $a_i < 0$, use $\boxed{-}$ instead of $\boxed{+}$.

Remark: Note how the calculator routine *repeats* the five-stroke sequence

$$\boxed{\times} \boxed{\text{RCL}} \boxed{+} a_i \boxed{=}$$

Section Exercises 4.3

In Exercises 1–20, divide the first polynomial by the second using long division.

1. $4x^3 - 7x^2 - 11x + 5, 4x + 5$
2. $6x^3 - 16x^2 + 17x - 6, 3x - 2$
3. $x^4 + 5x^3 + 6x^2 - x - 2, x + 2$
4. $6x^4 + 9x^3 + 10x^2 + x - 21, 2x + 3$
5. $3x^4 + 3x^3 - 10x^2 - 6x + 8, x^2 - 2$
6. $4x^5 - 8x^4 + 4x^3 + 7x^2 - 14x + 7, 4x^3 + 7$
7. $7x + 3, x + 2$
8. $8x - 5, 2x + 1$
9. $x^3, x^2 - 5$
10. $6x^3 + 10x^2 + x + 8, 2x^2 + 1$
11. $x^3 - 9, x^2 + 1$
12. $x^4 + 3x^2 + 1, x^2 - 2x + 3$
13. $x^5 + 7, x^3 - 1$
14. $5x^4, x^2 - 5x + 1$
15. $4x^4 + 5, 2x^2 + 6x - 3$
16. $11x^4 - 9x^3 + 33x^2 - 27x, 11x^2 - 9x$
17. $x^3 - 21x, 5 + 4x - x^2$
18. $x^3 - x + 3, x^2 + x - 2$
19. $2x^3 - 4x^2 - 15x + 5, x^2 - 2x - 8$
20. $x^4, (x - 1)^3$

In Exercises 21–40, use synthetic division to divide the first polynomial by the second.

21. $3x^3 - 17x^2 + 15x - 25, x - 5$
22. $5x^3 + 18x^2 + 7x - 6, x + 3$
23. $4x^3 - 9x + 8x^2 - 18, x + 2$
24. $9x^3 - 16x - 18x^2 + 32, x - 2$
25. $-x^3 + 75x - 250, x + 10$
26. $3x^3 - 16x^2 - 72, x - 6$
27. $5x^3 - 6x^2 + 8, x - 4$
28. $5x^3 + 6x + 8, x + 2$
29. $10x^4 - 50x^3 - 800, x - 6$
30. $x^5 - 13x^4 - 120x + 80, x + 3$
31. $x^5 - 32, x - 2$

32. $x^4 - 625, x - 5$
33. $x^3 + 512, x + 8$
34. $5x^3, x + 3$
35. $-3x^4, x - 2$
36. $-3x^4, x + 2$
37. $5 - 3x + 2x^2 - x^3, x + 1$
38. $180x - x^4, x - 6$
39. $4x^3 + 16x^2 - 23x - 15, x + \frac{1}{2}$
40. $3x^3 - 4x^2 + 5, x - \frac{3}{2}$
41. Use synthetic division to find (a) $f(1)$, (b) $f(-2)$, (c) $f(\frac{1}{2})$ and (d) $f(8)$ when

$$f(x) = 4x^3 - 13x + 10$$

42. Use synthetic division to find (a) $f(1)$, (b) $f(-2)$, (c) $f(5)$ and (d) $f(-10)$ when

$$f(x) = 0.4x^4 - 1.6x^3 + 0.7x^2 - 2$$

In Exercises 43–50, use synthetic division to show that r is a root of the third degree polynomial equation, and use the result to factor the polynomial completely.

43. $x^3 - 7x + 6 = 0, r = 2$
44. $x^3 - 28x - 48 = 0, r = -4$
45. $2x^3 - 15x^2 + 27x - 10 = 0, r = \frac{1}{2}$
46. $48x^3 - 80x^2 + 41x - 6 = 0, r = \frac{2}{3}$
47. $x^3 - 1.9x^2 + 1.1x - 0.2 = 0, r = 0.4$
48. $x^3 + 1.5x^2 - 7.21x - 0.735 = 0, r = -3.5$
49. $x^3 - 3x^2 + 2 = 0, r = 1 + \sqrt{3}$
50. $x^3 - x^2 - 13x - 3 = 0, r = 2 - \sqrt{5}$

In Exercises 51–54, use Horner's method to evaluate the polynomial at the indicated value of x.

51. $f(x) = x^3 - 6x^2 + 12x - 8$ at 5 and -4.5
52. $f(x) = 3x^4 + 6x^3 - 10x^2 - 7x + 2$ at -4.8 and 0.02
53. $f(x) = -5x^4 + 8.5x^3 + 10x - 3$ at 1.08 and -5.4
54. $f(x) = -2x^5 + 4x^3 - 6x^2 + 10$ at 4 and -3.7

Rational Zeros of Polynomial Functions 4.4

In Section 4.2 we saw that an nth degree polynomial function can have at most n real zeros. Although we do not have a general procedure for finding these real zeros, there is a procedure for finding the *rational* zeros of a polynomial function with rational coefficients. This procedure is called the

Rational Zero Test. Before looking at this test, let's look at the zeros of a few simple polynomial functions.

Function	*Real Zeros*	*Comment*
$f(x) = 3x + 4$	$x = -\frac{4}{3}$	One rational zero
$f(x) = x - \sqrt{2}$	$x = \sqrt{2}$	One irrational zero
$f(x) = 2x^2 - 8$	$x = -2, 2$	Two rational zeros
$f(x) = x^2 - 3$	$x = -\sqrt{3}, \sqrt{3}$	Two irrational zeros
$f(x) = x^2 + 1$	None	No real zeros

Several valuable observations can be made concerning the nature of the zeros:

1. Only four of these polynomial functions have rational coefficients. The Rational Zero Test does not apply to functions with irrational coefficients such as $\sqrt{2}$.

2. A polynomial function can have rational coefficients but no rational zeros.

3. The leading coefficient and constant term of a polynomial function are significant in determining the rational zeros.

THE RATIONAL ZERO TEST

If the polynomial function

$$f(x) = a_n x^n + a_{n-1} x^{n-1} + \cdots + a_2 x^2 + a_1 x + a_0$$

has *integer* coefficients, then every rational zero of f has the form

$$\text{rational zero} = \frac{p}{q}$$

where

$p = $ a factor of the constant term a_0

$q = $ a factor of the leading coefficient a_n

Proof:
We begin by assuming that p/q is a rational zero of f and that p/q is written in reduced form (that is, p and q have no common factors other than 1). Since $f(p/q) = 0$, we have

$$a_n \left(\frac{p}{q}\right)^n + a_{n-1} \left(\frac{p}{q}\right)^{n-1} + \cdots + a_2 \left(\frac{p}{q}\right)^2 + a_1 \left(\frac{p}{q}\right) + a_0 = 0$$

Multiplying by q^n, we have

$$a_n p^n + a_{n-1}p^{n-1}q + \cdots + a_2 p^2 q^{n-2} + a_1 p q^{n-1} + a_0 q^n = 0$$

Now we rewrite this equation in two convenient forms, one having p as a factor and one having q as a factor.

$$p(a_n p^{n-1} + a_{n-1}p^{n-2}q + \cdots + a_2 p q^{n-2} + a_1 q^{n-1}) = -a_0 q^n$$
$$q(a_{n-1}p^{n-1} + \cdots + a_2 p^2 q^{n-3} + a_1 p q^{n-2} + a_0 q^{n-1}) = -a_n p^n$$

From the first equation we see that p is a factor of $a_0 q^n$. However, since p and q have no factors in common, it follows that p must be a factor of a_0. Similarly, from the second equation we see that q is a factor of $a_n p^n$, and again, since p and q have no factors in common, it follows that q must be a factor of a_n.

To use the Rational Zero Test to find zeros of a polynomial function, we list all rational numbers whose numerators are factors of the constant term and whose denominators are factors of the leading coefficient. Once the list of *possible rational zeros* is formed, we use trial and error to determine which, if any, are actual zeros of the polynomial.

If the leading coefficient is 1, then the possible rational zeros are simply the factors of a_0. This situation is illustrated in Example 1.

EXAMPLE 1
Leading Coefficient is 1

Use the Rational Zero Test for each of the following functions:

(a) $f(x) = x^3 + x + 1$　　　　　　(b) $f(x) = x^4 - x^3 + x^2 - 3x - 6$

Solution:

(a) For

$$\boxed{1}x^3 + x + \boxed{1}$$

Factors of constant term: ± 1
Factors of leading coefficient: ± 1

the factors of the leading coefficient are ± 1, hence the possible rational zeros are simply the factors of the constant term.

Possible rational zeros $= \pm 1$

By testing these potential zeros, we see that neither works.

$$f(1) = (1)^3 + 1 + 1 = 3$$
$$f(-1) = (-1)^3 + (-1) + 1 = -1$$

Thus, we conclude that this polynomial function has *no* rational zeros.

(b) For

$$x^4 - x^3 + x^2 - 3x - 6$$

the leading coefficient is 1, hence the possible rational zeros are the factors of the constant term.

$$\text{Possible rational zeros} = \pm 1,\ \pm 2,\ \pm 3,\ \pm 6$$

By testing these potential zeros, you should conclude that $x = -1$ and $x = 2$ are the only two that work. Check the others to be sure.

The process of checking several possible rational roots can be tedious, and you can shorten the process in several different ways. First, a programmable calculator is useful. Second, the graph of the function can be used to estimate the location of the real zeros. A third shortcut is to find one zero, then divide the polynomial by the corresponding linear factor to obtain a polynomial of lower degree that may be more easily factorable. In particular, if you succeed in reducing the degree to two, then you can apply the quadratic formula.

Synthetic division makes this third shortcut even more convenient. For example, in part (b) of Example 1, after finding that $x = -1$ and $x = 2$ are zeros of

$$f(x) = x^4 - x^3 + x^2 - 3x - 6$$

we could save quite a bit of work by dividing both $(x + 1)$ and $(x - 2)$ into $f(x)$ as follows:

$$
\begin{array}{r|rrrrr}
-1 & 1 & -1 & 1 & -3 & -6 \\
   &   & -1 & 2 & -3 & 6 \\
\hline
   & 1 & -2 & 3 & -6 & 0 \\
\end{array}
$$

$$
\begin{array}{r|rrrr}
2 & 1 & -2 & 3 & -6 \\
  &   & 2 & 0 & 6 \\
\hline
  & 1 & 0 & 3 & 0 \\
\end{array}
$$

Thus, we have

$$f(x) = (x + 1)(x - 2)(x^2 + 3)$$

and since the factor $(x^2 + 3)$ has no real roots, we see that $x = -1$ and $x = 2$ are the only real zeros of f.

If the leading coefficient of the polynomial is something other than 1, the list of possible rational zeros can increase dramatically. In such cases, the shortcuts to the Rational Zero Test can be especially valuable.

EXAMPLE 2
Leading Coefficient Other Than 1

Find the rational roots of the following polynomial functions:

(a) $f(x) = 2x^3 + 3x^2 - 8x + 3$
(b) $f(x) = 10x^3 - 15x^2 - 16x + 12$

Solution:

(a) For $2x^3 + 3x^2 - 8x + 3$,

$$\text{Possible rational zeros} = \frac{\text{factors of 3}}{\text{factors of 2}}$$

$$= \frac{\pm 1, \pm 3}{\pm 1, \pm 2} = \pm 1, \pm 3, \pm \frac{1}{2}, \pm \frac{3}{2}$$

By synthetic division, we can determine that $x = 1$ is a zero *and* we can obtain the factored version of the polynomial.

$$
\begin{array}{r|rrrr}
1 & 2 & 3 & -8 & 3 \\
  &   & 2 & 5 & -3 \\
\hline
  & 2 & 5 & -3 & 0
\end{array}
$$

Now we see that

$$f(x) = (x - 1)(2x^2 + 5x - 3)$$
$$= (x - 1)(2x - 1)(x + 3) \qquad \textit{Factor quadratic}$$

which implies that the zeros of f are $x = 1, \frac{1}{2}$, and -3.

(b) For $10x^3 - 15x^2 - 16x + 12$,

$$\text{Possible rational zeros} = \frac{\text{factors of 12}}{\text{factors of 10}}$$

$$= \frac{\pm 1, \pm 2, \pm 3, \pm 4, \pm 6, \pm 12}{\pm 1, \pm 2, \pm 5, \pm 10}$$

With so many possibilities for zeros, it will be worth our time to stop and make a sketch of this function. From Figure 4.29, it looks as though three reasonable choices for zeros would be $x = -1$, $x = 0.5$ and $x = 2$. To test these choices, we use synthetic division as follows:

Test $x = -1$:

$$
\begin{array}{r|rrrr}
-1 & 10 & -15 & -16 & 12 \\
   &   & -10 & 25 & -9 \\
\hline
   & 10 & -25 & 9 & \boxed{3}
\end{array}
$$
$\implies$ *$x = -1$ is not a zero of f*

Test $x = 0.5$:

$$
\begin{array}{r|rrrr}
0.5 & 10 & -15 & -16 & 12 \\
    &   & 5 & -5 & -10.5 \\
\hline
    & 10 & -10 & -21 & \boxed{1.5}
\end{array}
$$
$\implies$ *$x = 0.5$ is not a zero of f*

Test $x = 2$:

$$
\begin{array}{r|rrrr}
2 & 10 & -15 & -16 & 12 \\
  &   & 20 & 10 & -12 \\
\hline
  & 10 & 5 & -6 & \boxed{0}
\end{array}
$$
$\implies$ *$x = 2$ is a zero of f*

$f(x) = 10x^3 - 15x^2 - 16x + 12$

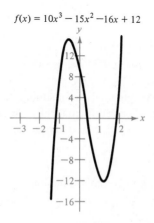

FIGURE 4.29

Thus, we have

$$f(x) = (x - 2)(10x^2 + 5x - 6)$$

and by the quadratic formula we find the two additional zeros to be

$$x = \frac{-5 + \sqrt{265}}{20} \approx 0.5639$$

and

$$x = \frac{-5 - \sqrt{265}}{20} \approx -1.0639$$

Remark: Remember that the last entry in the bottom row (of synthetic division) must be zero in order to conclude that the test value is a zero of f.

Upper and Lower Bounds for Rational Zeros

Occasionally, we gain additional information from synthetic division by recognizing certain patterns in the last row. In particular, if the test value is positive and each entry in the last row is positive (or zero), then we can conclude that the test value is an **upper bound** for the zeros of f. Similarly, if the test value is negative and the entries in the last row are alternately positive and negative (zero entries count as positive or negative in this test), then we can conclude that the test value is a **lower bound** for the zeros of f.

For example, in part (b) of Example 2, a test of $x = 3$ yields

Test $x = 3$:

3	10	-15	-16	12
		30	45	87
	10	15	29	99

Since the test value is positive and all entries in the last row are positive, we can conclude that $x = 3$ is an upper bound for the zeros of f. A test of $x = -2$ yields

Test $x = -2$:

-2	10	-15	-16	12
		-20	70	-108
	10	-35	54	-96

Since the test value is negative and the entries in the last row are alternately positive and negative, we can conclude that $x = -2$ is a lower bound for the zeros of f. This upper and lower bound information is useful because we can then restrict our test values to possible rational zeros that lie between -2 and 3.

The next example shows how to extend the Rational Zero Test to cover polynomial functions with rational (but not all integer) coefficients. The basic procedure consists of finding the lowest common denominator (LCD) of the coefficients and then factoring 1/LCD out of the polynomial.

EXAMPLE 3
Polynomial Functions with
Rational Coefficients

Find the rational roots of the polynomial function

$$f(x) = x^3 - \frac{1}{2}x^2 + \frac{1}{3}x - \frac{1}{6}$$

Solution:
Since the LCD of the coefficients is 6, we rewrite the function as follows:

$$f(x) = \frac{6}{6}x^3 - \frac{3}{6}x^2 + \frac{2}{6}x - \frac{1}{6}$$

$$= \frac{1}{6}(6x^3 - 3x^2 + 2x - 1)$$

Now, the zeros of f must coincide with the roots of

$$6x^3 - 3x^2 + 2x - 1 = 0$$

so

$$\text{Possible rational zeros} = \frac{\text{factors of } 1}{\text{factors of } 6}$$

$$= \frac{\pm 1}{\pm 1, \ \pm 2, \ \pm 3, \ \pm 6} = \pm 1, \ \pm\frac{1}{2}, \ \pm\frac{1}{3}, \ \pm\frac{1}{6}$$

By testing these potential zeros, we see that $x = \frac{1}{2}$ works.

$$
\begin{array}{r|rrrr}
0.5 & 6 & -3 & 2 & -1 \\
    &   & 3 & 0 & 1 \\
\hline
    & 6 & 0 & 2 & 0
\end{array}
$$

and we have

$$6x^3 - 3x^2 + 2x - 1 = \left(x - \frac{1}{2}\right)(6x^2 + 2) = (2x - 1)(3x^2 + 1)$$

Finally, since the factor $(3x^2 + 1)$ has no real roots, we conclude that $x = \frac{1}{2}$ is the only rational zero of f.

Section Exercises 4.4

In Exercises 1–30, use the Rational Zero Test and synthetic division as aids to finding the real zeros of the function.

1. $f(x) = x^3 - 6x^2 + 11x - 6$
2. $f(x) = x^3 - 7x - 6$
3. $f(x) = x^3 - 4x^2 - x + 4$
4. $f(x) = x^3 - 9x^2 + 20x - 12$
5. $f(x) = x^3 + 12x^2 + 21x + 10$
6. $f(x) = x^3 + 6x^2 + 12x + 8$
7. $f(x) = x^3 - 6x^2 + 12x - 8$
8. $f(x) = x^3 + 2x^2 - 7x + 4$
9. $f(x) = x^3 - 4x^2 + 5x - 2$
10. $f(x) = x^3 - 9x^2 + 27x - 27$
11. $f(x) = 2x^3 + 3x^2 - 1$
12. $f(x) = 3x^3 - 19x^2 + 33x - 9$
13. $f(x) = 4x^3 - 3x - 1$
14. $f(x) = 12x^3 - 4x^2 - 27x + 9$
15. $f(x) = 8x^3 - 6x^2 - 5x + 3$
16. $f(x) = 2x^3 - x^2 + 8x - 4$
17. $f(x) = 4x^3 + 3x^2 + 8x + 6$
18. $f(x) = 3x^3 - 2x^2 + 15x - 10$
19. $f(x) = x^4 - 3x^2 + 2$
20. $f(x) = x^4 - 7x^2 + 12$
21. $f(x) = x^4 - x^3 - 2x - 4$
22. $f(x) = x^4 - x^3 - 29x^2 - x - 30$

23. $f(x) = x^4 - 5x^2 + 4$
24. $f(x) = x^4 - 6x^3 + 11x^2 - 6x$
25. $f(x) = x^4 - 13x^2 - 12x$
26. $f(x) = 2x^4 + 7x^3 - 26x^2 + 23x - 6$
27. $f(x) = 2x^4 - 11x^3 - 6x^2 + 64x + 32$
28. $f(x) = x^5 - x^4 - 3x^3 + 5x^2 - 2x$
29. $f(x) = x^5 - 7x^4 + 10x^3 + 14x^2 - 24x$
30. $f(x) = 6x^4 - 11x^3 - 51x^2 + 99x - 27$

In Exercises 31–35, (a) list the possible rational zeros of the function, (b) sketch the graph of the function so that some of the possible zeros of part (a) can be disregarded, and (c) determine the real zeros of the polynomial function.

31. $f(x) = 32x^3 - 52x^2 + 17x + 3$
32. $f(x) = 6x^3 - x^2 - 13x + 8$
33. $f(x) = 4x^3 + 7x^2 - 11x - 18$
34. $f(x) = 6x^3 - 31x^2 - 7x + 60$
35. $f(x) = 4x^3 - 21x - 10$

In Exercises 36–40, find the rational zeros of the polynomial function.

36. $f(x) = x^3 - \frac{3}{2}x^2 - \frac{23}{2}x + 6$
37. $f(x) = x^3 - \frac{1}{4}x^2 - x + \frac{1}{4}$
38. $f(x) = x^3 + \frac{11}{6}x^2 - \frac{1}{2}x - \frac{1}{3}$
39. $f(x) = x^3 - \frac{2}{3}x^2 - \frac{5}{9}x + \frac{2}{9}$
40. $f(x) = x^3 + \frac{19}{4}x^2 + \frac{15}{8}x - \frac{9}{2}$

In Exercises 41–44, use synthetic division to determine if each value of x is an upper bound, lower bound, or neither of the zeros of f.

41. $f(x) = x^4 - 4x^3 + 15$
    (a) $x = 4$ (b) $x = -1$ (c) $x = 3$
42. $f(x) = 2x^3 - 3x^2 - 12x + 8$
    (a) $x = 2$ (b) $x = 4$ (c) $x = -1$
43. $f(x) = x^4 - 4x^3 + 16x - 16$
    (a) $x = -1$ (b) $x = -3$ (c) $x = 5$
44. $f(x) = 2x^4 - 8x + 3$
    (a) $x = 1$ (b) $x = 3$ (c) $x = -4$

Complex Zeros and the Fundamental Theorem of Algebra

4.5

We have already been using the fact that an nth degree polynomial function can have at most n real zeros. In this section we will improve upon that result to see that, in the complex number system, every nth degree polynomial function has *precisely* n zeros. This important result is derived from the **Fundamental Theorem of Algebra.** The proof of this theorem is outside the scope of this text, and we will be content with clearly stating and illustrating the theorem.

THE FUNDAMENTAL THEOREM OF ALGEBRA

> If f is a polynomial function of degree $n > 0$, then f has at least one zero in the complex number system.

Remark: We have stated the Fundamental Theorem in its standard form. Using the equivalency of zeros and factors, we can easily establish the following result.

NUMBER OF ZEROS OF A POLYNOMIAL FUNCTION

> If f is a polynomial function of degree $n > 0$, then f has precisely n zeros in the complex number system.

Proof:

Using the Fundamental Theorem, we know that f must have at least one zero c_1. In terms of factors, we know that $(x - c_1)$ is a factor of $f(x)$, and we have

$$f(x) = (x - c_1)f_1(x)$$

If $f_1(x)$ is of degree greater than zero, we can reapply the Fundamental Theorem to conclude that f must also have a zero c_2, which implies that

$$f(x) = (x - c_1)(x - c_2)f_2(x)$$

It is clear that the degree of $f_1(x)$ is $n - 1$, the degree of $f_2(x)$ is $n - 2$, and that we can repeatedly apply the Fundamental Theorem n times until we obtain

$$f(x) = a(x - c_1)(x - c_2) \cdots (x - c_n)$$

where a is the leading coefficient of the polynomial $f(x)$. Now, since $f(x)$ has precisely n linear factors, we apply the Factor Theorem to conclude that f has precisely the following n zeros:

$$c_1, c_2, \ldots, c_n$$

The n zeros of a polynomial can include zeros that are real or complex as well as zeros that are repeated. This is illustrated by the following functions:

Function	Degree	Factored Form	Zeros
$f(x) = x - 2$	1	$f(x) = (x - 2)$	$x = 2$
$f(x) = x^2 - 9$	2	$f(x) = (x - 3)(x + 3)$	$x = -3, 3$
$f(x) = x^2 + 4$	2	$f(x) = (x - 2i)(x + 2i)$	$x = -2i, 2i$
$f(x) = x^3 - 4x^2 + 5x - 2$	3	$f(x) = (x - 1)(x - 1)(x - 2)$	$x = 1, 1, 2$
$f(x) = x^4 - 1$	4	$f(x) = (x - 1)(x + 1)(x - i)(x + i)$	$x = -1, 1, -i, i$

Remark: Note that neither the Fundamental Theorem nor its corollary gives us any information on how to find the zeros of a polynomial function. To do that, we rely upon the techniques developed in earlier sections.

EXAMPLE 1
Finding the Zeros
of a Polynomial Function

Find all five zeros of the function

$$f(x) = x^5 + x^3 + 2x^2 - 12x + 8$$

and list the function in factored form.

Solution:
Using the Rational Zero Test, we can find the following three rational zeros of f:

$$
\begin{array}{r|rrrrrr}
1 & 1 & 0 & 1 & 2 & -12 & 8 \\
  &   & 1 & 1 & 2 & 4 & -8 \\
\hline
  & 1 & 1 & 2 & 4 & -8 & 0
\end{array}
$$

$$
\begin{array}{r|rrrrr}
1 & 1 & 1 & 2 & 4 & -8 \\
  &   & 1 & 2 & 4 & 8 \\
\hline
  & 1 & 2 & 4 & 8 & 0
\end{array}
\quad \Longrightarrow \quad \textit{1 is a repeated zero}
$$

$$
\begin{array}{r|rrrr}
-2 & 1 & 2 & 4 & 8 \\
   &   & -2 & 0 & -8 \\
\hline
   & 1 & 0 & 4 & 0
\end{array}
$$

Thus, we have

$$
\begin{aligned}
f(x) &= x^5 + x^3 + 2x^2 - 12x + 8 \\
&= (x - 1)(x - 1)(x + 2)(x^2 + 4)
\end{aligned}
$$

Finally, by factoring the quadratic factor, we have

$$
f(x) = (x - 1)(x - 1)(x + 2)(x - 2i)(x + 2i)
$$

which gives us the following five zeros:

$$
x = 1,\ 1,\ -2,\ 2i,\ -2i
$$

Notice that in Example 1 the two complex zeros are conjugates. That is, they are of the form

$$
a + bi \quad \text{and} \quad a - bi
$$

This is not a coincidence. In fact, the following result tells us that if a polynomial (with real coefficients) has one complex zero $(a + bi)$ then it *must* also have the conjugate $(a - bi)$ as a zero.

COMPLEX ZEROS OCCUR IN CONJUGATE PAIRS

If $a + bi$ $(b \neq 0)$ is a complex zero of a polynomial function f, with real coefficients, then the conjugate $a - bi$ is also a complex zero of f.

Proof:
We begin by letting $d(x)$ be defined by

$$
d(x) = [x - (a + bi)][x - (a - bi)] = x^2 - 2ax + (a^2 + b^2)
$$

By the Division Algorithm, we have

$$
f(x) = d(x) \cdot q(x) + r(x)
$$

where the degree of $r(x)$ is less than the degree of $d(x)$. This implies that

$$r(x) = cx + d$$

Since $f(x)$ and $d(x)$ have real coefficients, the division process for $f(x)/d(x)$ will yield only real coefficients in the quotient and remainder. In particular, we know that c and d are real numbers. Now, since $f(a + bi) = 0$ and $d(a + bi) = 0$, we have $r(a + bi) = 0$, which implies that

$$c(a + bi) + d = (ca + d) + (cb)i = 0$$

and we see that

$$cb = 0 \quad \text{and} \quad ca + d = 0$$

Since $b \neq 0$, it follows that $c = 0$, which in turn implies that $d = 0$. Thus, we have

$$f(x) = d(x) \cdot q(x) = [x - (a + bi)][x - (a - bi)] \cdot q(x)$$

from which it follows that $f(a - bi) = 0$.

The preceding result has the following important consequence relating to factors of a polynomial.

FACTORS OF A POLYNOMIAL

> Every polynomial of degree $n > 0$ with real coefficients can be written as the product of linear and quadratic factors with real coefficients, where the quadratic factors have no real zeros.

Proof:
This proof follows quite easily using complex conjugates. To begin, we know that if $f(x)$ is a polynomial of degree n, then it can be *completely* factored in the form

$$f(x) = A(x - c_1)(x - c_2)(x - c_3) \cdots (x - c_n)$$

If each c_i is real, there is nothing more to prove. If any c_i is complex, say

$$c_i = a + bi, \quad b \neq 0$$

then because the coefficients of $f(x)$ are real we know that the conjugate must also appear as a zero:

$$c_j = a - bi$$

By multiplying these two factors, we obtain the quadratic form

$$(x - c_i)(x - c_j) = [x - (a + bi)][x - (a - bi)]$$
$$= x^2 - 2ax + (a^2 + b^2)$$

where each coefficient is real. This completes the proof.

Remark: A quadratic factor with no real zeros is said to be **irreducible over the reals.** Be sure you see that this is not the same as being *irreducible* *over the rationals*. For example, the quadratic $x^2 + 1 = (x - i)(x + i)$ is irreducible over the reals (and the rationals). The quadratic $x^2 - 2 = (x - \sqrt{2})(x + \sqrt{2})$ is irreducible over the rationals, but *not* over the reals.

EXAMPLE 2
Factoring a Polynomial

Write the polynomial

$$f(x) = x^4 - x^2 - 20$$

(a) as the product of factors that are irreducible over the *rationals*
(b) as the product of linear factors and quadratic factors that are irreducible over the *reals*
(c) in completely factored form

Solution:

(a) We begin by factoring this fourth degree polynomial into the product of two quadratics, as follows:

$$x^4 - x^2 - 20 = (x^2 - 5)(x^2 + 4)$$

Both of these factors are irreducible over the rationals.

(b) By factoring over the reals, we have

$$x^4 - x^2 - 20 = (x + \sqrt{5})(x - \sqrt{5})(x^2 + 4)$$

Note that the quadratic factor is irreducible over the reals.

(c) In completely factored form, we have

$$x^4 - x^2 - 20 = (x + \sqrt{5})(x - \sqrt{5})(x + 2i)(x - 2i)$$

The polynomial in Example 2 was simple enough to factor directly over the reals and subsequently over the complex numbers. However, for more complicated polynomials, this is often not possible. In the next example we use synthetic division to factor polynomials with complex zeros.

EXAMPLE 3
Using Synthetic Division with Complex Zeros

Find all zeros of

$$f(x) = x^4 - 3x^3 + 6x^2 + 2x - 60$$

given that $1 + 3i$ is a zero.

Solution:
Since complex zeros occur in pairs, we divide by $1 + 3i$ and $1 - 3i$:

$1 + 3i$	1	-3	6	2	-60
		$1 + 3i$	$-11 - 3i$	$4 - 18i$	60
	1	$-2 + 3i$	$-5 - 3i$	$6 - 18i$	0
$1 - 3i$	1	$-2 + 3i$	$-5 - 3i$	$6 - 18i$	
		$1 - 3i$	$-1 + 3i$	$-6 + 18i$	
	1	-1	-6	0	

The resulting quadratic factors as

$$x^2 - x - 6 = (x - 3)(x + 2)$$

Therefore, the four zeros of

$$f(x) = x^4 - 3x^3 + 6x^2 + 2x - 60$$

are $-2, 3, 1 + 3i$, and $1 - 3i$.

Occasionally, we want to create polynomials with certain known zeros. For example, to create a polynomial that has 1, -2, and 5 as zeros, we can write

$$f(x) = (x - 1)(x + 2)(x - 5) = x^3 - 4x^2 - 7x + 10$$

Since complex zeros occur in conjugate pairs, we can create a polynomial given only one part of a conjugate complex pair. For example, suppose that we wanted to find a fourth degree polynomial with real coefficients that has $3 - i$ and $4i$ as complex zeros. Since the conjugate of each of these complex zeros must also be a zero of the polynomial, we write the following factored form:

$$f(x) = [x - (3 - i)][x - (3 + i)][x - (4i)][x - (-4i)]$$
$$= [x^2 - 6x + (9 + 1)][x^2 + 16]$$

which yields the polynomial

$$f(x) = x^4 - 6x^3 + 26x^2 - 96x + 160$$

Section Exercises 4.5

In Exercises 1–30, find all the zeros of the function and write the polynomial as a product of linear factors.

1. $f(x) = x^2 + 18x + 77$
2. $f(x) = x^2 - x + 56$
3. $f(x) = x^2 - 4x + 1$
4. $f(x) = x^2 + 10x + 23$
5. $f(x) = x^2 + 25$
6. $f(x) = x^2 + 144$
7. $f(x) = x^4 - 81$
8. $f(x) = x^4 - 625$
9. $f(x) = x^2 - 2x + 2$
10. $f(x) = x^3 - 3x^2 + 4x - 2$
11. $f(x) = x^3 - 6x^2 + 13x - 10$
12. $f(x) = x^3 - 2x^2 - 11x + 52$
13. $f(x) = x^3 - 3x^2 - 15x + 125$
14. $f(x) = x^3 + 11x^2 + 39x + 29$
15. $f(x) = x^3 + 24x^2 + 214x + 740$
16. $f(x) = 2x^3 - 5x^2 + 12x - 5$
17. $f(x) = 16x^3 - 20x^2 - 4x + 15$
18. $f(x) = 9x^3 - 15x^2 + 11x - 5$
19. $f(x) = x^3 - x + 6$

20. $f(x) = x^3 + 9x^2 + 27x + 27$
21. $f(x) = 5x^3 - 9x^2 + 28x + 6$
22. $f(x) = 3x^3 - 4x^2 + 8x + 8$
23. $f(x) = x^4 - 4x^3 + 8x^2 - 16x + 16$
24. $f(x) = x^4 + 6x^3 + 10x^2 + 6x + 9$
25. $f(x) = x^4 + 10x^2 + 9$
26. $f(x) = x^4 + 29x^2 + 100$
27. $f(x) = x^5 + x^4 - 13x^3 - 13x^2 + 36x + 36$
28. $f(x) = x^5 + 2x^4 + 10x^3 + 20x^2 + 9x + 18$
29. $f(x) = x^4 + 2x^3 - x^2 + 4x + 12$
30. $f(x) = x^5 - 8x^4 + 28x^3 - 56x^2 + 64x - 32$

In Exercises 31–40, find a polynomial with real coefficients that has the given zeros.

31. $1, 5i, -5i$
32. $4, 3i, -3i$
33. $2, 4 + i, 4 - i$
34. $6, -5 + 2i, -5 - 2i$
35. $i, -i, 6i, -6i$
36. $2, 2, 2, 4i, -4i$

37. $-5, -5, 1 + \sqrt{3}i$ 38. $\frac{2}{3}, -1, 3 + \sqrt{2}i$

39. $\frac{3}{4}, -2, -\frac{1}{2} + i$ 40. $0, 0, 4, 1 + i$

In Exercises 41–44, write the polynomial (a) as the product of factors that are irreducible over the *rationals*, (b) as the product of linear and quadratic factors that are irreducible over the *reals*, and (c) in completely factored form.

41. $f(x) = x^4 + 6x^2 - 27$
42. $f(x) = x^4 - 2x^3 - 3x^2 + 12x - 18$
    (Hint: One factor is $x^2 - 6$.)
43. $f(x) = x^4 - 4x^3 + 5x^2 - 2x - 6$
    (Hint: One factor is $x^2 - 2x - 2$.)

44. $f(x) = x^4 - 3x^3 - x^2 - 12x - 20$
    (Hint: One factor is $x^2 + 4$.)
45. Use synthetic division to show that $2i$ is a zero of the function

$$f(x) = 2x^4 - x^3 + 7x^2 - 4x - 4$$

46. Use synthetic division to show that $-3 + 2i$ is a zero of the function

$$f(x) = x^3 + 7x^2 + 19x + 13$$

Approximation Techniques for Zeros of Polynomials

4.6

By this time you should realize that one of the most important problems in algebra is that of finding zeros of polynomial functions. (An indication of this importance is given by the fact that the Fundamental Theorem of Algebra deals with the number of zeros of a polynomial.) Throughout history a great deal of time and effort have been devoted to methods for finding the zeros of polynomial functions. We have not looked at all of these methods, but we have looked at the basic ones. These include graphical methods, factorization methods, the Rational Zero Test, and reduction of degree.

In general, the higher the degree of a polynomial function, the more difficult it is to find its zeros. In practical applications involving polynomials of degree three or greater, we often must be content with an approximation technique for finding zeros. In this section we look at one of the simpler approximation methods known as the **Bisection Method.**

Most approximation methods are described by an **iterative process,** meaning that the method is applied repeatedly to obtain better and better approximations. This method is dependent upon finding two values for x—one at which the function is positive and one at which the function is negative. Then, by the Intermediate Value Theorem (Section 4.2), we know that the function has at least one zero between these two values. (See Figure 4.30.) In Figure 4.30, we can see that a zero must occur somewhere in the interval (a, b). To apply the Bisection Method, we cut this interval in half and consider the two intervals

$$\left(a, \frac{a + b}{2}\right) \quad \text{and} \quad \left(\frac{a + b}{2}, b\right)$$

Depending upon the value of $f(x)$ at the midpoint, we reapply the Bisection Method to one of these intervals. The process is illustrated in the next example.

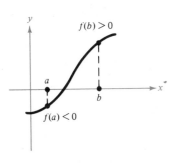

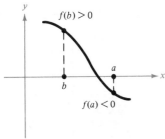

FIGURE 4.30

EXAMPLE 1
Using the Bisection Method

Use the Bisection Method to approximate the real zero of

$$f(x) = x^3 - x^2 + 1$$

to within 0.001 unit.

Solution:
We begin by making a sketch of f, as shown in Figure 4.31. In Example 3 of Section 4.2, we discovered that this function has a zero between -0.8 and -0.7, since

$$f(-0.8) = -0.152 < 0 \quad \text{and} \quad f(-0.7) = 0.167 > 0$$

Using the midpoint of this interval, we approximate the zero to be

$$c = \frac{-0.8 + (-0.7)}{2} = -0.75$$

Now the maximum error of this approximation can only be one-half of the interval. That is,

$$\text{max error} < \frac{-0.7 - (-0.8)}{2} = 0.05$$

If $f(-0.75) = 0$, then -0.75 is a zero of f.
If $f(-0.75) > 0$, then a zero occurs between -0.8 and -0.75.
If $f(-0.75) < 0$, then a zero occurs between -0.75 and -0.7.

Since $f(-0.75) = 0.015625$ is positive, we choose $(-0.8, -0.75)$ as our new interval. The results of several iterations are shown in Table 4.1. After the seventh iteration, the maximum error is less than 0.001, and we use

$$c = -0.7556 \approx -0.756 \qquad \textit{(to three-place accuracy)}$$

as our final approximation.

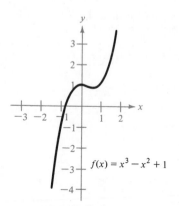

FIGURE 4.31

$f(x) = x^3 - x^2 + 1$

TABLE 4.1

Iteration	a	c	b	f(a)	f(c)	f(b)	Maximum Error
1	−0.8	−0.75	−0.7	−0.1520	0.0156	0.1670	0.05
2	−0.8	−0.775	−0.75	−0.1520	−0.0661	0.0156	0.025
3	−0.775	−0.7625	−0.75	−0.0661	−0.0247	0.0156	0.0125
4	−0.7625	−0.7563	−0.75	−0.0247	−0.0046	0.0156	0.0063
5	−0.7563	−0.7532	−0.75	−0.0046	0.0054	0.0156	0.0032
6	−0.7563	−0.7548	−0.7532	−0.0046	0.0003	0.0054	0.0016
7	−0.7563	−0.7556	−0.7548	−0.0046	−0.0023	0.0003	0.0008

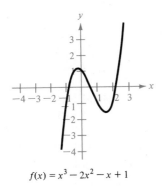

$f(x) = x^3 - 2x^2 - x + 1$

FIGURE 4.32

Remark: If a function is determined from its graph to have more than one real zero (see Figure 4.32), then we can use the Bisection Method to approximate each zero in turn by considering an interval appropriate to each.

If we continue the process in Table 4.1 indefinitely, the sequence of successively better approximations is said to **converge** to the zero of f. The convergence of the Bisection Method is relatively slow in that several iterations are usually necessary to obtain a very fine accuracy. However, the Bisection Method can be applied to *any* continuous function that crosses the x-axis. In calculus, you will encounter another approximation method, called **Newton's Method,** which usually has a more rapid convergence than the Bisection Method, but has the disadvantage of being less general.

A computer program written in BASIC is given below for the Bisection Method. The sample printout is for the function

$$f(x) = x^3 - 2x^2 - x + 1$$

and the final line gives one of the real zeros approximated to three decimal places. This particular function has three real zeros:

$$x \approx -0.802, \qquad x \approx 0.555, \qquad \text{and} \qquad x \approx 2.247$$

The graph of f is shown in Figure 4.32.

BASIC PROGRAM FOR BISECTION METHOD

```
 10  REM  DEFINE THE FUNCTION
 20          DEF FNY (X) = X^3 - 2*X^2 - X + 1
 30  REM  A = LEFT ENDPOINT, B = RIGHT ENDPOINT
 40          A = -1
 50          B = 0
 60          C = (A + B)/2
 70  REM  ME = MAXIMUM ERROR
 80          ME = (B - A)/2
 90          PRINT "A = ";A,"C = ";C,"B = ";B,"MAXIMUM ERROR = ";ME
100  REM  TEST THE SIZE OF THE ERROR
110          IF ME < .001 GOTO 150
120  REM  TEST THE SIGN OF F(C)
130          IF FNY(A)*FNY(C) < 0 THEN B = C: GOTO 60
140          IF FNY(B)*FNY(C) < 0 THEN A = C: GOTO 60
150          END
```

PRINTOUT FROM BASIC PROGRAM

A = −1	C = −.5	B = 0	MAXIMUM ERROR = .5
A = −1	C = −.75	B = −.5	MAXIMUM ERROR = .25
A = −1	C = −.875	B = −.75	MAXIMUM ERROR = .125
A = −.875	C = −.8125	B = −.75	MAXIMUM ERROR = .0625
A = −.8125	C = −.78125	B = −.75	MAXIMUM ERROR = .03125
A = −.8125	C = −.796875	B = −.78125	MAXIMUM ERROR = .015625
A = −.8125	C = −.804688	B = −.796875	MAXIMUM ERROR = .0078125
A = −.804688	C = −.800781	B = −.796875	MAXIMUM ERROR = .0039063
A = −.804688	C = −.802734	B = −.800781	MAXIMUM ERROR = .0019531
A = −.802734	C = −.801758	B = −.800781	MAXIMUM ERROR = .0009766

Section Exercises 4.6

In Exercises 1–8, use the Bisection Method to approximate the indicated real zero(s) of the function to within 0.001 unit.

1. $f(x) = x^3 + x - 1$ 2. $f(x) = x^5 + x - 1$

5. $f(x) = -x^3 + 3x^2 - x + 1$
6. $f(x) = x^3 - 3x - 1$

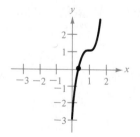
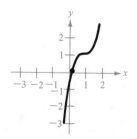
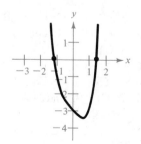
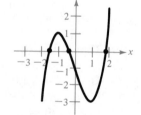
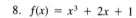
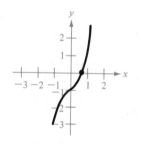
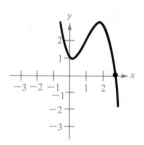
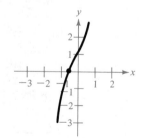

3. $f(x) = 2x^3 - 6x^2 + 6x - 1$
4. $f(x) = 4x^3 - 12x^2 + 12x - 3$

7. $f(x) = x^4 - x - 3$ 8. $f(x) = x^3 + 2x + 1$

9. Find the zero of the function

$$f(x) = 7x^4 - 42x^3 + 43x^2 + 216x - 324$$

accurate to within 0.001 unit in the interval $[1, 2]$.

10. Find the zero of the function

$$f(x) = 3x^4 - 12x^3 + 27x^2 + 4x - 4$$

accurate to within 0.001 unit in the interval $[0, 1]$.

Review Exercises / Chapter 4

In Exercises 1–10, analyze the function and sketch its graph.

1. $f(x) = (x + \frac{3}{2})^2 + 1$
2. $f(x) = (x - 4)^2 - 3$
3. $f(x) = -(x - 1)^3$
4. $f(x) = (x + 1)^3$
5. $f(x) = x^4 - x^3 - 2x^2$
6. $f(x) = -2x^3 - x^2 + x$
7. $f(x) = x^3 - 3x$
8. $f(x) = -x^3 + 3x - 2$
9. $f(x) = x(x + 3)^2$
10. $f(x) = x^4 - 4x^2$

In Exercises 11–15, use completion of the square to find the maximum or minimum value of the quadratic function.

11. $f(x) = x^2 - 2x$
12. $f(x) = x^2 + 8x + 10$
13. $f(x) = 6x - x^2$
14. $f(x) = 3 + 4x - x^2$
15. $f(x) = x^2 + 3x - 8$
16. Let x be the amount (in hundreds of dollars) a company spends on advertising, and let P be the profit. If

$$P = 230 + 20x - \frac{1}{2}x^2$$

what amount of advertising gives the maximum profit?
17. Find the number of units x that produce a maximum revenue R if $R = 900x - 0.1x^2$.
18. A real estate office handles 50 apartment units. When the rent is \$360 per month, all units are occupied. However, on the average, for each \$20 increase in rent, one unit becomes vacant. Each occupied unit requires an average of \$12 per month for service and repairs. What should be charged to realize the most profit?

In Exercises 19–24, divide the first polynomial by the second by using long division.

19. $x^4 - 3x^2 + 2, x^2 - 1$
20. $3x^4, x^2 - 1$
21. $3x^5 - 4x^3 + x^2 + 5, x^2 - 2x + 4$
22. $4x + 7, 3x - 2$
23. $3x^3 + 2x - 1, 2x^3 + x$
24. $-2x^4 - 4x^3 + 3x^2 - 8, x^2 + 4x + 4$

In Exercises 25–30, use synthetic division to divide the first polynomial by the second.

25. $\frac{1}{4}x^4 - 4x^3, x - 2$
26. $2x^3 + 2x^2 - x + 2, x - \frac{1}{2}$
27. $6x^4 - 4x^3 - 27x^2 + 18x, x - \frac{2}{3}$
28. $0.1x^3 + 0.3x^2 - 0.5, x - 5$
29. $2x^3 - 5x^2 + 12x - 5, x - (1 + 2i)$
30. $9x^3 - 15x^2 + 11x - 5, x - (\frac{1}{3} + \frac{2}{3}i)$

In Exercises 31–36, find all zeros of the function.

31. $f(x) = 4x^3 - 11x^2 + 10x - 3$
32. $f(x) = 10x^3 + 21x^2 - x - 6$
33. $f(x) = 6x^3 - 5x^2 + 24x - 20$
34. $f(x) = x^3 - 1.3x^2 - 1.7x + 0.6$
35. $f(x) = 6x^4 - 25x^3 + 14x^2 + 27x - 18$
36. $f(x) = 5x^4 + 126x^2 + 25$
37. Find a fourth degree polynomial with zeros -1, -1, $\frac{1}{3}$, $-\frac{1}{2}$.
38. Find a fourth degree polynomial with zeros $\frac{2}{3}$, 4, $\sqrt{3}i$, $-\sqrt{3}i$.
39. A spherical tank (see Figure 4.33) of radius 50 feet will be two-thirds full when the depth of the fluid is $y_0 + 50$ feet, and y_0 is the root of the equation

$$3y_0{}^3 - 22,500y_0 + 250,000 = 0$$

Use the Bisection Method to approximate y_0 to within 0.01 unit. (The equation was derived by calculus techniques.)

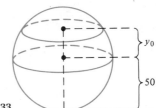

FIGURE 4.33

5 RATIONAL FUNCTIONS AND CONIC SECTIONS

Rational Functions and Their Graphs

5.1

Just as a rational number can be written as the quotient of two integers, so can a **rational function** be written as the quotient of two polynomials. Specifically, a function f is **rational** if it has the form

$$f(x) = \frac{P(x)}{Q(x)}, \qquad Q(x) \neq 0$$

where $P(x)$ and $Q(x)$ are polynomials.

Characteristics of Rational Functions

The domain of a rational function is the set of all real numbers other than those that make the denominator equal to zero. For example, the domain of

$$f(x) = \frac{x - 1}{(x - 2)(x + 1)}$$

is the set of all real numbers other than $x = 2$ and $x = -1$.

Note some special information about a rational function

$$f(x) = \frac{P(x)}{Q(x)}$$

1. Unless specified otherwise, we will assume that $P(x)$ and $Q(x)$ have no factors in common. That is, $P(x)/Q(x)$ is in *reduced form*.
2. $f(x) = 0$ if and only if $P(x) = 0$. That is, a rational function has zeros at the zeros of its numerator.
3. The zeros of $Q(x)$ make $f(x)$ undefined.

Though the function $f(x) = 1/x$ is one of the simplest rational functions, we can learn some important general characteristics of rational functions by studying its graph.

Symmetry: The graph has symmetry about the origin. Hence, we need only calculate points for positive x-values.

Domain: Note that $x = 0$ is not in the domain of f. Thus, we must pay special attention to the functional values when x is near zero.

Intercepts: Since $x = 0$ is not in the domain of f, there is no y-intercept. Furthermore, since the numerator of $f(x)$ is never zero, there are no x-intercepts.

Table of Values: Note the behavior of $f(x)$ as x approaches 0 and as x approaches very large values.

x	1	0.5	0.1	0.01	0.001	0.0001	0.00001	$\rightarrow$	0
$f(x)$	1	2	10	100	1000	10,000	100,000	$\rightarrow$	∞

x	1	2	10	100	1000	10,000	100,000	$\rightarrow$	∞
$f(x)$	1	0.5	0.1	0.01	0.001	0.0001	0.00001	$\rightarrow$	0

Using this table of values and information about symmetry and intercepts, we obtain the graph shown in Figure 5.1.

Note in Figure 5.1 and the accompanying table that the function values *increase without bound* as x approaches 0 from the right. Symbolically, we denote this behavior by

$$\frac{1}{x} \rightarrow \infty \quad \text{as} \quad x \rightarrow 0^+ \qquad \textit{(from the right)}$$

Similarly,

$$\frac{1}{x} \rightarrow -\infty \quad \text{as} \quad x \rightarrow 0^- \qquad \textit{(from the left)}$$

Under such conditions, the line $x = 0$ (the y-axis) is called a **vertical asymptote** of the graph of f.

When the function values become *arbitrarily close to zero* as x increases (or decreases) without bound, we denote this by

$$\frac{1}{x} \rightarrow 0 \quad \text{as} \quad x \rightarrow \infty \qquad \textit{(to the right)}$$

$$\frac{1}{x} \rightarrow 0 \quad \text{as} \quad x \rightarrow -\infty \qquad \textit{(to the left)}$$

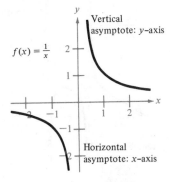

$f(x) = \frac{1}{x}$

Vertical asymptote: y–axis

Horizontal asymptote: x–axis

FIGURE 5.1

Under these conditions, the line $y = 0$ (the x-axis) is called a **horizontal asymptote** of the graph of f.

VERTICAL AND HORIZONTAL ASYMPTOTES

The line $x = a$ is a **vertical asymptote** of the graph of f if

$$f(x) \to \infty \quad \text{or} \quad f(x) \to -\infty$$

as $x \to a$, either from the right or from the left.

The line $y = b$ is a **horizontal asymptote** of the graph of f if

$$f(x) \to b$$

as $x \to \infty$ or $x \to -\infty$.

Remark: The graph of a function will never intersect its *vertical* asymptote. The graph of a function may intersect its horizontal asymptote any number of times. However, eventually (as $x \to \infty$ or $x \to -\infty$) the distance between the horizontal asymptote and the points on the graph must approach zero. Thus, an asymptote can be defined geometrically in the following way.

DEFINITION OF AN ASYMPTOTE

A line is called an **asymptote** to a curve if the distance between the line and a point (x, y) on the curve approaches zero as the distance between the origin and the point increases without bound.

We classify asymptotes as either vertical, horizontal, or slant, as shown in Figure 5.2.

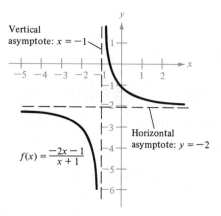
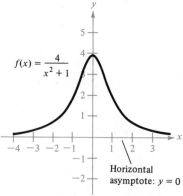
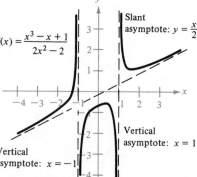

FIGURE 5.2   Asymptotes of Rational Functions

Asymptotes of Hyperbolas

The graph of $f(x) = 1/x$ (see Figure 5.1) is called an **hyperbola.** There are three basic types of rational functions whose graphs are hyperbolas.

$$f(x) = \frac{a}{Ax + B}, \quad g(x) = \frac{ax + b}{Ax + B}, \quad \text{and} \quad h(x) = \frac{ax^2 + bx + c}{Ax + B}$$

where $a \neq 0$ and $A \neq 0$. (We assume that $g(x)$ and $h(x)$ are given in reduced form.)

Each of these graphs has two asymptotes—a vertical asymptote and a horizontal (or slant) asymptote. We can find the vertical asymptote by setting the denominator equal to zero.

Vertical Asymptote: $Ax + B = 0$

$$x = -\frac{B}{A}$$

To find the horizontal (or slant) asymptote, it is helpful to write the function in the form

$$px + q + \frac{C}{Ax + B}$$

We can see that the line $y = px + q$ is an asymptote of the graph, since

$$\frac{C}{Ax + B} \to 0 \quad \text{as} \quad x \to \pm\infty$$

Horizontal or Slant Asymptote: $y = px + q$

Remark: The asymptote is horizontal if $p = 0$ and is a slant asymptote if $p \neq 0$.

You will see in Example 1 that if we find and plot the x- and y-intercepts and the asymptotes, we are well on the way to graphing a rational function.

EXAMPLE 1
Sketching the Graphs
of Rational Functions

Sketch the graphs of

(a) $f(x) = \dfrac{3}{x - 2}$ (b) $f(x) = \dfrac{2x - 1}{x}$

Solution:
For $f(x) = 3/(x - 2)$, we have the following.

(a) *y-intercept:* Since $f(0) = 3/(0 - 2) = -\frac{3}{2}$, the y-intercept occurs at $(0, -\frac{3}{2})$.

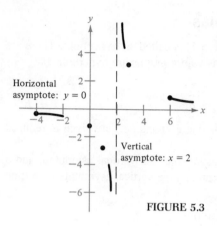

Horizontal asymptote: $y = 0$

Vertical asymptote: $x = 2$

FIGURE 5.3

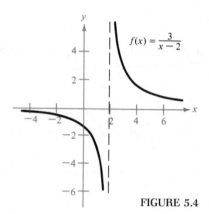

$f(x) = \dfrac{3}{x - 2}$

FIGURE 5.4

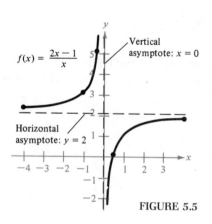

$f(x) = \dfrac{2x - 1}{x}$

Vertical asymptote: $x = 0$

Horizontal asymptote: $y = 2$

FIGURE 5.5

EXAMPLE 2
Graph with Slant Asymptote

x-intercepts: None.

Vertical Asymptote: $x - 2 = 0$ *Zero in denominator*

$x = 2$

Horizontal Asymptote: $y = 0$ $f(x) \rightarrow 0 \ as \ x \rightarrow \pm\infty$

Additional Points:

x	-4	1	3	5
$f(x)$	-0.5	-3	3	1

To sketch the graph of f, we begin by plotting the intercepts, asymptotes, and a few additional points, as shown in Figure 5.3. Then we can use these clues to make the final sketch shown in Figure 5.4.

(b) For $f(x) = (2x - 1)/x$, we have the following.

y-intercept: None.

x-intercept: $2x - 1 = 0$

$x = \dfrac{1}{2}$

Vertical Asymptote: $x = 0$

Horizontal Asymptote: By rewriting $f(x)$ as the sum of two fractions, we have

$$f(x) = \frac{2x - 1}{x} = 2 - \frac{1}{x}$$

In this form we see that $y = 2$ is a horizontal asymptote.

Additional Points:

x	-4	-1	4
$f(x)$	2.25	3	1.75

The graph of f is shown in Figure 5.5.

Sketch the graph of

$$f(x) = \frac{x^2 - x}{x + 1}$$

Solution:

$$y\text{-}intercept: \quad (0, 0).$$

$$x\text{-}intercepts: \quad x^2 - x = 0$$
$$x(x - 1) = 0$$
$$x = 0 \text{ or } 1$$

Vertical Asymptote: $x = -1$

Slant Asymptote: We begin by dividing the denominator into the numerator as follows:

$$
\begin{array}{r}
x \quad - 2 \\
x + 1 \overline{)\,x^2 - x } \\
\underline{x^2 + x } \\
-2x \\
\underline{-2x - 2} \\
2
\end{array}
$$

Thus, we have

$$f(x) = \frac{x^2 - x}{x + 1} = x - 2 + \frac{2}{x + 1}$$

Now, we can see that the graph of f has the line $y = x - 2$ as a slant asymptote. The graph of f is shown in Figure 5.6.

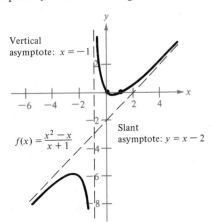

FIGURE 5.6

We can use the same approach for general rational functions that we have been using for functions that graph as hyperbolas. Use of the following guidelines should improve your skill in graphing rational functions of the form

$$f(x) = \frac{P(x)}{Q(x)}$$

For simplicity's sake, we assume that $P(x)$ and $Q(x)$ have no common factors.

GUIDELINES FOR GRAPHING RATIONAL FUNCTIONS

1. *Find and plot y-intercept* by evaluating $f(0)$.

2. *Find and plot x-intercept(s)* by setting the numerator $P(x)$ equal to zero and solving for x. Since $P(x)$ and $Q(x)$ have no common factors, all solutions to $P(x) = 0$ are in the domain of f and are legitimate x-intercepts of the graph.

3. *Find and plot the vertical asymptote(s)* by setting the denominator $Q(x)$ equal to zero and solving for x.

4. *Find and plot the horizontal (or slant) asymptotes* using the following guidelines:
   (a) If the degree of P is *less than* the degree of Q, then the graph has the x-axis ($y = 0$) as a horizontal asymptote.
   (b) If the degree of P is *equal* to the degree of Q, then the graph has a horizontal asymptote given by $y = a_n/b_m$, where a_n is the leading coefficient of P and b_m is the leading coefficient of Q.
   (c) If the degree of P is *one more than* the degree of Q, then the graph has a slant asymptote which can be determined by dividing $Q(x)$ into $P(x)$.
   (d) If the degree of P is *at least two more* than the degree of Q, then the graph has no horizontal or slant asymptotes.

5. *Plot a few additional points* both between and beyond any previously determined x-intercepts and vertical asymptotes. Remember to use any symmetry to the origin or y-axis.

6. *Complete the missing portions of the graph* in a manner that is consistent with the location of intercepts, asymptotes, and additional points.

Remark: The basic idea in applying step 4 is that we can determine the horizontal or slant asymptote of a rational function by comparing the degrees of the numerator and denominator. For instance, in part (a) of Example 1, the degree of the numerator was *less* than the degree of the denominator, thus yielding $y = 0$ as a horizontal asymptote. In part (b), the *equal* degrees yielded $y = 2$ as a horizontal asymptote. Then in Example 2, a slant asymptote was obtained because the degree of the numerator was *one more* than that of the denominator.

EXAMPLE 3
Sketching the Graph
of a Rational Function

Sketch the graph of

$$f(x) = \frac{2}{x^2 + 1}$$

Solution:

y-intercept: Since $f(0) = 2$, the y-intercept occurs at $(0, 2)$.

x-intercept: None.

Vertical Asymptotes: Since $x^2 + 1 = 0$ has no real solutions, there are no vertical asymptotes.

Horizontal Asymptotes: Since the degree of the numerator is less than the degree of the denominator, the *x*-axis is a horizontal asymptote. [Try to convince yourself that $f(x)$ approaches zero as *x* increases (or decreases) without bound.]

Additional Points: (Note that the graph is symmetrical with respect to the *y*-axis.)

x	1	2	3
$f(x)$	1	0.4	0.2

The graph of f is shown in Figure 5.7.

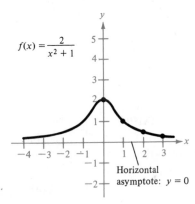

$$f(x) = \frac{2}{x^2 + 1}$$

Horizontal asymptote: $y = 0$

FIGURE 5.7

EXAMPLE 4
Sketching the Graph
of a Rational Function

Sketch the graph of

$$f(x) = \frac{x}{x^2 - x - 2}$$

Solution:

y-intercept: (0, 0).

x-intercept: (0, 0).

Vertical Asymptotes: $x^2 - x - 2 = 0$
$$(x - 2)(x + 1) = 0$$
$$x = -1 \text{ or } 2$$

Horizontal Asymptote: Since the degree of the numerator is less than the degree of the denominator, the *x*-axis (the line $y = 0$) is a horizontal asymptote.

Additional Points: Between and *beyond* the *x*-intercepts and vertical asymptotes.

x	-3	-0.5	1	3
$f(x)$	-0.3	0.4	-0.5	0.75

The graph of f is shown in Figure 5.8.

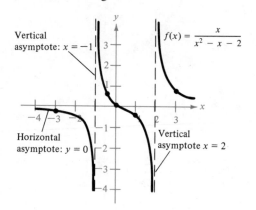

FIGURE 5.8

EXAMPLE 5
Sketching the Graph
of a Rational Function

Sketch the graph of

$$f(x) = \frac{2(x^2 - 9)}{x^2 - 4}$$

Solution: *y-intercept:* $(0, \frac{9}{2})$.

x-intercepts:
$$2(x^2 - 9) = 0$$
$$x^2 - 9 = 0$$
$$(x + 3)(x - 3) = 0$$
$$x = -3 \text{ or } 3$$

Vertical Asymptotes:
$$x^2 - 4 = 0$$
$$(x + 2)(x - 2) = 0$$
$$x = -2 \text{ or } 2$$

Horizontal Asymptote: Since the degree of the denominator is equal to the degree of the numerator, the horizontal asymptote is given by the ratio of the leading coefficients. That is, the horizontal asymptote is

$$y = \frac{2}{1} = 2$$

The graph has symmetry with respect to the *y*-axis and is shown in Figure 5.9.

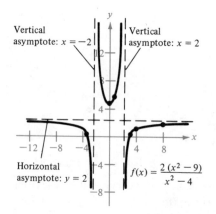

FIGURE 5.9

EXAMPLE 6
Sketching the Graph of a
Function with a Slant Asymptote

Sketch the graph of

$$f(x) = \frac{-x^3 + x^2 + 4}{x^2}$$

Solution:

y-intercept: None.

x-intercept: Using the rational zero test, we discover that $x = 2$ is a zero of the numerator. Thus, by factoring $(x - 2)$ out of the numerator, we have

$$-x^3 + x^2 + 4 = 0$$
$$-(x - 2)(x^2 + x + 2) = 0$$

Using the Quadratic Formula on the quadratic factor, we can determine that the only real zero of the numerator is $x = 2$.

Vertical Asymptote: $x = 0$

Slant Asymptote: Since the degree of the numerator exceeds the degree of the denominator by exactly 1, we divide to obtain

$$f(x) = \frac{-x^3 + x^2 + 4}{x^2} = -x + 1 + \frac{4}{x^2}$$

which implies that the slant asymptote is

$$y = -x + 1$$

The graph of f is shown in Figure 5.10.

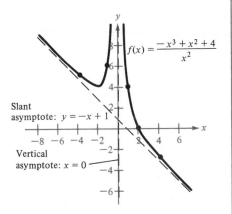

FIGURE 5.10

If the numerator and denominator of a rational function have a common factor, we must modify our sketching technique as demonstrated in the following example.

EXAMPLE 7
A Rational Function
with Common Factors

Sketch the graph of

$$g(x) = \frac{2x^2 - 7x + 3}{x^2 - 3x}$$

Solution:
We begin by factoring the numerator and denominator as follows:

$$g(x) = \frac{2x^2 - 7x + 3}{x^2 - 3x} = \frac{(2x - 1)(x - 3)}{x(x - 3)}$$

Now, for all points *other than* $x = 3$, the graph of g is the same as the graph of

$$f(x) = \frac{2x - 1}{x}$$

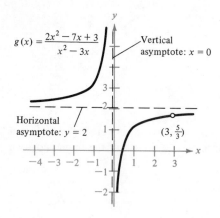

$g(x) = \dfrac{2x^2 - 7x + 3}{x^2 - 3x}$ Vertical asymptote: $x = 0$

Horizontal asymptote: $y = 2$ $(3, \tfrac{5}{3})$

FIGURE 5.11

We have already graphed this function in part (b) of Example 1. To sketch the graph of g, we must account for the fact that $x = 3$ is not in the domain of g. Graphically, we indicate that g is undefined when $x = 3$ by an *open dot* at the point $(3, \tfrac{5}{3})$, as shown in Figure 5.11.

Remark: In Example 7, the common factor $(x - 3)$ occurred just once as a factor of the numerator and once as a factor of the denominator. By canceling this factor, we formed a new function, since we introduced a new value to the domain. If $(x - 3)$ had occurred as a *multiple* factor of the denominator, then cancellation might not have added a new value to the domain, and in such cases we would not have formed a new function. For example, the following functions both exclude $x = 1$ from their domains and are therefore equal functions:

$$f(x) = \dfrac{x - 1}{(x - 1)^2} \qquad g(x) = \dfrac{1}{x - 1}$$

Section Exercises 5.1

In Exercises 1–10, match the rational function with the correct graph.

1. $f(x) = \dfrac{2}{x + 1}$

2. $f(x) = \dfrac{1}{x - 4}$

3. $f(x) = \dfrac{x + 1}{x}$

4. $f(x) = \dfrac{1 - 2x}{x}$

5. $f(x) = \dfrac{2 - x}{x - 1}$

6. $f(x) = \dfrac{x + 2}{x + 1}$

7. $f(x) = \dfrac{x - 2}{x - 1}$

8. $f(x) = -\dfrac{x + 2}{x + 1}$

9. $f(x) = \dfrac{x^2 + 1}{x}$

10. $f(x) = \dfrac{x^2 - 2x}{x - 1}$

(c)

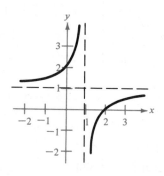

(d)

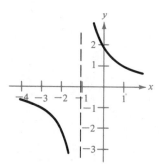

(a)

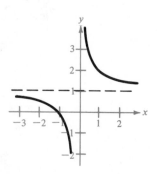

(b)

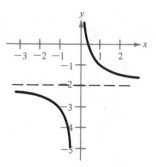

(e)

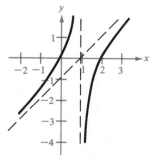

(f)

(g)

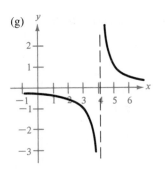

(h)

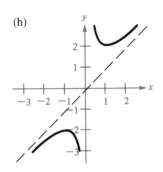

(i)

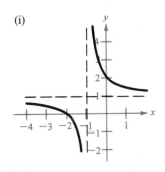

(j)

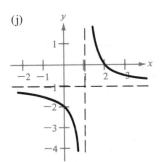

In Exercises 11–30, sketch the graph of the rational function, labeling all intercepts and asymptotes.

11. $f(x) = \dfrac{1}{x + 2}$

12. $f(x) = \dfrac{1}{x - 3}$

13. $f(x) = \dfrac{-1}{x + 2}$

14. $f(x) = \dfrac{1}{3 - x}$

15. $f(x) = \dfrac{1}{x + 2} + 2$

16. $f(x) = \dfrac{1}{x - 3} + 1$

17. $f(x) = \dfrac{x + 1}{x + 2}$

18. $f(x) = \dfrac{x - 2}{x - 3}$

19. $f(x) = \dfrac{2 + x}{1 - x}$

20. $f(x) = \dfrac{3 - x}{2 - x}$

21. $f(x) = \dfrac{3x + 1}{x}$

22. $f(x) = \dfrac{1 - 2x}{x}$

23. $f(x) = \dfrac{5 + 2x}{1 + x}$

24. $f(x) = \dfrac{1 - 3x}{1 - x}$

25. $f(x) = \dfrac{2x^2 + 1}{x}$

26. $f(x) = \dfrac{1 - x^2}{x}$

27. $f(x) = \dfrac{x^2 - x + 1}{x - 1}$

28. $f(x) = \dfrac{x^2}{x - 1}$

29. $f(x) = \dfrac{x^2 + 5x + 8}{x + 3}$

30. $f(x) = \dfrac{2x^2 + x}{x + 1}$

In Exercises 31–40, find all the asymptotes of the rational function.

31. $f(x) = \dfrac{1}{x^2}$

32. $f(x) = \dfrac{4}{(x - 2)^3}$

33. $f(x) = \dfrac{2 + x}{2 - x}$

34. $f(x) = \dfrac{1 - 5x}{1 + 2x}$

35. $f(x) = \dfrac{x^3}{x^2 - 1}$

36. $f(x) = \dfrac{2x^2}{x + 1}$

37. $f(x) = \dfrac{3x^2 + 1}{x^2 + 9}$

38. $f(x) = \dfrac{3x^2 + x - 5}{x^2 + 1}$

39. $f(x) = \dfrac{5x^4 - 2x^2 + 1}{x^2 + 1}$

40. $f(x) = \dfrac{x^5}{2x^3 + 1}$

In Exercises 41–60, sketch the graph of the rational function. As a sketching aid, examine the function for intercepts, symmetry, and asymptotes.

41. $f(x) = \dfrac{2 + x}{1 - x}$

42. $f(x) = \dfrac{x - 3}{x - 2}$

43. $f(x) = \dfrac{x^2}{x^2 + 9}$

44. $f(x) = 2 - \dfrac{3}{x^2}$

45. $f(x) = \dfrac{x^2}{x^2 - 9}$

46. $f(x) = \dfrac{x^2 + 1}{x}$

47. $f(x) = \dfrac{x^3}{x^2 - 1}$

48. $f(x) = \dfrac{x^3}{2x^2 - 8}$

49. $f(x) = \dfrac{x}{x^2 + 1}$

50. $f(x) = \dfrac{2x}{x^2 + x - 2}$

51. $f(x) = \dfrac{2x}{1 - x^2}$

52. $f(x) = \dfrac{2x^2 - 5x + 5}{x - 2}$

53. $f(x) = \dfrac{3x}{x^2 - x - 2}$

54. $f(x) = \dfrac{x^2 + x - 2}{x^2 - 2x + 1}$

55. $f(x) = \dfrac{x - 3}{x^2 - 9}$

56. $f(x) = \dfrac{x + 2}{x^2 - 3x - 10}$

57. $f(x) = \dfrac{x^2 - 25}{x + 5}$

58. $f(x) = \dfrac{2x^2 - x - 3}{x + 1}$

59. $f(x) = \dfrac{x^2}{1 + x^4}$

60. $f(x) = \dfrac{4}{1 + x^2}$

61. A right triangle is formed in the first quadrant by the x and y axes and a line segment through the point $(2, 3)$. (See Figure 5.12.)
    (a) Show that an equation of the line segment is

$$y = \frac{3(x - a)}{2 - a}$$

    (b) Show that the area of the triangle is

$$A = \frac{-3a^2}{2(2 - a)}$$

(c) Sketch the graph of the area function of part (b), and from the graph estimate the value of a ($a > 0$) so that the area is minimum.

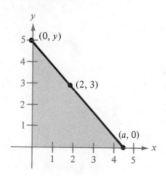

FIGURE 5.12

62. The cost of producing x units is $C = 0.2x^2 + 10x + 5$, and therefore the average cost per unit is

$$\overline{C} = \frac{C}{x} = 0.2x + 10 + \frac{5}{x}$$

Sketch the graph of the average cost function, and estimate the number of units that should be produced to minimize the average cost.

63. The region shown in Figure 5.13 is bounded by $y = x^2$, $y = 0$, and $x = 2$. It can be shown by techniques of calculus that the area A of this region is approximated by

$$f(n) = \frac{4}{3} \cdot \frac{2n^3 + 3n^2 + n}{n^3}$$

As n increases without bound, $f(n)$ approaches the exact area of the region [i.e., A can be obtained from the horizontal asymptote of $f(n)$]. Find A.

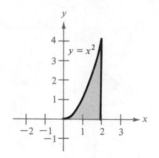

FIGURE 5.13   **FIGURE 5.14**

64. The region shown in Figure 5.14 is bounded by $y = x^3$, $y = 0$, $x = 1$, and $x = 4$. It can be shown by techniques of calculus that the area A of this region is approximated by

$$f(n) = 3 + \frac{27}{2} \cdot \frac{n(n + 1)}{n^2}$$
$$+ \frac{27}{2} \cdot \frac{n(n + 1)(2n + 1)}{n^3}$$
$$+ \frac{81}{4} \cdot \frac{n^2(n + 1)^2}{n^4}$$

As n increases without bound, $f(n)$ approaches the exact area of the region [i.e., A can be obtained from the horizontal asymptote of $f(n)$]. Find A.

Partial Fractions   5.2

In calculus it is often useful to rewrite a rational expression as the sum of simpler rational expressions. For example, the rational expression $(x + 7)/(x^2 - x - 6)$ can be written as the sum of two fractions with linear denominators, as follows:

$$\frac{x + 7}{x^2 - x - 6} = \frac{2}{x - 3} + \frac{-1}{x + 2}$$

The fractions on the right side are called **partial fractions,** and the entire right side is called the **partial fraction decomposition** of the left side.

In Chapter 4 we noted that it is theoretically possible to write any polynomial as the product of linear and irreducible quadratic factors. For instance,

$$x^5 + x^4 - x - 1 = (x - 1)(x + 1)^2(x^2 + 1)$$

where $(x - 1)$ is a linear factor, $(x + 1)^2$ is a repeated linear factor, and $(x^2 + 1)$ is an irreducible quadratic factor.

We can use this factorization to find the partial fraction decomposition of any rational expression having $x^5 + x^4 - x - 1$ as its denominator. Specifically, if $N(x)$ is a polynomial of degree less than five, then the partial fraction decomposition of $N(x)/(x^5 + x^4 - x - 1)$ has the form

$$\frac{N(x)}{x^5 + x^4 - x - 1} = \frac{N(x)}{(x - 1)(x + 1)^2(x^2 + 1)}$$

$$= \frac{A}{x - 1} + \frac{B}{x + 1} + \frac{C}{(x + 1)^2} + \frac{Dx + F}{x^2 + 1}$$

Note that the repeated linear factor $(x + 1)^2$ results in *two* fractions, one for $(x + 1)$ and one for $(x + 1)^2$. If $(x + 1)^3$ were a factor, then we would use three fractions: one for $(x + 1)$, one for $(x + 1)^2$, and one for $(x + 1)^3$. In general, the number of fractions resulting from a repeated linear factor is equal to the number of times the factor is repeated.

DECOMPOSITION OF $N(x)/D(x)$ INTO PARTIAL FRACTIONS

1. *Divide if improper:* If $N(x)/D(x)$ is an improper fraction, then divide the denominator into the numerator to obtain

   $$\frac{N(x)}{D(x)} = \text{(a polynomial)} + \frac{N_1(x)}{D(x)}$$

   and apply steps 2, 3, and 4 to the proper rational expression $N_1(x)/D(x)$.

2. *Factor denominator:* Completely factor the denominator into factors of the form

   $$(px + q)^m \quad \text{and} \quad (ax^2 + bx + c)^n$$

   where $(ax^2 + bx + c)$ is irreducible.

3. *Linear factors:* For *each* factor of the form $(px + q)^m$, the partial fraction decomposition must include the following sum of m fractions:

   $$\frac{A_1}{(px + q)} + \frac{A_2}{(px + q)^2} + \cdots + \frac{A_m}{(px + q)^m}$$

4. *Quadratic factors:* For *each* factor of the form $(ax^2 + bx + c)^n$, the partial fraction decomposition must include the following sum of n fractions:

   $$\frac{B_1x + C_1}{ax^2 + bx + c} + \frac{B_2x + C_2}{(ax^2 + bx + c)^2} + \cdots$$

   $$+ \frac{B_nx + C_n}{(ax^2 + bx + c)^n}$$

Algebraic techniques for determining the constants in the numerators are demonstrated in the examples that follow.

EXAMPLE 1
Distinct Linear Factors

Write the partial fraction decomposition for

$$\frac{x + 7}{x^2 - x - 6}$$

Solution:
Since

$$x^2 - x - 6 = (x - 3)(x + 2)$$

we include one partial fraction for each factor and write

$$\frac{x + 7}{x^2 - x - 6} = \frac{A}{x - 3} + \frac{B}{x + 2}$$

Multiplying this equation by the lowest common denominator (LCD), $(x - 3)(x + 2)$, leads to the **basic equation**

$$x + 7 = A(x + 2) + B(x - 3)$$

Since this equation is to be true for all x, we can substitute *convenient* values for x to obtain equations in A and B. These values are the ones that make particular factors zero. To solve for B, we let $x = -2$ and obtain

$$-2 + 7 = A(0) + B(-5)$$
$$5 = -5B$$
$$-1 = B$$

To solve for A, We let $x = 3$ and obtain

$$3 + 7 = A(5) + B(0)$$
$$10 = 5A$$
$$2 = A$$

Therefore, the decomposition is

$$\frac{x + 7}{x^2 - x - 6} = \frac{2}{x - 3} - \frac{1}{x + 2}$$

EXAMPLE 2
Repeated Linear Factors

Write the partial fraction decomposition for

$$\frac{5x^2 + 20x + 6}{x^3 + 2x^2 + x}$$

Solution:
Since

$$x^3 + 2x^2 + x = x(x^2 + 2x + 1) = x(x + 1)^2$$

we include one fraction for each power of x and $(x + 1)$ and write

$$\frac{5x^2 + 20x + 6}{x(x + 1)^2} = \frac{A}{x} + \frac{B}{x + 1} + \frac{C}{(x + 1)^2}$$

Multiplying by the LCD, $x(x + 1)^2$, leads to the *basic* equation

$$5x^2 + 20x + 6 = A(x + 1)^2 + Bx(x + 1) + Cx$$

Let $x = -1$ to eliminate the A and B terms:

$$5 - 20 + 6 = 0 + 0 - C$$
$$C = 9$$

Let $x = 0$ to eliminate the B and C terms:

$$6 = A(1) + 0 + 0$$
$$6 = A$$

We have exhausted the most convenient choices for x, so to find the value of B, we use *any other value* for x along with the calculated values of A and C. Thus, using $x = 1$, $A = 6$, and $C = 9$, we have

$$5 + 20 + 6 = A(4) + B(2) + C$$
$$31 = 6(4) + 2B + 9$$
$$-2 = 2B$$
$$-1 = B$$

Therefore,

$$\frac{5x^2 + 20x + 6}{x(x + 1)^2} = \frac{6}{x} - \frac{1}{x + 1} + \frac{9}{(x + 1)^2}$$

Remark: You need to make as many substitutions for x as there are unknowns $(A, B, C, \ldots)$.

EXAMPLE 3
Distinct Linear and
Quadratic Factors

Write the partial fraction decomposition for

$$\frac{2x^3 - 4x - 8}{(x^2 - x)(x^2 + 4)}$$

Solution:
Since

$$(x^2 - x)(x^2 + 4) = x(x - 1)(x^2 + 4)$$

we include one partial fraction for each factor and write

$$\frac{2x^3 - 4x - 8}{x(x - 1)(x^2 + 4)} = \frac{A}{x} + \frac{B}{x - 1} + \frac{Cx + D}{x^2 + 4}$$

Multiplying by the LCD, $[x(x - 1)(x^2 + 4)]$, yields the *basic* equation

$$2x^3 - 4x - 8 = A(x - 1)(x^2 + 4)$$
$$+ Bx(x^2 + 4) + (Cx + D)(x)(x - 1)$$

If $x = 1$, then

$$-10 = 0 + B(5) + 0$$
$$-2 = B$$

If $x = 0$, then

$$-8 = A(-1)(4) + 0 + 0$$
$$2 = A$$

At this point C and D are yet to be determined. We can find these remaining constants by comparing coefficients in the basic equation. We proceed as follows:

1. Expand the basic equation:

$$2x^3 - 4x - 8 = A(x - 1)(x^2 + 4) + Bx(x^2 + 4)$$
$$+ (Cx + D)(x)(x - 1)$$
$$= Ax^3 - Ax^2 + 4Ax - 4A + Bx^3$$
$$+ 4Bx + Cx^3 + Dx^2 - Cx^2 - Dx$$

2. Collect like terms:

$$2x^3 - 4x - 8 = (A + B + C)x^3 + (-A + D - C)x^2$$
$$+ (4A + 4B - D)x - 4A$$

3. Equate coefficients of like powers on opposite sides:

$$2x^3 + 0x^2 - 4x - 8 = (A + B + C)x^3 + (-A + D - C)x^2$$
$$+ (4A + 4B - D)x - 4A$$

Thus, we have

$$2 = A + B + C, \qquad 0 = -A + D - C,$$
$$-4 = 4A + 4B - D, \qquad -8 = -4A$$

Substituting the known values of $A = 2$ and $B = -2$, we have

$$2 = 2 - 2 + C \qquad -4 = 4(2) + 4(-2) - D$$
$$2 = C \qquad\qquad 4 = D$$

Finally, we have

$$\frac{2x^3 - 4x - 8}{x(x - 1)(x^2 + 4)} = \frac{A}{x} + \frac{B}{x - 1} + \frac{Cx + D}{x^2 + 4}$$

$$= \frac{2}{x} - \frac{2}{x - 1} + \frac{2x}{x^2 + 4} + \frac{4}{x^2 + 4}$$

In each of the first three examples, we began the solution of the basic equation by substituting values of x that made the linear factors zero. This method works well when the partial fraction decomposition involves *only* linear factors. However, if the decomposition involves a quadratic factor, then the alternative procedure (comparing coefficients in the basic equation) proves to be quite efficient. We suggest the latter procedure for partial fractions involving quadratic denominators. Both methods are outlined in the following summary.

GUIDELINES FOR SOLVING THE BASIC EQUATION

Linear Factors:

1. Substitute the *roots* of the distinct linear factors into the basic equation.

2. For repeated linear factors, use the coefficients determined in part 1 to rewrite the basic equation. Then substitute *other* convenient values for x and solve for the remaining coefficients.

Quadratic Factors:

1. Expand the basic equation.

2. Collect terms according to powers of x.

3. Equate the coefficients of like powers to obtain equations involving A, B, C, etc.

4. Use substitution to solve for A, B, C, ...

The second procedure for solving the basic equation is demonstrated in the next example.

EXAMPLE 4
Repeated Quadratic Factors

Write the partial fraction decomposition for

$$\frac{8x^3 + 13x}{(x^2 + 2)^2}$$

Solution:
We include one partial fraction for each power of $(x^2 + 2)$,

$$\frac{8x^3 + 13x}{(x^2 + 2)^2} = \frac{Ax + B}{x^2 + 2} + \frac{Cx + D}{(x^2 + 2)^2}$$

Multiplying by the LCD, $(x^2 + 2)^2$, yields the *basic* equation

$$
\begin{aligned}
8x^3 + 13x &= (Ax + B)(x^2 + 2) + Cx + D \\
&= Ax^3 + 2Ax + Bx^2 + 2B + Cx + D \\
&= Ax^3 + Bx^2 + (2A + C)x + (2B + D)
\end{aligned}
$$

Equating coefficients of like terms,

$$8x^3 + 0x^2 + 13x + 0 = Ax^3 + Bx^2 + (2A + C)x + (2B + D)$$

we have

$$8 = A, \quad 0 = B, \quad 13 = 2A + C, \quad 0 = 2B + D$$

Using the known values $A = 8$ and $B = 0$, we have

$$13 = 2A + C = 2(8) + C \qquad 0 = 2B + D = 2(0) + D$$
$$-3 = C \qquad\qquad\qquad 0 = D$$

Finally, we conclude that

$$\frac{8x^3 + 13x}{(x^2 + 2)^2} = \frac{8x}{x^2 + 2} + \frac{-3x}{(x^2 + 2)^2}$$

When working the exercises, you should keep in mind that for *improper* rational expressions like

$$\frac{N(x)}{D(x)} = \frac{2x^3 + x^2 - 7x + 7}{x^2 + x - 2}$$

you must first divide to obtain the form

$$\frac{N(x)}{D(x)} = (\text{polynomial}) + \frac{N_1(x)}{D(x)}$$

The proper rational expression $N_1(x)/D(x)$ is then decomposed into its partial fractions by the usual methods.

Section Exercises 5.2

In Exercises 1–30, write the partial fraction decomposition for the rational expression.

1. $\dfrac{1}{x^2 - 1}$

2. $\dfrac{1}{4x^2 - 9}$

3. $\dfrac{3}{x^2 + x - 2}$

4. $\dfrac{x + 1}{x^2 + 4x + 3}$

5. $\dfrac{5 - x}{2x^2 + x - 1}$

6. $\dfrac{3x^2 - 7x - 2}{x^3 - x}$

7. $\dfrac{x^2 + 12x + 12}{x^3 - 4x}$

8. $\dfrac{x^3 - x + 3}{x^2 + x - 2}$

9. $\dfrac{2x^3 - 4x^2 - 15x + 5}{x^2 - 2x - 8}$

10. $\dfrac{x + 2}{x(x - 4)}$

11. $\dfrac{4x^2 + 2x - 1}{x^2(x + 1)}$

12. $\dfrac{2x - 3}{(x - 1)^2}$

13. $\dfrac{x^4}{(x - 1)^3}$

14. $\dfrac{4x^2 - 1}{2x(x + 1)^2}$

15. $\dfrac{3x}{(x - 3)^2}$

16. $\dfrac{6x^2 + 1}{x^2(x - 1)^3}$

17. $\dfrac{x^2 - 1}{x(x^2 + 1)}$

18. $\dfrac{x}{(x - 1)(x^2 + x + 1)}$

19. $\dfrac{x^2}{x^4 - 2x^2 - 8}$

20. $\dfrac{2x^2 + x + 8}{(x^2 + 4)^2}$

21. $\dfrac{x}{16x^4 - 1}$

22. $\dfrac{x^2 - 4x + 7}{(x + 1)(x^2 - 2x + 3)}$

23. $\dfrac{x^2 + x + 2}{(x^2 + 2)^2}$

24. $\dfrac{x^3}{(x + 2)^2(x - 2)^2}$

25. $\dfrac{x^2 + 5}{(x + 1)(x^2 - 2x + 3)}$

26. $\dfrac{x - 1}{x^3 + x^2}$

27. $\dfrac{x + 1}{x^3 + x}$

28. $\dfrac{x^2 - x}{x^2 + x + 1}$

29. $\dfrac{1}{y(L - y)}$, L is a constant

30. $\dfrac{1}{(x + 1)(n - x)}$, n is a positive integer

Conic Sections   5.3

Conic sections have a rich historical background, going back to early Greek mathematics. Initially, interest in conics centered around geometrical problems. With the advent of scientific discovery in the seventeenth century, the broad applicability of conics became apparent, and they played a prominent role in the early history of calculus.

The name **conic section,** or simply **conic,** refers to the description of a conic as the intersection of a double-napped cone and a plane. Notice from Figure 5.15 that in the formation of the four basic conics, the intersecting plane does not pass through the vertex of the cone. When the plane does pass through the vertex, we call the resulting figure a **degenerate conic,** as shown in Figure 5.16.

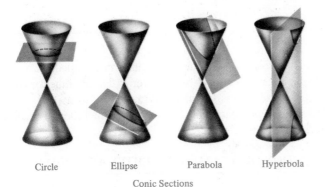

Circle   Ellipse   Parabola   Hyperbola

Point   Line   Two intersecting lines

Conic Sections

Degenerate Conics

FIGURE 5.15

FIGURE 5.16

There are several ways to approach a study of conics. We could begin by defining the conics in terms of the intersections of planes and cones, as the Greeks did. Or we could define them algebraically, in terms of the general second-degree equation

$$Ax^2 + Bxy + Cy^2 + Dx + Ey + F = 0$$

However, we will use a third approach, in which each of the conics is defined as a *locus* (collection) of points satisfying a certain geometric property. For example, in Section 3.2, we saw how the definition of a circle as *the collection of all points (x, y) that are equidistant from a fixed point (h, k)* led easily to the standard equation of a circle,

$$(x - h)^2 + (y - k)^2 = r^2$$

For the sake of simplicity, we will restrict our study of conics in this section to parabolas with vertices at the origin and ellipses and hyperbolas with centers at the origin. In the following section, we will look at the more general cases.

DEFINITION OF A PARABOLA

A **parabola** is set of all points (x, y) that are equidistant from a fixed line (**directrix**) and a fixed point (**focus**) not on the line. (See Figure 5.17.)

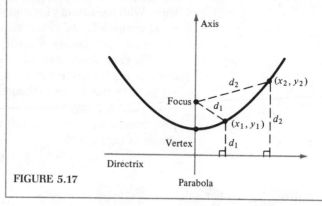

FIGURE 5.17

The midpoint between the focus and the directrix is called the **vertex,** and the line passing through the focus and the vertex is called the **axis** of the parabola. Note that a parabola is symmetric with respect to its axis.

Using this definition of a parabola, we can derive the following standard form of the equation of a parabola.

STANDARD EQUATION OF A PARABOLA (VERTEX AT ORIGIN)

The **standard form of the equation of a parabola** with vertex at $(0, 0)$ and directrix $y = -p$ is

$$x^2 = 4py \qquad\qquad \textit{(Vertical axis)}$$

For directrix $x = -p$, the equation is

$$y^2 = 4px \qquad\qquad \textit{(Horizontal axis)}$$

The focus lies on the axis p units (directed distance) from the vertex.

Proof:

Since the two cases are similar, we give a proof for the first case only. Suppose the directrix ($y = -p$) is parallel to the x-axis. In Figure 5.18 we assume $p > 0$, and since p is the directed distance from the vertex to the

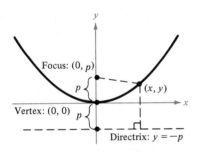

FIGURE 5.18

focus, it follows that the focus lies above the vertex. Since by definition the point (x, y) is equidistant from $(0, p)$ and $y = -p$, we can apply the distance formula to obtain

$$\sqrt{(x - 0)^2 + (y - p)^2} = y + p$$

Squaring both sides of this equation yields

$$x^2 + (y - p)^2 = (y + p)^2$$
$$x^2 + y^2 - 2py + p^2 = y^2 + 2py + p^2$$
$$x^2 = 4py$$

EXAMPLE 1
Finding the Focus of a Parabola

Find the focus of the parabola whose equation is

$$y = -2x^2$$

Solution:
Since the squared term in the equation involves x, we know that the axis is vertical, and we have the standard form

$$x^2 = 4py$$

Writing the given equation in this form, we have

$$x^2 = -\frac{1}{2}y$$

$$x^2 = 4\left(-\frac{1}{8}\right)y$$

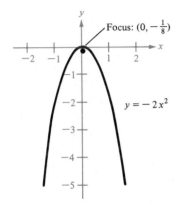

FIGURE 5.19

Thus, we have $p = -\frac{1}{8}$. Since p is negative, the parabola opens downward (see Figure 5.19), and the focus of the parabola is

$$(0, p) = \left(0, -\frac{1}{8}\right)$$

EXAMPLE 2
A Parabola with a Horizontal Axis

Write the standard equation of the parabola with its vertex at the origin and its focus at $(2, 0)$.

Solution:
The axis of the parabola is horizontal, passing through $(0, 0)$ and $(2, 0)$. (See Figure 5.20.) Thus, we consider the form

$$y^2 = 4px$$

Since $p = 2$, the equation is

$$y^2 = 4(2)x = 8x$$

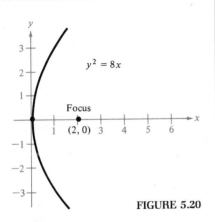

FIGURE 5.20

The term *parabola* is a technical term used in mathematics to describe a special type of U-shaped curve with explicit properties. Similarly, the term *ellipse* describes only a very special type of oval curve having the following properties.

DEFINITION OF AN ELLIPSE

An **ellipse** is the set of all points (x, y) the sum of whose distances from two distinct fixed points (**foci**) is constant. (See Figure 5.21.)

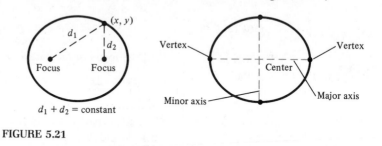

FIGURE 5.21

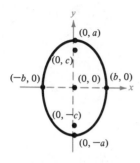

FIGURE 5.22

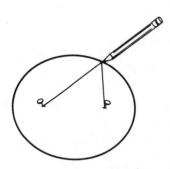

Major axis is horizontal.
Minor axis is vertical.

FIGURE 5.23

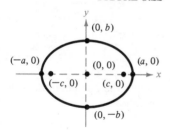

Major axis is vertical.
Minor axis is horizontal.

FIGURE 5.24

One way to visualize this definition of an ellipse is to consider two thumbtacks placed at the foci (see Figure 5.22). If we fasten the ends of a fixed length of string to the thumbtacks and draw the string taut with a pencil, then the traceable path of the pencil will be an ellipse.

Referring to Figure 5.21, the line through the foci intersects the ellipse at two points called the **vertices.** The chord joining the vertices is called the **major axis,** and its midpoint is called the **center** of the ellipse. The chord perpendicular to the major axis at the center is called the **minor axis** of the ellipse.

STANDARD EQUATION OF AN ELLIPSE (WITH CENTER AT ORIGIN)

The **standard form of the equation of an ellipse** with center at origin and major and minor axes of lengths $2a$ and $2b$, respectively, is

$$\frac{x^2}{a^2} + \frac{y^2}{b^2} = 1$$

(Horizontal major axis, Figure 5.23)

or

$$\frac{x^2}{b^2} + \frac{y^2}{a^2} = 1$$

(Vertical major axis, Figure 5.24)

The foci lie on the major axis, c units from the center, where a, b, and c are related by the equation

$$c^2 = a^2 - b^2$$

EXAMPLE 3
Finding the Standard
Equation of an Ellipse

Find the standard form of the equation of the ellipse with a major axis of length 6, having foci at $(-2, 0)$ and $(2, 0)$, as shown in Figure 5.25.

Solution:
Since the foci occur at $(-2, 0)$ and $(2, 0)$, the center of the ellipse is $(0, 0)$, and the major axis is horizontal. Thus, the ellipse has an equation of the form

$$\frac{x^2}{a^2} + \frac{y^2}{b^2} = 1$$

Furthermore, since we are given the length of the major axis, we have

$$2a = 6$$
$$a = 3$$

and since we know that c is the distance from the center to either focus we have

$$c = 2$$

Finally, we have

$$b^2 = a^2 - c^2 = 3^2 - 2^2 = 9 - 4 = 5$$

which gives us the equation

$$\frac{x^2}{9} + \frac{y^2}{5} = 1$$

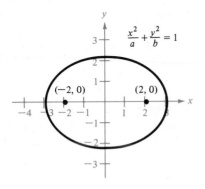

FIGURE 5.25

EXAMPLE 4
Finding the Foci and
Vertices of an Ellipse

Find the foci and vertices of the ellipse given by

$$4x^2 + y^2 = 36$$

Solution:
We begin by writing the equation in standard form:

$$4x^2 + y^2 = 36$$
$$\frac{4x^2}{36} + \frac{y^2}{36} = \frac{36}{36}$$
$$\frac{x^2}{3^2} + \frac{y^2}{6^2} = 1$$

Since the denominator of the y^2 term is larger than the denominator of the x^2 term, we conclude that the major axis is vertical. Thus, the standard form is

$$\frac{x^2}{b^2} + \frac{y^2}{a^2} = 1$$

and we have $a = 6$ and $b = 3$. To find c, we write

$$c^2 = a^2 - b^2 = 36 - 9 = 27$$
$$c = \sqrt{27} = 3\sqrt{3}$$

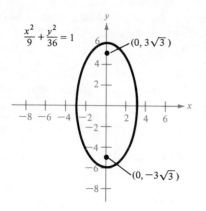

$$\frac{x^2}{9} + \frac{y^2}{36} = 1$$

FIGURE 5.26

Finally, we conclude that the foci are

$$(0, \ -3\sqrt{3}) \qquad \text{and} \qquad (0, \ 3\sqrt{3})$$

and the vertices are

$$(0, \ -6) \qquad \text{and} \qquad (0, \ 6)$$

See Figure 5.26.

Remark: In Example 4, note that once we have written the equation in standard form we can easily sketch the ellipse by locating the endpoints of the major and minor axes. Since 3^2 is the denominator of the x^2 term, we move 3 units to the *right and left* of the center to locate the endpoints of the horizontal axis. Similarly, since 6^2 is the denominator of the y^2 term, we move 6 units *up and down* from the center to locate the endpoints of the vertical axis.

The definition of a **hyperbola** parallels that of an ellipse. The distinction is that for an ellipse the *sum* of the distances between the foci and a point on the ellipse is fixed, while for a hyperbola the *difference* of these distances is fixed.

DEFINITION OF A HYPERBOLA

A **hyperbola** is the set of all points (x, y), the difference of whose distances from two distinct fixed points (**foci**) is constant. (See Figure 5.27.)

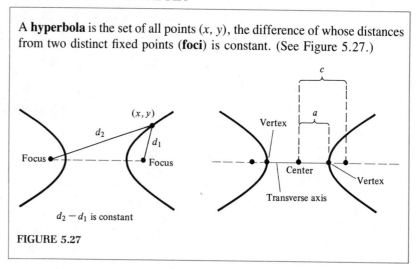

FIGURE 5.27

One distinguishing feature of hyperbolas is that their graphs have two disconnected parts called **branches.** Referring to Figure 5.27, the line through the two foci intersects the hyperbola at two points called **vertices.** The line segment connecting the vertices is called the **transverse axis,** and the midpoint of the transverse axis is called the **center** of the hyperbola.

STANDARD EQUATION OF A HYPERBOLA (CENTER AT ORIGIN)

The **standard form of the equation of a hyperbola** with center at $(0, 0)$ is

$$\frac{x^2}{a^2} - \frac{y^2}{b^2} = 1$$

(Transverse axis is horizontal, Figure 5.28)

$$\frac{y^2}{a^2} - \frac{x^2}{b^2} = 1$$

(Transverse axis is vertical, Figure 5.29)

where the vertices and foci are, respectively, a and c units from the center and $b^2 = c^2 - a^2$.

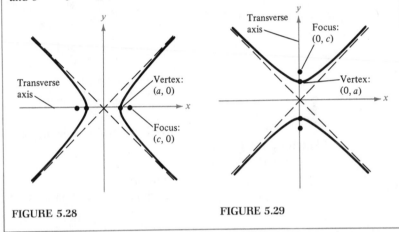

FIGURE 5.28   FIGURE 5.29

EXAMPLE 5

Finding the Standard Equation of a Hyperbola

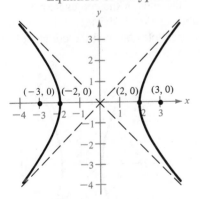

FIGURE 5.30

Find the standard form of the equation of the hyperbola with foci at $(-3, 0)$ and $(3, 0)$ and vertices at $(-2, 0)$ and $(2, 0)$, as shown in Figure 5.30.

Solution:
Since the foci are 3 units from the center, we have $c = 3$, and since the vertices are 2 units from the center, we have $a = 2$. Thus, it follows that

$$b^2 = c^2 - a^2 = 3^2 - 2^2 = 9 - 4 = 5$$

Finally, since the transverse axis is horizontal, the standard equation has the form

$$\frac{x^2}{a^2} - \frac{y^2}{b^2} = 1$$

Substituting the values of a^2 and b^2, we have

$$\frac{x^2}{4} - \frac{y^2}{5} = 1$$

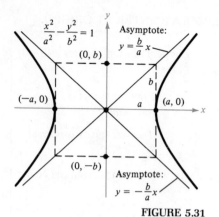

FIGURE 5.31

An important aid in sketching the graph of a hyperbola is the determination of its **asymptotes** (see Figure 5.31). Each hyperbola has two asymptotes that intersect at the center of the hyperbola. Furthermore, the asymptotes pass through the corners of a rectangle of dimensions $2a$ by $2b$. The line segment of length $2b$, joining $(0, b)$ and $(0, -b)$, is referred to as the **conjugate axis** of the hyperbola.

ASYMPTOTES OF A HYPERBOLA (WITH CENTER AT ORIGIN)

The asymptotes of a hyperbola with center at $(0, 0)$ are

$$y = \frac{b}{a} x \quad \text{and} \quad y = -\frac{b}{a} x \qquad \text{\textit{(Transverse axis is horizontal)}}$$

or

$$y = \frac{a}{b} x \quad \text{and} \quad y = -\frac{a}{b} x \qquad \text{\textit{(Transverse axis is vertical)}}$$

(See Figure 5.31.)

EXAMPLE 6
Sketching the Graph
of a Hyperbola

Sketch the graph of the hyperbola whose equation is

$$4x^2 - y^2 = 16$$

Solution:
Rewriting this equation in standard form, we have

$$\frac{4x^2}{16} - \frac{y^2}{16} = \frac{16}{16}$$

$$\frac{x^2}{2^2} - \frac{y^2}{4^2} = 1$$

From this, we conclude that the transverse axis is horizontal and the vertices occur at $(-2, 0)$ and $(2, 0)$. Moreover, the ends of the conjugate axis occur at $(0, -4)$ and $(0, 4)$, and we are able to sketch the rectangle shown in Figure 5.32. Finally, by drawing the asymptotes through the corners of this rectangle, we complete the sketch shown in Figure 5.33.

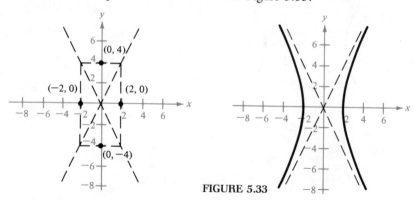

FIGURE 5.32   **FIGURE 5.33**

EXAMPLE 7
Finding the Standard
Equation of a Hyperbola

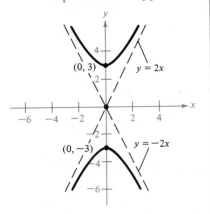

FIGURE 5.34

Find the standard form of the equation of the hyperbola having vertices at $(0, -3)$ and $(0, 3)$ and with asymptotes $y = -2x$ and $y = 2x$, as shown in Figure 5.34.

Solution:
Since the transverse axis is vertical, we have asymptotes of the form

$$y = \frac{a}{b}x \quad \text{and} \quad y = -\frac{a}{b}x$$

Thus, we have

$$\frac{a}{b} = 2$$

and since $a = 3$, we can determine that $b = \frac{3}{2}$. Now we see that the hyperbola has the following equation:

$$\frac{y^2}{3^2} - \frac{x^2}{(\frac{3}{2})^2} = 1$$

Section Exercises 5.3

In Exercises 1–10, match the equation with the correct graph.

1. $\dfrac{x^2}{1} + \dfrac{y^2}{4} = 1$

2. $x^2 = 4y$

3. $\dfrac{x^2}{1} - \dfrac{y^2}{4} = 1$

4. $y^2 = -4x$

5. $\dfrac{x^2}{4} + \dfrac{y^2}{1} = 1$

6. $\dfrac{y^2}{4} - \dfrac{x^2}{1} = 1$

7. $x^2 = -6y$

8. $x^2 + 5y^2 = 10$

9. $x^2 - 5y^2 = -5$

10. $y^2 - 8x = 0$

(c)

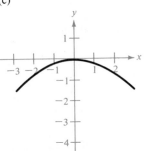

(d)

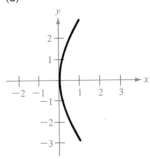

(a)

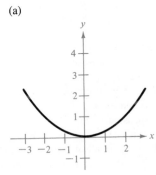

(b)

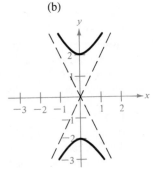

(e)

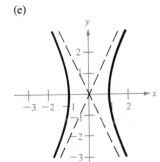

(f)

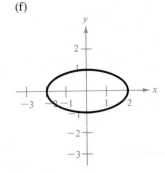

(g)

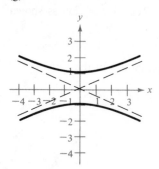

(h)

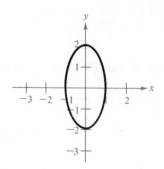

(i)

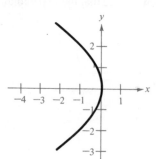

(j)

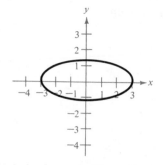

In Exercises 11–18, find the vertex, focus, and directrix of the parabola and sketch its graph.

11. $y = 4x^2$
12. $y = 2x^2$
13. $y^2 = -6x$
14. $y^2 = 3x$
15. $x^2 + 8y = 0$
16. $x + y^2 = 0$
17. $y^2 - 8x = 0$
18. $x^2 + 12y = 0$

In Exercises 19–28, find an equation of the specified parabola.

19. Vertex $(0, 0)$; focus $(0, -\frac{3}{2})$
20. Vertex $(0, 0)$; focus $(\frac{3}{4}, 0)$
21. Vertex $(0, 0)$; focus $(-\frac{2}{3}, 0)$
22. Vertex $(0, 0)$; focus $(0, -\frac{5}{2})$
23. Vertex $(0, 0)$; directrix $y = -1$
24. Vertex $(0, 0)$; directrix $x = 3$
25. Focus $(0, -4)$; directrix $y = 4$
26. Focus $(2, 0)$; directrix $x = -2$
27. Vertex $(0, 0)$; axis horizontal, passes through the point $(4, 6)$
28. Vertex $(0, 0)$; axis vertical, passes through the point $(-2, -2)$

In Exercises 29–36, find the center, foci, and vertices of the ellipse and sketch the graph.

29. $\dfrac{x^2}{25} + \dfrac{y^2}{16} = 1$
30. $\dfrac{x^2}{144} + \dfrac{y^2}{169} = 1$
31. $\dfrac{x^2}{16} + \dfrac{y^2}{25} = 1$
32. $\dfrac{x^2}{169} + \dfrac{y^2}{144} = 1$
33. $\dfrac{x^2}{9} + \dfrac{y^2}{5} = 1$
34. $\dfrac{x^2}{28} + \dfrac{y^2}{64} = 1$
35. $5x^2 + 3y^2 = 15$
36. $x^2 + 4y^2 = 4$

In Exercises 37–44, find an equation of the specified ellipse.

37. Center $(0, 0)$; focus $(5, 0)$; vertex $(6, 0)$
38. Center $(0, 0)$; vertex $(2, 0)$; minor axis of length 3
39. Vertices $(5, 0), (-5, 0)$; focus $(2, 0)$
40. Vertices $(0, 8), (0, -8)$; focus $(0, 4)$
41. Vertices $(0, -2), (0, 2)$; minor axis of length 2
42. Foci $(-2, 0), (2, 0)$; major axis of length 8
43. Foci $(0, 5), (0, -5)$; sum of distances from the foci to any point on the ellipse is 14
44. Center $(0, 0)$; major axis vertical; curve passes through points $(0, 4)$ and $(2, 0)$

In Exercises 45–52, find the center, vertices, and foci of the hyperbola and sketch its graph, using asymptotes as an aid.

45. $x^2 - y^2 = 1$
46. $\dfrac{x^2}{9} - \dfrac{y^2}{16} = 1$
47. $\dfrac{y^2}{1} - \dfrac{x^2}{4} = 1$
48. $\dfrac{y^2}{9} - \dfrac{x^2}{1} = 1$
49. $\dfrac{y^2}{25} - \dfrac{x^2}{144} = 1$
50. $\dfrac{x^2}{36} - \dfrac{y^2}{4} = 1$
51. $2x^2 - 3y^2 = 6$
52. $3y^2 - 5x^2 = 15$

In Exercises 53–60, find an equation of the specified hyperbola.

53. Center $(0, 0)$; one vertex $(0, 2)$; one focus $(0, 4)$
54. Center $(0, 0)$; one vertex $(3, 0)$; one focus $(5, 0)$
55. Vertices $(-1, 0), (1, 0)$; asymptotes $y = \pm 3x$
56. Vertices $(0, -3), (0, 3)$; asymptotes $y = \pm 3x$
57. Vertices $(0, 3), (0, -3)$; passing through the point $(-2, 5)$
58. Asymptotes $y = \pm \frac{3}{4}x$; focus $(10, 0)$
59. For any point on the hyperbola, the difference of its distances from $(-4, 0)$ and $(4, 0)$ is 6.
60. For any point on the hyperbola, the difference of its distances from the points $(0, -\frac{3}{2})$ and $(0, \frac{3}{2})$ is 2.
61. Each cable of a particular suspension bridge is suspended (in the shape of a parabola) between two towers that are 400 feet apart and 50 feet above the roadway. (See Figure 5.35.) If the cables touch the roadway midway between the

towers, find an equation for the parabolic shape of each cable.

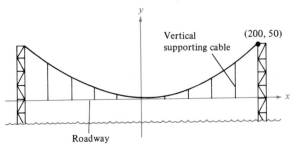

FIGURE 5.35

62. A fireplace arch is to be constructed in the shape of a semi-ellipse. The opening is to have a height of 2 feet at the center and a width of 5 feet along the base (see Figure 5.36). The contractor will first draw the form of the ellipse by the method shown in Figure 5.22. Where should the tacks be placed and how long a piece of string should be used?

63. A line segment through a focus with endpoints on the ellipse and perpendicular to the major axis is called a **latus rectum**

of the ellipse. Therefore, an ellipse has two latus recta. Knowing the length of the latus recta is helpful in sketching an ellipse, because it yields another locus point on the curve. (See Figure 5.37.) Show that the length of each latus rectum is $2b^2/a$.

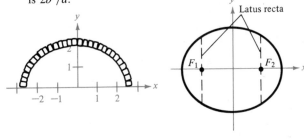

FIGURE 5.36   **FIGURE 5.37**

64. Sketch the graph of each ellipse, making use of the end-points of the latus recta (see Exercise 63).

   (a) $\dfrac{x^2}{4} + \dfrac{y^2}{1} = 1$ (b) $6x^2 + 4y^2 = 1$

   (c) $5x^2 + 3y^2 = 15$

65. Use the definition of an ellipse to derive its standard form.

66. Use the definition of a hyperbola to derive its standard form.

Conic Sections and Translation   5.4

In Section 5.3, we looked at conic sections whose graphs were in standard position. In this section, we will study the equations of conic sections that have been shifted vertically or horizontally in the plane. The following summary lists the standard forms of the four basic conics.

STANDARD FORMS OF EQUATIONS OF CONICS

	Standard Position	**With Vertex at (h, k)**	
Parabola			
Vertical axis	$x^2 = 4py$	$(x - h)^2 = 4p(y - k)$	
Horizontal axis	$y^2 = 4px$	$(y - k)^2 = 4p(x - h)$	

	Standard Position	**With Center at (h, k)**	
Circle	$x^2 + y^2 = r^2$	$(x - h)^2 + (y - k)^2 = r^2$	
Ellipse			
Horizontal major axis	$\dfrac{x^2}{a^2} + \dfrac{y^2}{b^2} = 1$	$\dfrac{(x - h)^2}{a^2} + \dfrac{(y - k)^2}{b^2} = 1$	
Vertical major axis	$\dfrac{x^2}{b^2} + \dfrac{y^2}{a^2} = 1$	$\dfrac{(x - h)^2}{b^2} + \dfrac{(y - k)^2}{a^2} = 1$	
Hyperbola			
Horizontal transverse axis	$\dfrac{x^2}{a^2} - \dfrac{y^2}{b^2} = 1$	$\dfrac{(x - h)^2}{a^2} - \dfrac{(y - k)^2}{b^2} = 1$	
Vertical transverse axis	$\dfrac{y^2}{a^2} - \dfrac{x^2}{b^2} = 1$	$\dfrac{(y - k)^2}{a^2} - \dfrac{(x - h)^2}{b^2} = 1$	

We use the completing the square procedure to rewrite equations of conics in standard form.

EXAMPLE 1

Finding the Standard Form of a Parabola

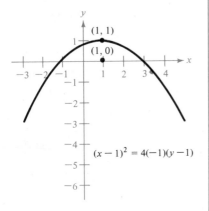

FIGURE 5.38

Find the vertex and focus of the parabola given by

$$x^2 - 2x + 4y - 3 = 0$$

Solution:

$x^2 - 2x + 4y - 3 = 0$	*Given equation*
$x^2 - 2x = -4y + 3$	*Group terms*
$x^2 - 2x + 1 = -4y + 3 + 1$	*Add 1 to both sides*
$(x - 1)^2 = -4y + 4$	*Completed square form*
$(x - 1)^2 = 4(-1)(y - 1)$	*Standard form*
$(x - h)^2 = 4p(y - k)$	

In standard form, we see that $h = 1$, $k = 1$, and $p = -1$. Since the axis is vertical and p is negative, we see that the parabola opens downward.

Vertex: $(h, k) = (1, 1)$

Focus: $(h, k + p) = (1, 0)$

The graph of this parabola is shown in Figure 5.38.

Remark: Note in Example 1 that p is the directed distance from the vertex to the focus. Since the axis of this parabola is vertical and since $p = -1$, we know that the focus is 1 unit *below* the vertex.

EXAMPLE 2

Sketching an Ellipse

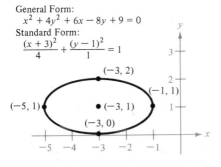

FIGURE 5.39

Sketch the graph of the ellipse whose equation is

$$x^2 + 4y^2 + 6x - 8y + 9 = 0$$

Solution:

$x^2 + 4y^2 + 6x - 8y + 9 = 0$	*Given equation*
$(x^2 + 6x + \ ) + (4y^2 - 8y + \ ) = -9$	*Group terms*
$(x^2 + 6x + \ ) + 4(y^2 - 2y + \ ) = -9$	*Factor 4 out of y-terms*
$(x^2 + 6x + 9) + 4(y^2 - 2y + 1) = -9 + 9 + 4(1)$	*Add 9 and 4 to both sides*
$(x + 3)^2 + 4(y - 1)^2 = 4$	*Completed square form*
$\dfrac{(x + 3)^2}{4} + \dfrac{(y - 1)^2}{1} = 1$	*Standard form*

Now we see that the center occurs at $(h, k) = (-3, 1)$. Since the denominator of the x-term is $4 = a^2 = 2^2$, we locate the endpoints of the major axis 2 units to the right and left of the center. Similarly, since the denominator of the y-term is $1 = b^2 = 1^2$, we locate the endpoints of the minor axis 1 unit up and down from the center. The graph of this ellipse is shown in Figure 5.39.

EXAMPLE 3
Sketching the Graph
of a Hyperbola

Sketch the graph of the hyperbola given by the equation

$$y^2 - 4x^2 + 4y + 24x - 41 = 0$$

Solution:

$$y^2 - 4x^2 + 4y + 24x - 41 = 0 \qquad \textit{Given equation}$$
$$(y^2 + 4y + \quad) - (4x^2 - 24x + \quad) = 41 \qquad \textit{Group terms}$$
$$(y^2 + 4y + \quad) - 4(x^2 - 6x + \quad) = 41 \qquad \textit{Factor 4 out of x-terms}$$
$$(y^2 + 4y + 4) - 4(x^2 - 6x + 9) = 41 + 4 - 4(9)$$

Add 4, subtract 4(9)

$$(y + 2)^2 - 4(x - 3)^2 = 9 \qquad \textit{Completed square form}$$
$$\frac{(y + 2)^2}{9} - \frac{4(x - 3)^2}{9} = 1 \qquad \textit{Divide both sides by 9}$$
$$\frac{(y + 2)^2}{9} - \frac{(x - 3)^2}{\frac{9}{4}} = 1 \qquad \textit{Change 4 to } \frac{1}{1/4}$$
$$\frac{(y + 2)^2}{3^2} - \frac{(x - 3)^2}{(\frac{3}{2})^2} = 1 \qquad \textit{Standard form}$$

Now we see that the transverse axis is vertical and the center lies at $(h, k) = (3, -2)$. Since the denominator of the y-term is $a^2 = 3^2$, we know that the vertices occur 3 units above and below the center:

$$(3, -5) \qquad \text{and} \qquad (3, 1)$$

To sketch the hyperbola, we sketch a rectangle whose top and bottom pass through the vertices. Since the denominator of the x-term is $b^2 = (\frac{3}{2})^2$, we locate the sides of the rectangle $\frac{3}{2}$ units to the right and left of center as shown in Figure 5.40. Finally, we sketch the asymptotes by drawing lines through the opposite corners of the rectangle. Using these asymptotes, we complete the graph of the hyperbola, as shown in Figure 5.40.

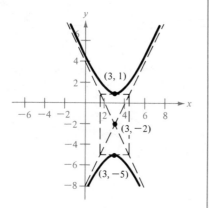

FIGURE 5.40

In Example 3, had we been asked to find the foci of the hyperbola, we would have proceeded as follows:

$$c^2 = a^2 + b^2 = 9 + \frac{9}{4} = \frac{45}{4}$$

Since the transverse axis is vertical, the foci lie c units above and below the center:

$$\left(3, 1 + \frac{3\sqrt{5}}{2}\right) \qquad \text{and} \qquad \left(3, 1 - \frac{3\sqrt{5}}{2}\right)$$

One interesting application of conic sections involves the orbits of comets in our solar system. Of the 566 comets identified prior to 1960, 211 have elliptical orbits, 290 have parabolic orbits, and 65 have hyperbolic orbits. For example Halley's Comet has an elliptical orbit, and we can predict the

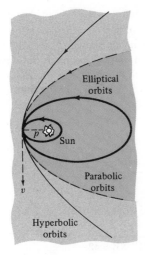

reappearance of this comet every 75 years. The center of the sun is a focus point of each of these orbits, and each orbit has a vertex at the point where the comet is closest to the sun, as shown in Figure 5.41.

If p is the distance between the vertex and the focus, and v is the speed of the comet at the vertex, then the orbit is:

an *ellipse* if $\quad v < \sqrt{\dfrac{2GM}{p}}$

a *parabola* if $v = \sqrt{\dfrac{2GM}{p}}$

a *hyperbola* if $v > \sqrt{\dfrac{2GM}{p}}$

where M is the mass of the sun and $G \approx 6.67(10^{-8})$ cm^3/(gm · sec^2).

FIGURE 5.41

Section Exercises 5.4

In Exercises 1–12, find the vertex, focus, and directrix of the parabola.

1. $(x - 1)^2 + 8(y + 2) = 0$
2. $(x + 3) + (y - 2)^2 = 0$
3. $(y + \frac{1}{2})^2 = 2(x - 5)$ 4. $(x + \frac{1}{2})^2 = 4(y - 3)$
5. $y = \frac{1}{4}(x^2 - 2x + 5)$ 6. $y = -\frac{1}{6}(x^2 + 4x - 2)$
7. $4x - y^2 - 2y - 33 = 0$
8. $y^2 + x + y = 0$
9. $y^2 + 6y + 8x + 25 = 0$
10. $x^2 - 2x + 8y + 9 = 0$
11. $y^2 - 4y - 4x = 0$ 12. $y^2 - 4x - 4 = 0$

In Exercises 13–20, find an equation of the specified parabola.

13. Vertex (3, 2); focus (1, 2)
14. Vertex (−1, 2); focus (−1, 0)
15. Vertex (0, 4); directrix $y = 2$
16. Vertex (−2, 1); directrix $x = 1$
17. Focus (2, 2); directrix $x = -2$
18. Focus (0, 0); directrix $y = 4$
19.   20.

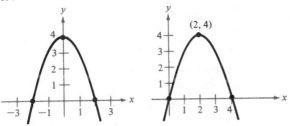

In Exercises 21–28, find the center, foci, and vertices of the ellipse.

21. $\dfrac{(x - 1)^2}{9} + \dfrac{(y - 5)^2}{25} = 1$
22. $(x + 2)^2 + 4(y + 4)^2 = 1$
23. $9x^2 + 4y^2 + 36x - 24y + 36 = 0$
24. $9x^2 + 4y^2 - 36x + 8y + 31 = 0$
25. $16x^2 + 25y^2 - 32x + 50y + 31 = 0$
26. $9x^2 + 25y^2 - 36x - 50y + 61 = 0$
27. $12x^2 + 20y^2 - 12x + 40y - 37 = 0$
28. $36x^2 + 9y^2 + 48x - 36y + 43 = 0$

In Exercises 29–36, find an equation for this specified ellipse.

29. Vertices (0, 2), (4, 2); minor axis of length 2
30. Foci (0, 0), (4, 0); major axis of length 8
31. Foci (0, 0), (0, 8); major axis of length 16
32. Center (2, −1); one vertex at (2, $\frac{1}{2}$); minor axis of length 2
33. Vertices (3, 1), (3, 9); minor axis of length 6
*34. Center (3, 2); $e = \frac{1}{3}$; foci (1, 2), (5, 2)
*35. Center (0, 4); $e = \frac{1}{2}$; vertices (−4, 4), (4, 4)
36. Vertices (5, 0), (5, 12); endpoints of the minor axis (0, 6), (10, 6)

*For an ellipse the **eccentricity**, e, is defined as $e = c/a$. It measures the flatness of the ellipse.

In Exercises 37–46, find the center, vertices, and foci of the hyperbola and sketch its graph, using asymptotes as an aid.

37. $\dfrac{(x-1)^2}{4} - \dfrac{(y+2)^2}{1} = 1$

38. $\dfrac{(x+1)^2}{144} - \dfrac{(y-4)^2}{25} = 1$

39. $(y+6)^2 - (x-2)^2 = 1$

40. $\dfrac{(y-1)^2}{\frac{1}{4}} - \dfrac{(x+3)^2}{\frac{1}{9}} = 1$

41. $9x^2 - y^2 - 36x - 6y + 18 = 0$

42. $x^2 - 9y^2 + 36y - 72 = 0$

43. $9y^2 - x^2 + 2x + 54y + 62 = 0$

44. $16y^2 - x^2 + 2x + 64y + 63 = 0$

45. $x^2 - 9y^2 + 2x - 54y - 80 = 0$

46. $9x^2 - y^2 + 54x + 10y + 55 = 0$

In Exercises 47–50, find an equation for the specified hyperbola.

47. Vertices $(0, 2)$, $(6, 2)$; asymptotes $y = \frac{2}{3}x$ and $y = 4 - \frac{2}{3}x$

48. Vertices $(2, 3)$, $(2, -3)$; foci $(2, 5)$, $(2, -5)$

49. Vertices $(2, 3)$, $(2, -3)$; passing through the point $(0, 5)$

50. For any point on the hyperbola, the difference of its distances from $(2, 2)$ and $(10, 2)$ is 6.

In Exercises 51–58, classify the graph of each equation as a circle, a parabola, an ellipse, or a hyperbola.

51. $x^2 + y^2 - 6x + 4y + 9 = 0$

52. $x^2 + 4y^2 - 6x + 16y + 21 = 0$

53. $4x^2 - y^2 - 4x - 3 = 0$

54. $y^2 - 4y - 4x = 0$

55. $4x^2 + 3y^2 + 8x - 24y + 51 = 0$

56. $4y^2 - 2x^2 - 4y - 8x - 15 = 0$

57. $25x^2 - 10x - 200y - 119 = 0$

58. $4x^2 + 4y^2 - 16y + 15 = 0$

59. An earth satellite in a 100-mile high circular orbit around the earth has a velocity of approximately 17,500 miles per hour. If this velocity is multipled by $\sqrt{2}$, then the satellite will have the minimum velocity necessary to escape the earth's gravity and it will follow a parabolic path with the center of the earth as the focus. (See Figure 5.42.)

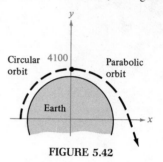

FIGURE 5.42

(a) Find the escape velocity of the satellite.

(b) Find an equation of its path (assume the radius of the earth is 4000 miles).

60. A large parabolic antenna is described as the surface formed by revolving the parabola $y = x^2/200$ on the interval $0 \le x \le 100$ about the y-axis. The receiving and transmitting equipment is positioned at the focus. Find the coordinates of the focus.

61. The earth moves in an elliptical orbit with the sun at one of the foci. If the length of half of the major axis is 93 million miles and the eccentricity is 0.017, find the least and greatest distances of the earth from the sun.

62. Show that the equation of an ellipse can be written

$$\frac{(x-h)^2}{a^2} + \frac{(y-k)^2}{a^2(1-e^2)} = 1$$

Note that as e approaches zero, with a remaining fixed, the ellipse approaches a circle of radius a.

Review Exercises / Chapter 5

In Exercises 1–16, analyze the equation and sketch its graph.

1. $y = \dfrac{x}{x^2 + 1}$

2. $y = \dfrac{2x}{x^2 + 4}$

3. $y = \dfrac{x^2}{x^2 + 1}$

4. $y = \dfrac{2x^2}{x^2 + 4}$

5. $y = \dfrac{x}{x^2 - 1}$

6. $y = \dfrac{2x}{x^2 - 4}$

7. $y = \dfrac{2x^2}{x^2 - 4}$

8. $y = \dfrac{x^2 + 1}{x + 1}$

9. $y = \dfrac{2x^3}{x^2 + 1}$

10. $xy - 4 = 0$

11. $5y^2 - 4x^2 = 20$

12. $y^2 - 12y - 8x + 20 = 0$

13. $x^2 - 6x + 2y + 9 = 0$

14. $4x^2 + y^2 - 16x + 15 = 0$
15. $x^2 + y^2 - 2x - 4y + 5 = 0$
16. $16x^2 + 16y^2 - 16x + 24y - 3 = 0$

In Exercises 17–22, write the partial fraction decomposition for the rational expression.

17. $\dfrac{x^2}{x^2 + 2x - 15}$

18. $\dfrac{9}{x^2 - 9}$

19. $\dfrac{x^2 + 2x}{x^3 - x^2 + x - 1}$

20. $\dfrac{4x - 2}{3(x - 1)^2}$

21. $\dfrac{3x^3 + 4x}{(x^2 + 1)^2}$

22. $\dfrac{4x^2}{(x - 1)(x^2 + 1)}$

In Exercises 23–32, find an equation of the specified conic.

23. Hyperbola: Vertices $(0, \pm 1)$; foci $(0, \pm 3)$

24. Hyperbola: Vertices $(2, 2)$, $(-2, 2)$; foci $(4, 2)$, $(-4, 2)$
25. The ellipse such that the sum of the distances from any of the points on its graph to the points $(0, 0)$ and $(4, 0)$ is 10.
26. Parabola: Vertex $(4, 2)$; focus $(4, 0)$
27. Parabola: Vertex $(2, 0)$; focus $(0, 0)$
28. Ellipse: Vertices $(2, 0)$, $(2, 4)$; foci $(2, 1)$, $(2, 3)$
29. Hyperbola: Foci $(\pm 4, 0)$; asymptotes $y = \pm 2x$
30. The hyperbola such that the absolute value of the difference of the distances from any point on its graph to the points $(4, 0)$ and $(-4, 0)$ is 4.
31. Parabola: Vertex $(0, 2)$; passes through point $(-1, 0)$; vertical axis
32. Ellipse: Vertices $(0, 1)$, $(4, 1)$; endpoints of minor axis $(2, 0)$, $(2, 2)$

CHAPTER 6 EXPONENTIAL AND LOGARITHMIC FUNCTIONS

Exponential Functions

6.1

The functions we have dealt with so far have primarily been polynomials and rational functions. In this chapter we will study two new functions: the exponential function and its inverse, the logarithmic function.

Exponential functions are widely used in describing economic and physical phenomena such as compound interest, growth (or decline) of population sizes, and decay of radioactive material. Logarithms have in the past been used extensively to simplify complicated arithmetic calculations. With the advent of electronic computing devices, their computational usefulness has diminished. Nevertheless, properties of logarithms and logarithmic functions still play a critical role in mathematics and its real-world applications.

Consider the familiar functions

$$f(x) = x^3 \quad \text{and} \quad g(x) = \sqrt{x} = x^{1/2}$$

which involve a *variable raised to a constant power*. If we interchange the roles of the variable and the constant, we get new functions

$$F(x) = 3^x \quad \text{and} \quad G(x) = \left(\frac{1}{2}\right)^x$$

having constant bases and variable exponents. Such functions are called **exponential functions.**

EXPONENTIAL FUNCTION

The **exponential function with base** a is denoted by

$$f(x) = a^x$$

where $a > 0$, $a \neq 1$, and x is any real number.

From our study of Sections 1.4 and 1.5, we know how to interpret real numbers that are raised to integer or rational powers. Now, we need to interpret forms with *irrational* exponents, such as $a^{\sqrt{2}}$ and a^{π}. A technical definition of such forms is beyond the scope of this text. For our purposes it is sufficient to think of

$$a^{\sqrt{2}} \qquad (\text{where } \sqrt{2} \approx 1.414213)$$

as that value which has the successively closer approximations

$$a^{1.4}, \; a^{1.41}, \; a^{1.414}, \; a^{1.4142}, \; a^{1.41421}, \ldots$$

Consequently, for exponential functions like

$$f(x) = a^x$$

we assume in this text that a^x exists and that the properties of exponents (Section 1.4) hold for all real numbers x. Hence, the graphs of exponential functions will be continuous, with no holes or jumps.

EXAMPLE 1
Graphs of $y = a^x$, $a > 1$

On the same coordinate plane, sketch graphs of

(a) $y = 2^x$ (b) $y = (\tfrac{3}{2})^x$ (c) $y = 4^x$

Solution:
Table 6.1 lists some values for each function, and Figure 6.1 shows their graphs.

TABLE 6.1

x	-2	-1	0	1	2	3
(a) 2^x	1/4	1/2	1	2	4	8
(b) $(3/2)^x$	4/9	2/3	1	3/2	9/4	27/8
(c) 4^x	1/16	1/4	1	4	16	64

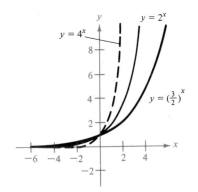

FIGURE 6.1

EXAMPLE 2
Graph of $y = a^{-x} = (1/a)^x$, $a > 1$

On the same coordinate plane, sketch graphs of

(a) $y = (\frac{1}{4})^x$ (b) $y = 2^{-x}$

Solution:
Note in part (b) that we can write

$$2^{-x} = \frac{1}{2^x} = \left(\frac{1}{2}\right)^x$$

Table 6.2 lists some values for each function, and Figure 6.2 shows their graphs.

TABLE 6.2

x	-3	-2	-1	0	1	2
(a) $(1/4)^x$	64	16	4	1	1/4	1/16
(b) 2^{-x}	8	4	2	1	1/2	1/4

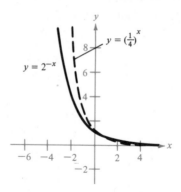

FIGURE 6.2

Exponential Graphs

From Figures 6.1 and 6.2 we can make several important observations about the graphs of $y = a^x$ and $y = a^{-x}$. These are summarized in Figures 6.3 and 6.4.

   Graph of $y = a^x$, $a > 1$

1. Domain: all real x

2. Range: all positive reals

3. Intercept: (0, 1)

4. a^x *increases* as x increases

5. x-axis is horizontal asymptote to *left*

6. Reflection (in y-axis) of graph of $y = a^{-x}$

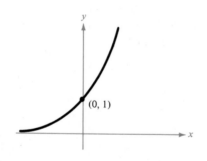

FIGURE 6.3

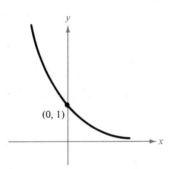

FIGURE 6.4

Graph of $y = a^{-x}$, $a > 1$

1. Domain: all real x
2. Range: all positive reals
3. Intercept: $(0, 1)$
4. a^{-x} *decreases* as x increases
5. x-axis is horizontal asymptote to *right*
6. Reflection (in y-axis) of graph of $y = a^x$

These observations can be used to establish two additional properties of exponential functions not listed back in Section 1.4. These properties are useful in solving equations that involve exponential functions.

SPECIAL PROPERTIES OF EXPONENTIAL FUNCTIONS

For $a > 0$ and $a \neq 1$:
1. If $a^x = a^y$, then $x = y$.
2. If $a^x = b^x$, and $x \neq 0$, then $a = b$.

EXAMPLE 3
Solving Exponential Equations

Solve for x in each of the following:

(a) $(\frac{1}{2})^x = 64$ (b) $2^5 = (x + 3)^5$ (c) $27 = x^{3/2}$

Solution:

(a) Since

$$\left(\frac{1}{2}\right)^x = \frac{1}{2^x} = 2^{-x} \quad \text{and} \quad 64 = 2^6$$

we have

$$2^{-x} = 2^6$$

Thus we conclude that $x = -6$.

(b) Since $a^x = b^x$ implies $a = b$, we have

$$2^5 = (x + 3)^5 \quad \text{implies} \quad 2 = x + 3$$

Thus,

$$-1 = x$$

(c) Rewriting the given equation, we have

$$(27)^{2/3} = (x^{3/2})^{2/3} \qquad \textit{Raise to reciprocal power}$$
$$(\sqrt[3]{27})^2 = x$$
$$3^2 = 9 = x$$

Applications

Many physical and economic phenomena can be described by exponential functions. One of the most familiar is the exponential growth of an investment on which compound interest is earned. Consider a principal P that is invested at an annual rate r compounded once a year. If the interest is added to the principal at the end of one year, then the new principal P_1 is

$$P_1 = P + Pr = P(1 + r)$$

This pattern of multiplying the previous principal by $1 + r$ is then repeated each successive year. Table 6.3 shows how the results lead to an exponential function.

TABLE 6.3
Compound Interest Formula

Time in Years	Balance After Each Compounding
0	$P = P$
1	$P_1 = P(1 + r)$
2	$P_2 = P_1(1 + r) = P(1 + r)(1 + r) = P(1 + r)^2$
3	$P_3 = P_2(1 + r) = P(1 + r)^2(1 + r) = P(1 + r)^3$
.	.
.	.
.	.
n	$B = P_n = P(1 + r)^n$

EXAMPLE 4
Finding the Balance for Compound Interest

Suppose that $9000 is invested at an annual rate of 8.5%, compounded annually. Find the balance in the account to the nearest dollar at the end of 3 years.

Solution:
In this case, $P = 9000$, $r = 8.5\% = 0.085$, and $n = 3$. Using the formula

$$B = P(1 + r)^n$$

we have

$$B = 9000(1 + 0.085)^3 = 9000(1.085)^3$$
$$B \approx 9000(1.2773) \approx \$11{,}496$$

EXAMPLE 5
Radioactive Decay

Suppose a 5-gram sample of a radioactive substance decays at a rate such that it loses half of its mass each day.

(a) Verify that after t days its mass is

$$M(t) = \frac{5}{2^t} = 5(2^{-t})$$

(b) Sketch a graph of M.

Solution:

(a) We need to show that $M(0)$ is equal to the original mass 5, and that the mass on a given day is $\frac{1}{2}$ the mass of the previous day. That is, we must show that

$$M(t + 1) = \frac{1}{2} M(t)$$

For $t = 0$, we have

$$M(0) = \frac{5}{2^0} = \frac{5}{1} = 5$$

Furthermore, replacing t by $t + 1$, we get

$$M(t + 1) = \frac{5}{2^{t+1}} = \frac{5}{2 \cdot 2^t} = \frac{1}{2}\left(\frac{5}{2^t}\right) = \frac{1}{2} M(t)$$

(b) Using the values in the following table, we obtain the graph shown in Figure 6.5.

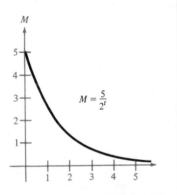

$M = \frac{5}{2^t}$

FIGURE 6.5

t	0	1	2	3	4
$M(t)$	5	$\frac{5}{2}$	$\frac{5}{4}$	$\frac{5}{8}$	$\frac{5}{16}$

Remark: The radioactive substance in Example 5 is said to have a **half-life** of one day.

Exponential Functions and Calculators

Scientific calculators have an exponential key, $\boxed{y^x}$, that is used to raise a number to a power. Note the key stroke sequences for the following examples:

Exponential	Key Strokes	Display
$(1.085)^3$	1.085 $\boxed{y^x}$ 3 $\boxed{=}$	1.27729
$12^{5/7}$	12 $\boxed{y^x}$ $\boxed{(}$ 5 $\boxed{\div}$ 7 $\boxed{)}$ $\boxed{=}$	5.89989
$29^{-8.6}$	29 $\boxed{y^x}$ 8.6 $\boxed{+/-}$ $\boxed{=}$	2.6508 − 13 (scientific notation)

The Natural Base e

Our next example uses an exponential function similar to the one in Example 4 to identify an important irrational number that is the base for the exponential and natural logarithmic functions used in calculus. We denote this number by the letter e. To five decimal places, the value of e is

$$e \approx 2.71828 \ldots$$

This base arises in many natural phenomena such as the population growth shown in Example 7.

EXAMPLE 6
An Estimation of the
Irrational Number e

Evaluate the exponential expression

$$\left(1 + \frac{1}{n}\right)^n$$

for several large values of n to see that it approaches the value $e \approx 2.71828$ as n increases without bound.

Solution:
With a calculator we use the key strokes

$$n \boxed{1/x} \boxed{+} 1 \boxed{=} \boxed{y^x} n \boxed{=}$$

to obtain the values shown in the table that follows.

n	10	100	1000	10,000	100,000	1,000,000
$\left(1 + \dfrac{1}{n}\right)^n$	2.59374	2.70481	2.71692	2.71815	2.71827	2.71828

From these values it seems reasonable to conclude that

$$\left(1 + \frac{1}{n}\right)^n \rightarrow e \quad \text{as} \quad n \rightarrow \infty$$

EXAMPLE 7
Population Growth

In Professor Jacob's research, the number of fruit flies in a certain experimental population at time t hours is given by the exponential equation

$$Q(t) = 20e^{0.03t}$$

(a) Find the initial number of fruit flies in the population.
(b) How large is the population of fruit flies after 72 hours?
(c) Sketch a graph of Q.

Solution:

(a) To find the initial population, we evaluate $Q(t)$ at $t = 0$.

$$Q(0) = 20e^{0.03(0)} = 20e^0 = 20(1) = 20$$

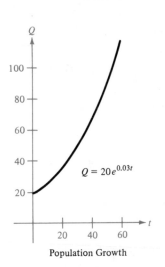

$Q = 20e^{0.03t}$

Population Growth

FIGURE 6.6

(b) After 72 hours, the population size is

$$Q(72) = 20e^{(0.03)(72)} = 20e^{2.16} \approx 173 \quad \textit{(Nearest integer)}$$

(c) The next table shows some values of $Q(t)$ (to the nearest integer), and Figure 6.6 shows the graph of Q.

t	0	5	10	20	40	60
$20e^{0.03t}$	20	23	27	36	66	121

Remark: Many populations in nature have a growth pattern described by the function

$$Q(t) = ce^{kt}$$

where c is the original population, $Q(t)$ is the population at time t, and k is a constant that affects the rate of growth.

We complete this section with another look at compound interest. This time we consider interest that is compounded n times per year. In this case we have the amount of deposit P, the annual rate of interest, r, the number of compoundings per year, n and the number of years, t. The rate per compounding period is therefore r/n, and the balance after t years is

$$B = P\left(1 + \frac{r}{n}\right)^{tn}$$

If we let the number of compoundings, n, increase without bound, we have what is called **continuous compounding.** In the formula for n compoundings per year, let $m = n/r$. Then we have

$$B = P\left(1 + \frac{r}{n}\right)^{nt} = P\left(1 + \frac{1}{m}\right)^{mrt} = P\left[\left(1 + \frac{1}{m}\right)^{m}\right]^{rt}$$

Now as m gets larger and larger, we know from Example 6 that

$$\left[\left(1 + \frac{1}{m}\right)^{m}\right] \to e$$

Hence, for continuous compounding, it follows that

$$P\left[\left(1 + \frac{1}{m}\right)^{m}\right]^{rt} \to P[e]^{rt}$$

and we write

$$B = Pe^{rt}$$

FORMULAS FOR COMPOUND INTEREST

The balance in an account with principal P, interest rate r, and time in years t is given by the following formulas:

1. For n compoundings per year:

$$B = P\left(1 + \frac{r}{n}\right)^{nt}$$

2. For continuous compoundings:

$$B = Pe^{rt}$$

EXAMPLE 8
Compounding n Times and Continuously

Suppose \$12,000 is invested at an annual rate of 9%. Find the balance after 5 years if it is compounded

(a) quarterly   (b) continuously

Solution:

(a) For quarterly compoundings, we have $n = 4$. Thus, in 5 years at 9% we have a balance of

$$B = P\left(1 + \frac{r}{n}\right)^{nt} = 12{,}000\left(1 + \frac{0.09}{4}\right)^{4(5)}$$

$$B = 12{,}000(1.0225)^{20} = 12{,}000(1.56051) = \$18{,}726.12$$

(b) Compounded continuously, the balance is

$$B = Pe^{rt} = 12{,}000e^{0.09(5)} = 12{,}000e^{0.45}$$
$$= 12{,}000(1.56831) = \$18{,}819.72$$

Note that continuous compounding yields

$$\$18{,}819.72 - \$18{,}726.12 = \$93.60$$

more than quarterly compounding.

Section Exercises 6.1

In Exercises 1–10, match the given exponential function with one of the accompanying graphs.

1. $f(x) = 3^x$
2. $f(x) = -3^x$
3. $f(x) = 3^{-x}$
4. $f(x) = 3^{2x}$
5. $f(x) = 3^{x+1}$
6. $f(x) = 3^x + 1$
7. $f(x) = 3^{x/2}$
8. $f(x) = 3^{x-2}$
9. $f(x) = 3^x - 4$
10. $f(x) = 3^x + 3^{-x}$

(a)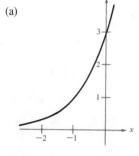

(b)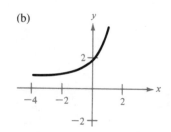

(c)

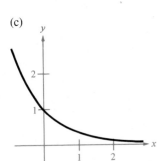

(d)

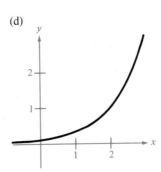

(e)

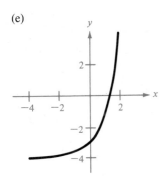

(f)

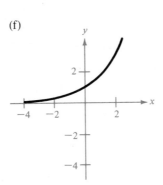

(g)

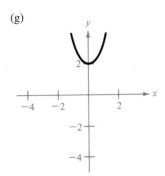

(h)

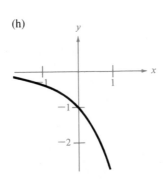

(i)

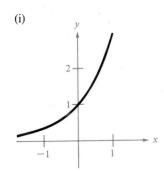

(j)

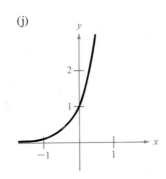

In Exercises 11–20, solve for x.

11. $3^x = 81$
12. $5^{x+1} = 125$
13. $(\frac{1}{3})^{x-1} = 27$
14. $(\frac{1}{5})^{2x} = 625$
15. $4^3 = (x + 2)^3$
16. $4^2 = (x + 2)^2$
17. $x^{3/4} = 8$
18. $(x + 3)^{4/3} = 16$
19. $4(2^x) = 1$
20. $e^{-2x} = e^5$

In Exercises 21–37, sketch the graph of the given exponential function.

21. $f(x) = 4^x$
22. $g(x) = 5^x$
23. $h(x) = (\frac{1}{4})^x = 4^{-x}$
24. $f(x) = (\frac{1}{5})^x = 5^{-x}$
25. $f(x) = (\frac{1}{4})^{-x}$
26. $g(x) = (\frac{1}{5})^{-x}$
27. $y = 2^{-x^2}$
28. $y = 3^{-x^2}$
29. $y = 3^{|x|}$
30. $y = 3^{-|x|}$
31. $s(t) = 2^{-t} + 3$
32. $s(t) = \dfrac{3^{-t}}{4}$
33. $f(x) = e^{2x}$
34. $h(x) = e^{x-2}$
35. $A(t) = 500e^{0.15t}$
36. $N(t) = 1000e^{-0.2t}$
37. $g(x) = \dfrac{10}{1 + e^{-x}}$

38. If $1000 is invested at 10% interest, find the amount after 10 years if the interest is compounded
    (a) annually   (b) semiannually   (c) quarterly
    (d) monthly   (e) daily   (f) continuously

39. If $2500 is invested at 12% interest, find the amount after 20 years if the interest is compounded
    (a) annually   (b) semiannually   (c) quarterly
    (d) monthly   (e) daily   (f) continuously

40. The demand equation for a certain product is given by

$$p = 5000\left(1 - \frac{4}{4 + e^{-0.002x}}\right)$$

Find the price of the product if the demand is
    (a) $x = 100$ units   (b) $x = 500$ units

41. The demand equation for a certain product is given by

$$p = 500 - 0.5e^{0.004x}$$

Find the price of the product if the demand is
    (a) $x = 1000$ units   (b) $x = 1500$ units

42. Find the amount of money that should be deposited in an account paying 12% interest compounded continuously to produce a final balance of $100,000 in
    (a) 1 year   (b) 10 years
    (c) 20 years   (d) 50 years

43. A certain type of bacteria increases according to the model

$$P(t) = 100e^{0.2197t}$$

where t is time in hours. Find
    (a) $P(0)$ (b) $P(5)$ (c) $P(10)$

44. The population of a town increases according to the model

$$P(t) = 2500e^{0.0293t}$$

where t is time in years, with $t = 0$ corresponding to 1980. Use the model to approximate the population in
(a) 1985   (b) 1990   (c) 2000

45. The half-life of radioactive radium is approximately 1600 years, and its decay is approximated by the model

$$Q(t) = Q_0e^{-0.0004t}$$

where t is time in years, $Q(t)$ is the quantity at time t, and Q_0 is the initial quantity. What percentage of a given amount remains after 100 years?

If the average time between successive occurrences of some event is λ, then the probability of waiting less than t units of time between successive occurrences can sometimes be approximated by the model

$$F(t) = 1 - e^{-t/\lambda}$$

Use this model in Exercises 46 and 47.

46. Trucks arrive at a terminal at an average of 3 per hour (therefore $\lambda = 20$ minutes). If a truck has just arrived, find the probability that the next arrival will be within
(a) 10 minutes   (b) 30 minutes   (c) 1 hour

47. The average time between incoming calls at a switchboard is 3 minutes. If a call has just come in, find the probability that the next call will be within
(a) $\frac{1}{2}$ minute   (b) 2 minutes   (c) 5 minutes

48. A certain automobile gets 28 miles per gallon of gasoline for speeds up to 50 miles per hour. Over 50 miles per hour, the miles per gallon drops at the rate of 12% for each 10 miles per hour. If s is the speed (in miles per hour) and y is the miles per gallon, then

$$y = 28e^{0.6-0.012s}, \qquad s \geq 50$$

Use this function to complete the following table:

Speed	50	55	60	65	70
Miles per gallon					

49. A solution of a certain drug contained 500 units per milliliter when prepared. It was analyzed after 40 days and found to contain 300 units per milliliter. Assuming that the rate of decomposition is proportional to the amount present, the equation giving the amount A after t days is

$$A = 500e^{-0.013t}$$

Use this model to find A when $t = 60$.

50. In calculus it is shown that

$$e^x \approx 1 + x + \frac{x^2}{2} + \frac{x^3}{6} + \frac{x^4}{24}$$

Use this equation to approximate
(a) e $(x = 1)$ (b) $e^{1/2}$ $(x = \frac{1}{2})$ (c) $e^{-1/2}$ $(x = -\frac{1}{2})$

Logarithms   6.2

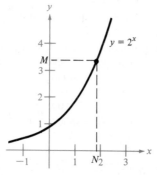

FIGURE 6.7

We noted in Figure 6.1 that the graph of $y = 2^x$ is always increasing and that the range for this exponential function is the set of positive real numbers. This means that to every real number ($M > 0$) in the range there corresponds a unique real number N in the domain, such that $2^N = M$. For instance, if $M = 3.42$, then from Figure 6.7 we can see that there exists a real number N such that

$$2^N = 3.42$$

If $2^N = 3.42$, we call N the *logarithm of 3.42 to base 2*. Or in equation form, we write

$$N = \log_2 3.42$$

In general, we use the following definition of a logarithm.

DEFINITION OF LOGARITHM TO BASE b

Let $b > 0$ and $b \neq 1$. If M is any positive real number, then there exists a unique real number N such that

$$b^N = M$$

This number N is called the **logarithm of M to base b** and is denoted by

$$N = \log_b M$$

Remark: This definition of logarithms gives us the following equivalence:

$$N = \log_b M \quad \text{if and only if} \quad b^N = M$$

From the definition of logarithms, it should be clear that **A LOGARITHM IS AN EXPONENT.** Keep this fact in mind as you work with logarithms. For instance, to evaluate $\log_2 16$ you can ask yourself the question

2 to what power yields 16?

Since $16 = 2^4$, it follows that

$$\log_2 16 = 4$$

In Example 1, we show how to use the equivalent logarithmic and exponential forms

$$\log_b M = N \quad \text{and} \quad b^N = M$$

to evaluate logarithms.

EXAMPLE 1
Evaluating Logarithms

Evaluate the following (if possible).

(a) $\log_4 2$ (b) $\log_3(\frac{1}{81})$ (c) $\log_4 8$ (d) $\log_5(-2)$

Solution:

Logarithmic Form	Exponential Form	Value of N
(a) Let $N = \log_4 2$	$4^N = 2$	$N = \dfrac{1}{2} = \log_4 2$
(b) Let $N = \log_3\left(\dfrac{1}{81}\right)$	$3^N = \dfrac{1}{81} = \dfrac{1}{3^4}$	$N = -4 = \log_3\left(\dfrac{1}{81}\right)$
(c) Let $N = \log_4 8$	$4^N = 8 = (\sqrt{4})^3$	$N = \dfrac{3}{2} = \log_4 8$
(d) Let $N = \log_5(-2)$	$5^N = -2$	Not possible*

*It is not possible to find a value of N such that $5^N = -2$, since $5^N > 0$ for all N. Thus, $\log_5(-2)$ is not defined.

EXAMPLE 2
Converting from Logarithmic to
Exponential Form and Vice Versa

In each of the following, convert all logarithmic equations to exponential form and exponential equations to logarithmic form. Find the value of x in each case.

(a) $\log_5 x = 2$

(b) $36^x = 6$

(c) $(\frac{1}{27}) = x^{-3}$

(d) $\log_3(x - 1) = 2$

(e) $x = 16^{3/2}$

(f) $\log_8(\frac{1}{2}) = x$

Solution:
Using the equivalent forms

$$\log_b M = N \quad \text{and} \quad b^N = M$$

we can readily make up the following table of solutions. In most cases the value of x is more easily obtained from the exponential form.

Logarithmic Form	Exponential Form	Value of x
(a) $\log_5 x = 2$	$5^2 = x$	$x = 25$
(b) $\log_{36} 6 = x$	$36^x = 6$	$\sqrt{36} = 6, x = \dfrac{1}{2}$
(c) $\log_x\left(\dfrac{1}{27}\right) = -3$	$\dfrac{1}{27} = x^{-3}$	$\dfrac{1}{27} = \dfrac{1}{x^3}, x = 3$
(d) $\log_3(x - 1) = 2$	$3^2 = x - 1$	$x = 1 + 3^2 = 10$
(e) $\log_{16} x = \dfrac{3}{2}$	$16^{3/2} = x$	$x = (\sqrt{16})^3 = 64$
(f) $\log_8\left(\dfrac{1}{2}\right) = x$	$8^x = \dfrac{1}{2}$	$\dfrac{1}{2} = \dfrac{1}{\sqrt[3]{8}} = 8^{-1/3}$
		$x = -\dfrac{1}{3}$

Logarithms are powerful as computational aids primarily because they have fundamental properties that convert complex multiplications, divisions, and exponentiations into simpler additions, subtractions, and multiplications, respectively. The following properties of logarithms can be derived from the properties of exponents that we studied back in Section 1.4.

PROPERTIES OF LOGARITHMS

If $b > 0$, $b \neq 1$, and c is any real number, then the following properties hold:

Property	**Comments**
1. $\log_b b = 1$	1. True because $b^1 = b$.
2. $\log_b 1 = 0$	2. True because $b^0 = 1$.
3. $\log_b(MN) = \log_b M + \log_b N$	3. The log of a product converts to a sum.

4. $\log_b\left(\dfrac{M}{N}\right) = \log_b M - \log_b N$ 4. The log of a quotient converts to a difference.

5. $\log_b(M^c) = c \log_b M$ 5. The log of an exponential converts to a product.

6. $\log_b(b^c) = c$ 6. Inverse property.

7. $b^{(\log_b c)} = c$ 7. Inverse property.

8. $\log_b M = \log_b N$ implies $M = N$ 8. One-to-one correspondence.

Proof:

All of these properties follow quite directly from the corresponding properties of exponents. As an example of this correspondence, we list a proof of Property 3. We leave the others up to you.

To prove Property 3, let

$$u = \log_b M \qquad \text{and} \qquad v = \log_b N$$

Using the equivalent forms

$$b^u = M \qquad \text{and} \qquad b^v = N$$

we can multiply to get

$$b^u(b^v) = M(N) \qquad \text{or} \qquad b^{u+v} = MN$$

Or, equivalently, we have

$$\log_b(MN) = u + v = \log_b M + \log_b N$$

Remark: Note that we have no general property that can be used to simplify $\log_b(M \pm N)$.

EXAMPLE 3
Using Properties of Logs

Given $\log_5 2 \approx .4307$, $\log_5 3 \approx .6826$, and $\log_5 7 \approx 1.2091$, use properties of logs to evaluate each of the following:

(a) $\log_5 6$ (b) $\log_5(\frac{7}{27})$ (c) $\log_5 \sqrt[3]{49}$

Solution:

(a) By Property 3,

$$\log_5 6 = \log_5(2 \cdot 3) = \log_5 2 + \log_5 3$$
$$\approx 0.4307 + 0.6826 = 1.1133$$

(b) $\log_5(\frac{7}{27}) = \log_5 7 - \log_5 27$ *Property 4*

            $= \log_5 7 - \log_5(3^3)$ *Prime factoring*

            $= \log_5 7 - 3 \log_5 3$ *Property 5*

            $\approx 1.2091 - 3(0.6826) = -0.8387$

(c) $\log_5 \sqrt[3]{49} = \log_5(7^2)^{1/3}$ *Rational exponents*

$\quad\quad\quad\quad = \log_5(7^{2/3})$ *Property of exponents*

$\quad\quad\quad\quad = \frac{2}{3}\log_5 7$ *Property 5*

$\quad\quad\quad\quad \approx \frac{2}{3}(1.2091) = 0.80607$

The base of a logarithm can be any positive real number other than 1. For computational purposes, the most widely used base is 10, since our number system uses base 10. For calculus, the most appropriate base is the irrational number e. Logarithms to these bases are given the following special names.

COMMON AND NATURAL LOGARITHMS

Logarithms to base 10 are called **common logarithms** and are denoted by

$$\log_{10} x = \log x$$

Logarithms to base e are called **natural logarithms** and are denoted by

$$\log_e x = \ln x$$

Remark: The eight *properties of logarithms* hold for both common and natural logarithms. On most scientific calculators $\boxed{\text{LOG}}$ and $\boxed{\text{ln}}$ are the keys for common and natural logarithms, respectively.

Sometimes you need to work with logarithms to bases other than 10 or e. In such cases the following *change of base formula* is useful.

CHANGE OF BASE FORMULA

For $a > 0$, $b > 0$, and $a, b \neq 1$:

$$\log_b x = \frac{\log_a x}{\log_a b} = \frac{1}{\log_a b}(\log_a x)$$

Remark: One way of looking at the change of base formula is that logarithms to base b are simply constant multiples of logarithms to base a. The constant multiplier is $1/(\log_a b)$.

EXAMPLE 4
Changing Bases

Change to *common* logarithms, then evaluate using the $\boxed{\text{LOG}}$ key.

(a) $\log_4 30$ (b) $\ln 100$

Change to *natural* logarithms, then evaluate using the $\boxed{\text{ln}}$ key.

(c) $\log e$ (d) $\log_3 x$

Solution:
In all four cases, we apply the change of base formula as indicated.

(a) $\log_4 30 = \dfrac{\log 30}{\log 4} = \dfrac{1.47712}{0.60206} \approx 2.4534$

(b) $\ln 100 = \dfrac{\log 100}{\log e} = \dfrac{2}{\log e} \approx \dfrac{2}{\log 2.71828} \approx 4.6052$

(c) $\log e = \log_{10} e = \dfrac{\ln e}{\ln 10} = \dfrac{1}{\ln 10} \approx 0.4343$

(d) $\log_3 x = \dfrac{\ln x}{\ln 3} \approx \dfrac{\ln x}{1.0986}$

In Example 4, we were asked to find the logarithm of a given number. In science, we are often faced with the reverse problem. For example, we might be asked to solve the following equation for N:

$$\log_b N = u$$

The key to solving this equation lies in Property 7 of logarithms. Using the left and right sides of the equation as exponents for the base b, we have

$$b^{(\log_b N)} = b^u$$
$$N = b^u \qquad\qquad \textit{Property 7}$$

Since we obtain the number N by eliminating a logarithm, we call N the **antilogarithm of u.**

DEFINITION OF ANTILOGARITHM

If

$$\log_b N = u$$

then N is called the **antilogarithm of u for base b.** Specifically, the antilogarithm of u has the value

$$N = b^u$$

EXAMPLE 5
Finding Antilogarithms
with a Calculator

Find the antilogarithm for each of the following:

(a) $\log N = 3.718$　　　(b) $\log N = -2.04$　　　(c) $\ln N = 0.147$

Solution:

(a) Since $\log N = 3.718$ is equivalent to $10^{3.718} = N$, the calculator sequence 10 $\boxed{y^x}$ 3.718 $\boxed{=}$ yields the antilogarithm value $N = 5223.96$.

Note: Some scientific calculators have an antilogarithm key (base 10) that finds N directly from 3.718.

(b) Again using base 10, we have $\log N = -2.04$ or $10^{-2.04} = N$. Thus, the calculator sequence 10 $\boxed{y^x}$ 2.04 $\boxed{+/-}$ $\boxed{=}$ yields $N = 0.0091201$.

(c) In this case, $\ln N = 0.147$ is equivalent to $e^{0.147} = N$. The calculator sequence .147 $\boxed{e^x}$ $\boxed{=}$ yields $N = 1.15835$.

EXAMPLE 6
Magnitude of Earthquakes

On the Richter Scale, the magnitude R of an earthquake of intensity I is given by

$$R = \log\left(\frac{I}{I_0}\right)$$

where I_0 is the minimum intensity used for comparison. If we let $I_0 = 1$, find the intensity per unit of area for the following earthquakes. (Intensity is a measure of the wave energy of an earthquake.)

(a) San Francisco in 1906, $R = 8.3$
(b) Mexico City in 1978, $R = 7.85$
(c) Predicted in 1985, $R = 6.3$

Solution:

(a) In this case, we have $8.3 = \log I$. We solve for I using the antilogarithm of 8.3 to the base 10.

$$I = 10^{8.3} \approx 199,526,000$$

(b) For Mexico City, we have $7.85 = \log I$.

$$I = 10^{7.85} \approx 70,794,600$$

(c) For $R = 6.3$, we have $6.3 = \log I$.

$$I = 10^{6.3} \approx 1,995,260$$

Note that a drop of 2 units on the Richter Scale (from 8.3 to 6.3) represents an intensity change by a factor of

$$\frac{1}{10^2} = \frac{1}{100}$$

Logarithmic Expressions

In addition to being of assistance in computing with logarithms, the *properties of logarithms* are valuable aids for rewriting logarithmic expressions in forms that simplify the operations of algebra and calculus. Examples 7 and 8 illustrate some cases.

EXAMPLE 7
Rewriting Logarithmic
Expressions

Use the *properties of logarithms* to rewrite each of the following as the sum and/or difference of logarithms:

(a) $\log \dfrac{x^2 y}{5}$

(b) $\ln \dfrac{\sqrt{3x-5}}{7x^3}$

(c) $\ln\left(\dfrac{5x^2}{y}\right)^3$

Solution:

(a) $\log \dfrac{x^2 y}{5} = \log x^2 y - \log 5$

$\qquad = \log x^2 + \log y - \log 5$

$\qquad = 2\log x + \log y - \log 5$

(b) $\ln \dfrac{\sqrt{3x-5}}{7x^3} = \ln(3x-5)^{1/2} - \ln 7x^3$

$\qquad = \ln(3x-5)^{1/2} - \ln 7 - \ln x^3$

$\qquad = \dfrac{1}{2}\ln(3x-5) - \ln 7 - 3\ln x$

(c) $\ln\left(\dfrac{5x^2}{y}\right)^3 = 3\ln\dfrac{5x^2}{y}$

$\qquad = 3(\ln 5x^2 - \ln y)$

$\qquad = 3(\ln 5 + \ln x^2 - \ln y)$

$\qquad = 3(\ln 5 + 2\ln x - \ln y)$

EXAMPLE 8
Rewriting Logarithmic
Expressions

Rewrite as a single logarithm:

(a) $\frac{1}{2}\log x - 3\log(x+1)$

(b) $2\ln(x+2) - \frac{1}{3}(\ln x + \ln y)$

Solution:

(a) In this case, we use the *properties of logarithms* in the reverse (right to left) direction.

$$\frac{1}{2}\log x - 3\log(x+1) = \log x^{1/2} - \log(x+1)^3$$

$$= \log \frac{\sqrt{x}}{(x+1)^3}$$

(b) Again, in reverse direction we have

$$2\ln(x+2) - \frac{1}{3}\left(\ln x + \ln y\right) = \ln(x+2)^2 - (\ln x^{1/3} + \ln y^{1/3})$$

$$= \ln(x+2)^2 - \ln(xy)^{1/3}$$

$$= \ln \frac{(x+2)^2}{\sqrt[3]{xy}}$$

Logarithms Using Tables

Although calculators are more efficient than tables in computations with logarithms, it is instructive to see how to work with tables of logarithms. Using base 10, we note first that every positive real number can be written as a product $c \times 10^k$, where $1 \leq c < 10$. For example,

$$1986 = 1.986 \times 10^3, \qquad 5.37 = 5.37 \times 10^0,$$
$$0.0439 = 4.39 \times 10^{-2}$$

Suppose we apply the properties of logarithms to the number 1986:

$$1986 = 1.986 \times 10^3$$
$$\log 1986 = \log(1.986 \times 10^3)$$
$$= \log(1.986) + \log(10^3)$$
$$= \log(1.986) + 3 \log(10)$$
$$= \log(1.986) + 3$$

In general, for any positive real number x (expressible as $x = c \times 10^k$), its logarithm has the **standard form**

$$\log x = \log c + \log 10^k = \log c + k$$

where $1 \leq c < 10$. We call $\log c$ the **mantissa** and k the **characteristic** of $\log x$. Since $\log x$ increases as x increases, it follows, from $1 \leq c < 10$, that

$$\log 1 \leq \log c < \log 10$$
$$0 \leq \log c < 1$$

which means that the *mantissa* always lies between 0 and 1.

The common logarithm table in Appendix C gives four-decimal-place approximations of the *mantissa* for the logarithm of every three-digit number between 1.00 and 9.99. The next example shows how to use these tables to evaluate logarithms.

EXAMPLE 9
Evaluating Logarithms
with Tables

Use the tables in Appendix C to approximate

(a) log 85.6   (b) log 0.000329

Solution:

(a) Since $85.6 = 8.56 \times 10^1$, the characteristic is 1. By the tables, the mantissa is $\log 8.56 = 0.9325$. Therefore,

$$\log 85.6 = (\text{mantissa}) + (\text{characteristic})$$
$$= \log 8.56 + 1$$
$$= 0.9325 + 1 = 1.9325$$

(b) Since $0.000329 = 3.29 \times 10^{-4}$, the characteristic is -4. Using tables for the mantissa, we obtain

$$\log 0.000329 = \log 3.29 + (-4) = 0.5172 - 4 = -3.4828$$

EXAMPLE 10
Finding Antilogarithms
with Tables

Use tables to approximate the value of x if

(a) $\log x = 2.7582$ (b) $\log x = -3.6364$

Solution:

(a) In standard form, we have

$$\log x = 0.7582 + 2 = \log c + k$$

By locating the mantissa 0.7582 in the common logarithm table, we see that it corresponds to $\log 5.73$. Thus, since $k = 2$, we have

$$x = c \times 10^k = 5.73 \times 10^2 = 573$$

(b) To obtain a nonnegative mantissa, we add and subtract 4 to get the standard form

$$\log x = 4 - 3.6364 - 4 = 0.3636 - 4$$

From the logarithm table we find that $0.3636 = \log 2.31$. Thus, since $k = -4$, we have

$$x = c \times 10^k = 2.31 \times 10^{-4} = 0.000231$$

To do numerical computations with logarithms, we combine the properties of logarithms with the procedures shown in Examples 9 and 10.

EXAMPLE 11
Computations Using
Logarithm Tables

Use tables of logarithms to approximate the value of

$$N = \frac{(1.9)^3}{\sqrt{82.7}}$$

Solution:
Using properties of logarithms as in Example 7, we write

$$\log N = 3 \log 1.9 - \frac{1}{2} \log 82.7$$

$$= 3(0.2788) - \frac{1}{2}\left(1.9175\right)$$

$$= 0.8364 - 0.95875$$

$$\approx -0.12235$$

Adding and subtracting 1, we obtain the standard form

$$\log N \approx (1 - 0.12195) - 1 = 0.87805 - 1$$

Using the mantissa value closest to 0.87805 in the table, we have $0.8779 = \log 7.55$. Therefore,

$$\log N \approx 0.8779 - 1$$

$$N = c \times 10^k \approx 7.55 \times 10^{-1} = 0.755$$

Remark: A slightly more accurate approximation could be obtained by using *linear interpolation* (Section 3.5) to find the antilogarithm for 0.87805.

Section Exercises 6.2

In Exercises 1–10, write the equation in exponential form.

1. $4 = \log_2 16$
2. $3 = \log_4 64$
3. $-2 = \log_5(\frac{1}{25})$
4. $-3 = \log_2(\frac{1}{8})$
5. $\frac{1}{2} = \log_{16} 4$
6. $\frac{2}{3} = \log_{27} 9$
7. $0 = \log_7 1$
8. $3 = \log 1000$
9. $2 = \ln e^2$
10. $a = \log_b c$

In Exercises 11–20, write the equation in logarithmic form.

11. $5^3 = 125$
12. $8^2 = 64$
13. $81^{1/4} = 3$
14. $9^{3/2} = 27$
15. $6^{-2} = \frac{1}{36}$
16. $10^{-3} = 0.001$
17. $e^3 = 20.0855\ldots$
18. $e^0 = 1$
19. $u^v = w$
20. $e^x = 4$

In Exercises 21–34, find the value of the unknown.

21. $\log 1000 = x$
22. $\log 0.1 = x$
23. $\log_4(\frac{1}{64}) = x$
24. $\log_5 25 = x$
25. $\log_3 x = -1$
26. $\log_2 x = -4$
27. $\log_b 27 = 3$
28. $\log_b 2 = \frac{1}{5}$
29. $\ln e^x = 3$
30. $e^{\ln x} = 4$
31. $x^2 - x = \log_5 25$
32. $3x + 5 = \log_2 64$
33. $x - 3 = \log_2 32$
34. $x - x^2 = \log_4(\frac{1}{64})$

In Exercises 35–50, evaluate the logarithm using the properties of logarithms, given $\log_b 2 = 0.3562$, $\log_b 3 = 0.5646$, and $\log_b 5 = 0.8271$.

35. $\log_b 6$
36. $\log_b 15$
37. $\log_b(\frac{3}{2})$
38. $\log_b(\frac{5}{3})$
39. $\log_b 25$
40. $\log_b 18 \ (18 = 2 \cdot 3^2)$
41. $\log_b \sqrt{2}$
42. $\log_b(\frac{9}{2})$
43. $\log_b 40$
44. $\log_b \sqrt[3]{75}$
45. $\log_b(\frac{1}{4})$
46. $\log_b(3b^2)$
47. $\log_b \sqrt{5b}$
48. $\log_b 1$
49. $\log_b\left[\dfrac{(4.5)^3}{\sqrt{3}}\right]$
50. $\log_b\left(\dfrac{4}{15^2}\right)$

In Exercises 51–60, use the properties of logarithms to write the expressions as a sum, difference, or multiple of logarithms.

51. $\log_2 xyz$
52. $\ln\left(\dfrac{xy}{z}\right)$
53. $\ln\sqrt{a-1}$
54. $\log_3\left(\dfrac{x^2-1}{x^3}\right)^3$
55. $\ln z(z-1)^2$
56. $\log_4 \sqrt{\dfrac{x^2}{y^3}}$
57. $\log_b\left(\dfrac{x^2}{y^2 z^3}\right)$
58. $\log_b\left(\dfrac{\sqrt{x}y^4}{z^4}\right)$
59. $\ln\sqrt[3]{x\sqrt{y}}$
60. $\ln\left(\dfrac{x}{\sqrt{x^2+1}}\right)$

In Exercises 61–65, write the expression as a single logarithm.

61. $\log_3(x-2) - \log_3(x+2)$
62. $3\ln x + 2\ln y - 4\ln z$
63. $\frac{1}{3}[2\ln(x+3) + \ln x - \ln(x^2-1)]$
64. $2[\ln x - \ln(x+1) - \ln(x-1)]$
65. $2\ln 3 - \frac{1}{2}\ln(x^2+1)$

In Exercises 66–73, evaluate the logarithm using the change of base formula and your calculator. Do the problem twice, once with common logarithms and once with natural logarithms.

66. $\log_3 7$
67. $\log_7 4$
68. $\log_{1/2} 4$
69. $\log_4(0.55)$
70. $\log_9(0.4)$
71. $\log_{20} 125$
72. $\log_{15} 1250$
73. $\log_{1/3}(0.015)$

In Exercises 74 and 75, use a calculator to evaluate the logarithms.

74. $\ln\left(\dfrac{1+\sqrt{3}}{2}\right)$
75. $\ln(\sqrt{5}-2)$

76. The time (in hours) necessary for a certain object to cool $10°$ is

$$t = \frac{10\,\ln(\frac{1}{2})}{\ln(\frac{3}{4})}$$

Find t.

77. The population of a town will double in

$$t = \frac{10\,\ln 2}{\ln 67 - \ln 50}$$

years. Find t.

78. The work (in foot-pounds) done in compressing an initial volume of 9 cubic feet at a pressure of 15 pounds per square inch to a volume of 3 cubic feet is

$$W = 19440(\ln 9 - \ln 3)$$

Find W.

In Exercises 79 and 80, use the Richter Scale (Example 6) for measuring the magnitude of earthquakes.

79. Find the magnitude R of an earthquake of intensity I (let $I_0 = 1$):
    (a) $I = 80,500,000$ (b) $I = 48,275,000$
80. Find the intensity I of an earthquake measuring R on the Richter Scale (let $I_0 = 1$):
    (a) Columbia in 1906, $R = 8.6$
    (b) Los Angeles in 1971, $R = 6.7$

Acidity (pH) is a measure of the hydrogen ion concentration H^+ (measured in moles of hydrogen per liter of solution) of a solution. The formula for pH is

$$pH = -\log[H^+]$$

Use this formula in Exercises 81–84.

81. Compute $[H^+]$ for a solution in which pH $= 5.8$.
82. Compute $[H^+]$ for a solution in which pH $= 3.2$.
83. If $[H^+] = 2.3 \times 10^{-5}$, find the pH.
84. If $[H^+] = 11.3 \times 10^{-6}$, find the pH.

In Exercises 85–87, find the logarithm by using the table in Appendix C.

85. (a) log 417   (b) log 41.7   (c) log 0.0417

86. (a) log 985   (b) log 9.85   (c) log 0.985
87. (a) log 4385   (b) log 43.85   (c) log 0.004385

In Exercises 88–90, find N (antilogarithm) by using the table in Appendix C.

88. (a) $\log N = 4.3979$ (b) $\log N = 1.3979$
    (c) $\log N = -1.6021$
89. (a) $\log N = 2.6294$ (b) $\log N = 0.6294$
    (c) $\log N = -2.3706$
90. (a) $\log N = 6.1335$ (b) $\log N = -3.8665$
    (c) $\log N = 8.1335 - 10$

In Exercises 91–100, use logarithms to approximate the quantity to four-digit accuracy.

91. $\dfrac{(86.4)(8.09)}{38.6}$ 92. $\dfrac{1243}{(42.8)(67.9)}$

93. $(1.05)^{15}$ 94. $(0.2313)^6$

95. $\sqrt[3]{86.5}$ 96. $\sqrt[4]{(4.705)(18.86)}$

97. $\left(\dfrac{18.60}{59.25}\right)^{0.75}$ 98. $(48.8)^{0.75} + (75.4)^{0.45}$

99. $\dfrac{(3165)^{0.19}}{(525)^{0.45}}$ 100. $(5.85 \times 10^4)(16.4 \times 10^5)$

Logarithmic Functions   6.3

Having studied logarithms and their properties, we are now ready to look at a new family of functions known as **logarithmic functions.**

DEFINITION OF A LOGARITHMIC FUNCTION

> The function f defined by
>
> $$f(x) = \log_a x$$
>
> for all positive real numbers x is called the **logarithmic function** to the base a.

Since logarithms are exponents, we would expect logarithmic functions to have a close relationship to the exponential functions studied in Section 6.1. To see the connection between these functions, consider the fact that

$$y = \log_a x \qquad \text{if and only if} \qquad x = a^y$$

This suggests that for $a > 0$ and $a \neq 1$, the logarithmic function $f(x) = \log_a x$ is the **inverse** of the exponential function $g(x) = a^x$ and vice versa. (See Section 3.6.) Furthermore, this means that their graphs are reflections of each other in the line $y = x$. This is shown in Example 1.

EXAMPLE 1
Graphing a Logarithmic
Function and Its Inverse

On the same set of axes, sketch the graphs of

$$f(x) = 2^x \qquad \text{and} \qquad g(x) = \log_2 x$$

Solution:
For $f(x) = 2^x$, we make up the following table of values:

x	-2	-1	0	1	2	3
$f(x) = 2^x$	$\dfrac{1}{4}$	$\dfrac{1}{2}$	1	2	4	8

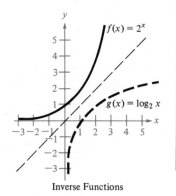

Inverse Functions

FIGURE 6.8

By plotting these points and connecting them with a smooth curve, we have the graph shown in Figure 6.8. Finally, to sketch the graph of $g(x) = \log_2 x$, we use the fact that g is the inverse of f. Notice in Figure 6.8 that the graph of g is obtained by reflecting the graph of f in the line $y = x$.

In Figure 6.8, note how the graph of $g(x) = \log_2 x$ falls sharply as x nears zero from the right. This observation and the fact that g is not defined at $x = 0$ suggest that the y-axis is a *vertical asymptote* for the graph of g. Actually, all logarithmic functions of the form

$$f(x) = \log_b x, \qquad b > 1$$

have the y-axis as a vertical asymptote, as shown in Figure 6.9. Study the general properties of logarithmic functions as we have listed them.

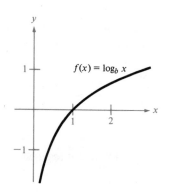

FIGURE 6.9

Graph of $y = \log_b x$, $b > 1$

1. Domain: All positive reals
2. Range: All reals
3. Intercept: (1, 0)
4. $\log_b x$ increases as x increases
5. y-axis is a vertical asymptote
6. Reflection of the graph of $y = b^x$

EXAMPLE 2
Graphing Logarithmic Functions

On the same set of axes, sketch graphs of

(a) $f(x) = \log x$
(b) $g(x) = \ln x$

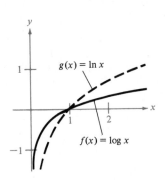

FIGURE 6.10

Solution:

(a) Since

$$f(x) = \log x = \log_{10} x$$

we make up the following table of values. Note that the first three entries in the table can be found without the use of a calculator.

	Without Calculator			With Calculator		
x	$\dfrac{1}{10}$	1	10	2	5	8
$\log_{10} x$	-1	0	1	0.301	0.699	0.903

The graph of f is shown in Figure 6.10.

(b) Since

$$g(x) = \ln x = \log_e x$$

we make up the following table of values. When sketching functions involving the natural logarithm, it is helpful to remember the following approximations:

$$\frac{1}{e} \approx .37, \qquad e \approx 2.72, \qquad e^2 \approx 7.39$$

	Without Calculator				With Calculator		
x	$1/e$	1	e	e^2	2	5	8
$\ln x$	-1	0	1	2	0.693	1.609	2.079

The graph of g is shown in Figure 6.10.

The graphs of logarithmic functions to bases other than 10 or e can be readily obtained from the graph of $f(x) = \log x$ (or $\ln x$) by use of the change of base formula

$$\log_b x = \frac{\log_{10} x}{\log_{10} b} = \left(\frac{1}{\log_{10} b}\right) \log_{10} x$$

This rule shows us that $\log_b x$ is simply a constant multiple of $\log_{10} x$. Hence, a graph of $g(x) = \log_b x$ can be obtained from the graph of $\log_{10} x$ *by multiplying ordinates* (y-values) *by the factor* $1/\log_{10} b$. For instance, the graph of

$$g(x) = \log_4 x$$

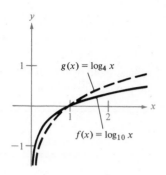

FIGURE 6.11

can be obtained from the graph of $f(x) = \log x$ by stretching away from the x-axis by the factor $(1/\log 4) \approx 1.661$, as shown in Figure 6.11. That is, for each x,

$$g(x) = (1.661)f(x)$$

Since logarithms are defined only for positive real numbers, the domain is important in sketching the graphs of logarithmic functions. For instance, Property 5 of logarithms guarantees that

$$\ln(x - 1)^2 = 2 \ln(x - 1)$$

but only for positive values for $(x - 1)$, that is, for $x > 1$. Consequently, the graphs of

$$f(x) = \ln(x - 1)^2 \qquad \text{and} \qquad g(x) = 2 \ln(x - 1)$$

are not the same because they have different domains, as we will see in the next example.

EXAMPLE 3
Shifts in the Graphs of
Logarithmic Functions

Sketch graphs of

(a) $f(x) = \ln(x - 1)^2$ (b) $g(x) = 2 \ln(x - 1)$

Solution:
As we pointed out in the preceding discussion, these two functions have the same values for $x > 1$. The domains of f and g are as follows:

Function	Domain
$f(x) = \ln(x - 1)^2$ $g(x) = 2 \ln(x - 1)$	all real $x \neq 1$ $x > 1$

Using the $\boxed{\text{LN}}$ key on a calculator, we obtain the points shown in the following table. Note the symmetry with respect to the line $x = 1$.

x	-2	-1	0	0.5	1	1.5	2	3	4
$\ln(x - 1)^2$	2.20	1.39	0	-1.39	undef.	-1.39	0	1.39	2.20

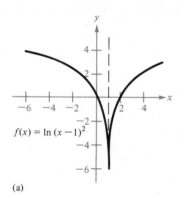

(a)

The graph of f is shown in Figure 6.12(a). To obtain the graph of g, we simply retrace that portion of the graph of f that lies to the right of the line $x = 1$, as shown in Figure 6.12(b). Note that in both cases the x-intercept is $(2, 0)$, a shift of one to the right compared to the graph of $y = \ln x$ in Figure 6.10. Note also that the vertical asymptote falls at $x = 1$, where both functions are undefined.

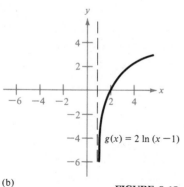

(b)

FIGURE 6.12

EXAMPLE 4
Application of a
Logarithmic Function

Students in a precalculus class were given a final exam and then were retested monthly. On equivalent exams given each month thereafter, the average scores decreased according to the *human memory model*

$$F(t) = 75 - 15 \log(t + 1)$$

(a) What was the average score on the original ($t = 0$) exam?
(b) What was the average score at the end of $t = 2$ months? What was the average score at the end of $t = 6$ months?
(c) How long will it take for the average score to drop below 45?
(d) Sketch a graph of F.

Solution:

(a) The original average was

$$\begin{aligned}
F(0) &= 75 - 15 \log(0 + 1) \\
&= 75 - 15 \log 1 = 75 - 15(0) \\
&= 75
\end{aligned}$$

(b) After 2 months, the average was

$$\begin{aligned}
F(2) &= 75 - 15 \log 3 \approx 75 - 15(.4771) \\
&\approx 67.8
\end{aligned}$$

After 6 months, the average was

$$\begin{aligned}
F(6) &= 75 - 15 \log 7 \approx 75 - 15(.8451) \\
&\approx 62.3
\end{aligned}$$

(c) To find how long it would take for the average score to drop to 45, we set $F(t) = 45$ and solve for t.

$$\begin{aligned}
F(t) = 75 - 15 \log(t + 1) &= 45 \\
-15 \log(t + 1) &= -30 \\
\log(t + 1) &= 2
\end{aligned}$$

Using the antilogarithm to the base 10, we have

$$\begin{aligned}
t + 1 &= 10^2 \\
t &= 100 - 1 \\
&= 99 \text{ months}
\end{aligned}$$

(d) Several points are shown in the following table, and the graph of F is shown in Figure 6.13.

t	0	1	2	6	12
$F(t)$	75	70.5	67.8	62.3	58.3

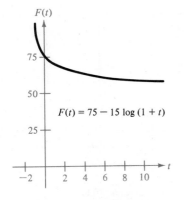

$F(t) = 75 - 15 \log (1 + t)$

FIGURE 6.13

Section Exercises 6.3

In Exercises 1–8, sketch the graph of the function and its inverse on the same set of axes.

1. $f(x) = 3^x$, $g(x) = \log_3 x$
2. $f(x) = 5^x$, $g(x) = \log_5 x$
3. $f(x) = (\frac{1}{2})^x$, $g(x) = \log_{1/2} x$
4. $f(x) = (\frac{1}{3})^x$, $g(x) = \log_{1/3} x$
5. $f(x) = e^x$, $g(x) = \ln x$
6. $f(x) = e^{2x}$, $g(x) = \ln \sqrt{x}$
7. $f(x) = e^{x-1}$, $g(x) = 1 + \ln x$
8. $f(x) = e^x - 1$, $g(x) = \ln(x + 1)$

In Exercises 9–18, use the graph of $y = \ln x$ to match the given function with one of the accompanying graphs.

9. $f(x) = 4 + \ln x$
10. $f(x) = \ln \dfrac{1}{x} = -\ln x$
11. $f(x) = \ln(x - 2)$
12. $f(x) = \ln(x + 5)$
13. $f(x) = \ln x^3 = 3 \ln x$
14. $f(x) = \ln \sqrt{x} = \frac{1}{2} \ln x$
15. $f(x) = -\ln(x + 1)$
16. $f(x) = \ln 2x$
17. $f(x) = \ln x^2$
18. $f(x) = \ln |x - 3|$

(a)

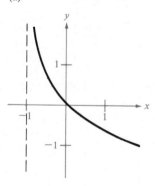

(b)

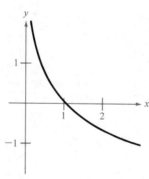

(c)

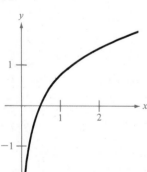

(d)

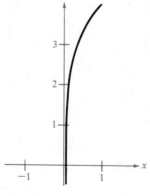

(e)

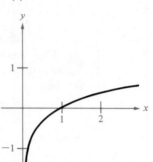

(f)

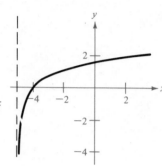

(g)

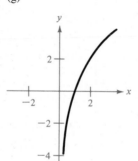

(h)

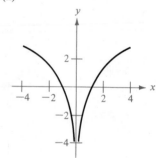

(i)

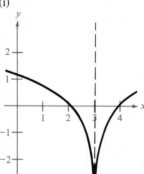

(j)

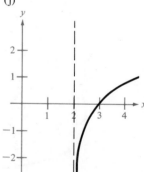

In Exercises 19–26, determine the domain of the function.

19. $f(x) = \log(2x - 1)$
20. $f(x) = \log_4(1 - x)$
21. $g(x) = \ln(-x)$
22. $h(x) = \ln(x - 2) + \ln x$
23. $h(x) = \log_2 \sqrt{4 - x^2}$
24. $g(x) = \log_b(b^{-x})$
25. $f(x) = \log_b |x + 2|$
26. $f(x) = \log_b(x^2 - 4)$

27. The intensity level β, in decibels, of a sound wave is defined by

$$\beta(I) = 10 \log \frac{I}{I_0}$$

where I_0 is an arbitrary intensity of 10^{-16} watts per square centimeter, corresponding roughly to the faintest sound that can be heard. Determine $\beta(I)$ if
    (a) $I = 10^{-14}$ watts per square centimeter (whisper)
    (b) $I = 10^{-9}$ watts per square centimeter (busy street corner)
    (c) $I = 10^{-6.5}$ watts per square centimeter (air hammer)
    (d) $I = 10^{-4}$ watts per square centimeter (threshold of pain)
    (Intensity is the power transported per unit area of a surface perpendicular to the direction of propagation of the sound wave.)

28. Students in a mathematics class were given an exam and then retested monthly with an equivalent exam. With each passing month the average score for the class decreased according to the human memory model

$$F(t) = 80 - 17 \log(t + 1)$$

    (a) What was the average score on the original exam $(t = 0)$?
    (b) What was the average score after 4 months?
    (c) How long will it take for the average score to drop below 53?

29. A principal P invested at $9\frac{1}{2}\%$, compounded continuously, increases to an amount KP after t years, where t is given by

$$t = \frac{\ln K}{0.095}$$

    (a) Complete the following table for the given function:

K	1	2	3	4	6	8	10	12
t								

    (b) Use the table of part (a) to graph this function.

30. The time in years for the world population to double if it is increasing at a continuous rate of r is given by

$$t = \frac{\ln 2}{r}$$

Use this function to complete the following table:

r	0.005	0.010	0.015	0.020	0.025	0.030
t						

31. Use a calculator to demonstrate that

$$\frac{\ln x}{\ln y} \neq \ln\left(\frac{x}{y}\right) = \ln x - \ln y$$

by completing the following table:

x	y	$\dfrac{\ln x}{\ln y}$	$\ln \dfrac{x}{y}$	$\ln x - \ln y$
1	2			
3	4			
10	5			
4	0.5			

32. (a) Use a calculator to complete the following table for the function

$$f(x) = \frac{\ln x}{x}$$

x	1	5	10	10^2	10^4	10^6
$f(x)$						

    (b) Use the table of part (a) to determine what $f(x)$ approaches as x increases without bound.

Exponential and Logarithmic Equations

6.4

In the first three sections of this chapter, we focused our study mainly on the definitions, properties, graphs, and applications of exponential and logarithmic functions. Here we concentrate on algebraic procedures for solving *equations* involving these functions. In Section 6.1, we solved some simple exponential equations such as $(\frac{1}{2})^x = 64$ by writing each term as a rational power of the base, in this case

$$2^{-x} = 64 = 2^6 \quad \Rightarrow \quad x = -6$$

However, this method does not work as well for equations like

$$2^x = 7$$

since it is not easy to write 7 as a power of 2. As a general guideline, the following suggestions work well for solving exponential and logarithmic equations.

SUGGESTED RULES FOR SOLVING EXPONENTIAL AND LOGARITHMIC EQUATIONS

> To solve an exponential equation, first take the logarithm of both sides. Then try solving for the unknown.
>
> To solve a logarithmic equation, first rewrite it in exponential form, then try solving for the unknown.

We can use the preceding suggestions to prove the *change of base formula* given in Section 6.2:

$$\log_b x = \frac{\log_a x}{\log_a b}$$

Let $y = \log_b x$, then the equivalent exponential form is $b^y = x$. Therefore,

$$\log_a(b^y) = \log_a x \qquad \textit{Take } \log_a \textit{ of both sides}$$

$$y \log_a b = \log_a x \qquad \textit{Property 5}$$

$$y = \frac{\log_a x}{\log_a b} \qquad \textit{Solve for y}$$

Finally,

$$\log_b x = \frac{\log_a x}{\log_a b} \qquad \textit{Back substitute for y}$$

EXAMPLE 1
Solving Exponential Equations

Solve for x:

(a) $5^x = 72$

(b) $3^{-x} = 0.026$

Solution:

(a) To solve for x, we take the common logarithm of both sides (natural logarithms would work just as well) to get

$$\log(5^x) = \log 72$$
$$x \log 5 = \log 72 \qquad \textit{Property 5}$$
$$x = \frac{\log 72}{\log 5} \qquad \textit{Divide by log 5}$$

The calculator keystroke sequence

72 $\boxed{\text{LOG}}$ $\boxed{\div}$ 5 $\boxed{\text{LOG}}$ $\boxed{=}$

yields $x \approx 2.6572$.

(b) Taking the common logarithm of both sides, we get

$$\log(3^{-x}) = \log 0.026$$
$$-x \log 3 = \log 0.026 \qquad \textit{Property 5}$$
$$x = -\frac{\log 0.026}{\log 3} \approx 3.3221 \quad \textit{Divide by log 3}$$

When both members of an equation are exponential functions, a procedure similar to that shown in Example 1 can still be used. However, the algebra is a bit more complicated. Watch how it is done in our next example.

EXAMPLE 2
Solving Exponential Equations

Solve for x:

(a) $4^{2x+3} = 7^{x+2}$

(b) $\dfrac{e^x - e^{-x}}{2} = 5$

Solution:

(a) Taking the common logarithm of both sides, we obtain

$$\log(4^{2x+3}) = \log(7^{x+2})$$
$$(2x + 3) \log 4 = (x + 2) \log 7 \qquad \textit{Property 5}$$
$$2x \log 4 + 3 \log 4 = x \log 7 + 2 \log 7 \qquad \textit{Distributive Law}$$
$$2x \log 4 - x \log 7 = 2 \log 7 - 3 \log 4 \qquad \textit{Collect like terms}$$
$$x(2 \log 4 - \log 7) = 2 \log 7 - 3 \log 4 \qquad \textit{Factor out x}$$
$$x = \frac{2 \log 7 - 3 \log 4}{2 \log 4 - \log 7} \qquad \textit{Divide}$$
$$x \approx -0.3231$$

(b) Some preliminary algebra is helpful.

$$\frac{e^x - e^{-x}}{2} = 5 \qquad \textit{Given}$$

$$e^x - e^{-x} = 10 \qquad \textit{Multiply by 2}$$

$$e^{2x} - 1 = 10e^x \qquad \textit{Multiply by } e^x$$

$$(e^x)^2 - 10(e^x) - 1 = 0 \qquad \textit{Quadratic form}$$

Let $u = e^x$, then we obtain the quadratic equation

$$u^2 - 10u - 1 = 0$$

whose solution, by the Quadratic Formula, is

$$u = \frac{10 \pm \sqrt{100 + 4}}{2} = \frac{10 \pm 2\sqrt{26}}{2} = 5 \pm \sqrt{26}$$

Now, since $u = e^x$, we have

$$u = e^x = 5 \pm \sqrt{26}$$

But e^x is never negative, so the only valid solution is

$$e^x = 5 + \sqrt{26}$$

Taking the *natural* logarithm of both sides, we obtain

$$\ln(e^x) = \ln(5 + \sqrt{26})$$

$$x \ln e = \ln(5 + \sqrt{26}) \qquad \textit{Property 5}$$

$$x(1) = \ln(5 + \sqrt{26}) \qquad \textit{Property 1}$$

$$x = \ln(5 + \sqrt{26}) \approx 2.3124$$

In our next example, the *properties of logarithms* play a vital role in the solution procedure.

EXAMPLE 3
Solving Logarithmic Equations

Solve for x:

(a) $\ln(x - 2) + \ln(2x - 1) = 2 \ln x$
(b) $\log (5x + 3) - \log(x - 1) = 2$

Solution:

(a) $\ln(x - 2) + \ln(2x - 1) = 2 \ln x$ *Given*

      $\ln(x - 2)(2x - 1) = \ln x^2$ *Properties 3 and 5*

         $(x - 2)(2x - 1) = x^2$ *Property 8*

           $2x^2 - 5x + 2 = x^2$

             $x^2 - 5x + 2 = 0$ *Collect like terms*

Using the Quadratic Formula, we obtain

$$x = \frac{5 \pm \sqrt{17}}{2}$$

Since $(x - 2)$ is negative for $x = (5 - \sqrt{17})/2$, the *only valid solution* is

$$x = \frac{5 + \sqrt{17}}{2} \approx 4.5616$$

(b) $\log(5x + 3) - \log(x - 1) = 2$

$$\log\left(\frac{5x + 3}{x - 1}\right) = 2$$

$$\frac{5x + 3}{x - 1} = 10^2$$

$$5x + 3 = 100(x - 1)$$

$$x = \frac{103}{95}$$

We complete this section with two practical applications of exponential equations.

EXAMPLE 4
Compound Interest

An amount of $11,000 is invested in a trust fund at an annual interest rate of 9.5%, compounded continuously.

(a) How long will it take for the initial investment to double in value?

(b) What interest rate is required for the initial investment to double in 10 years' time?

Solution:

(a) For continuous compounding, we have the formula

$$B = Pe^{rt}$$

In this case, $P = 11,000$, $r = 0.095$, and $B = 2P = 22,000$. Thus, we have

$$22,000 = 11,000(e^{0.095t})$$

$$2 = e^{0.095t}$$

$$\ln 2 = \ln(e^{0.095t}) \qquad \textit{Take ln of both sides}$$

$$\ln 2 = 0.095t(\ln e) = 0.095t$$

$$t = \frac{\ln 2}{0.095} \approx 7.3 \text{ years}$$

(b) In this case, $P = 11,000$, $B = 22,000$, and $t = 10$. Thus, we have

$$22,000 = 11,000(e^{10r})$$

$$2 = e^{10r}$$

$$\ln 2 = \ln(e^{10r}) = 10r(\ln e) = 10r$$

$$r = \frac{\ln 2}{10} \approx 0.0693 \approx 6.93\%.$$

EXAMPLE 5
Logistic Curve

On a college campus of 5000 students, a single student returned from vacation with a contagious flu virus. The spread of this disease through the student body is given by

$$s(t) = \frac{5,000}{1 + 4,000\,e^{-0.8t}}$$

where $s(t)$ is the total number infected after t days.

(a) How many are infected after 5 days?
(b) If the college will close if 40% of the students are ill, after how many days will it close?

Solution:

(a) After 5 days, the number infected is

$$s(5) = \frac{5000}{1 + 4000e^{-0.8(5)}}$$

$$= \frac{5000}{1 + 4000e^{-4}} \approx 67 \qquad \textit{(To nearest integer)}$$

One possible calculator key stroke sequence for finding $s(5)$ is

4 $\boxed{+/-}$ $\boxed{e^x}$ $\boxed{\times}$ 4000 $\boxed{+}$ 1 $\boxed{=}$ $\boxed{1/x}$ $\boxed{\times}$ 5000 $\boxed{=}$

(b) In this case, the number infected is

$$(0.40)(5000) = 2000$$

Therefore, we solve for t in the equation

$$2000 = \frac{5000}{1 + 4000e^{-0.8t}}$$

obtaining

$$2000 + 8,000,000\,e^{-0.8t} = 5000$$

$$e^{-0.8t} = \frac{3000}{8,000,000} = \frac{3}{8000}$$

$$\ln(e^{-0.8t}) = \ln 3 - \ln 8000$$

$$-0.8t = \ln 3 - \ln 8000$$

$$t = \frac{\ln 3 - \ln 8000}{-0.8} \approx 9.86$$

Hence, in 10 days, at least 40% of the students will be infected and the college will close.

Section Exercises 6.4

In Exercises 1–30, solve for the unknown.

1. $4^x = \frac{1}{16}$

2. $7^x = \frac{1}{49}$

3. $3^x = 243$

4. $8^x = 4$

5. $4^x = 12$ (Use common logarithms)

6. $6^{-x} = 52$ (Use common logarithms)

7. $(8.5)^{-x} = 360$ (Use natural logarithms)

8. $(\frac{3}{4})^x = 0.15$ (Use natural logarithms)

9. $e^{0.09t} = 3$

10. $e^{0.125t} = 8$

11. $\left(1 + \frac{0.10}{12}\right)^{12t} = 2$

12. $\left(1 + \frac{0.065}{365}\right)^{365t} = 4$

13. $\frac{10,000}{1 + 19e^{-t/5}} = 2000$

14. $80e^{-t/2} + 20 = 70$

15. $\left(\frac{1}{1.0775}\right)^N = 0.2247$

16. $3^{2x+1} = 5^{x+2}$

17. $10^{7-x} = 5^{x+1}$

18. $4^{x^2} = 100$

19. $\log_3 x + \log_3(x - 2) = 1$

20. $\log_{10}(x + 3) - \log_{10}x = 1$

21. $x^2 - 2x = \log_5 125$

22. $x - x^2 = \log_4 \frac{1}{16}$

23. $\log_2(x + 5) - \log_2(x - 2) = 3$

24. $\log\sqrt{x + 2} = 1$

25. $\ln x^2 = (\ln x)^2$

26. $\frac{e^x + e^{-x}}{2} = 2$

27. $\frac{e^x + e^{-x}}{e^x - e^{-x}} = 2$

28. $\frac{e^x - e^{-x}}{e^x + e^{-x}} = \frac{1}{4}$

29. $\ln(e^{x^2}) = 4$

30. $\log_b(b^{2x-1}) = 5$

31. $1000 is deposited into a fund at an annual percentage rate of 11%. Find the time for the investment to double if the interest is compounded
    (a) annually   (b) monthly   (c) daily   (d) continuously

32. Repeat Exercise 31, using a percentage rate of $10\frac{1}{2}$% and finding the time it will take for the investment to triple.

33. Complete the following table for the time t necessary for P dollars to triple if interest is compounded continuously at the rate r.

r	2%	4%	6%	8%	10%	12%
t						

34. The demand equation for a certain product is given by

$$P = 500 - 0.5(e^{0.004x})$$

Find the demand x if the price charged is
(a) $P = $350   (b) $P = $300

35. The demand equation for a certain product is given by

$$P = 5000\left(1 - \frac{4}{4 + e^{-0.002x}}\right)$$

Find the demand x if the price charged is
(a) $P = $600   (b) $P = $400

36. The population growth of a city is given by $105,300e^{0.015t}$, where t is the time in years with $t = 0$ corresponding to 1985. According to this model, in what year will the city have a population of 150,000?

37. The yield V (in millions of cubic feet per acre) for a forest at age t is given by

$$V = 6.7e^{-48.1/t}$$

Find the time necessary to have a yield of
(a) 1.3 million cubic feet   (b) 2 million cubic feet

38. In a group project in learning theory, a mathematical model for the proportion P of correct responses after n trials was found to be

$$P = \frac{0.83}{1 + e^{-0.2n}}$$

After how many trials will 60% of the responses be correct?

39. A certain lake is stocked with 500 fish, and their population increases according to the **logistics curve**

$$p(t) = \frac{10,000}{1 + 19e^{-t/5}}$$

where t is measured in months. After how many months will the fish population be 2000?

Newton's Law of Cooling states that the rate of change in the temperature of an object is proportional to the difference between its temperature and the temperature of its environment. If $T(t)$ is the temperature of the object at time t, T_0 is the initial temperature, and T_e is the constant temperature of the environment, then

$$T(t) = T_e + (T_0 - T_e)e^{-kt}$$

Use this model in Exercises 40 and 41.

40. An object in a room at 70° cools from 350° to 150° in 45 minutes. Find

(a) the temperature of the object as a function of time
(b) the temperature after it has cooled for 1 hour
(c) the time necessary for the object to cool to 80°

41. Using Newton's Law of Cooling, determine the reading on a thermometer 5 minutes after it is taken from a room at 72° Fahrenheit to the outdoors where the temperature is 20°, if the reading dropped to 48° after 1 minute.

Review Exercises / Chapter 6

In Exercises 1–15, sketch the graph of the function.

1. $f(x) = 6^x$
2. $f(x) = 0.3^x$
3. $g(x) = 6^{-x}$
4. $g(x) = 0.3^{-x}$
5. $h(x) = e^{-x/2}$
6. $h(x) = 2 - e^{-x/2}$
7. $s(t) = \dfrac{4}{1 + e^{-t}}$
8. $s(t) = 4e^{-2/t},\ t > 0$
9. $f(x) = \ln(x - 3)$
10. $f(x) = \ln |x|$
11. $f(x) = \ln x + 3$
12. $f(x) = \dfrac{\ln x}{4}$
13. $g(x) = \log_2 x$
14. $g(x) = \log_5 x$
15. $h(x) = \ln(e^{x-1})$

In Exercises 16–20, use the properties of logarithms to write the expression as a sum, difference, or multiple of logarithms.

16. $\ln \left| \dfrac{x - 1}{x + 1} \right|$

17. $\ln \left| \dfrac{x^2 + 1}{x} \right|$

18. $\ln \sqrt{\dfrac{x^2 + 1}{x^4}}$

19. $\ln[(x^2 + 1)(x - 1)]$

20. $\ln \sqrt[5]{\dfrac{4x^2 - 1}{4x^2 + 1}}$

In Exercises 21–25, write the expression as a single logarithm.

21. $\frac{1}{2} \ln |2x - 1| - 2 \ln |x + 1|$
22. $5 \ln |x - 2| - \ln |x + 2| - 3 \ln |x|$
23. $2[\ln x + \frac{1}{3} \ln \sqrt{y}]$
24. $\frac{1}{2} \ln (x^2 + 4x) - \ln 2 - \ln x$
25. $\ln 3 + \frac{1}{3} \ln (4 - x^2) - \ln x$

In Exercises 26–30, determine if the statement or equation is true or false.

26. The domain of the function $f(x) = \ln |x|$ is the set of all real numbers.

27. $\ln (x + y) = \ln x + \ln y$

28. $\dfrac{\ln x}{\ln y} = \ln x - \ln y$

29. $\ln \sqrt{x^4 + 2x^2} = \ln(|x|\sqrt{x^2 + 2})$

30. $e^{x-1} = \dfrac{e^x}{e}$

31. The demand equation for a certain product is given by

$$p = 500 - 0.5e^{0.004x}$$

Find the demand x if the price charged is
(a) $p = \$450$ (b) $p = \$400$

32. In a typing class, the average number of words per minute typed after t weeks of lessons was found to be

$$N = \dfrac{157}{1 + 5.4e^{-0.12t}}$$

Find the time necessary to type
(a) 50 words per minute   (b) 75 words per minute

In Exercises 33–36, find the exponential growth function $y = Ce^{kt}$ that passes through the two points.

33.
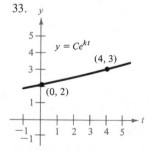

34.
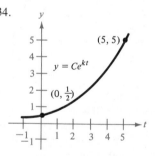

35.
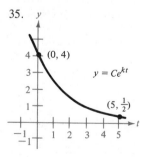

36.
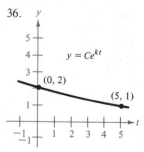

37. A deposit of $750 is made in a savings account for which the interest is compounded continuously. The balance will double in $7\frac{3}{4}$ years.
    (a) What is the annual percentage rate for this account?
    (b) Find the balance in the account after 10 years.

38. A deposit of $10,000 is made in a savings account for which the interest is compounded continuously. The balance will double in 5 years.
    (a) What is the annual percentage rate for this account?
    (b) Find the balance after 1 year.

39. The management at a certain factory has found that the maximum number of units a worker can produce in a day is 30. The learning curve for the number of units N produced per day after a new employee has worked t days is given by

    $$N = 30(1 - e^{kt})$$

    After 20 days on the job, a particular worker produced 19 units.
    (a) Find the learning curve for this worker.
    (b) How many days should pass before this worker is producing 25 units per day?

40. The management in Exercise 39 requires that a new employee be producing at least 20 units per day after 30 days on the job.
    (a) Find the learning curve describing this minimum requirement.

(b) Find the number of days before a minimal achiever is producing 25 units per day.

41. The sales S (in thousands of units) of a new product after it is on the market t years is given by

    $$S = 30(1 - e^{kt})$$

    (a) Find S as a function of t if 5000 units have been sold after 1 year.
    (b) How many units have been sold after 5 years?
    (c) Sketch a graph of this sales function.

42. Complete the following table for the function $f(x) = e^{-x}$:

x	0	1	2	5	10	15
e^{-x}	1	0.36788				

This table demonstrates that for $k > 0$, e^{-kx} approaches 0 as x increases.

43. Use the result of Exercise 42 to determine what each function ($k > 0$) approaches as x increases.
    (a) $f(x) = 30(1 - e^{-0.025x})$
    (b) $f(x) = \dfrac{50}{2 + 3e^{-0.2x}}$
    (c) $f(x) = (100 - a)e^{-kx} + a$
    (d) $f(x) = 5000\left(1 - \dfrac{4}{4 + e^{-0.002x}}\right)$

CHAPTER

7 | SYSTEMS OF EQUATIONS AND INEQUALITIES

Systems of Equations

7.1

Recall from Chapter 3 that the graph of an equation in two variables consists of all points in the plane that satisfy the equation. We call such points **solution points** (or simply **solutions**) of the equation. For example, the equation

$$x + y = 4$$

is satisfied when $x = 3$ and $y = 1$. Thus, $(3, 1)$ is a solution point of the equation. Some additional solution points are $(-4, 8)$, $(0, 4)$, $(-2, 6)$, $(6, -2)$, and $(5, -1)$. On the other hand, $(2, 3)$ is not a solution point of the equation $x + y = 4$, since $x = 2$ and $y = 3$ yields

$$2 + 3 = 5 \neq 4$$

Keep in mind that in this chapter we are concerned only with the real number solutions to an equation.

Typically, a single equation in two variables will have infinitely many solutions. However, on occasion equations in two variables have only one solution or no solutions. For instance, the equation

$$x^2 + y^2 = 0$$

has $(0, 0)$ as its only solution, and the equation

$$x^2 + y^2 = -1$$

has no solutions.

In this section we will study techniques for finding solutions to systems of two or more equations in two or more variables. That is, we show how to find points that satisfy two or more equations simultaneously.

For example, consider the following problem:

Find two numbers whose sum is four and whose difference is two.

Using x and y to denote the numbers, we can translate the verbal statement into two equations, as follows:

Sum is four: $x + y = 4$	*Equation 1*
Difference is two: $x - y = 2$	*Equation 2*

We now have what is called a **system of two equations involving two variables, x and y.** By a **solution** of this system, we mean an ordered pair (x, y) that satisfies *both* equations. This particular system happens to have only one solution, and we can find it by the **substitution method,** shown in Example 1.

EXAMPLE 1

Solving a System of Two Equations in Two Variables

Find the solution(s) to the system of equations

$$x + y = 4 \qquad \text{and} \qquad x - y = 2$$

Solution:

We have

$x + y = 4$	*Equation 1*
$x - y = 2$	*Equation 2*

We can solve for y in Equation 1 to get

$$y = 4 - x$$

Substituting this value for y into Equation 2, we obtain the equation

$$x - (4 - x) = 2$$
$$x = 3$$

Finally, by backsubstituting $x = 3$ into the equation $y = 4 - x$, we get

$$y = 4 - x$$
$$y = 4 - 3 = 1$$

Thus, the solution to the given system is $x = 3$ and $y = 1$.

In the solution to Example 1, we can identify four basic steps in the **method of substitution,** shown schematically below.

The Method of Substitution

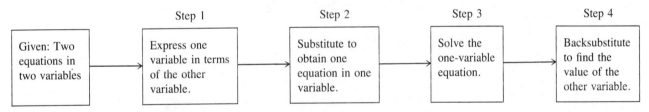

	Step 1	Step 2	Step 3	Step 4
Given: Two equations in two variables	Express one variable in terms of the other variable.	Substitute to obtain one equation in one variable.	Solve the one-variable equation.	Backsubstitute to find the value of the other variable.

EXAMPLE 2
Solving a System by Substitution:
One-Solution Case

Find the solution(s) to the following system:

$$x - y = 2 \qquad \text{\textit{Equation 1}}$$

$$\frac{x}{2} + 3y = 1 \qquad \text{\textit{Equation 2}}$$

Solution:

(1) Solve for x in Equation 1:

$$x = 2 + y$$

(2) Substitute into Equation 2:

$$\frac{2 + y}{2} + 3y = 1$$

(3) Solve for y:

$$2 + y + 6y = 2$$
$$7y = 0 \implies y = 0$$

(4) Backsubstitute $y = 0$ to get

$$x = 2 + y = 2 + 0 = 2$$

The solution is

$$x = 2, \qquad y = 0$$

EXAMPLE 3
Solving a System by Substitution:
Two-Solution Case

Find the solution(s) to the system

$$x^2 - 2x + y^2 = 3 \qquad \text{\textit{Equation 1}}$$
$$2x + 3y = -4 \qquad \text{\textit{Equation 2}}$$

Solution:

(1) Solve for y in Equation 2:

$$3y = -4 - 2x$$

$$y = \frac{-4 - 2x}{3}$$

(2) Substitute into Equation 1:

$$x^2 - 2x + \left(\frac{-4 - 2x}{3}\right)^2 = 3$$

(3) Solve for x:

$$x^2 - 2x + \frac{16 + 16x + 4x^2}{9} = 3$$

$$9x^2 - 18x + 16 + 16x + 4x^2 = 27$$

$$13x^2 - 2x - 11 = 0$$
$$(13x + 11)(x - 1) = 0$$
$$x = 1 \text{ or } -\frac{11}{13}$$

(4) Backsubstitute $x = 1$:

$$y = \frac{-4 - 2(1)}{3} = \frac{-4 - 2}{3} = \frac{-6}{3} = -2$$

Backsubstitute $x = -\frac{11}{13}$:

$$y = \frac{-4 - 2(-11/13)}{3} = \frac{-52 + 22}{39} = \frac{-30}{39} = -\frac{10}{13}$$

The solutions are

$$x = 1, \qquad y = -2$$
$$x = -\frac{11}{13}, \qquad y = -\frac{10}{13}$$

EXAMPLE 4
Solving a System by Substitution:
No-Solution Case

Find the solution(s) to the system

$$x^2 + y^2 = 1 \qquad \qquad \textit{Equation 1}$$
$$x + y = 3 \qquad \qquad \textit{Equation 2}$$

Solution:

(1) Solve for y in Equation 2:

$$y = 3 - x$$

(2) Substitute into Equation 1:

$$x^2 + (3 - x)^2 = 1$$

(3) Solve for x:

$$x^2 + 9 - 6x + x^2 = 1$$
$$2x^2 - 6x + 8 = 0$$
$$x = \frac{6 \pm \sqrt{36 - 64}}{12}$$

Since the discriminant is negative, there are no real solutions for x, and hence this system has no (real) solutions.

Graphical Approach to Finding Solutions

From Examples 2, 3, and 4, we can see that a system of two equations in two unknowns can have exactly one solution, more than one solution, or no solutions. In practice, we can gain valuable insight about the number of solutions for a system of equations by graphing each of the equations on the same coordinate plane. We call the solutions of the system the **points of intersection** of the graphs. For instance, in Figure 7.1 we see that the two equations in Example 2 graph as two lines with a single point of intersection. Similarly, the two equations in Example 3 graph as a circle and a line, and it makes sense that there are two points of intersection.

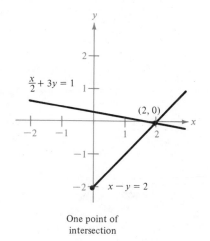

One point of
intersection

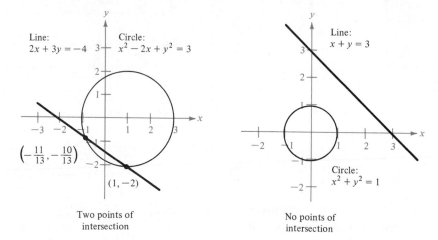

Two points of
intersection

No points of
intersection

FIGURE 7.1   *Points of Intersection*

When solving a system of equations in two unknowns, we can first use a graphical approach to get a good estimate for both the number and values of the solutions. Then to get exact solutions we use substitution.

FINDING POINTS OF INTERSECTION

1. Sketch the graph of each equation in the system.
2. Estimate the points of intersection of the graphs. Occasionally you can guess the exact coordinates from the sketch, but *be sure to test the points in each equation.*
3. Use substitution to find the exact solutions, and check to see that each of these solutions is represented on your sketch.

EXAMPLE 5
Finding Points of Intersection

Find the points of intersection of the graphs of

$$y = \ln x \quad \text{and} \quad x + y = 1$$

Solution:
The graph of each equation is shown in Figure 7.2. From this sketch it is clear that there is only one point of intersection. Also, it appears that $(1, 0)$

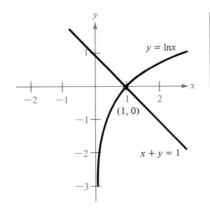

FIGURE 7.2

is the solution point, and we confirm this by checking these coordinates in *both* equations.

When $x = 1$ and $y = 0$, the equation $y = \ln x$ is true, since $0 = \ln(1)$. When $x = 1$ and $y = 0$, the equation $x + y = 1$ is true, since $1 + 0 = 1$. Hence $(1, 0)$ is the single point of intersection.

Remark: Example 5 shows us the value of a graphical approach to solving systems of equations in two variables. Notice what would have happened if we had tried only the substitution method in Example 5. By substituting $y = \ln x$ into $x + y = 1$, we obtain

$$x + \ln x = 1$$

You would be hard pressed to solve this equation for x using analytical techniques.

Economics Application

Many applications involve finding the point(s) of intersection of two graphs. A common one from business is called **break-even analysis.** The marketing of a new product typically requires a substantial investment to develop and produce the product. When enough units have been sold so that the total revenue has offset the total cost, we say that the sale of the product has reached the **break-even point.** We denote the **total cost** of producing x units of a product by C, and the **total revenue** received from selling x units of the product by R. Thus, we can find the break-even point by setting the cost C equal to the revenue R and solving for x. In other words, the break-even point corresponds to the point of intersection of the cost and revenue curves.

EXAMPLE 6
An Application:
Break-Even Analysis

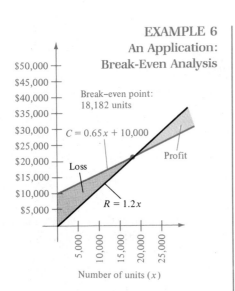

Number of units (x)

FIGURE 7.3

Roger Fisher is setting up a small business to manufacture and market an item he has developed. Roger has invested $10,000 in equipment and can produce each item for $0.65. If he can sell each item for $1.20, how many items must he sell before he breaks even?

Solution:
The total cost of producing x units is

$$C = 0.65x + 10,000$$

and the revenue obtained by selling x units is

$$R = 1.2x$$

Since the break-even point occurs when $R = C$, we have

$$1.2x = 0.65x + 10,000$$
$$0.55x = 10,000$$
$$x = \frac{10,000}{0.55} \approx 18,182 \text{ units}$$

Note in Figure 7.3 that sales less than the break-even point correspond to an overall loss, while sales greater than the break-even point correspond to a profit.

Solving Systems of Three Equations in Three Variables

We can extend the method of substitution to systems of equations involving more than two variables. The basic strategy is to reduce the system from three equations in three variables to a new system of two equations in two variables. The resulting new system is then solved with the usual substitution method.

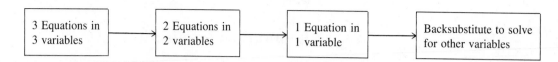

EXAMPLE 7
Solving a System Involving Three Variables

Find all solutions to the following system of equations:

$$x \ - 2y + z = -4 \qquad \textit{Equation 1}$$
$$2x + \ y - z = \ \ 5 \qquad \textit{Equation 2}$$
$$x^2 + \ y + z = \ \ 2 \qquad \textit{Equation 3}$$

Solution:

(1) Solve for z in Equation 1:

$$z = -x + 2y - 4$$

(2) Substitute this expression for z into the other two equations:

$$2x + y - z = 5 \qquad \textit{Equation 2}$$
$$2x + y - (-x + 2y - 4) = 5$$
$$3x - y = 1$$
$$x^2 + y + z = 2 \qquad \textit{Equation 3}$$
$$x^2 + y + (-x + 2y - 4) = 2$$
$$x^2 - x + 3y = 6$$

Now, we have succeeded in reducing the original system of three equations in three variables to a new system of two equations in two variables:

$$3x - \ y = 1$$
$$x^2 - \ x + 3y = 6$$

(3) By using substitution on this system, we have

$$3x - y = 1 \quad \Longrightarrow \quad y = 3x - 1$$

and substitution yields

$$x^2 - x + 3y = 6$$
$$x^2 - x + 3(3x - 1) = 6$$
$$x^2 + 8x - 9 = 0$$
$$(x - 1)(x + 9) = 0$$
$$x = 1 \text{ or } -9$$

(4) Backsubstitute to find y and z. If $x = 1$, then

$$y = 3x - 1 = 3(1) - 1 = 3 - 1 = 2$$
$$z = -x + 2y - 4 = -1 + 2(2) - 4 = -1 + 4 - 4 = -1$$

If $x = -9$, then

$$y = 3x - 1 = 3(-9) - 1 = -27 - 1 = -28$$
$$z = -x + 2y - 4$$
$$= -(-9) + 2(-28) - 4 = 9 - 56 - 4 = -51$$

Thus, we see that the original system has two solutions:

$$x = 1, \qquad y = 2, \qquad z = -1$$

and

$$x = -9, \qquad y = -28, \qquad z = -51$$

Section Exercises 7.1

In Exercises 1–10, find all solutions to the given system by the method of substitution.

1. $2x + y = 4$
   $-x + y = 1$

2. $x - y = -5$
   $x + 2y = 4$

3. $x - y = -3$
   $x^2 - y = -1$

4. $3x - y = -2$
   $x^3 - y = 0$

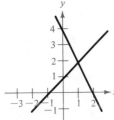 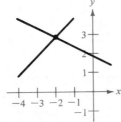 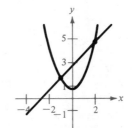 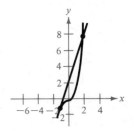

5. $x + 3y = 15$
   $x^2 + y^2 = 25$

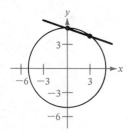

6. $x - y = 0$
     $x^3 - 5x + y = 0$

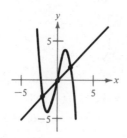

7. $x^2 \qquad - y = 0$
   $x^2 - 4x + y = 0$

8. $x^2 + y = 1$
   $y = x^4 - 2x^2 + 1$

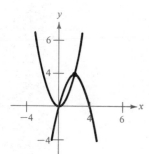

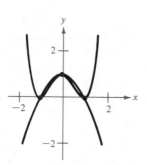

9. $x - 3y = -4$
   $x^2 - y^3 = 0$

10. $y = x^3 - 3x^2 + 3$
       $2x + y = 3$

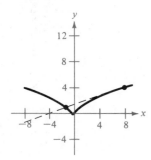

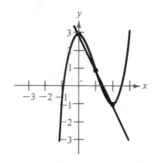

In Exercises 11–30, find all solutions to the system by the method of substitution.

11. $y = x$
      $5x - 3y = 10$

12. $x + 2y = 1$
     $5x - 4y = -23$

13. $2x - y + 2 = 0$
    $4x + y - 5 = 0$

14. $6x - 3y - 4 = 0$
    $x + 2y - 4 = 0$

15. $30x - 40y - 33 = 0$
    $10x + 20y - 21 = 0$

16. $1.5x + 0.8y = 2.3$
    $0.3x - 0.2y = 0.1$

17. $0.2x + 0.5y = 8$
    $x + \quad y = 20$

18. $\frac{1}{2}x + \frac{3}{4}y = 10$
    $\frac{3}{2}x - \quad y = 4$

19. $y = 2x$
    $y = x^2 + 1$

20. $x + y = 4$
    $x^2 - y = 2$

21. $3x - 7y + 6 = 0$
    $x^2 - y^2 = 4$

22. $x^2 + y^2 = 25$
    $2x + y = 10$

23. $x^2 + y^2 = 5$
    $x - y = 1$

24. $y = x^3 - 2x^2 + x - 1$
    $y = \quad - x^2 + 3x - 1$

25. $y = x^4 - 2x^2 + 1$
    $y = 1 - x^2$

26. $x^2 + y = 4$
    $2x - y = 1$

27. $x - y + z = 2$
    $2x + y - 3z = -5$
    $5x + y + z = 10$

28. $x - 2y - 2z = 9$
    $3x - y + z = 13$
    $2x + 3y + z = 8$

29. $xy - 1 = 0$
    $2x - 4y + 7 = 0$

30. $x - 2y = 1$
    $y = \sqrt{x - 1}$

In Exercises 31–38, find all points of intersection of the graphs of the given pair of equations.

31. $x + y = 4$
      $x^2 + y^2 - 4x = 0$

32. $3x - 2y = 0$
    $x^2 - y^2 = 4$

33. $2x - y + 3 = 0$
      $x^2 + y^2 - 4x = 0$

34. $x - y + 3 = 0$
      $x^2 - 4x + 7 = y$

35. $y = e^x$
      $x - y + 1 = 0$

36. $x^2 + y^2 = 8$
          $y = x^2$

37. $y = -\ln x$
    $y = \ln(2 - x)$

38. $x - y = 3$
    $x - y^2 = 1$

39. Andrea Jones is setting up a small business and has invested $16,000 to produce an item that will sell for $5.95. If each unit can be produced for $3.45, how many units must Andrea sell in order to break even?

40. Find two numbers whose sum is 20 and product is 96.

41. The sum of two numbers is 12 and the sum of their squares is 80. Find the numbers.

42. What are the dimensions of a rectangle if its perimeter is 40 miles and its area is 96 square miles?

In certain optimization problems in calculus, systems of equations (as found in Exercises 43–46) must be solved. The variable λ is called a Lagrange Multiplier. In the following exercises, find x, y, and λ satisfying the given system.

43.
$$y + \lambda = 0$$
$$x + \lambda = 0$$
$$x + y - 10 = 0$$

44.
$$2x + \lambda = 0$$
$$2y + \lambda = 0$$
$$x + y - 4 = 0$$

45.
$$2x - 2x\lambda = 0$$
$$-2y + \lambda = 0$$
$$y - x^2 = 0$$

46.
$$2 + 2y + 2\lambda = 0$$
$$2x + 1 + \lambda = 0$$
$$2x + y - 100 = 0$$

Systems of Linear Equations in Two Variables

7.2

In Section 2.1, we looked at equations of the form

$$ax + b = 0$$

where a and b are real numbers and $a \neq 0$. We call this type of equation a **linear equation in one variable.** We can generalize equations of this type to any number of variables:

$$ax + b = 0 \qquad \textit{Linear equation in one variable}$$

$$ax + by + c = 0 \qquad \textit{Linear equation in two variables}$$

$$ax + by + cz + d = 0 \qquad \textit{Linear equation in three variables}$$

$$a_n x_n + a_{n-1} x_{n-1} + \cdots + a_1 x_1 + a_0 = 0 \qquad \textit{Linear equation in n variables}$$

Note that we usually revert to a subscript notation for linear equations involving a large number of variables.

In Section 7.1 we used the method of substitution to solve various systems of equations. Now we restrict our study to linear systems and introduce a second method, which is called the method of **elimination.**

THE METHOD OF ELIMINATION

Given a linear system in two variables x and y:

1. Obtain coefficients for x (or y) that differ only in sign by multiplying all terms of one or both equations by suitably chosen constants.
2. Add the two equations to obtain a linear equation in (at most) one variable.
3. Solve the resulting linear equation in one variable.
4. Backsubstitute this solution into either of the original equations to solve for the other variable.
5. Check your solution in *both* of the original equations.

Remark: Note that the name *elimination* is derived from the fact that in step 2 one of the original variables is eliminated from the system.

EXAMPLE 1
Solving a Linear System
by Elimination

Solve the system

$$2x - 3y = -7 \qquad \text{\textit{Equation 1}}$$
$$3x + y = -5 \qquad \text{\textit{Equation 2}}$$

Solution:

(1) In Equation 2, if we change the coefficient of y to 3 by multiplying through by 3, then the coefficients of y in Equations 1 and 2 will differ only in sign. Multiply Equation 2 by 3:

$$3x + y = -5 \quad \Longrightarrow \quad 9x + 3y = -15$$

(2) Add equations to eliminate y:

$$
\begin{array}{r}
2x - 3y = -7 \\
9x + 3y = -15 \\
\hline
11x = -22
\end{array}
$$

(3) Solve for x:

$$11x = -22$$
$$x = -2$$

(4) Backsubstitute (into Equation 1) to solve for y:

$$2x - 3y = -7$$
$$2(-2) - 3y = -7$$
$$-3y = -3$$
$$y = 1$$

(5) Check the solution $x = -2$, $y = 1$.

$$2(-2) - 3(1) = -4 - 3 = -7 \quad \text{\textit{Equation 1}}$$
$$3(-2) + 1 = -6 + 1 = -5 \quad \text{\textit{Equation 2}}$$

EXAMPLE 2
Solving a Linear System
by Elimination

Solve the system

$$4x + 3y = 6 \qquad \text{\textit{Equation 1}}$$
$$5x + 2y = 11 \qquad \text{\textit{Equation 2}}$$

Solution:

(1) To eliminate y, we first change the coefficients of y to 6 and -6, as follows. Multiply Equation 1 by 2:

$$4x + 3y = 6 \quad \Longrightarrow \quad 8x + 6y = 12$$

Multiply Equation 2 by -3:

$$5x + 2y = 11 \qquad \Rightarrow \qquad -15x - 6y = -33$$

(2) Add equations to eliminate y:

$$
\begin{array}{r}
8x + 6y = 12 \\
-15x - 6y = -33 \\
\hline
-7x = -21
\end{array}
$$

(3) Solve for x:

$$-7x = -21$$
$$x = 3$$

(4) Backsubstitute (into Equation 1) to solve for y:

$$8x + 6y = 12$$
$$8(3) + 6y = 12$$
$$6y = -12$$
$$y = -2$$

(5) Check the solution $x = 3$, $y = -2$.

$$4(3) + 3(-2) = 12 - 6 = 6 \qquad \textit{Equation 1}$$
$$5(3) + 2(-2) = 15 - 4 = 11 \qquad \textit{Equation 2}$$

In Examples 1 and 2, both of the systems had exactly one solution. By sketching the graphs of these systems, we see that each one corresponds to two intersecting lines with a single point of intersection, as shown in Figure 7.4.

As we observed in Section 7.1, it is possible for a system of equations to have infinitely many solutions, no solution, or exactly one solution. In the case of two linear equations in two variables, we can interpret this result graphically as follows.

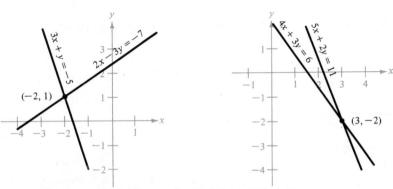

FIGURE 7.4   Linear Systems with a Single Point of Intersection

GRAPHIC INTERPRETATION OF SOLUTIONS

For a system of two linear equations in two variables, the number of solutions is given by one of the following:

Number of Solutions	**Graphical Interpretation**
Infinitely many solutions	The lines are identical.
No solution	The lines are parallel.
Exactly one solution	The lines intersect at one point.

Figure 7.5 illustrates the three cases described in this graphical interpretation of solutions.

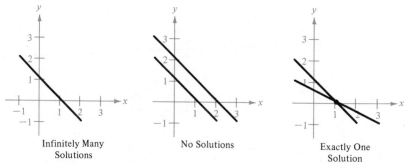

Infinitely Many Solutions   No Solutions   Exactly One Solution

FIGURE 7.5

We say that a system of two equations in two variables is **consistent** if it has at least one solution and is **inconsistent** if it has no solution. Moreover, a consistent system is said to be **dependent** if it has infinitely many solutions.

EXAMPLE 3
An Inconsistent System

Find all solutions to the following system:

$$x - 2y = 3 \qquad \textit{Equation 1}$$
$$-2x + 4y = 1 \qquad \textit{Equation 2}$$

Solution:

(1) Multiply Equation 1 by 2:

$$x - 2y = 3 \quad \Longrightarrow \quad 2x - 4y = 6$$

(2) Add equations to eliminate x:

$$\begin{array}{r} 2x - 4y = 6 \\ -2x + 4y = 1 \\ \hline 0 = 7 \end{array}$$

From this absurd result, we conclude that there is no solution and the system is *inconsistent*. Figure 7.6 shows that these two equations correspond to two parallel lines that have no point of intersection.

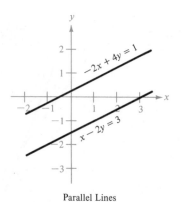

Parallel Lines

FIGURE 7.6

EXAMPLE 4
A Dependent System

Find all solutions to the following system:

$$2x - y = 1 \qquad \textit{Equation 1}$$
$$-4x + 2y = -2 \qquad \textit{Equation 2}$$

Solution:

(1) Multiply Equation 2 by $\frac{1}{2}$:

$$-4x + 2y = -2 \qquad \Rightarrow \qquad -2x + y = -1$$

(2) Add equations to eliminate x:

$$
\begin{array}{r}
2x - y = 1 \\
-2x + y = -1 \\
\hline
0 = 0
\end{array}
$$

Since *both* variables are eliminated *and* we obtained a valid equation, we conclude that the equations are dependent. Hence, there are infinitely many solutions to this system. Graphically, we interpret this result to mean that both equations graph as the same line, and thus *every* point on the line is a valid solution to the system. To represent this infinite solution set, we first solve for y:

$$2x - y = 1 \qquad \Rightarrow \qquad y = 2x - 1$$

Now, let x be any real number a; then y is given by $y = 2a - 1$. In other words, every ordered pair of the form

$$(a, 2a - 1)$$

is a solution to the system. For instance, if $a = 1$, then we see that $(1, 1)$ is a solution. Similarly, if $a = 2$, we see that $(2, 3)$ is a solution.

It is important to properly interpret the results obtained with the elimination method.

1. If the elimination method leads to an equation of the form

$$0x + 0y = c, \qquad c \neq 0$$

the given system is *inconsistent* and has *no solution.*

2. If the elimination method leads to an equation of the form

$$0x + 0y = 0$$

the given system is *dependent* and has *infinitely many solutions.*

Systems of linear equations have a wide variety of applications. The following example shows how a system of linear equations may be used to solve a problem involving speed.

EXAMPLE 5
An Application Involving Rates

An airplane flying into a headwind travels the 2000-mile flying distance between two cities in 4 hours and 24 minutes. On the return flight, the same distance is traveled in 4 hours. Find the ground speed of the plane and the speed of the wind, assuming that both remain constant.

Solution:

The two unknown quantities in this problem are the speeds of the wind and the plane. We let

$$r_1 = \text{rate of plane (in miles per hour)}$$

$$r_2 = \text{rate of wind (in miles per hour)}$$

For the trip into the wind, the resulting speed is given by the difference of r_1 and r_2. For the trip with the wind, the resulting speed is given by the sum of the two speeds. (See Figure 7.7.) Using the formula

$$\text{distance} = (\text{rate})(\text{time})$$

we can write the following equations:

$$2000 = (r_1 - r_2)\left(4 + \frac{24}{60}\right) \qquad \textit{Equation 1}$$

$$2000 = (r_1 + r_2)(4) \qquad \textit{Equation 2}$$

These two equations simplify to the following system:

$$5000 = 11r_1 - 11r_2 \qquad \textit{Equation 1}$$

$$500 = r_1 + r_2 \qquad \textit{Equation 2}$$

By elimination, the solution is

$$r_1 = \frac{5250}{11} \approx 477.27 \text{ miles per hour}$$

$$r_2 = \frac{250}{11} \approx 22.73 \text{ miles per hour}$$

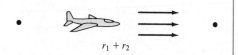

$r_1 - r_2$

$r_1 + r_2$

FIGURE 7.7

Section Exercises 7.2

In Exercises 1–10, solve the linear system by elimination. Label each line on the graph with the appropriate equation.

1. $2x + y = 4$
   $x - y = 2$

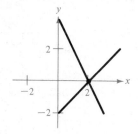

2. $x + 3y = 2$
   $-x + 2y = 3$

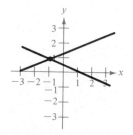

3. $x - y = 0$
    $3x - 2y = -1$

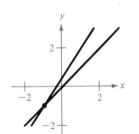

4. $2x - y = 2$
    $4x + 3y = 24$

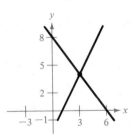

5. $x - y = 1$
    $-2x + 2y = 5$

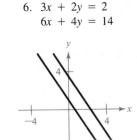

6. $3x + 2y = 2$
    $6x + 4y = 14$

7. $\dfrac{x}{2} - \dfrac{y}{3} = 1$
    $-2x + \frac{4}{3}y = -4$

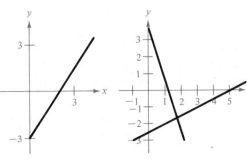

8. $x - 2y = 5$
    $6x + 2y = 7$

9. $9x - 3y = -1$
    $\frac{1}{5}x + \frac{2}{5}y = -\frac{1}{3}$

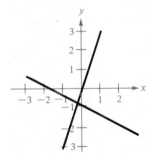

10. $5x + 3y = 18$
    $2x - 7y = -1$

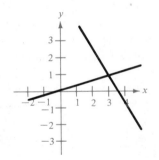

In Exercises 11–30, solve the system by elimination.

11. $x + 2y = 4$
    $x - 2y = 1$

12. $3x - 5y = 2$
    $2x + 5y = 13$

13. $2x + 3y = 18$
    $5x - y = 11$

14. $x + 7y = 12$
    $3x - 5y = 10$

15. $3x + 2y = 10$
    $2x + 5y = 3$

16. $8r + 16s = 20$
    $16r + 50s = 55$

17. $2u + v = 120$
    $u + 2v = 120$

18. $5u + 6v = 24$
    $3u + 5v = 18$

19. $6r - 5s = 3$
    $10s - 12r = 5$

20. $1.8x + 1.2y = 4$
    $9x + 6y = 3$

21. $2.5x - 3y = 1.5$
    $10x - 12y = 6$

22. $\frac{2}{3}x + \frac{1}{6}y = \frac{2}{3}$
    $4x + y = 4$

23. $\dfrac{x}{4} + \dfrac{y}{6} = 1$
    $x - y = 3$

24. $\dfrac{x - 1}{2} + \dfrac{y + 2}{3} = 4$
    $x - 2y = 5$

25. $\dfrac{x + 3}{4} + \dfrac{y - 1}{3} = 1$
    $2x - y = 12$

26. $0.02x - 0.05y = -0.19$
    $0.03x + 0.04y = 0.52$

27. $0.05x - 0.03y = 0.21$
    $0.07x + 0.02y = 0.16$

28. $3b + 3m = 7$
    $3b + 5m = 3$

29. $4b + 3m = 3$
    $3b + 11m = 13$

30. $\dfrac{12}{x} - \dfrac{12}{y} = 7$ (Hint: Let $X = 1/x$ and $Y = 1/y$.
    $\dfrac{3}{x} + \dfrac{4}{y} = 0$ Then solve the linear system in X and Y.)

In Exercises 31–38, solve the problem using a system of linear equations.

31. An airplane flying into a headwind travels the 1800-mile flying distance between two cities in 3 hours and 36 minutes. On the return flight, the distance is traveled in 3 hours. Find the ground speed of the plane and the speed of the wind, assuming that both remain constant.

32. Two planes start from the same airport and fly in opposite directions. The second plane starts one-half hour after the first plane, but its speed is 50 miles per hour faster. Find the ground speed of each plane if 2 hours after the first plane starts its flight the planes are 2000 miles apart.

33. Ten gallons of a 30% acid solution is obtained by mixing some 20% solution with some 50% solution. How much of each must be used?

34. Ten pounds of mixed nuts at a food co-op are to sell for $5.95 per pound. The mixture is obtained from two kinds of nuts, with one variety priced at $4.29 per pound and the other at $6.55 per pound. How much of each must be used?

35. Five hundred tickets were sold for a certain performance of a play. The tickets for adults and children sold for $2.50 and $2.00, respectively, and the receipts for the performance were $1187.50. How many of each kind of ticket were sold?

36. A woman invested an inheritance of $12,000 in two corporate bonds that pay 10.5% and 12% simple interest. If she receives $1380 in yearly interest, how much is invested in each bond?

37. The perimeter of a rectangle is 40 feet, and the length is 4 feet greater than the width. Find the rectangle's dimensions.

38. The sum of the digits of a given two-digit number is 12. If the digits are reversed, the number is increased by 36. Find the number.

The *least squares regression line*, $y = ax + b$, for the points

$$(x_1, y_1), (x_2, y_2), \ldots, (x_n, y_n)$$

is obtained by solving the linear system

$$nb + \left(\sum_{i=1}^{n} x_i\right)a = \sum_{i=1}^{n} y_i$$

$$\left(\sum_{i=1}^{n} x_i\right)b + \left(\sum_{i=1}^{n} x_i^2\right)a = \sum_{i=1}^{n} x_i y_i$$

for a and b, the slope and y-intercept. This gives the *best-fitting* line to the given points. For $n = 4$, recall that

$$\sum_{i=1}^{n} x_i = \sum_{i=1}^{4} x_i = x_1 + x_2 + x_3 + x_4$$

In Exercises 39–42, (a) find the least squares regression line, and (b) plot the given points and sketch the least squares regression line on the same axes.

39. $(-2, 0), (0, 1), (2, 3)$
40. $(-3, 0), (-1, 1), (1, 1), (3, 2)$
41. $(0, 4), (1, 3), (1, 1), (2, 0)$
42. $(1, 0), (2, 0), (3, 0), (3, 1), (4, 1), (4, 2), (5, 2), (6, 2)$

Systems of Linear Equations in More Than Two Variables

7.3

Triangular Form

The method of elimination can be applied to systems of equations in more than two variables. When we use this procedure, our goal is to transform the original system into an *equivalent* system in **triangular form.** (This may take several steps.) The following two systems are in triangular form:

$$
\begin{aligned}
x - 2y + 3z &= 9 \\
y + 2z &= 3 \\
z &= 2
\end{aligned}
\qquad
\begin{aligned}
2x - y + z - 2w &= 6 \\
2y + z - w &= 3 \\
z + w &= 1 \\
w &= -1
\end{aligned}
$$

Backsubstitution works well with a system in triangular form, since the last equation in the system gives us the value of the nth variable. By substituting this value back into the next to last equation, we find the value of an additional variable, and we continue backsubstituting until the value of each variable has been found.

    Transforming a linear system into triangular form usually involves a *chain* of equivalent systems, each of which is obtained by using one or more of the following rules.

OPERATIONS THAT LEAD TO EQUIVALENT
SYSTEMS OF EQUATIONS

1. Interchange any two equations.
2. Multiply all terms of an equation by a nonzero constant.
3. Replace an equation by the sum of itself and a constant multiple of any other equation in the system.

Study Example 1 to see how we use backsubstitution to solve a system in triangular form.

EXAMPLE 1
Using Backsubstitution on a
System in Triangular Form

Solve the following system:

$$x - 2y + 3z = 9$$
$$y + 2z = 3$$
$$z = 2$$

Solution:

$$x - 2y + 3z = 9 \qquad \text{\textit{Equation 1}}$$
$$y + 2z = 3 \qquad \text{\textit{Equation 2}}$$
$$z = 2 \qquad \text{\textit{Equation 3}}$$

Substituting the value of z from the third equation into the second equation yields

$$y + 2z = 3$$
$$y + 2(2) = 3$$
$$y = -1$$

Next, substituting the values for both y and z into the first equation yields

$$x - 2y + 3z = 9$$
$$x - 2(-1) + 3(2) = 9$$
$$x = 1$$

Thus, the solution is $x = 1$, $y = -1$, and $z = 2$.

In transforming a system into triangular form, the goal is to reduce the size of the system by pairing equations and eliminating one of the variables. In the next example, note the operations used to obtain an equivalent triangular system.

EXAMPLE 2
Rewriting a System
in Triangular Form

Solve the following system:

$$x - 2y + 3z = 9 \qquad \textit{Equation 1}$$
$$-x + 3y - z = -6 \qquad \textit{Equation 2}$$
$$2x - 5y + 5z = 17 \qquad \textit{Equation 3}$$

Solution:

Though there are several ways to begin, we can eliminate x by adding the first two equations and by adding the first equation (multiplied by -2) to the third equation.

$$
\begin{array}{r}
x - 2y + 3z = 9 \\
-x + 3y - z = -6 \\
\hline
y + 2z = 3
\end{array}
\qquad
\begin{array}{r}
-2x + 4y - 6z = -18 \\
2x - 5y + 5z = 17 \\
\hline
-y - z = -1
\end{array}
$$

Now, we continue the process of elimination on these two new equations:

$$
\begin{array}{r}
y + 2z = 3 \\
-y - z = -1 \\
\hline
z = 2
\end{array}
$$

Finally, by choosing one equation from each system, we obtain the following triangular system:

$$
\begin{aligned}
x - 2y + 3z &= 9 \\
y + 2z &= 3 \\
z &= 2
\end{aligned}
$$

This is the same system we solved in Example 1, and we conclude that the solution is $x = 1$, $y = -1$, and $z = 2$.

From Example 2, we see a general procedure to follow.

THE METHOD OF ELIMINATION

1. Reduce the original system to triangular form by successively adding pairs of equations (or multiples of equations).
2. Backsubstitute to solve the triangular system of equations.
3. Check your solution in *each* of the original equations.

As with a system of linear equations in two variables, the solution(s) to a system of linear equations in more than two variables must fall into one and only one of the following categories:

1. There are infinitely many solutions (consistent system).
2. There is no solution (inconsistent system).
3. There is exactly one solution (consistent system).

When a system of equations has no solution, we simply state that it is inconsistent. If a system has exactly one solution, we list the value of each variable. However, for systems that have infinitely many solutions, we encounter a certain awkwardness in listing the solutions. For example, we might give the solutions to a system in three variables as

$$(a, a + 1, 2a), \qquad \text{where } a \text{ is any real number}$$

This means that for each value of a, we have a valid solution to the system. A few of the many possible solutions are $(-1, 0, -2)$, $(0, 1, 0)$, $(1, 2, 2)$, $(2, 3, 4)$, and so on.

Now consider the solutions represented by

$$(b - 1, b, 2b - 2), \qquad \text{where } b \text{ is any real number}$$

A few possible solutions are $(-1, 0, -2)$, $(0, 1, 0)$, $(1, 2, 2)$, $(2, 3, 4)$, and so on. Both descriptions result in the same collection of solutions. Thus, when comparing descriptions of infinite solution sets, bear in mind that there is more than one way to describe such sets.

EXAMPLE 3
A System with Infinitely Many Solutions

Solve the following system:

$$
\begin{array}{rl}
x + y - 3z = -1 & \qquad \textit{Equation 1} \\
-x + 2y \phantom{{}- 3z} = 1 & \qquad \textit{Equation 2} \\
\hline
3y - 3z = 0 & \qquad \textit{Equation 3}
\end{array}
$$

Solution:
We begin by adding the first two equations to eliminate x:

$$
\begin{array}{r}
x + y - 3z = -1 \\
-x + 2y \phantom{{}- 3z} = 1 \\
\hline
3y - 3z = 0
\end{array}
$$

Multiplying this new equation by $-\frac{1}{3}$ and adding it to Equation 3, we have

$$
\begin{array}{r}
-y + z = 0 \\
y - z = 0 \\
\hline
0 = 0
\end{array}
$$

This means that Equation 3 is *dependent* on Equations 1 and 2 in the sense that it gives us no additional information about the variables. Thus, the original system is equivalent to

$$
\begin{array}{r}
x + y - 3z = -1 \\
y - z = 0
\end{array}
$$

We solve for y in terms of z:

$$y - z = 0 \qquad \Longrightarrow \qquad y = z$$

and backsubstitute to solve for x, also in terms of z:

$$x + y - 3z = -1$$
$$x + z - 3z = -1$$
$$x = 2z - 1$$

Finally, by letting z be any real number a, we have the following solution:

$$(2a - 1, a, a)$$

If we let $z = b/2$, the solution has the alternative form

$$\left(b - 1, \frac{b}{2}, \frac{b}{2}\right)$$

Nonsquare Systems

So far we have only considered **square** systems, for which the number of equations is equal to the number of variables. In a **nonsquare** system, the number of equations differs from the number of variables. In Chapter 11 we will prove that a system of linear equations cannot have a unique solution unless there are at least as many equations as there are variables in the system. Typically, systems with fewer equations than variables have either infinitely many solutions or no solutions.

EXAMPLE 4
A System with Fewer Equations than Variables

Solve the following system:

$$x - 3y + z = 2 \qquad \textit{Equation 1}$$
$$2x + y - z = 1 \qquad \textit{Equation 2}$$

Solution:
To eliminate x, we multiply Equation 1 by (-2) and add the result to Equation 2:

$$
\begin{array}{r}
-2x + 6y - 2z = -4 \\
2x + y - z = 1 \\
\hline
7y - 3z = -3
\end{array}
$$

Thus, the following system is equivalent to the original system:

$$x - 3y + z = 2$$
$$7y - 3z = -3$$

Now we solve for both y and x in terms of z:

$$7y = 3z - 3$$
$$y = \frac{3z - 3}{7}$$

Backsubstitution into Equation 1 yields

$$x - 3\left(\frac{3z - 3}{7}\right) + z = 2$$

$$x = \frac{2z + 5}{7}$$

Finally, by letting z be any real number a, we have the solution

$$\left(\frac{2a + 5}{7}, \frac{3a - 3}{7}, a \right)$$

Homogeneous Systems

A system of linear equations in which the constant term in each equation is zero is called **homogeneous.** For example,

$$a_1 x + b_1 y + c_1 z = 0$$
$$a_2 x + b_2 y + c_2 z = 0$$
$$a_3 x + b_3 y + c_3 z = 0$$

The *trivial* (or obvious) solution to this homogeneous system is $(0, 0, 0)$. That means if all variables are given the value zero, then the equations must be satisfied. Often homogeneous systems will have nontrivial solutions also, and we can find these in the same way we find solutions for nonhomogeneous systems.

Applications

We conclude this section with two applications involving systems of linear equations in three variables.

EXAMPLE 5
An Application: Moving Object

The height at time t of an object that is moving in a (vertical) line with constant acceleration a is given by the **position equation**

$$s = \frac{1}{2} at^2 + v_0 t + s_0$$

The height s is measured in feet, t is measured in seconds, v_0 is the initial velocity (at time $t = 0$), and s_0 is the initial height. Find the values of a, v_0, and s_0 if at 1 second, $s = 52$ feet; at 2 seconds, $s = 52$ feet; and at 3 seconds, $s = 20$ feet.

Solution:
By substituting the three values of t and s into the equation for the height, we obtain three linear equations.

When t = 1: $\dfrac{1}{2} a(1^2) + v_0(1) + s_0 = 52$

When t = 2: $\dfrac{1}{2} a(2^2) + v_0(2) + s_0 = 52$

When t = 3: $\dfrac{1}{2} a(3^2) + v_0(3) + s_0 = 20$

This system can be rewritten

$$a + 2v_0 + 2s_0 = 104 \qquad \textit{Equation 1}$$
$$2a + 2v_0 + s_0 = 52 \qquad \textit{Equation 2}$$
$$9a + 6v_0 + 2s_0 = 40 \qquad \textit{Equation 3}$$

By multiplying Equation 1 by (-2) and by (-9) and adding the results to the second and third equations respectively, we eliminate a.

$$
\begin{array}{rcl}
-2a - 4v_0 - 4s_0 &=& -208 \\
2a + 2v_0 + s_0 &=& 52 \\
\hline
-2v_0 - 3s_0 &=& -156
\end{array}
\qquad
\begin{array}{rcl}
-9a - 18v_0 - 18s_0 &=& -936 \\
9a + 6v_0 + 2s_0 &=& 40 \\
\hline
-12v_0 - 16s_0 &=& -896
\end{array}
$$

Now, by multiplying the first of these new equations by 3 and dividing the second by -2, we have

$$
\begin{array}{rcl}
-6v_0 - 9s_0 &=& -468 \\
6v_0 + 8s_0 &=& 448 \\
\hline
-s_0 &=& -20 \\
s_0 &=& 20 \text{ feet}
\end{array}
$$

By backsubstitution, we have

$$-2v_0 - 3s_0 = -156$$
$$-2v_0 - 3(20) = -156$$
$$v_0 = 48 \text{ feet per second}$$

and finally,

$$a + 2v_0 + 2s_0 = 104$$
$$a + 2(48) + 2(20) = 104$$
$$a = -32 \text{ feet per second squared}$$

This means that the position equation for this object is

$$s = -16t^2 + 48t + 20$$

In the next example we show how to fit a parabola through three given points in the plane. This procedure can be generalized to fit an nth degree polynomial function to $n + 1$ points in the plane. The only restriction to the procedure is that (since we are trying to fit a *function* to the points) every point must have a distinct x-coordinate.

EXAMPLE 6
An Application: Curve Fitting

Find an equation for the quadratic function

$$f(x) = ax^2 + bx + c$$

whose graph passes through the points $(-1, 3)$, $(1, 1)$, and $(2, 6)$.

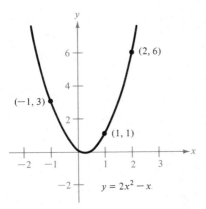

FIGURE 7.8

Solution:

Since the graph of f passes through the points $(-1, 3)$, $(1, 1)$, and $(2, 6)$, we have

$$f(-1) = a(-1)^2 + b(-1) + c = 3$$
$$f(1) = a(1)^2 + b(1) + c = 1$$
$$f(2) = a(2)^2 + b(2) + c = 6$$

This produces the following system of linear equations in the variables a, b, and c:

$$a - b + c = 3 \qquad \text{Equation 1}$$
$$a + b + c = 1 \qquad \text{Equation 2}$$
$$4a + 2b + c = 6 \qquad \text{Equation 3}$$

The solution to this system turns out to be

$$a = 2, \qquad b = -1, \qquad \text{and} \qquad c = 0$$

Thus, the equation of the parabola passing through the three given points is

$$f(x) = 2x^2 - x$$

as shown in Figure 7.8.

Section Exercises 7.3

In Exercises 1–26, solve the systems of equations.

1. $\quad x + y + z = 6$
   $\quad 2x - y + z = 3$
   $\quad 3x \quad\;\; - z = 0$

2. $\quad x + y + z = 2$
   $\quad -x + 3y + 2z = 8$
   $\quad 4x + y \qquad = 4$

3. $\quad 4x + y - 3z = 11$
   $\quad 2x - 3y + 2z = 9$
   $\quad x + y + z = -3$

4. $\quad 2x \qquad + 2z = 2$
   $\quad 5x + 3y \qquad = 4$
   $\quad 3y - 4z = 4$

5. $\qquad 6y + 4z = -12$
   $\quad 3x + 3y \qquad = 9$
   $\quad 2x \qquad - 3z = 10$

6. $\quad 2x + 4y + z = -4$
   $\quad x - 2y + 3z = \frac{13}{2}$
   $\quad 2x - y + \frac{1}{2}z = 3$

7. $\quad 3x - 2y + 4z = 1$
   $\quad x + y - 2z = 3$
   $\quad 2x - 3y + 6z = 8$

8. $\quad 5x - 3y + 2z = 3$
   $\quad 2x + 4y - z = 7$
   $\quad x - 11y + 4z = 3$

9. $\quad 3x + 3y + 5z = 1$
   $\quad 3x + 5y + 9z = 0$
   $\quad 5x + 9y + 17z = 0$

10. $\quad 2x + y + 3z = 1$
    $\quad 2x + 6y + 8z = 3$
    $\quad 6x + 8y + 18z = 5$

11. $\quad x + 2y - 7z = -4$
    $\quad 2x + y + z = 13$
    $\quad 3x + 9y - 36z = -33$

12. $\quad 2x + y - 3z = 4$
    $\quad 4x \qquad + 2z = 10$
    $\quad -2x + 3y - 13z = -8$

13. $\quad x \qquad + 4z = 13$
    $\quad 4x - 2y + z = 7$
    $\quad 2x - 2y - 7z = -19$

14. $\quad 4x - y + 5z = 11$
    $\quad x + 2y - z = 5$
    $\quad 5x - 8y + 13z = 7$

15. $\quad x - 2y + 5z = 2$
    $\quad 3x + 2y - z = -2$

16. $\quad x - 3y + 2z = 18$
    $\quad 5x - 13y + 12z = 80$

17. $\quad 2x - 3y + z = -2$
    $\quad -4x + 9y \qquad = 7$

18. $\quad 2x + 3y + 3z = 7$
    $\quad 4x + 18y + 15z = 44$

19. $\quad x \qquad + 4z = 1$
    $\quad x + y + 10z = 10$
    $\quad 2x - y + 2z = -5$

20. $\quad x + y + z + w = 6$
    $\quad 2x + 3y \qquad - w = 0$
    $\quad -3x + 4y + z + 2w = 4$
    $\quad x + 2y - z + w = 0$

21. $\quad x \qquad\qquad + 3w = 4$
    $\quad 2y - z - w = 0$
    $\quad 3y \qquad - 2w = 1$
    $\quad 2x - y + 4z \qquad = 5$

22. $\begin{aligned} 3x - 2y - 6z &= -4 \\ -3x + 2y + 6z &= 1 \\ x - y - 5z &= -3 \end{aligned}$

23. $\begin{aligned} 4x + 3y + 17z &= 0 \\ 5x + 4y + 22z &= 0 \\ 4x + 2y + 19z &= 0 \end{aligned}$ 24. $\begin{aligned} 2x + 3y \quad\;\; &= 0 \\ 4x + 3y - z &= 0 \\ 8x + 3y + 3z &= 0 \end{aligned}$

25. $\begin{aligned} 5x + 5y - z &= 0 \\ 10x + 5y + 2z &= 0 \\ 5x + 15y - 9z &= 0 \end{aligned}$ 26. $\begin{aligned} 12x + 5y + z &= 0 \\ 12x + 4y - z &= 0 \end{aligned}$

In Exercises 27–30, find the equation of the parabola

$$y = ax^2 + bx + c$$

passing through the given points.

27. $(0, -4), (1, 1), (2, 10)$ 28. $(0, 5), (1, 6), (2, 5)$
29. $(1, 0), (3, 0), (2, -1)$ 30. $(1, 2), (2, 1), (3, -4)$

In Exercises 31–34, find an equation of the circle

$$x^2 + y^2 + Dx + Ey + F = 0$$

passing through the given points.

31. $(0, 0), (2, -2), (4, 0)$ 32. $(0, 0), (0, 6), (-3, 3)$
33. $(3, -1), (-2, 4), (6, 8)$ 34. $(0, 0), (0, 2), (3, 0)$

In Exercises 35–38, use the given information to find a, v_0, and s_0 in the position equation

$$s = \tfrac{1}{2}at^2 + v_0 t + s_0$$

35. At $t = 1$ second, $s = 128$ feet; at $t = 2$ seconds, $s = 80$ feet; and at $t = 3$ seconds, $s = 0$.
36. At $t = 1$ second, $s = 48$ feet; at $t = 2$ seconds, $s = 64$ feet; and at $t = 3$ seconds, $s = 48$ feet.
37. At $t = 1$ second, $s = 452$ feet; at $t = 3$ seconds, $s = 260$ feet; and at $t = 4$ seconds, $s = 116$ feet.
38. At $t = 2$ seconds, $s = 132$ feet; at $t = 3$ seconds, $s = 100$ feet; and at $t = 4$ seconds, $s = 36$ feet.

In Exercises 39–42, use a system of linear equations to decompose each rational fraction into partial fractions. (See Section 5.3.)

39. $\dfrac{1}{x^3 - x} = \dfrac{A}{x} + \dfrac{B}{x - 1} + \dfrac{C}{x + 1}$

40. $\dfrac{x^2 + 4x - 1}{x^2 - x} = 1 + \dfrac{5x - 1}{x^2 - x} = 1 + \dfrac{A}{x} + \dfrac{B}{x - 1}$

41. $\dfrac{x^2 - 3x - 3}{x(x - 2)(x + 3)} = \dfrac{A}{x} + \dfrac{B}{x - 2} + \dfrac{C}{x + 3}$

42. $\dfrac{12}{x(x - 2)(x + 3)} = \dfrac{A}{x} + \dfrac{B}{x - 2} + \dfrac{C}{x + 3}$

43. A small company that manufactures products A and B has an order for 15 units of product A and 16 units of product B. The company has trucks of three different sizes that can haul the products, as shown in the following table:

		Product	
		A	B
	Large	6	3
Truck	Medium	4	4
	Small	0	3

How many trucks of each size are needed to deliver the order? (Give *two* possible solutions.)

44. A chemist needs 10 liters of a 25% acid solution. He has three solutions containing the acid, in which the concentration is 10%, 20%, and 50%, respectively. How many liters of each solution should the chemist use if he wants to use
(a) as little as possible of the 50% solution?
(b) as much as possible of the 50% solution?
(c) two liters of the 50% solution?

The Least Squares Regression Parabola, $y = ax^2 + bx + c$, for the points

$$(x_1, y_1), (x_2, y_2), \ldots, (x_n, y_n)$$

is obtained by solving the linear system

$$nc + \left(\sum_{i=1}^{n} x_i\right)b + \left(\sum_{i=1}^{n} x_i^2\right)a = \sum_{i=1}^{n} y_i$$

$$\left(\sum_{i=1}^{n} x_i\right)c + \left(\sum_{i=1}^{n} x_i^2\right)b + \left(\sum_{i=1}^{n} x_i^3\right)a = \sum_{i=1}^{n} x_i y_i$$

$$\left(\sum_{i=1}^{n} x_i^2\right)c + \left(\sum_{i=1}^{n} x_i^3\right)b + \left(\sum_{i=1}^{n} x_i^4\right)a = \sum_{i=1}^{n} x_i^2 y_i$$

for a, b, and c. This gives the *best-fitting* parabola to the given points. In Exercises 45–48, (a) find the least squares regression parabola, and (b) plot the given points and sketch the least squares parabola on the same axes.

45. $(-2, 0), (-1, 0), (0, 1), (1, 2), (2, 5)$
46. $(-4, 5), (-2, 6), (2, 6), (4, 2)$
47. $(0, 0), (2, 2), (3, 6), (4, 12)$
48. $(0, 10), (1, 9), (2, 6), (3, 0)$

Systems of Inequalities and Linear Programming

7.4

The following statements are **inequalities in two variables:**

$$3x - 2y < 4 \qquad \text{and} \qquad 2x^2 + 3y^2 \geq 6$$

A **solution** to an inequality in two variables is an ordered pair (x, y) that satisfies the inequality. The **graph** of an inequality in two variables is the collection of all solutions. To sketch the graph of an inequality in two variables, begin by sketching the graph of the corresponding *equality* (or equation). This graph is made with a dashed line for the strict inequalities $<$ or $>$ and a solid line for the inequalities $\leq$ or $\geq$. The graph will normally divide the plane into two or more regions. In each such region, one of the following must be true:

1. All points in the region are solutions of the inequality, or
2. No points in the region are solutions of the inequality.

This means that we can determine whether or not the points in an entire region satisfy the inequality by simply testing *one* point in the region.

SKETCHING THE GRAPH OF AN INEQUALITY IN TWO VARIABLES

1. Replace the inequality sign by an equality sign, and sketch the graph of the resulting equation. (Use a dashed line for $<$ or $>$ and a solid line for $\leq$ or $\geq$.)
2. Test one point in each of the regions formed by the graph in step 1. If the point satisfies the inequality, then shade the entire region to denote that every point in the region satisfies the inequality.

EXAMPLE 1
Sketching the Graph of an Inequality

Sketch the graphs of the following inequalities:

(a) $3x - 2y < 6$ (b) $y \geq x^2 - 1$ (c) $xy \leq 1$

Solution:

(a) We begin by sketching the graph of the equation

$$3x - 2y = 6$$

as shown in Figure 7.9. Note that we use a dashed line to indicate that the points on the line *do not* satisfy the original inequality. Next, we choose two convenient test points, one from each of the regions determined in Figure 7.9. By testing these points, we have the following:

Test Point $(0, 0)$: $3(0) - 2(0) = 0 < 6$

Test Point $(0, -4)$: $3(0) - 2(-4) = 8 > 6$

Thus, we conclude that every point in the half-plane above the line

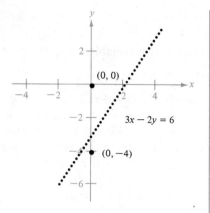

Graph of Equality

FIGURE 7.9

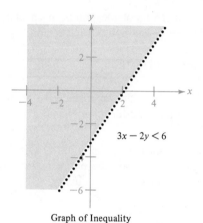

Graph of Inequality

FIGURE 7.10

satisfies the inequality and every point in the half-plane below (or on) the line does not satisfy the inequality, as shown in Figure 7.10.

(b) The graph of the equality is the parabola shown in Figure 7.11. By testing $(0, 0)$ and $(0, -2)$, we see that the points that satisfy the inequality are those lying above (or on) the parabola, as shown.

(c) The graph of the equality is the hyperbola shown in Figure 7.12. By testing $(-2, -2)$, $(0, 0)$, and $(2, 2)$, we see that the points that satisfy the inequality are those lying between (or on) the two branches of the hyperbola, as shown.

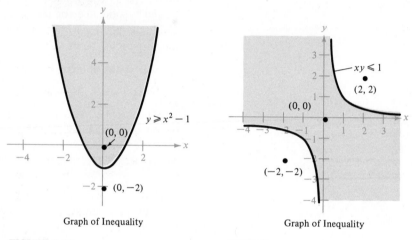

Graph of Inequality

FIGURE 7.11

Graph of Inequality

FIGURE 7.12

Remark: Remember that in general we can add or subtract variable quantities to both sides of an inequality, but we *cannot* multiply or divide both sides of an inequality by a variable quantity. For example,

$$xy \leq 1 \qquad \textit{is not} \text{ equivalent to} \qquad y \leq \frac{1}{x}$$

LINEAR INEQUALITIES IN TWO VARIABLES

A **linear inequality in two variables** has one of the following forms:

$$ax + by < c \qquad ax + by \leq c$$
$$ax + by > c \qquad ax + by \geq c$$

The graph of a linear inequality is a half-plane.

Remark: Since the graph of a linear inequality must be a half-plane, it is actually unnecessary to test points in both half-planes. *However,* because this test is relatively simple, we suggest that you test a point in each half-plane as a check on your conclusion.

EXAMPLE 2
Graphing Linear Inequalities
in Two Variables

Sketch the following linear inequalities in the plane:

(a) $x - y \leq 2$ (b) $x > -2$ (c) $y \leq 3$

Solution:

(a) By replacing $\leq$ with $=$, we obtain the equation

$$x - y = 2 \qquad \text{or} \qquad y = x - 2$$

which is the line shown in Figure 7.13. Since the origin $(0, 0)$ satisfies the inequality, we see that the graph consists of the half-plane lying above this line, as shown. (Check a point in the other half-plane.)

(b) The equation $x = -2$ graphs as a vertical line. The points that satisfy the inequality $x > -2$ are the points lying to the right of this line, as shown in Figure 7.14.

(c) The equation $y = 3$ graphs as a horizontal line. The points that satisfy the inequality $y \leq 3$ are the points lying below (or on) this line, as shown in Figure 7.15.

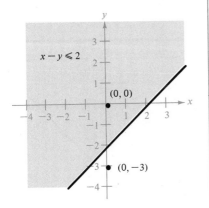

FIGURE 7.13

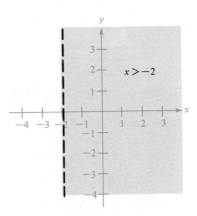

FIGURE 7.14

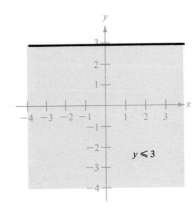

FIGURE 7.15

Systems of Inequalities

Graphing the solution of a *system* of inequalities in two variables is accomplished by graphing each inequality in the system and then hunting for the region in the plane that is common to every graph in the system.

EXAMPLE 3
Solving a System of Inequalities

Sketch the graph (and label the vertices) of the region containing all points that satisfy the following system:

$$x - y \leq 2$$
$$x > -2$$
$$y \leq 3$$

Solution:

We have already sketched the graph of each of these inequalities (in Example 2). To find the region common to all three graphs, we superimpose all three graphs on the same coordinate plane, as shown in Figure 7.16. To find the vertices of this triangular region, we solve the three systems of equations obtained by taking pairs of the equations that represent the borders of the region.

Inequality	**Equation of Border**
$x - y \leq 2$	$x - y = 2$
$x > -2$	$x = -2$
$y \leq 3$	$y = 3$

Vertex 1: $(-2, -4)$
Obtained by solving the system

$x - y = 2$
$x = -2$

Vertex 2: $(5, 3)$
Obtained by solving the system

$x - y = 2$
$y = 3$

Vertex 3: $(-2, 3)$
Obtained by solving the system

$x = -2$
$y = 3$

FIGURE 7.16

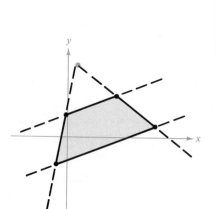

Border lines can intersect at a point that is not a vertex.

FIGURE 7.17

For the triangular region shown in Example 3, we found the vertices by intersecting each pair of border lines. With more complicated regions, two border lines can sometimes intersect at a point that is not a vertex of the region, as shown in Figure 7.17. In order to keep track of which points of intersection are actually vertices of the region, we suggest that you make a careful sketch of the region and refer to your sketch as you find each point of intersection.

In Example 3 (and in Figure 7.17), we dealt with systems of *linear* inequalities. Since the border equations for linear systems are straight lines, we can see that the enclosed region will always be a polygon. If one or more of the inequalities in the system are nonlinear, then the region will not (usually) be a polygon. Example 4 illustrates this more general type of region.

EXAMPLE 4
Solving a System of Inequalities

Sketch the graph (and label the vertices) of the region containing all points that satisfy the following system:

$$x^2 + y^2 \geq 25$$
$$7x - y \leq 25$$

Solution:
As shown in Figure 7.18, the points that satisfy the inequality $x^2 + y^2 \geq 25$ are the points lying outside (or on) a circle of radius 5 centered at the origin. The points satisfying the inequality $7x - y \leq 25$ are the points lying above the line given by $7x - y = 25$.

To find the points of intersection, we can use substitution. We begin by solving for y in the linear *equation:*

$$7x - y = 25 \quad \Longrightarrow \quad y = 7x - 25$$

Then we substitute this value of y into the equation of the circle and solve for x:

$$x^2 + y^2 = 25$$
$$(7x - 25)^2 + x^2 = 25$$
$$49x^2 - 350x + 625 + x^2 - 25 = 0$$
$$50x^2 - 350x + 600 = 0$$
$$x^2 - 7x + 12 = 0$$
$$(x - 3)(x - 4) = 0$$
$$x = 3 \text{ or } 4$$

Finally, we backsubstitute to solve for y. If $x = 3$, then

$$y = 7(3) - 25 = -4$$

If $x = 4$, then

$$y = 7(4) - 25 = 3$$

Thus, the two points of intersection are $(3, -4)$ and $(4, 3)$, as shown in Figure 7.18.

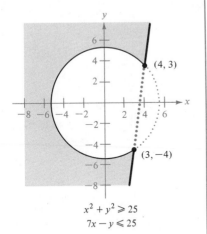

$$x^2 + y^2 \geqslant 25$$
$$7x - y \leqslant 25$$

FIGURE 7.18

Three types of unusual situations may arise in systems of inequalities:

1. Sometimes no points are common to a system. For example, the system

$$x + y > 3$$
$$x + y < -1$$

obviously has no solution points, since the quantity $(x + y)$ cannot be both less than -1 and greater than 3 for any values of x and y. (See Figure 7.19.)

2. A system of inequalities may describe a region that is unbounded in certain directions. For example, the system

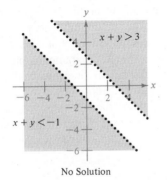

$x + y > 3$

$x + y < -1$

No Solution

FIGURE 7.19

$$x + y < 3$$
$$x + 2y > 3$$

has solutions that form an *infinite wedge*, as shown in Figure 7.20.

3. One or more of the inequalities in the system might be redundant. For example, the two systems

$$x + y < 2 \qquad x + y < 2$$
$$x + y > 0 \qquad x > 0$$
$$x > 0 \qquad y > 0$$
$$y > 0$$

have the same solution region. We can see in Figure 7.21 that the inequality $x + y > 0$ is redundant, since it adds no further restriction to that given by the other three inequalities.

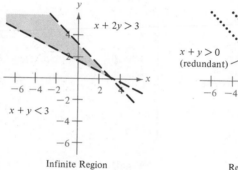

Infinite Region   Redundant Inequality

FIGURE 7.20   FIGURE 7.21

Linear Programming

In many applications in business or economics, we are concerned with a process called **optimization.** For example, an optimization problem can involve finding the minimum cost, the maximum profit, or the minimum use of resources. There are many different types of optimization problems, and many of these require calculus techniques. One type that does not require calculus is called **linear programming.** A linear programming problem consists of a linear **objective function** and a set of linear inequalities called **constraints.** The objective function gives the quantity that is to be minimized (or maximized), and the constraints determine the allowable solution points.

For example, consider the following linear programming problem, in which you are asked to maximize the value of C subject to the given constraints:

Objective Function: $C = 3x + 2y$

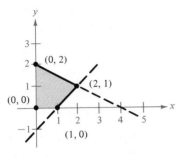

FIGURE 7.22

Constraints:
$$x \geq 0$$
$$y \geq 0$$
$$x + 2y \leq 4$$
$$x - y \leq 1$$

These constraints form the boundaries for the region shown in Figure 7.22. Every point in the region satisfies all four constraints, and since there are infinitely many points, it is not evident how we should go about finding the point that gives a maximum value of C. Fortunately, it can be shown that the *solution to a linear programming problem must occur at one of the vertices of the region bounded by the constraints.* This means that we can find the maximum value by testing C at each of the vertices. In this particular case we have the following:

At $(0, 0)$: $C = 3(0) + 2(0) = 0$
At $(1, 0)$: $C = 3(1) + 2(0) = 3$
At $(2, 1)$: $C = 3(2) + 2(1) = 8$ *(Maximum value of C)*
At $(0, 2)$: $C = 3(0) + 2(2) = 4$

Thus, the maximum value of C occurs when $x = 2$ and $y = 1$.

EXAMPLE 5
An Application of Linear Programming

A manufacturer wants to maximize the profit for two products. The first product yields a profit of \$1.50 per unit, and the second product yields a profit of \$2.00 per unit. Market tests and available resources have indicated the following constraints: (1) The combined production level should not exceed 1200 units per month, (2) the demand for product II is at most half of that for product I, and (3) the production level of product I can exceed three times the production level of product II by at most 600 units.

Solution:
If we let x be the number of units of product I and y be the number of units of product II, then the objective function (for the combined profit) is given by

$$P = 1.5x + 2y$$

The three given constraints can be interpreted as linear inequalities,

$$x + y \leq 1200 \qquad\qquad\qquad\qquad\qquad (1)$$

$$y \leq \frac{x}{2} \qquad \text{or} \qquad x - 2y \geq 0 \qquad (2)$$

$$x \leq 3y + 600 \qquad \text{or} \qquad x - 3y \leq 600 \qquad (3)$$

Since neither x nor y can be negative, we also have the following two additional constraints:

$$x \geq 0 \quad \text{and} \quad y \geq 0$$

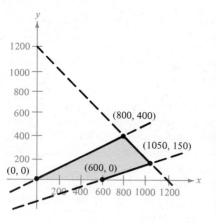

FIGURE 7.23

Figure 7.23 shows the region determined by these linear inequalities. To find the maximum profit, we test the value of P at the vertices of the region.

$$At\ (0,\ 0):\quad P = 1.5(0) + 2(0) = 0$$
$$At\ (600,\ 0):\quad P = 1.5(600) + 2(0) = \$900$$
$$At\ (1050,\ 150):\quad P = 1.5(1050) + 2(150) = \$1875$$
$$At\ (800,\ 400):\quad P = 1.5(800) + 2(400) = \$2000$$

Thus, the maximum profit is $2000 and occurs when the monthly production consists of 800 units of product I and 400 units of product II.

Section Exercises 7.4

In Exercises 1–10, match the inequality with the correct graph.

1. $x > 3$
2. $y \leq 2$
3. $2x + 3y \leq 6$
4. $2x - y \geq -2$
5. $x^2 + y^2 < 4$
6. $(x - 2)^2 + (y - 3)^2 > 4$
7. $xy > 2$
8. $y \leq 4 - x^2$
9. $y^2 - x^2 \leq 4$
10. $y \leq e^{-x^2}$

(a)

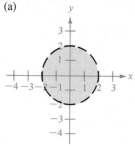

(b)

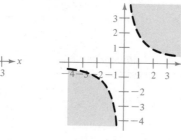

(c)

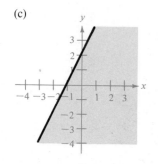

(d)

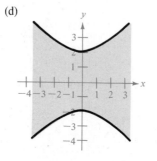

(e)

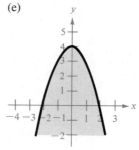

(f)

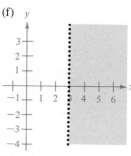

(g)

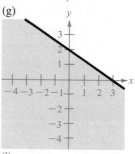

(h)

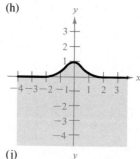

(i)

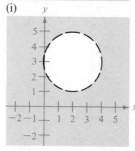

(j)

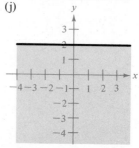

In Exercises 11–20, sketch the graph of the inequality.

11. $y \geq -1$

12. $x \leq 4$

13. $y < 2 - x$

14. $y > 2x - 4$

15. $2y - x \geq 4$

16. $5x + 3y \geq -15$

17. $(x + 1)^2 + (y - 2)^2 < 9$

18. $y^2 - x < 0$

19. $y \leq \dfrac{1}{1 + x^2}$

20. $y < \ln x$

In Exercises 21–40, sketch the graph of the solution of the system of inequalities.

21. $\begin{aligned} x + y &\leq 1 \\ -x + y &\leq 1 \\ y &\geq 0 \end{aligned}$

22. $\begin{aligned} 3x + 2y &< 6 \\ x &> 0 \\ y &> 0 \end{aligned}$

23. $\begin{aligned} x + y &\leq 5 \\ x &\geq 2 \\ y &\geq 0 \end{aligned}$

24. $\begin{aligned} 2x + y &\geq 2 \\ y &\leq 1 \\ x &\leq 2 \end{aligned}$

25. $\begin{aligned} -3x + 2y &< 6 \\ x + 4y &> -2 \\ 2x + y &< 3 \end{aligned}$

26. $\begin{aligned} x - 7y &> -36 \\ 5x + 2y &> 5 \\ 6x - 5y &> 6 \end{aligned}$

27. $\begin{aligned} 2x + y &> 2 \\ 6x + 3y &< 2 \end{aligned}$

28. $\begin{aligned} x - 2y &< -6 \\ 5x - 3y &> -9 \end{aligned}$

29. $\begin{aligned} x &\geq 1 \\ x - 2y &\leq 3 \\ 3x + 2y &\geq 9 \\ x + y &\leq 6 \end{aligned}$

30. $\begin{aligned} x - y^2 &> 0 \\ x - y &< 2 \end{aligned}$

31. $\begin{aligned} x^2 + y^2 &\leq 9 \\ x^2 + y^2 &\geq 1 \end{aligned}$

32. $\begin{aligned} x^2 + y^2 &\leq 25 \\ 4x - 3y &\leq 0 \end{aligned}$

33. $\begin{aligned} x &> y^2 \\ x &< y + 2 \end{aligned}$

34. $\begin{aligned} x &< 2y - y^2 \\ 0 &< x + y \end{aligned}$

35. $\begin{aligned} y &\leq \sqrt{3x} + 1 \\ y &\geq x + 1 \end{aligned}$

36. $\begin{aligned} y &< -x^2 + 2x + 3 \\ y &> x^2 - 4x + 3 \end{aligned}$

37. $\begin{aligned} y &< x^3 - 2x + 1 \\ y &> -2x \\ x &\leq 1 \end{aligned}$

38. $\begin{aligned} y &\geq x^4 - 2x^2 + 1 \\ y &\leq 1 - x^2 \end{aligned}$

39. $\begin{aligned} x^2 y &\geq 1 \\ 0 &< x \leq 4 \\ y &\leq 4 \end{aligned}$

40. $\begin{aligned} y &\leq e^{-x^2/2} \\ y &\geq 0 \\ -2 &\leq x \leq 2 \end{aligned}$

41. A furniture company can sell all the tables and chairs that it produces. Each table requires 1 hour in the assembly center and $1\frac{1}{3}$ hours in the finishing center. Each chair requires $1\frac{1}{2}$ hours in the assembly center and $1\frac{1}{2}$ hours in the finishing center. The company's assembly center is available 12 hours per day, and its finishing center is available 15 hours per day. If x is the number of tables produced per day and y is the number of chairs, find a system of inequalities describing all possible production levels. Sketch the graph of the system.

42. A store sells two models of a certain brand of computer.

Because of the demand, it is necessary to stock at least twice as many units of model A as units of model B. The cost to the store for the two models is $800 and $1200, respectively. The management does not want more than $20,000 in computer inventory at any time, and it wants at least 4 model A computers and 2 model B computers in inventory at all times. Devise a system of inequalities describing all possible inventory levels, and sketch the graph of the system.

In Exercises 43–48, derive a set of inequalities to describe the region.

43. Rectangle with vertices at (2, 1), (5, 1), (5, 7), and (2, 7).

44. Triangle with vertices at (0, 0), (5, 0), and (2, 3).

45. Triangle with vertices at (−1, 0), (1, 0), and (0, 1).

46. Segment of a circle

47. Segment of a circle

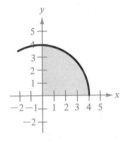
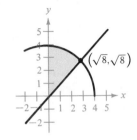

48. Parabolic region

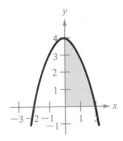

In Exercises 49–52, find the minimum and maximum values of *both* objective functions, subject to the given constraints.

49. Constraints: $\begin{aligned} x &\geq 0 \\ y &\geq 0 \\ x + 3y &\leq 15 \\ 4x + y &\leq 16 \end{aligned}$

Objective function:

(a) $C = 3x + 2y$

(b) $C = 5x + \dfrac{y}{2}$

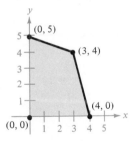

50. Constraints: $x \geq 0$
$2x + 3y \geq 6$
$3x - 2y \leq 9$
$x + 5y \leq 20$

Objective function:
(a) $C = 4x + 3y$
(b) $C = x + 6y$

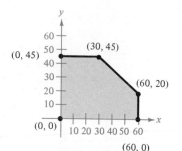

(0, 4)   (5, 3)
(0, 2)
(3, 0)

51. Constraints: $x \geq 0$
$y \geq 0$
$x \leq 60$
$y \leq 45$
$5x + 6y \leq 420$

Objective function:
(a) $C = 10x + 7y$
(b) $C = 25x + 30y$

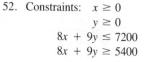

(0, 45)   (30, 45)
(60, 20)
(0, 0)   (60, 0)

52. Constraints: $x \geq 0$
$y \geq 0$
$8x + 9y \leq 7200$
$8x + 9y \geq 5400$

Objective function:
(a) $C = 50x + 35y$
(b) $C = 16x + 18y$

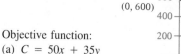

(0, 800)
(0, 600)
(900, 0)
(675, 0)

53. A merchant plans to sell two models of home computers that will cost $250 and $400, respectively, and he estimates that the total monthly demand will not exceed 250 units. Find the number of units of each model that the merchant should stock in order to maximize profit if he does not want to invest more than $70,000 in computer inventory and if his profit on the two models is $45 (for the model costing $250) and $50 (for the model costing $400).

54. A fruit grower has 150 acres of land available to raise two crops, A and B. It takes 1 day to trim an acre of A and 2 days to trim an acre of B, and there are 240 days per year available for trimming. It takes $\frac{3}{10}$ day to pick an acre of A and $\frac{1}{10}$ day to pick an acre of B, and there are 30 days per year available for picking. Find the number of acres of each fruit that should be planted to maximize profit, if the profit for crops A and B is $140 per acre and $235 per acre, respectively.

55. Find the minimum value of the objective function $C = 4x + 5y$ subject to the constraints

$$x + y \geq 8$$
$$4x + 3y \geq 27$$
$$3x + 5y \geq 30$$
$$x \geq 0$$
$$y \geq 0$$

56. A farmer mixes two brands of cattle feed. Brand X costs $25 per bag and contains 2 units of nutritional element A, 2 units of element B, and 2 units of element C. Brand Y costs $20 per bag and contains 1 unit of nutritional element A, 9 units of element B, and 3 units of element C. Find the number of bags of each brand that should be mixed to produce a mixture having a minimum cost per bag. The minimum requirements of nutrients A, B, and C are 12 units, 36 units, and 24 units, respectively.

Review Exercises / Chapter 7

In Exercises 1–20, find all solutions to the system of equations.

1. $x + y = 2$
$x - y = 0$

2. $2x = 3(y - 1)$
$y = x$

3. $x^2 - y^2 = 9$
$x - y = 1$

4. $x^2 + y^2 = 169$
$3x + 2y = 39$

5. $y = 2x^2$
$y = x^4 - 2x^2$

6. $x = y + 3$
$x = y^2 + 1$

7. $y^2 - 2y + x = 0$
$x + y = 0$

8. $y = 2x^2 - 4x + 1$
$y = x^2 - 4x + 3$

9. $2x - y = 2$
$6x + 8y = 39$

10. $40x + 30y = 24$
$20x - 50y = -14$

11. $0.2x + 0.3y = 0.14$
$0.4x + 0.5y = 0.20$

12. $12x + 42y = -17$
$30x - 18y = 19$

13. $\dfrac{x}{2} - \dfrac{y}{3} = 0$
$3x + 2(y + 5) = 10$

14. $\dfrac{x}{3} + \dfrac{4y}{7} = 3$
$2x + 3y = 15$

15. $x + 2y + 6z = 4$
$-3x + 2y - z = -4$
$4x + 2z = 16$

16. $2x + 3y + 4z = 21$
$5x + y - 2z = -17$
$8x + 9y + z = -12$

17. $\begin{aligned} x - 2y + z &= -6 \\ 2x - 3y &= -7 \\ -x + 3y - 3z &= 11 \end{aligned}$

18. $\begin{aligned} 2x + 6z &= -9 \\ 3x - 2y + 11z &= -16 \\ 3x - y + 7z &= -11 \end{aligned}$

19. $\begin{aligned} 2x + 5y - 19z &= 34 \\ 3x + 8y - 31z &= 54 \end{aligned}$

20. $\begin{aligned} 2x + y + z + 2w &= -1 \\ 5x - 2y + z - 3w &= 0 \\ -x + 3y + 2z + 2w &= 1 \\ 3x + 2y + 3z - 5w &= 12 \end{aligned}$

In Exercises 21–28, sketch the graph of the solutions of the systems of inequalities.

21. $\begin{aligned} x + 2y &\le 160 \\ 3x + y &\le 180 \\ x &\ge 0 \\ y &\ge 0 \end{aligned}$

22. $\begin{aligned} 2x + 3y &\le 24 \\ 2x + y &\le 16 \\ x &\ge 0 \\ y &\ge 0 \end{aligned}$

23. $\begin{aligned} 3x + 2y &\ge 24 \\ x + 2y &\ge 12 \\ x &\ge 2 \\ x &\le 15 \\ y &\le 15 \end{aligned}$

24. $\begin{aligned} 2x + y &\ge 16 \\ x + 3y &\ge 18 \\ 0 \le x &\le 25 \\ 0 \le y &\le 25 \end{aligned}$

25. $\begin{aligned} y &< x + 1 \\ y &> x^2 - 1 \end{aligned}$

26. $\begin{aligned} y &\le 6 - 2x - x^2 \\ y &\ge x + 6 \end{aligned}$

27. $\begin{aligned} 2x - 3y &\ge 0 \\ 2x - y &\le 8 \\ y &\ge 0 \end{aligned}$

28. $\begin{aligned} x^2 + y^2 &\le 9 \\ (x - 3)^2 + y^2 &\le 9 \end{aligned}$

29. A mixture of 6 parts of chemical A, 8 parts of chemical B, and 13 parts of chemical C is required to kill a certain destructive crop insect. Commercial spray X contains 1, 2, and 2 parts, respectively, of these chemicals. Commercial spray Y contains only chemical C. Commercial spray Z contains chemicals A, B, and C in equal amounts. How much of each type of commercial spray is needed to get the desired mixture?

CHAPTER

8 | MATRICES AND DETERMINANTS

Matrix Solutions of Systems of Linear Equations

8.1

In mathematics we always look for a valid shortcut to solving problems. In this section we will look at a streamlined technique for solving systems of linear equations. Let's reconsider the system of equations in Example 2 of Section 7.3. (Note that the equations are arranged so that the variables line up vertically. This is important.)

$$x - 2y + 3z = 9$$
$$-x + 3y - z = -6$$
$$2x - 5y + 5z = 17$$

If you look back at the solution of Example 2, you will see that variables x, y, and z served mainly to keep track of the position of the coefficients. The actual decisions in the solution were based on the values of the coefficients. By writing only the coefficients, we obtain the following two-dimensional array, called a **matrix:**

$$\begin{bmatrix} 1 & -2 & 3 & 9 \\ -1 & 3 & -1 & -6 \\ 2 & -5 & 5 & 17 \end{bmatrix}$$

This particular matrix has three **rows** (numbers in horizontal lines) and four **columns** (numbers in vertical lines).

DEFINITION OF A MATRIX

If m and n are positive integers, then an $m \times n$ **matrix** is an array of the form

$$\begin{bmatrix} a_{11} & a_{12} & a_{13} & \cdots & a_{1n} \\ a_{21} & a_{22} & a_{23} & \cdots & a_{2n} \\ a_{31} & a_{32} & a_{33} & \cdots & a_{3n} \\ \cdot & \cdot & \cdot & & \cdot \\ \cdot & \cdot & \cdot & & \cdot \\ \cdot & \cdot & \cdot & & \cdot \\ a_{m1} & a_{m2} & a_{m3} & \cdots & a_{mn} \end{bmatrix}$$

where each **element a_{ij} of the matrix** is a real number.

An $m \times n$ matrix (read "m by n") has m rows and n columns and is said to be of *order* $m \times n$. If $m = n$, the matrix is **square** of order n. The element a_{ij} is located in the ith row and jth column. We call i the **row subscript** and j the **column subscript.** The elements a_{11}, a_{22}, a_{33}, . . . are called the **main diagonal elements.**

A matrix derived from a system of linear equations is called the **augmented matrix of the system.** Moreover, the matrix that is derived from the coefficients of the system (and does not include the constant terms) is called the **coefficient matrix of the system.** For example,

System

$$\begin{aligned} x - 4y + 3z &= 5 \\ -x + 3y - z &= -3 \\ 2x \quad\quad - 4z &= 6 \end{aligned}$$

Coefficient Matrix

$$\begin{bmatrix} 1 & -4 & 3 \\ -1 & 3 & -1 \\ 2 & 0 & -4 \end{bmatrix}$$

Augmented Matrix

$$\begin{bmatrix} 1 & -4 & 3 & 5 \\ -1 & 3 & -1 & -3 \\ 2 & 0 & -4 & 6 \end{bmatrix}$$

Remark: Note the use of 0 for the missing y-variable in both matrices, and note the fourth column of constant terms from the system in the augmented matrix.

Recall that two systems of linear equations are equivalent if they possess the same solutions. Moreover, we can construct an equivalent system in *triangular form* by adding equations, multiplying equations by constants, or interchanging the position of equations. In matrix terminology, these three types of changes correspond to what we call **elementary row operations.** Thus, an elementary row operation on an augmented matrix (corresponding to a given system of linear equations) produces a new augmented matrix corresponding to a new (but equivalent) system of linear equations.

ELEMENTARY ROW OPERATIONS

If an augmented matrix (of a given system of linear equations) is changed by any of the following operations, it produces an augmented matrix of an *equivalent* system of linear equations:

1. Interchange any two rows of the matrix.
2. Multiply every element in a row by the same nonzero constant.
3. Replace a row by the sum of that row and a multiple of any other row.

Though each of the three elementary row operations is simple to perform, there is a considerable amount of arithmetic involved. It is easy to make a careless error, and we suggest that you use the following scheme to designate these operations:

Interchange two rows:

$$\begin{bmatrix} 0 & 1 & 3 & 4 \\ -1 & 2 & 0 & 3 \\ 2 & -3 & 4 & 1 \end{bmatrix} \quad \begin{matrix} \uparrow R_2 \\ \downarrow R_1 \end{matrix} \quad \begin{bmatrix} -1 & 2 & 0 & 3 \\ 0 & 1 & 3 & 4 \\ 2 & -3 & 4 & 1 \end{bmatrix}$$

Multiply a row by a (nonzero) constant ($\frac{1}{2}R_1$ is new R_1):

$$\begin{bmatrix} 2 & -4 & 6 & -2 \\ 1 & 3 & -3 & 0 \\ 5 & -2 & 1 & 2 \end{bmatrix} \quad \frac{1}{2}R_1 \quad \begin{bmatrix} 1 & -2 & 3 & -1 \\ 1 & 3 & -3 & 0 \\ 5 & -2 & 1 & 2 \end{bmatrix}$$

Add a row to another row ($R_1 + R_2$ is new R_2):

$$\begin{bmatrix} 1 & 2 & -4 & 3 \\ -1 & 3 & -2 & -1 \\ 2 & 1 & 5 & -2 \end{bmatrix} \quad R_1 + R_2 \quad \begin{bmatrix} 1 & 2 & -4 & 3 \\ 0 & 5 & -6 & 2 \\ 2 & 1 & 5 & -2 \end{bmatrix}$$

Add a multiple of a row to another row ($-2R_1 + R_3$ is new R_3):

$$\begin{bmatrix} 1 & 2 & -4 & 3 \\ -1 & 3 & -2 & -1 \\ 2 & 1 & 5 & -2 \end{bmatrix} \quad -2R_1 + R_3 \quad \begin{bmatrix} 1 & 2 & -4 & 3 \\ -1 & 3 & -2 & -1 \\ 0 & -3 & 13 & -8 \end{bmatrix}$$

Note that we write the elementary row operations beside the row that we are changing. To emphasize this, you may want to use the following notation (written to the left of the changed row):

$$\tfrac{1}{2}R_1 \to R_1, \qquad R_1 + R_2 \to R_2, \qquad -2R_1 + R_3 \to R_3$$

EXAMPLE 1
Using Elementary Row Operations

Use matrices and elementary row operations to solve the following system:

$$\begin{aligned} x - 2y + 3z &= 9 \\ -x + 3y - z &= -6 \\ 2x - 5y + 5z &= 17 \end{aligned}$$

Solution:
We begin by forming the augmented matrix of this system.

$$\begin{bmatrix} 1 & -2 & 3 & 9 \\ -1 & 3 & -1 & -6 \\ 2 & -5 & 5 & 17 \end{bmatrix}$$

Our goal is to apply elementary row operations to obtain a matrix that corresponds to a system in *triangular form*. This means that we need to change the elements below the main diagonal to zeros (by means of elementary row operations). Study the following steps carefully to see how this is done.

$$\begin{bmatrix} 1 & -2 & 3 & 9 \\ -1 & 3 & -1 & -6 \\ 2 & -5 & 5 & 17 \end{bmatrix} \quad R_1 + R_2 \quad \begin{bmatrix} 1 & -2 & 3 & 9 \\ 0 & 1 & 2 & 3 \\ 2 & -5 & 5 & 17 \end{bmatrix}$$

$$-2R_1 + R_3 \quad \begin{bmatrix} 1 & -2 & 3 & 9 \\ 0 & 1 & 2 & 3 \\ 0 & -1 & -1 & -1 \end{bmatrix}$$

$$R_2 + R_3 \quad \begin{bmatrix} 1 & -2 & 3 & 9 \\ 0 & 1 & 2 & 3 \\ 0 & 0 & 1 & 2 \end{bmatrix}$$

Now, we write the system of linear equations corresponding to this final matrix and use backsubstitution to find the solution:

$$x - 2y + 3z = 9$$
$$y + 2z = 3$$
$$z = 2$$

This is the same system (in triangular form) that we solved in Example 1 in Section 7.3. The solution is

$$x = 1, \quad y = -1, \quad \text{and} \quad z = 2$$

The final matrix obtained in Example 1 is said to be in **echelon form.** The general procedure followed is

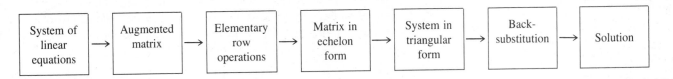

We can eliminate the need for backsubstitution by continuing to apply elementary row operations to the echelon (triangular) form so as to transform it into a matrix in **reduced echelon form.** To be in this form, a matrix must have the following properties.

$$\begin{bmatrix} 1 & 0 & 0 & 0 & a \\ 0 & 1 & 0 & 0 & b \\ 0 & 0 & 1 & 0 & c \\ 0 & 0 & 0 & 1 & d \end{bmatrix}$$

FIGURE 8.1
*Reduced Echelon Form for 4 × 5
Augmented Matrix*

REDUCED ECHELON FORM

1. All rows consisting entirely of zeros are placed at the bottom of the matrix.

2. The first nonzero entry in any row is 1 (called a leading 1).

3. Nonzero rows are arranged so that leading 1's occur farther to the right in each successive row.

4. Each column with a leading 1 has zeros as all other entries.

Once an augmented matrix is written in reduced echelon form, we can read the solution directly from the matrix. For example, in Figure 8.1, we can see that $x = a$, $y = b$, $z = c$, and $w = d$. This procedure of finding the solution to a linear system by writing the augmented matrix in reduced echelon form is called **Gauss-Jordan elimination**, summarized as follows.

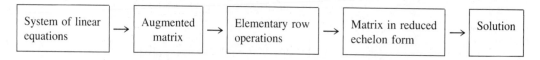

Had we applied Gauss-Jordan elimination to the system in Example 1, we would have obtained the reduced echelon form

$$\begin{bmatrix} 1 & 0 & 0 & 1 \\ 0 & 1 & 0 & -1 \\ 0 & 0 & 1 & 2 \end{bmatrix}$$

from which we obtain a system of equations that corresponds precisely to the solution $x = 1$, $y = -1$, and $z = 2$.

Remark: The order in which the elementary row operations are performed in the Gauss-Jordan elimination process is important. We move from *left to right by columns*, changing to zero all elements directly above or below the leading 1's.

The power of the Gauss-Jordan elimination method and the convenience of our notation become increasingly evident with linear systems involving a large number of variables. In the next example, note how several row operations are performed as single steps in the solution.

EXAMPLE 2
**The Gauss-Jordan
Elimination Method**

Use Gauss-Jordan elimination to solve the following system:

$$\begin{aligned}
y + z - 2w &= -3 \\
x + 2y - z \quad\;\;\; &= 2 \\
2x + 4y + z - 3w &= -2 \\
x - 4y - 7z - w &= -19
\end{aligned}$$

Solution:

$$\begin{bmatrix} 0 & 1 & 1 & -2 & -3 \\ 1 & 2 & -1 & 0 & 2 \\ 2 & 4 & 1 & -3 & -2 \\ 1 & -4 & -7 & -1 & -19 \end{bmatrix}$$

$$\begin{matrix} R_2 \\ R_1 \end{matrix} \begin{bmatrix} 1 & 2 & -1 & 0 & 2 \\ 0 & 1 & 1 & -2 & -3 \\ 2 & 4 & 1 & -3 & -2 \\ 1 & -4 & -7 & -1 & -19 \end{bmatrix} \qquad \textit{Interchange rows}$$

$$\begin{matrix} \\ \\ -2R_1 + R_3 \\ -R_1 + R_4 \end{matrix} \begin{bmatrix} 1 & 2 & -1 & 0 & 2 \\ 0 & 1 & 1 & -2 & -3 \\ 0 & 0 & 3 & -3 & -6 \\ 0 & -6 & -6 & -1 & -21 \end{bmatrix} \qquad \textit{Zeros in column 1}$$

$$\begin{matrix} -2R_2 + R_1 \\ \\ \\ 6R_2 + R_4 \end{matrix} \begin{bmatrix} 1 & 0 & -3 & 4 & 8 \\ 0 & 1 & 1 & -2 & -3 \\ 0 & 0 & 3 & -3 & -6 \\ 0 & 0 & 0 & -13 & -39 \end{bmatrix} \qquad \textit{Zeros in column 2}$$

$$\begin{matrix} \\ \\ \frac{1}{3}R_3 \\ -\frac{1}{13}R_4 \end{matrix} \begin{bmatrix} 1 & 0 & -3 & 4 & 8 \\ 0 & 1 & 1 & -2 & -3 \\ 0 & 0 & 1 & -1 & -2 \\ 0 & 0 & 0 & 1 & 3 \end{bmatrix} \qquad \textit{Ones on diagonal}$$

$$\begin{matrix} 3R_3 + R_1 \\ -R_3 + R_2 \\ \\ \\ \end{matrix} \begin{bmatrix} 1 & 0 & 0 & 1 & 2 \\ 0 & 1 & 0 & -1 & -1 \\ 0 & 0 & 1 & -1 & -2 \\ 0 & 0 & 0 & 1 & 3 \end{bmatrix} \qquad \textit{Zeros in column 3}$$

$$\begin{matrix} -R_4 + R_1 \\ R_4 + R_2 \\ R_4 + R_3 \\ \\ \end{matrix} \begin{bmatrix} 1 & 0 & 0 & 0 & -1 \\ 0 & 1 & 0 & 0 & 2 \\ 0 & 0 & 1 & 0 & 1 \\ 0 & 0 & 0 & 1 & 3 \end{bmatrix} \qquad \textit{Zeros in column 4}$$

Thus, the solution is

$$x = -1, \qquad y = 2, \qquad z = 1, \qquad \text{and} \qquad w = 3$$

Remark: We suggest that you repeat Example 2 using the echelon form
and backsubstitution to see which method you prefer.

EXAMPLE 3

A System with Infinitely Many Solutions

Solve the following system:

$$x - y + 2z = 4$$
$$2x + 4y + 4z = -1$$
$$x + y + 2z = 1$$

Solution:

$$\begin{bmatrix} 1 & -1 & 2 & 4 \\ 2 & 4 & 4 & -1 \\ 1 & 1 & 2 & 1 \end{bmatrix} \quad \begin{matrix} -2R_1 + R_2 \\ -R_1 + R_3 \end{matrix} \quad \begin{bmatrix} 1 & -1 & 2 & 4 \\ 0 & 6 & 0 & -9 \\ 0 & 2 & 0 & -3 \end{bmatrix}$$

$$\frac{1}{3}R_2 \quad \begin{bmatrix} 1 & -1 & 2 & 4 \\ 0 & 2 & 0 & -3 \\ 0 & 2 & 0 & -3 \end{bmatrix}$$

$$-R_2 + R_3 \quad \begin{bmatrix} 1 & -1 & 2 & 4 \\ 0 & 2 & 0 & -3 \\ 0 & 0 & 0 & 0 \end{bmatrix}$$

Now, since one of the rows has all zero elements, we can convert back to the corresponding system of linear equations and write the solution following the procedure given in Section 7.3.

$$x - y + 2z = 4$$
$$2y = -3$$

From this system, $y = -\frac{3}{2}$ and

$$x - \left(-\frac{3}{2}\right) + 2z = 4$$

$$x = \frac{-4z + 5}{2}$$

Letting z be any real number a, we have the following solution:

$$\left(\frac{-4a + 5}{2}, \frac{-3}{2}, a\right)$$

Remark: Note in Example 3 that we used the combined method of reducing to triangular form and then applying backsubstitution. As a general procedure, we have found this combined method to be more efficient than the Gauss-Jordan elimination method (involving the fully reduced echelon form).

EXAMPLE 4

A System with No Solutions

Solve the following system:

$$2x + 3y + z = 2$$
$$x - 3y + 2z = 4$$
$$x + z = 6$$

Solution:

$$\begin{bmatrix} 2 & 3 & 1 & 2 \\ 1 & -3 & 2 & 4 \\ 1 & 0 & 1 & 6 \end{bmatrix} \quad \begin{matrix} \curvearrowright R_2 \\ \curvearrowleft R_1 \end{matrix} \quad \begin{bmatrix} 1 & -3 & 2 & 4 \\ 2 & 3 & 1 & 2 \\ 1 & 0 & 1 & 6 \end{bmatrix}$$

$$\begin{matrix} -2R_1 + R_2 \\ -R_1 + R_3 \end{matrix} \quad \begin{bmatrix} 1 & -3 & 2 & 4 \\ 0 & 9 & -3 & -6 \\ 0 & 3 & -1 & 2 \end{bmatrix}$$

$$\tfrac{1}{3}R_2 \quad \begin{bmatrix} 1 & -3 & 2 & 4 \\ 0 & 3 & -1 & -2 \\ 0 & 3 & -1 & 2 \end{bmatrix}$$

$$-R_2 + R_3 \quad \begin{bmatrix} 1 & -3 & 2 & 4 \\ 0 & 3 & -1 & -2 \\ 0 & 0 & 0 & 4 \end{bmatrix}$$

Now, by converting back to a system of linear equations, we have

$$x - 3y + 2z = 4$$
$$3y - z = -2$$
$$0 = 4$$

Since the last of these equations is an absurdity, we conclude that the original system is *inconsistent* and has no solutions.

Section Exercises 8.1

1. Given the matrix

$$\begin{bmatrix} 1 & 2 & 3 \\ 2 & -1 & -4 \\ 3 & 1 & -1 \end{bmatrix}$$

perform the following *sequence* of elementary row operations to put the matrix in reduced echelon form:
   (a) Add (-2) times Row 1 to Row 2. (Only Row 2 should change.)
   (b) Add (-3) times Row 1 to Row 3. (Only Row 3 should change.)
   (c) Add (-1) times Row 2 to Row 3.
   (d) Multiply Row 2 by $(-\tfrac{1}{5})$.
   (e) Add (-2) times Row 2 to Row 1.

2. Given the matrix

$$\begin{bmatrix} 7 & 1 \\ 0 & 2 \\ -3 & 4 \\ 4 & 1 \end{bmatrix}$$

perform the following *sequence* of elementary row operations to put the matrix in reduced echelon form:
   (a) Add Row 3 to Row 4. (Only Row 4 should change.)
   (b) Interchange Rows 1 and 4. (Note that the first element in the matrix is now 1, and it was obtained without introducing fractions.)
   (c) Add (3) times Row 1 to Row 3.
   (d) Add (-7) times Row 1 to Row 4.
   (e) Multiply Row 2 by $\tfrac{1}{2}$.
   (f) Add the appropriate multiple of Row 2 to Rows 1, 3, and 4.

In Exercises 3–10, put the matrix in reduced echelon form.

3. $\begin{bmatrix} 1 & 1 & 0 & 5 \\ -2 & -1 & 2 & -10 \\ 3 & 6 & 7 & 14 \end{bmatrix}$

4. $\begin{bmatrix} 1 & 2 & -1 & 3 \\ 3 & 7 & -5 & 14 \\ -2 & -1 & -3 & 8 \end{bmatrix}$

5. $\begin{bmatrix} 1 & -1 & -1 & 1 \\ 5 & -4 & 1 & 8 \\ -6 & 8 & 18 & 0 \end{bmatrix}$

6. $\begin{bmatrix} 1 & -3 & 0 & -7 \\ -3 & 10 & 1 & 23 \\ 4 & -10 & 2 & -24 \end{bmatrix}$

7. $\begin{bmatrix} 3 & 3 & 3 \\ -1 & 0 & -4 \\ 2 & 4 & -2 \\ 5 & 6 & 12 \end{bmatrix}$

8. $\begin{bmatrix} 1 & 3 & 2 \\ 5 & 15 & 9 \\ 2 & 6 & 10 \end{bmatrix}$

9. $\begin{bmatrix} 1 & 2 & 3 & -5 \\ 1 & 2 & 4 & -9 \\ -2 & -4 & -4 & 3 \\ 4 & 8 & 11 & -14 \end{bmatrix}$

10. $\begin{bmatrix} 1 & -3 \\ -1 & 8 \\ 0 & 4 \\ -2 & 10 \end{bmatrix}$

In Exercises 11–30, use matrices to solve the system of equations.

11. $x + 2y = 7$
$2x + y = 8$

12. $2x + 6y = 16$
$2x + 3y = 7$

13. $-3x + 5y = -22$
$3x + 4y = 4$
$4x - 8y = 32$

14. $x + 2y = 0$
$x + y = 6$
$3x - 2y = 8$

15. $8x - 4y = 7$
$5x + 2y = 1$

16. $2x - y = -0.1$
$3x + 2y = 1.6$

17. $-x + 2y = 1.5$
$2x - 4y = 3$

18. $x - 3y = 5$
$-2x + 6y = -10$

19. $x - 3z = -2$
$3x + y - 2z = 5$
$2x + 2y + z = 4$

20. $2x - y + 3z = 24$
$2y - z = 14$
$7x - 5y = 6$

21. $x + y - 5z = 3$
$x - 2z = 1$
$2x - y - z = 0$

22. $2x + 3z = 3$
$4x - 3y + 7z = 5$
$8x - 9y + 15z = 9$

23. $x + 2y + z = 8$
$3x + 7y + 6z = 26$

24. $4x + 12y - 7z - 20w = 22$
$3x + 9y - 5z - 28w = 30$

25. $3x + 3y + 12z = 6$
$x + y + 4z = 2$
$2x + 5y + 20z = 10$
$-x + 2y + 8z = 4$

26. $2x + 10y + 2z = 6$
$x + 5y + 2z = 6$
$x + 5y + z = 3$
$-3x - 15y - 3z = -9$

27. $2x + y - z + 2w = -6$
$3x + 4y + w = 1$
$x + 5y + 2z + 6w = -3$
$5x + 2y - z - w = 3$

28. $x_1 + x_2 - 2x_3 + 3x_4 + 2x_5 = 9$
$3x_1 + 3x_2 - x_3 + x_4 + x_5 = 5$
$2x_1 + 2x_2 - x_3 + x_4 - 2x_5 = 1$
$4x_1 + 4x_2 + x_3 - 3x_5 = 4$
$8x_1 + 5x_2 - 2x_3 - x_4 + 2x_5 = 3$

29. $x_1 - x_2 + 2x_3 + 2x_4 + 6x_5 = 6$
$3x_1 - 2x_2 + 4x_3 + 4x_4 + 12x_5 = 14$
$x_2 - x_3 - x_4 - 3x_5 = -3$
$2x_1 - 2x_2 + 4x_3 + 5x_4 + 15x_5 = 10$
$2x_1 - 2x_2 + 4x_3 + 4x_4 + 13x_5 = 13$

30. $2x_1 - 3x_2 + x_3 + 5x_4 = 1$

31. A small corporation borrowed $1,500,000 to expand its product line. Some of the money was borrowed at 8%, some at 9%, and some at 12%. How much was borrowed at each rate if the annual interest was $133,000 and the amount borrowed at 8% was 4 times the amount borrowed at 12%?

32. A grocer wishes to mix three kinds of nuts costing $3.50, $4.50, and $6.00 per pound, to obtain 50 pounds of a mixture costing $4.95 per pound. How many pounds of each variety should the grocer use if half of the mixture is composed of the two cheapest varieties?

33. Find a, b, and c for the quadratic function $f(x) = ax^2 + bx + c$, such that $f(1) = 1$, $f(-3) = 17$, and $f(2) = -\frac{1}{2}$.

34. Find a, b, c, and d for the cubic function $f(x) = ax^3 + bx^2 + cx + d$, such that $f(1) = 3$, $f(2) = 19$, $f(-1) = -11$, and $f(-2) = -33$.

35. Find D, E, and F for the circle $x^2 + y^2 + Dx + Ey + F = 0$ if the circle passes through the points $(1, 1)$, $(3, 3)$, and $(4, 2)$.

36. The sum of three positive numbers is 33. Find the three numbers if the second is 3 greater than the first and the third is 4 times the first.

The Algebra of Matrices

8.2

In Section 8.1 we looked at a matrix as if it were simply an array of real numbers. In reality, there is much more to matrices than that. In fact, there is a rich mathematical theory of matrices, and there are numerous practical applications of this theory. In this and the next section, we will study some of the basic portions of this theory. A more comprehensive study of the theory of matrices can be gained by taking a complete course in linear algebra.

In this section, we will look at the basic properties of **matrix addition and multiplication.** Throughout this chapter, when we refer to a matrix, we will mean a *real* matrix, whose elements are real numbers. Matrices can be represented in three convenient ways.

MATRIX NOTATION

1. A matrix can be denoted by an uppercase letter of the alphabet, such as A, B, C, ...
2. A matrix can be denoted by a representative element enclosed in brackets, such as $[a_{ij}]$, $[b_{ij}]$, $[c_{ij}]$, ...
3. A matrix can be denoted by an array of real numbers.

We say that two matrices are equal if and only if their corresponding elements are equal.

DEFINITION OF EQUALITY OF MATRICES

Two matrices $A = [a_{ij}]$ and $B = [b_{ij}]$ are equal if and only if

$$a_{ij} = b_{ij}$$

for every i and j. This definition implies that two matrices can be equal to each other only if they have the same order ($m \times n$).

We can add two matrices (of the same order) by adding their corresponding elements.

DEFINITION OF ADDITION OF MATRICES

If $A = [a_{ij}]$ and $B = [b_{ij}]$ are of the same order $m \times n$, then we define their sum to be the $m \times n$ matrix given by

$$A + B = [a_{ij} + b_{ij}]$$

We do not define the sum of two matrices of different orders.

Since addition of matrices is defined in terms of addition of real numbers, it is not surprising that many of the properties of addition of real numbers carry over to matrices.

PROPERTIES OF MATRIX ADDITION

If A, B, and C are matrices of order $m \times n$, then the following statements are true:

1. *Additive Identity:* The matrix whose elements are all zero is the additive identity, and we denote this matrix by 0.

$$A + 0 = A = 0 + A$$

2. *Commutative:* $A + B = B + A$

3. *Associative:* $A + (B + C) = (A + B) + C$

4. *Additive Inverse:* The additive inverse of $A = [a_{ij}]$ is given by $-A = [-a_{ij}]$ and has the property that

$$A + (-A) = 0 = -A + A$$

5. *Subtraction of Matrices:* The difference of two matrices is defined to be

$$A - B = A + (-B)$$

Proof: The proof of each of these five properties is quite straightforward and follows directly from the definition of addition of matrices together with the corresponding property of real numbers. For example, to prove the first property, we can use the fact that 0 is the additive identity in the real number system. Thus we have

$$A + 0 = \begin{bmatrix} a_{11} & a_{12} & \cdots & a_{1n} \\ a_{21} & a_{22} & \cdots & a_{2n} \\ \vdots & \vdots & & \vdots \\ a_{m1} & a_{m2} & \cdots & a_{mn} \end{bmatrix} + \begin{bmatrix} 0 & 0 & \cdots & 0 \\ 0 & 0 & \cdots & 0 \\ \vdots & \vdots & & \vdots \\ 0 & 0 & & 0 \end{bmatrix}$$

$$= \begin{bmatrix} a_{11} + 0 & a_{12} + 0 & \cdots & a_{1n} + 0 \\ a_{21} + 0 & a_{22} + 0 & \cdots & a_{2n} + 0 \\ \vdots & \vdots & & \vdots \\ a_{m1} + 0 & a_{m2} + 0 & \cdots & a_{mn} + 0 \end{bmatrix}$$

$$= \begin{bmatrix} a_{11} & a_{12} & \cdots & a_{1n} \\ a_{21} & a_{22} & \cdots & a_{2n} \\ \cdot & \cdot & & \cdot \\ \cdot & \cdot & & \cdot \\ \cdot & \cdot & & \cdot \\ a_{m1} & a_{m2} & \cdots & a_{mn} \end{bmatrix} = A$$

EXAMPLE 1
Addition of Matrices

Perform the following operations:

(a) $\begin{bmatrix} 0 & 1 & -2 \\ 1 & 2 & 3 \end{bmatrix} + \begin{bmatrix} 1 & -1 & 3 \\ 4 & -3 & 0 \end{bmatrix}$

(b) $\begin{bmatrix} 0 & 0 \\ 0 & 0 \end{bmatrix} + \begin{bmatrix} 1 & 3 \\ -1 & 2 \end{bmatrix}$

(c) $\begin{bmatrix} 3 & 1 & 4 \\ 2 & -1 & 2 \\ 0 & 5 & 3 \end{bmatrix} - \begin{bmatrix} 1 & 4 & -2 \\ 2 & 1 & 3 \\ 2 & -4 & 1 \end{bmatrix}$

(d) $\begin{bmatrix} 1 & -2 \\ 3 & 0 \end{bmatrix} + \begin{bmatrix} -1 & 2 \\ -3 & 0 \end{bmatrix}$

Solution:

(a) $\begin{bmatrix} 0 & 1 & -2 \\ 1 & 2 & 3 \end{bmatrix} + \begin{bmatrix} 1 & -1 & 3 \\ 4 & -3 & 0 \end{bmatrix}$

$$= \begin{bmatrix} 0+1 & 1-1 & -2+3 \\ 1+4 & 2-3 & 3+0 \end{bmatrix} = \begin{bmatrix} 1 & 0 & 1 \\ 5 & -1 & 3 \end{bmatrix}$$

(b) $\begin{bmatrix} 0 & 0 \\ 0 & 0 \end{bmatrix} + \begin{bmatrix} 1 & 3 \\ -1 & 2 \end{bmatrix} = \begin{bmatrix} 0+1 & 0+3 \\ 0-1 & 0+2 \end{bmatrix} = \begin{bmatrix} 1 & 3 \\ -1 & 2 \end{bmatrix}$

(c) $\begin{bmatrix} 3 & 1 & 4 \\ 2 & -1 & 2 \\ 0 & 5 & 3 \end{bmatrix} - \begin{bmatrix} 1 & 4 & -2 \\ 2 & 1 & 3 \\ 2 & -4 & 1 \end{bmatrix}$

$$= \begin{bmatrix} 3-1 & 1-4 & 4+2 \\ 2-2 & -1-1 & 2-3 \\ 0-2 & 5+4 & 3-1 \end{bmatrix} = \begin{bmatrix} 2 & -3 & 6 \\ 0 & -2 & -1 \\ -2 & 9 & 2 \end{bmatrix}$$

(d) $\begin{bmatrix} 1 & -2 \\ 3 & 0 \end{bmatrix} + \begin{bmatrix} -1 & 2 \\ -3 & 0 \end{bmatrix} = \begin{bmatrix} 1-1 & -2+2 \\ 3-3 & 0+0 \end{bmatrix} = \begin{bmatrix} 0 & 0 \\ 0 & 0 \end{bmatrix}$

We refer to the multiplication of a matrix by a real number as **scalar multiplication,** to distinguish it from the multiplication of two matrices.

DEFINITION OF SCALAR MULTIPLICATION

If $A = [a_{ij}]$ is an $m \times n$ matrix and c is a real number, then the product cA is defined to be the $m \times n$ matrix given by

$$cA = [ca_{ij}] = \begin{bmatrix} ca_{11} & ca_{12} & ca_{13} & \cdots & ca_{1n} \\ ca_{21} & ca_{22} & ca_{23} & \cdots & ca_{2n} \\ ca_{31} & ca_{32} & ca_{33} & \cdots & ca_{3n} \\ \cdot & \cdot & \cdot & & \cdot \\ \cdot & \cdot & \cdot & & \cdot \\ \cdot & \cdot & \cdot & & \cdot \\ ca_{m1} & ca_{m2} & ca_{m3} & \cdots & ca_{mn} \end{bmatrix}$$

PROPERTIES OF SCALAR MULTIPLICATION

If $A = [a_{ij}]$ and $B = [b_{ij}]$ are $m \times n$ matrices and c and d are real numbers, then the following statements are true:

1. Left Distributive: $c(A + B) = cA + cB$
2. Right Distributive: $(c + d)A = cA + dA$
3. Associative: $(cd)A = c(dA)$

EXAMPLE 2
Scalar Multiplication

Given

$$A = \begin{bmatrix} 1 & 2 & 4 \\ -3 & 0 & -1 \\ 2 & 1 & 2 \end{bmatrix} \qquad B = \begin{bmatrix} 2 & 0 & 0 \\ 1 & -4 & 3 \\ -1 & 3 & 2 \end{bmatrix}$$

perform the following operations:

(a) $3A$
(b) $A - 2B$

Solution:

(a) $3A = 3 \begin{bmatrix} 1 & 2 & 4 \\ -3 & 0 & -1 \\ 2 & 1 & 2 \end{bmatrix} = \begin{bmatrix} 3(1) & 3(2) & 3(4) \\ 3(-3) & 3(0) & 3(-1) \\ 3(2) & 3(1) & 3(2) \end{bmatrix}$

$$= \begin{bmatrix} 3 & 6 & 12 \\ -9 & 0 & -3 \\ 6 & 3 & 6 \end{bmatrix}$$

(b) $A - 2B = \begin{bmatrix} 1 & 2 & 4 \\ -3 & 0 & -1 \\ 2 & 1 & 2 \end{bmatrix} - 2 \begin{bmatrix} 2 & 0 & 0 \\ 1 & -4 & 3 \\ -1 & 3 & 2 \end{bmatrix}$

$$= \begin{bmatrix} 1 & 2 & 4 \\ -3 & 0 & -1 \\ 2 & 1 & 2 \end{bmatrix} - \begin{bmatrix} 4 & 0 & 0 \\ 2 & -8 & 6 \\ -2 & 6 & 4 \end{bmatrix}$$

$$= \begin{bmatrix} 1-4 & 2-0 & 4-0 \\ -3-2 & 0+8 & -1-6 \\ 2+2 & 1-6 & 2-4 \end{bmatrix}$$

$$= \begin{bmatrix} -3 & 2 & 4 \\ -5 & 8 & -7 \\ 4 & -5 & -2 \end{bmatrix}$$

EXAMPLE 3
Solving a Matrix Equation

Solve the following equation for X:

$$3X + A = B$$

where $A = \begin{bmatrix} 1 & -2 \\ 0 & 3 \end{bmatrix}$ and $B = \begin{bmatrix} -3 & 4 \\ 2 & 1 \end{bmatrix}$.

Solution: We have

$$3X + A = B$$

$$X = \frac{1}{3}(B - A)$$

Now, using the given values of A and B, we have

$$X = \frac{1}{3}\left(\begin{bmatrix} -3 & 4 \\ 2 & 1 \end{bmatrix} - \begin{bmatrix} 1 & -2 \\ 0 & 3 \end{bmatrix} \right)$$

$$= \frac{1}{3} \begin{bmatrix} -4 & 6 \\ 2 & -2 \end{bmatrix} = \begin{bmatrix} -\frac{4}{3} & 2 \\ \frac{2}{3} & -\frac{2}{3} \end{bmatrix}$$

The third matrix operation we will introduce in this section is matrix multiplication. Although more complicated than addition or scalar multiplication, this operation is very important and has numerous practical applications.

DEFINITION OF MATRIX MULTIPLICATION

If $A = [a_{ij}]$ is an $m \times n$ matrix and $B = [b_{ij}]$ is an $n \times p$ matrix, then the product AB is an $m \times p$ matrix

$$AB = [c_{ij}]$$

where $c_{ij} = a_{i1}b_{1j} + a_{i2}b_{2j} + a_{i3}b_{3j} + \cdots + a_{in}b_{nj}$.

Remark: This definition says that the element in the ith row and jth column of the product is obtained by multiplying the elements in the ith row of A by the corresponding elements in the *j*th column of *B* and adding the results. The following diagram should make this clearer.

To obtain the element in *i*th row and *j*th column of the product, use the *i*th row of *A* and the *j*th row of *B*.

$$
\begin{bmatrix}
a_{11} & a_{12} & a_{13} & \cdots & a_{1n} \\
a_{21} & a_{22} & a_{23} & \cdots & a_{2n} \\
\cdot & \cdot & \cdot & & \cdot \\
\cdot & \cdot & \cdot & & \cdot \\
a_{i1} & a_{i2} & a_{i3} & \cdots & a_{in} \\
\cdot & \cdot & \cdot & & \cdot \\
\cdot & \cdot & \cdot & & \cdot \\
\cdot & \cdot & \cdot & & \cdot \\
a_{m1} & a_{m2} & a_{m3} & \cdots & a_{mn}
\end{bmatrix}
\begin{bmatrix}
b_{11} & b_{12} & \cdots & b_{1j} & \cdots & b_{1p} \\
b_{21} & b_{22} & \cdots & b_{2j} & \cdots & b_{2p} \\
b_{31} & b_{32} & \cdots & b_{3j} & \cdots & b_{3p} \\
\cdot & \cdot & & \cdot & & \cdot \\
\cdot & \cdot & & \cdot & & \cdot \\
b_{n1} & b_{n2} & \cdots & b_{nj} & \cdots & b_{np}
\end{bmatrix}
=
\begin{bmatrix}
c_{11} & c_{12} & \cdots & c_{1j} & \cdots & c_{1p} \\
c_{21} & c_{22} & \cdots & c_{2j} & \cdots & c_{2p} \\
\cdot & \cdot & & \cdot & & \cdot \\
\cdot & \cdot & & \cdot & & \cdot \\
c_{i1} & c_{i2} & \cdots & c_{ij} & \cdots & c_{ip} \\
\cdot & \cdot & & \cdot & & \cdot \\
\cdot & \cdot & & \cdot & & \cdot \\
c_{m1} & c_{m2} & \cdots & c_{mj} & \cdots & c_{mp}
\end{bmatrix}
$$

$$a_{i1}b_{1j} + a_{i2}b_{2j} + a_{i3}b_{3j} + \cdots + a_{in}b_{nj} = c_{ij}$$

Remark: In order for the product of two matrices to be defined, the number of columns of the first matrix must equal the number of rows of the second matrix.

$$
\begin{matrix}
A & B & = & C \\
m \times n & n \times p & & m \times p
\end{matrix}
$$

equal

order of *AB*

EXAMPLE 4
Finding the Product
of Two Matrices

Find the following products:

(a) $\begin{bmatrix} 1 & 2 \\ 0 & -1 \end{bmatrix} \begin{bmatrix} 3 & -1 \\ 2 & 1 \end{bmatrix}$ (b) $\begin{bmatrix} 1 & 0 & 3 \\ 2 & -1 & -2 \end{bmatrix} \begin{bmatrix} -2 & 4 & 2 \\ 1 & 0 & 0 \\ -1 & 1 & -1 \end{bmatrix}$

(c) $\begin{bmatrix} 3 & 4 \\ -2 & 5 \end{bmatrix} \begin{bmatrix} 1 & 0 \\ 0 & 1 \end{bmatrix}$ (d) $\begin{bmatrix} 1 & 2 \\ 1 & 1 \end{bmatrix} \begin{bmatrix} -1 & 2 \\ 1 & -1 \end{bmatrix}$

Solution:

(a) $\begin{bmatrix} 1 & 2 \\ 0 & -1 \end{bmatrix} \begin{bmatrix} 3 & -1 \\ 2 & 1 \end{bmatrix} = \begin{bmatrix} (1)(3) + (2)(2) & (1)(-1) + (2)(1) \\ (0)(3) + (-1)(2) & (0)(-1) + (-1)(1) \end{bmatrix}$

$= \begin{bmatrix} 7 & 1 \\ -2 & -1 \end{bmatrix}$

(b)

$$(1)(-2) + (0)(1) + (3)(-1) = -2 + 0 - 3$$

$$\begin{bmatrix} 1 & 0 & 3 \\ 2 & -1 & -2 \end{bmatrix} \begin{bmatrix} -2 & 4 & 2 \\ 1 & 0 & 0 \\ -1 & 1 & -1 \end{bmatrix}$$

$$= \begin{bmatrix} -2 + 0 - 3 & 4 + 0 + 3 & 2 + 0 - 3 \\ -4 - 1 + 2 & 8 + 0 - 2 & 4 + 0 + 2 \end{bmatrix}$$

$$= \begin{bmatrix} -5 & 7 & -1 \\ -3 & 6 & 6 \end{bmatrix}$$

(c) $\begin{bmatrix} 3 & 4 \\ -2 & 5 \end{bmatrix} \begin{bmatrix} 1 & 0 \\ 0 & 1 \end{bmatrix} = \begin{bmatrix} 3 + 0 & 0 + 4 \\ -2 + 0 & 0 + 5 \end{bmatrix} = \begin{bmatrix} 3 & 4 \\ -2 & 5 \end{bmatrix}$

(d) $\begin{bmatrix} 1 & 2 \\ 1 & 1 \end{bmatrix} \begin{bmatrix} -1 & 2 \\ 1 & -1 \end{bmatrix} = \begin{bmatrix} -1 + 2 & 2 - 2 \\ -1 + 1 & 2 - 1 \end{bmatrix} = \begin{bmatrix} 1 & 0 \\ 0 & 1 \end{bmatrix}$

In part (c) of Example 4, notice that the matrix with 1's on its main diagonal and zeros elsewhere acts as an identity with respect to matrix multiplication. This and several other properties of matrix multiplication are summarized as follows.

THEOREM: PROPERTIES OF MATRIX MULTIPLICATION

For each of the following properties, assume that the matrices A, B, and C are of the appropriate orders.

1. Associative: $A(BC) = (AB)C$
2. Distributive: $A(B + C) = AB + AC$
   $(A + B)C = AC + BC$
3. Identity Matrix of Order n:

$$I_n = \begin{bmatrix} 1 & 0 & 0 & \cdots & 0 \\ 0 & 1 & 0 & \cdots & 0 \\ 0 & 0 & 1 & \cdots & 0 \\ \cdot & \cdot & \cdot & & \cdot \\ \cdot & \cdot & \cdot & & \cdot \\ \cdot & \cdot & \cdot & & \cdot \\ 0 & 0 & 0 & & 1 \end{bmatrix}$$

If A is of order $m \times n$ and B is of order $n \times p$, then we have

$$AI_n = A \quad \text{and} \quad I_n B = B$$

Remark: Note that we did *not* list a commutative property for matrix multiplication, because the product of two matrices is rarely commutative.

EXAMPLE 5
Noncommutativity of Matrix Multiplication

Given the matrices

$$A = \begin{bmatrix} 1 & 3 \\ 2 & -1 \end{bmatrix}, \quad B = \begin{bmatrix} 2 & -1 \\ 0 & 2 \end{bmatrix},$$

$$C = \begin{bmatrix} 1 & 0 & 2 \\ 3 & -2 & 1 \end{bmatrix}, \quad D = \begin{bmatrix} -1 & 0 \\ 3 & 1 \\ 2 & 4 \end{bmatrix}$$

find the following products (if they exist):

(a) AB and BA (b) AC and CA (c) CD and DC

Solution:

(a) $AB = \begin{bmatrix} 1 & 3 \\ 2 & -1 \end{bmatrix}\begin{bmatrix} 2 & -1 \\ 0 & 2 \end{bmatrix} = \begin{bmatrix} 2 + 0 & -1 + 6 \\ 4 + 0 & -2 - 2 \end{bmatrix} = \begin{bmatrix} 2 & 5 \\ 4 & -4 \end{bmatrix}$

$BA = \begin{bmatrix} 2 & -1 \\ 0 & 2 \end{bmatrix}\begin{bmatrix} 1 & 3 \\ 2 & -1 \end{bmatrix} = \begin{bmatrix} 2 - 2 & 6 + 1 \\ 0 + 4 & 0 - 2 \end{bmatrix} = \begin{bmatrix} 0 & 7 \\ 4 & -2 \end{bmatrix}$

Note that $AB \neq BA$.

(b) $AC = \begin{bmatrix} 1 & 3 \\ 2 & -1 \end{bmatrix}\begin{bmatrix} 1 & 0 & 2 \\ 3 & -2 & 1 \end{bmatrix} = \begin{bmatrix} 1 + 9 & 0 - 6 & 2 + 3 \\ 2 - 3 & 0 + 2 & 4 - 1 \end{bmatrix}$

$$= \begin{bmatrix} 10 & -6 & 5 \\ -1 & 2 & 3 \end{bmatrix}$$

The product CA is *not defined,* since C is of order 2×3 and A is of order 2×2.

(c) $CD = \begin{bmatrix} 1 & 0 & 2 \\ 3 & -2 & 1 \end{bmatrix}\begin{bmatrix} -1 & 0 \\ 3 & 1 \\ 2 & 4 \end{bmatrix} = \begin{bmatrix} -1 + 0 + 4 & 0 + 0 + 8 \\ -3 - 6 + 2 & 0 - 2 + 4 \end{bmatrix}$

$$= \begin{bmatrix} 3 & 8 \\ -7 & 2 \end{bmatrix}$$

$DC = \begin{bmatrix} -1 & 0 \\ 3 & 1 \\ 2 & 4 \end{bmatrix}\begin{bmatrix} 1 & 0 & 2 \\ 3 & -2 & 1 \end{bmatrix} = \begin{bmatrix} -1 + & 0 & 0 + 0 & -2 + 0 \\ 3 + & 3 & 0 - 2 & 6 + 1 \\ 2 + & 12 & 0 - 8 & 4 + 4 \end{bmatrix}$

$$= \begin{bmatrix} -1 & 0 & -2 \\ 6 & -2 & 7 \\ 14 & -8 & 8 \end{bmatrix}$$

Remark: You should not conclude from Example 5 that the product of two matrices is never commutative. Occasionally, it can happen that the products AB and BA are the same. (For example, try reversing the order of the product in part (d) of Example 4.) What you should conclude from Example 5 is that we cannot presume commutativity of matrix multiplication and we must compute both orders to see if they are equal or not.

Section Exercises 8.2

In Exercises 1–6, find (a) $A + B$, (b) $A - B$, (c) $3A$, and (d) $3A - 2B$.

1. $A = \begin{bmatrix} 1 & -1 \\ 2 & -1 \end{bmatrix}$, $B = \begin{bmatrix} 2 & -1 \\ -1 & 8 \end{bmatrix}$

2. $A = \begin{bmatrix} 1 & 2 \\ 2 & 1 \end{bmatrix}$, $B = \begin{bmatrix} -3 & -2 \\ 4 & 2 \end{bmatrix}$

3. $A = \begin{bmatrix} 6 & -1 \\ 2 & 4 \\ -3 & 5 \end{bmatrix}$, $B = \begin{bmatrix} 1 & 4 \\ -1 & 5 \\ 1 & 10 \end{bmatrix}$

4. $A = \begin{bmatrix} 2 & 1 & 1 \\ -1 & -1 & 4 \end{bmatrix}$, $B = \begin{bmatrix} 2 & -3 & 4 \\ -3 & 1 & -2 \end{bmatrix}$

5. $A = \begin{bmatrix} 2 & 2 & -1 & 0 & 1 \\ 1 & 1 & -2 & 0 & -1 \end{bmatrix}$,
$B = \begin{bmatrix} 1 & 1 & -1 & 1 & 0 \\ -3 & 4 & 9 & -6 & -7 \end{bmatrix}$

6. $A = \begin{bmatrix} 3 \\ 2 \\ -1 \end{bmatrix}$, $B = \begin{bmatrix} -4 \\ 6 \\ 2 \end{bmatrix}$

In Exercises 7–12, find (a) AB and (b) BA.

7. $A = \begin{bmatrix} 1 & 2 \\ 4 & 2 \end{bmatrix}$, $B = \begin{bmatrix} 2 & -1 \\ -1 & 8 \end{bmatrix}$

8. $A = \begin{bmatrix} 2 & -1 \\ 1 & 4 \end{bmatrix}$, $B = \begin{bmatrix} 0 & 0 \\ 3 & -3 \end{bmatrix}$

9. $A = \begin{bmatrix} 3 & -1 \\ 1 & 3 \end{bmatrix}$, $B = \begin{bmatrix} 1 & -3 \\ 3 & 1 \end{bmatrix}$

10. $A = \begin{bmatrix} 1 & -1 \\ 1 & 1 \end{bmatrix}$, $B = \begin{bmatrix} 1 & 3 \\ -3 & 1 \end{bmatrix}$

11. $A = \begin{bmatrix} 1 & -1 & 7 \\ 2 & -1 & 8 \\ 3 & 1 & -1 \end{bmatrix}$, $B = \begin{bmatrix} 1 & 1 & 2 \\ 2 & 1 & 1 \\ 1 & -3 & 2 \end{bmatrix}$

12. $A = \begin{bmatrix} 3 & 2 & 1 \end{bmatrix}$, $B = \begin{bmatrix} 2 \\ 3 \\ 0 \end{bmatrix}$

In Exercises 13–20, find AB, if possible.

13. $A = \begin{bmatrix} 2 & 1 \\ -3 & 4 \\ 1 & 6 \end{bmatrix}$, $B = \begin{bmatrix} 0 & -1 & 0 \\ 4 & 0 & 2 \\ 8 & -1 & 7 \end{bmatrix}$

14. $A = \begin{bmatrix} 0 & -1 & 0 \\ 4 & 0 & 2 \\ 8 & -1 & 7 \end{bmatrix}$, $B = \begin{bmatrix} 2 & 1 \\ -3 & 4 \\ 1 & 6 \end{bmatrix}$

15. $A = \begin{bmatrix} -1 & 3 \\ 4 & -5 \\ 0 & 2 \end{bmatrix}$, $B = \begin{bmatrix} 1 & 2 \\ 0 & 7 \end{bmatrix}$

16. $A = \begin{bmatrix} 1 & 0 & 0 \\ 0 & 4 & 0 \\ 0 & 0 & -2 \end{bmatrix}$, $B = \begin{bmatrix} 3 & 0 & 0 \\ 0 & -1 & 0 \\ 0 & 0 & 5 \end{bmatrix}$

17. $A = \begin{bmatrix} 5 & 0 & 0 \\ 0 & -8 & 0 \\ 0 & 0 & 7 \end{bmatrix}$, $B = \begin{bmatrix} \frac{1}{5} & 0 & 0 \\ 0 & -\frac{1}{8} & 0 \\ 0 & 0 & \frac{1}{2} \end{bmatrix}$

18. $A = \begin{bmatrix} 0 & 0 & 5 \\ 0 & 0 & -3 \\ 0 & 0 & 4 \end{bmatrix}$, $B = \begin{bmatrix} 6 & -11 & 4 \\ 8 & 16 & 4 \\ 0 & 0 & 0 \end{bmatrix}$

19. $A = \begin{bmatrix} 6 \\ -2 \\ 1 \\ 6 \end{bmatrix}$, $B = \begin{bmatrix} 10 & 12 \end{bmatrix}$

20. $A = \begin{bmatrix} 1 & 0 & 3 & -2 & 4 \\ 6 & 13 & 8 & -17 & 20 \end{bmatrix}$, $B = \begin{bmatrix} 1 & 6 \\ 4 & 2 \end{bmatrix}$

If

$$f(x) = a_0 + a_1 x + a_2 x^2 + \cdots + a_n x^n$$

then for an $n \times n$ matrix A, we define $f(A)$ to be

$$f(A) = a_0 I_n + a_1 A + a_2 A^2 + \cdots + a_n A^n$$

In Exercises 21–24, find $f(A)$.

21. $f(x) = x^2 - 5x + 2$, $A = \begin{bmatrix} 2 & 0 \\ 4 & 5 \end{bmatrix}$

22. $f(x) = x^2 - 7x + 6$, $A = \begin{bmatrix} 5 & 4 \\ 1 & 2 \end{bmatrix}$

23. $f(x) = x^3 - 10x^2 + 31x - 30$, $A = \begin{bmatrix} 3 & 1 & 4 \\ 0 & 2 & 6 \\ 0 & 0 & 5 \end{bmatrix}$

24. $f(x) = x^2 - 10x + 24$, $A = \begin{bmatrix} 8 & -4 \\ 2 & 2 \end{bmatrix}$

25. In matrix multiplication, if $AC = BC$, then A is *not* necessarily equal to B. Illustrate this using the matrices

$$A = \begin{bmatrix} 1 & 2 & 3 \\ 0 & 5 & 4 \\ 3 & -2 & 1 \end{bmatrix}, \quad B = \begin{bmatrix} 4 & -6 & 3 \\ 5 & 4 & 4 \\ -1 & 0 & 1 \end{bmatrix}$$

and

$$C = \begin{bmatrix} 0 & 0 & 0 \\ 0 & 0 & 0 \\ 4 & -2 & 3 \end{bmatrix}$$

26. In matrix multiplication, if $AB = 0$, then it is *not* necessarily true that $A = 0$ or $B = 0$. Illustrate this using the matrices

$$A = \begin{bmatrix} 3 & 3 \\ 4 & 4 \end{bmatrix} \quad \text{and} \quad B = \begin{bmatrix} 1 & -1 \\ -1 & 1 \end{bmatrix}$$

27. Explain why the following nonequalities are valid for matrices:

$$(A + B)(A - B) \neq A^2 - B^2$$

and

$$(A + B)(A + B) \neq A^2 + 2AB + B^2$$

28. A certain corporation has four factories, each of which produces two products. The number of units of product i produced at factory j in one day is represented by a_{ij} in the matrix

$$A = \begin{bmatrix} 100 & 90 & 70 & 30 \\ 40 & 20 & 60 & 60 \end{bmatrix}$$

Use scalar multiplication (multiply by 1.10) to give the production levels if production is increased by 10%.

29. A fruit grower raises two crops, which he ships to three outlets. The number of units of product i that are shipped to outlet j is represented by a_{ij} in the matrix

$$A = \begin{bmatrix} 100 & 75 & 75 \\ 125 & 150 & 100 \end{bmatrix}$$

The profit on one unit of product i is represented by b_{1i} in the matrix

$$B = [\$3.75 \quad \$7.00]$$

Find the product BA, and state what each entry of the product represents.

30. The matrix

$$P = \begin{array}{c} \text{From } R \\ \text{From } D \\ \text{From } I \end{array} \begin{bmatrix} \overset{\text{To } R}{0.75} & \overset{\text{To } D}{0.15} & \overset{\text{To } I}{0.10} \\ 0.20 & 0.60 & 0.20 \\ 0.30 & 0.40 & 0.30 \end{bmatrix}$$

is called a stochastic matrix. Each p_{ij} ($i \neq j$) represents the proportion of the voting population that changes from party i to party j, and p_{ii} represents the proportion that remains loyal to the party from one election to the next. Find P^2. (This matrix gives the transition probabilities from the first election to the third.)

The Inverse of a Matrix   8.3

There are many similarities between the algebra of real numbers and the algebra of matrices. For example, compare the following solutions:

Real Numbers **(Solve for x)**	**$m \times n$ Matrices** **(Solve for X)**
$x + a = b$	$X + A = B$
$x + a + (-a) = b + (-a)$	$X + A + (-A) = B + (-A)$
$x + 0 = b - a$	$X + 0 = B - A$
$x = b - a$	$X = B - A$

The solution of equations involving multiplication is a bit more complicated. For example, we can solve the real number equation $ax = b$ for x and obtain $x = b/a = a^{-1}b$ only if $a \neq 0$. Here, a^{-1} is the *multiplicative inverse of a*, with the property that $(a^{-1})a = 1$. We use a similar definition for the multiplicative inverse of a matrix.

DEFINITION OF THE INVERSE OF A MATRIX

If A is a square matrix of order n, then A^{-1} is called the **inverse of A** if it has the property that

$$A(A^{-1}) = (A^{-1})A = I_n$$

where I_n is the identity matrix of order n.

Remark: The symbol A^{-1} is read "A inverse." We do not use the reciprocal notation $1/A$ because matrix division is not defined.

EXAMPLE 1
The Inverse of a Matrix

Show that B is the inverse of A.

(a) $A = \begin{bmatrix} -1 & 2 \\ -1 & 1 \end{bmatrix}$, $B = \begin{bmatrix} 1 & -2 \\ 1 & -1 \end{bmatrix}$

(b) $A = \begin{bmatrix} 2 & 3 & 1 \\ 3 & 3 & 1 \\ 2 & 4 & 1 \end{bmatrix}$, $B = \begin{bmatrix} -1 & 1 & 0 \\ -1 & 0 & 1 \\ 6 & -2 & -3 \end{bmatrix}$

Solution:

(a) $AB = \begin{bmatrix} -1 & 2 \\ -1 & 1 \end{bmatrix}\begin{bmatrix} 1 & -2 \\ 1 & -1 \end{bmatrix} = \begin{bmatrix} -1+2 & 2-2 \\ -1+1 & 2-1 \end{bmatrix} = \begin{bmatrix} 1 & 0 \\ 0 & 1 \end{bmatrix}$

$BA = \begin{bmatrix} 1 & -2 \\ 1 & -1 \end{bmatrix}\begin{bmatrix} -1 & 2 \\ -1 & 1 \end{bmatrix} = \begin{bmatrix} -1+2 & 2-2 \\ -1+1 & 2-1 \end{bmatrix} = \begin{bmatrix} 1 & 0 \\ 0 & 1 \end{bmatrix}$

(b) $AB = \begin{bmatrix} 2 & 3 & 1 \\ 3 & 3 & 1 \\ 2 & 4 & 1 \end{bmatrix}\begin{bmatrix} -1 & 1 & 0 \\ -1 & 0 & 1 \\ 6 & -2 & -3 \end{bmatrix} = \begin{bmatrix} 1 & 0 & 0 \\ 0 & 1 & 0 \\ 0 & 0 & 1 \end{bmatrix}$

$BA = \begin{bmatrix} -1 & 1 & 0 \\ -1 & 0 & 1 \\ 6 & -2 & -3 \end{bmatrix}\begin{bmatrix} 2 & 3 & 1 \\ 3 & 3 & 1 \\ 2 & 4 & 1 \end{bmatrix} = \begin{bmatrix} 1 & 0 & 0 \\ 0 & 1 & 0 \\ 0 & 0 & 1 \end{bmatrix}$

Both matrices in Example 1 are square. If a matrix is not square, then it does not possess an inverse. But not all square matrices possess inverses. If a matrix does have an inverse, then we say that it is **invertible** (or **nonsingular**). A square matrix that has no inverse is said to be **singular.** We can use a system of equations to determine whether a matrix has an inverse.

EXAMPLE 2
A Singular Matrix

Show that the following matrix has no inverse:

$$A = \begin{bmatrix} 1 & -2 \\ -2 & 4 \end{bmatrix}$$

Solution:
To find an inverse for A, we try to solve the matrix equation

$$AB = I$$

$$\begin{bmatrix} 1 & -2 \\ -2 & 4 \end{bmatrix} \begin{bmatrix} a & b \\ c & d \end{bmatrix} = \begin{bmatrix} 1 & 0 \\ 0 & 1 \end{bmatrix}$$

Through matrix multiplication, this equation results in two systems of linear equations:

$$a - 2c = 1 \qquad b - 2d = 0$$
$$-2a + 4c = 0 \qquad -2b + 4d = 1$$

Neither of these systems has a solution. (Check this to convince yourself that this is true.) Thus, A has no inverse.

Example 2 shows us one way to find the inverse of a matrix if it exists. However, this procedure is somewhat cumbersome, and we recommend the following alternative procedure.

A PROCEDURE FOR FINDING THE INVERSE OF A MATRIX

If A is a square matrix of order n, we can find its inverse as follows:

1. Write the $n \times 2n$ matrix that consists of the given matrix A on the left augmented with the identity matrix I_n.

$$[A : I_n] = \begin{bmatrix} a_{11} & a_{12} & a_{13} & \cdots & a_{1n} & 1 & 0 & 0 & \cdots & 0 \\ a_{21} & a_{22} & a_{23} & \cdots & a_{2n} & 0 & 1 & 0 & \cdots & 0 \\ a_{31} & a_{32} & a_{33} & \cdots & a_{3n} & 0 & 0 & 1 & \cdots & 0 \\ \cdot & \cdot & \cdot & & \cdot & \cdot & \cdot & \cdot & & \cdot \\ \cdot & \cdot & \cdot & & \cdot & \cdot & \cdot & \cdot & & \cdot \\ \cdot & \cdot & \cdot & & \cdot & \cdot & \cdot & \cdot & & \cdot \\ a_{n1} & a_{n2} & a_{n3} & \cdots & a_{nn} & 0 & 0 & 0 & \cdots & 1 \end{bmatrix}$$

2. Transform A to I_n using elementary row operations *on the entire matrix* $[A : I_n]$. The result will be a matrix of the form

$$[I_n : B]$$

The matrix B is the inverse of A. That is, $B = A^{-1}$.

3. Check your work by multiplying to see that

$$A(A^{-1}) = I_n = (A^{-1})A$$

EXAMPLE 3
Finding the Inverse of a Matrix

Find the inverse of the following matrix:

$$A = \begin{bmatrix} 2 & 3 & 1 \\ 3 & 3 & 1 \\ 2 & 4 & 1 \end{bmatrix}$$

Solution:

We begin by adjoining the identity matrix, I_3, as follows:

$$\begin{bmatrix} 2 & 3 & 1 & 1 & 0 & 0 \\ 3 & 3 & 1 & 0 & 1 & 0 \\ 2 & 4 & 1 & 0 & 0 & 1 \end{bmatrix}$$

Now, using elementary row operations, we attempt to rewrite this matrix in the form

$$\begin{bmatrix} 1 & 0 & 0 & b_{11} & b_{12} & b_{13} \\ 0 & 1 & 0 & b_{21} & b_{22} & b_{23} \\ 0 & 0 & 1 & b_{31} & b_{32} & b_{33} \end{bmatrix}$$

We have

$$\begin{bmatrix} 2 & 3 & 1 & 1 & 0 & 0 \\ 3 & 3 & 1 & 0 & 1 & 0 \\ 2 & 4 & 1 & 0 & 0 & 1 \end{bmatrix} \begin{smallmatrix} R_2 \\ R_1 \end{smallmatrix} \begin{bmatrix} 3 & 3 & 1 & 0 & 1 & 0 \\ 2 & 3 & 1 & 1 & 0 & 0 \\ 2 & 4 & 1 & 0 & 0 & 1 \end{bmatrix}$$

$$\begin{aligned} -R_2 + R_1 \\ \\ -R_2 + R_3 \end{aligned} \begin{bmatrix} 1 & 0 & 0 & -1 & 1 & 0 \\ 2 & 3 & 1 & 1 & 0 & 0 \\ 0 & 1 & 0 & -1 & 0 & 1 \end{bmatrix}$$

$$\begin{smallmatrix} R_3 \\ R_2 \end{smallmatrix} \begin{bmatrix} 1 & 0 & 0 & -1 & 1 & 0 \\ 0 & 1 & 0 & -1 & 0 & 1 \\ 2 & 3 & 1 & 1 & 0 & 0 \end{bmatrix}$$

$$-2R_1 + R_3 \begin{bmatrix} 1 & 0 & 0 & -1 & 1 & 0 \\ 0 & 1 & 0 & -1 & 0 & 1 \\ 0 & 3 & 1 & 3 & -2 & 0 \end{bmatrix}$$

$$-3R_2 + R_3 \begin{bmatrix} 1 & 0 & 0 & -1 & 1 & 0 \\ 0 & 1 & 0 & -1 & 0 & 1 \\ 0 & 0 & 1 & 6 & -2 & -3 \end{bmatrix}$$

Thus, the inverse of A is given by

$$A^{-1} = B = \begin{bmatrix} -1 & 1 & 0 \\ -1 & 0 & 1 \\ 6 & -2 & -3 \end{bmatrix}$$

In part (b) of Example 1, we already verified that this is the correct inverse of A.

The process shown in Example 3 is general in the sense that it can be used both to determine if a matrix has an inverse and to find the inverse if it exists. For example, if we applied this process to the singular matrix in Example 2, we would obtain a row (in that part of the matrix occupied by A) with all zero elements:

$$[A : I] = \begin{bmatrix} 1 & -2 & 1 & 0 \\ -2 & 4 & 0 & 1 \end{bmatrix} \quad 2R_2 + R_1 \quad \begin{bmatrix} 1 & -2 & 1 & 0 \\ 0 & 0 & 2 & 1 \end{bmatrix}$$

This means that the matrix has no inverse. In general, when reducing the matrix $[A : I_n]$, we know A is not invertible if we arrive at a point at which all possible candidates for the next main diagonal element are zero.

One of the many practical applications involving the inverse of a matrix is solving a system of linear equations. Note how the system

$$2x + 3y + z = -1$$
$$3x + 3y + z = 1$$
$$2x + 4y + z = -2$$

can be written as a matrix equation using matrix multiplication.

$$\underset{A}{\begin{bmatrix} 2 & 3 & 1 \\ 3 & 3 & 1 \\ 2 & 4 & 1 \end{bmatrix}} \underset{X}{\begin{bmatrix} x \\ y \\ z \end{bmatrix}} = \underset{B}{\begin{bmatrix} -1 \\ 1 \\ -2 \end{bmatrix}}$$

To solve this system, we need A^{-1} to obtain

$$X = (A^{-1})B$$

For a *single* system of equations, this procedure would not be worth the effort. However, we often encounter several systems in which each system has the same coefficients but the right-hand constants differ. In this situation, it is convenient to find the inverse of the coefficient matrix and use this inverse to solve each system. This procedure is demonstrated in the next example.

EXAMPLE 4
Solving a System of Equations Using an Inverse

Use an inverse matrix to solve the following systems:

(a) $2x + 3y + z = -1$
    $3x + 3y + z = 1$
    $2x + 4y + z = -2$

(b) $2x + 3y + z = 4$
    $3x + 3y + z = 8$
    $2x + 4y + z = 5$

(c) $2x + 3y + z = 2$
    $3x + 3y + z = -1$
    $2x + 4y + z = 4$

Solution:
The coefficient matrix for each system is the same:

$$A = \begin{bmatrix} 2 & 3 & 1 \\ 3 & 3 & 1 \\ 2 & 4 & 1 \end{bmatrix}$$

From Example 3, we know that the inverse of this matrix is

$$A^{-1} = \begin{bmatrix} -1 & 1 & 0 \\ -1 & 0 & 1 \\ 6 & -2 & -3 \end{bmatrix}$$

(a) We let B be the following column matrix:

$$B = \begin{bmatrix} -1 \\ 1 \\ -2 \end{bmatrix}$$

Now, the solution to the matrix equation $AX = B$ is

$$X = (A^{-1})B = \begin{bmatrix} -1 & 1 & 0 \\ -1 & 0 & 1 \\ 6 & -2 & -3 \end{bmatrix} \begin{bmatrix} -1 \\ 1 \\ -2 \end{bmatrix}$$

$$= \begin{bmatrix} 1 + 1 + 0 \\ 1 + 0 - 2 \\ -6 - 2 + 6 \end{bmatrix} = \begin{bmatrix} 2 \\ -1 \\ -2 \end{bmatrix}$$

Thus, the solution to the system of linear equations is

$$x = 2, \quad y = -1, \quad \text{and} \quad z = -2$$

(b) In this case,

$$B = \begin{bmatrix} 4 \\ 8 \\ 5 \end{bmatrix}$$

Therefore, the solution to the matrix equation $AX = B$ is

$$X = (A^{-1})B = \begin{bmatrix} -1 & 1 & 0 \\ -1 & 0 & 1 \\ 6 & -2 & -3 \end{bmatrix} \begin{bmatrix} 4 \\ 8 \\ 5 \end{bmatrix}$$

$$= \begin{bmatrix} -4 + 8 + 0 \\ -4 + 0 + 5 \\ 24 - 16 - 15 \end{bmatrix} = \begin{bmatrix} 4 \\ 1 \\ -7 \end{bmatrix}$$

Thus, the solution to the system of linear equations is

$$x = 4, \quad y = 1, \quad \text{and} \quad z = -7$$

(c) Here we have

$$B = \begin{bmatrix} 2 \\ -1 \\ 4 \end{bmatrix}$$

Therefore, the solution to the matrix equation $AX = B$ is

$$X = (A^{-1})B = \begin{bmatrix} -1 & 1 & 0 \\ -1 & 0 & 1 \\ 6 & -2 & -3 \end{bmatrix} \begin{bmatrix} 2 \\ -1 \\ 4 \end{bmatrix}$$

$$X = (A^{-1})B = \begin{bmatrix} -2 - 1 + 0 \\ -2 + 0 + 4 \\ 12 + 2 - 12 \end{bmatrix} = \begin{bmatrix} -3 \\ 2 \\ 2 \end{bmatrix}$$

Thus, the solution to the system of linear equations is

$$x = -3, \qquad y = 2, \qquad \text{and} \qquad z = 2$$

Section Exercises 8.3

In Exercises 1–5, show that B is the inverse of A.

1. $A = \begin{bmatrix} 1 & 2 \\ 3 & 4 \end{bmatrix}$, $B = \begin{bmatrix} -2 & 1 \\ \frac{3}{2} & -\frac{1}{2} \end{bmatrix}$

2. $A = \begin{bmatrix} 1 & -1 \\ 2 & 3 \end{bmatrix}$, $B = \begin{bmatrix} \frac{3}{5} & \frac{1}{5} \\ -\frac{2}{5} & \frac{1}{5} \end{bmatrix}$

3. $A = \begin{bmatrix} -2 & 2 & 3 \\ 1 & -1 & 0 \\ 0 & 1 & 4 \end{bmatrix}$, $B = \frac{1}{3}\begin{bmatrix} -4 & -5 & 3 \\ -4 & -8 & 3 \\ 1 & 2 & 0 \end{bmatrix}$

4. $A = \begin{bmatrix} 2 & -17 & 11 \\ -1 & 11 & -7 \\ 0 & 3 & -2 \end{bmatrix}$, $B = \begin{bmatrix} 1 & 1 & 2 \\ 2 & 4 & -3 \\ 3 & 6 & -5 \end{bmatrix}$

5. $A = \begin{bmatrix} 1 & 2 & 0 & 0 \\ 2 & 1 & 2 & 0 \\ 0 & 2 & 1 & 2 \\ 0 & 0 & 2 & 1 \end{bmatrix}$,

$B = \frac{1}{5}\begin{bmatrix} -7 & 6 & 4 & -8 \\ 6 & -3 & -2 & 4 \\ 4 & -2 & -3 & 6 \\ -8 & 4 & 6 & -7 \end{bmatrix}$

In Exercises 6–30, find the inverse of the matrix (if it exists).

6. $\begin{bmatrix} 1 & 2 \\ 3 & 7 \end{bmatrix}$

7. $\begin{bmatrix} 1 & -2 \\ 2 & -3 \end{bmatrix}$

8. $\begin{bmatrix} -7 & 33 \\ 4 & -19 \end{bmatrix}$

9. $\begin{bmatrix} -1 & 1 \\ -2 & 1 \end{bmatrix}$

10. $\begin{bmatrix} 11 & 1 \\ -1 & 0 \end{bmatrix}$

11. $\begin{bmatrix} 2 & 4 \\ 4 & 8 \end{bmatrix}$

12. $\begin{bmatrix} 2 & 3 \\ 1 & 4 \end{bmatrix}$

13. $\begin{bmatrix} 2 & 7 \\ -3 & -9 \end{bmatrix}$

14. $\begin{bmatrix} -2 & 5 \\ 6 & -15 \end{bmatrix}$

15. $\begin{bmatrix} 1 & 1 & 1 \\ 3 & 5 & 4 \\ 3 & 6 & 5 \end{bmatrix}$

16. $\begin{bmatrix} 1 & 2 & 2 \\ 3 & 7 & 9 \\ -1 & -4 & -7 \end{bmatrix}$

17. $\begin{bmatrix} 1 & 2 & -1 \\ 3 & 7 & -10 \\ -5 & -7 & -15 \end{bmatrix}$

18. $\begin{bmatrix} 10 & 5 & -7 \\ -5 & 1 & 4 \\ 3 & 2 & -2 \end{bmatrix}$

19. $\begin{bmatrix} 1 & -2 & -1 & -2 \\ 3 & -5 & -2 & -3 \\ 2 & -5 & -2 & -5 \\ -1 & 4 & 4 & 11 \end{bmatrix}$

20. $\begin{bmatrix} 4 & 8 & -7 & 14 \\ 2 & 5 & -4 & 6 \\ 0 & 2 & 1 & -7 \\ 3 & 6 & -5 & 10 \end{bmatrix}$

21. $\begin{bmatrix} 1 & 1 & 2 \\ 3 & 1 & 0 \\ -2 & 0 & 3 \end{bmatrix}$

22. $\begin{bmatrix} 3 & 2 & 2 \\ 2 & 2 & 2 \\ -4 & 4 & 3 \end{bmatrix}$

23. $\begin{bmatrix} 0.1 & 0.2 & 0.3 \\ -0.3 & 0.2 & 0.2 \\ 0.5 & 0.4 & 0.4 \end{bmatrix}$

24. $\begin{bmatrix} 2 & 0 & 0 \\ 0 & 3 & 0 \\ 0 & 0 & 5 \end{bmatrix}$

25. $\begin{bmatrix} -8 & 0 & 0 & 0 \\ 0 & 1 & 0 & 0 \\ 0 & 0 & 4 & 0 \\ 0 & 0 & 0 & -5 \end{bmatrix}$

26. $\begin{bmatrix} 1 & 3 & -2 & 0 \\ 0 & 2 & 4 & 6 \\ 0 & 0 & -2 & 1 \\ 0 & 0 & 0 & 5 \end{bmatrix}$

27. $\begin{bmatrix} 1 & 0 & 0 \\ 3 & 4 & 0 \\ 2 & 5 & 5 \end{bmatrix}$

28. $\begin{bmatrix} 1 & 0 & 0 \\ 3 & 0 & 0 \\ 2 & 5 & 5 \end{bmatrix}$

29. $\begin{bmatrix} 1 & 0 & 3 & 0 \\ 0 & 2 & 0 & 4 \\ 1 & 0 & 3 & 0 \\ 0 & 2 & 0 & 4 \end{bmatrix}$

30. $\begin{bmatrix} 1/a & 0 \\ 0 & a \end{bmatrix}$

31. Use an inverse matrix to solve the following systems (see Exercise 9):

(a) $-x + y = 4$
     $-2x + y = 0$

(b) $-x + y = -3$
     $-2x + y = 5$

(c) $-x + y = 0$
      $-2x + y = 7$

32. Use an inverse matrix to solve the following systems (see Exercise 12):

(a) $2x + 3y = 5$ (b) $2x + 3y = 0$
    $x + 4y = 10$ $x + 4y = 3$

(c) $2x + 3y = 1$
    $x + 4y = -2$

33. Use an inverse matrix to solve the following systems (see Exercise 22):

(a) $3x + 2y + 2z = 0$
      $2x + 2y + 2z = 5$
      $-4x + 4y + 3z = 2$

(b) $3x + 2y + 2z = -1$
      $2x + 2y + 2z = 2$
      $-4x + 4y + 3z = 0$

(c) $3x + 2y + 2z = 0$
      $2x + 2y + 2z = 0$
      $-4x + 4y + 3z = 1$

34. Use an inverse matrix to solve the following systems (see Exercise 19):

(a) $x_1 - 2x_2 - x_3 - 2x_4 = 0$
      $3x_1 - 5x_2 - 2x_3 - 3x_4 = 1$
      $2x_1 - 5x_2 - 2x_3 - 5x_4 = -1$
      $-x_1 + 4x_2 + 4x_3 + 11x_4 = 2$

(b) $x_1 - 2x_2 - x_3 - 2x_4 = 1$
      $3x_1 - 5x_2 - 2x_3 - 3x_4 = -2$
      $2x_1 - 5x_2 - 2x_3 - 5x_4 = 0$
      $-x_1 + 4x_2 + 4x_3 + 11x_4 = -3$

Determinants   8.4

With every *square* matrix, we can associate a real number called its **determinant.** Determinants have many uses, and we will encounter a few of these in the next two sections. In this section we concentrate on procedures for computing (or evaluating) the determinant of a matrix. From this point on, we assume that all matrices under discussion are square matrices.

THE DETERMINANT OF A MATRIX OF ORDER 2

The **determinant** of the matrix

$$A = \begin{bmatrix} a_{11} & a_{12} \\ a_{21} & a_{22} \end{bmatrix}$$

is given by

$$|A| = \begin{vmatrix} a_{11} & a_{12} \\ a_{21} & a_{22} \end{vmatrix} = a_{11}a_{22} - a_{12}a_{21}$$

Remark: A convenient method for remembering this formula is shown in the following diagram. Note that the determinant is given by the difference of the products of the two diagonals of the matrix.

$$|A| = \begin{vmatrix} a_{11} & a_{12} \\ a_{21} & a_{22} \end{vmatrix} = a_{11}a_{22} - a_{12}a_{21}$$

EXAMPLE 1
The Determinant
of a Matrix of Order 2

Find the determinants of the following matrices:

(a) $A = \begin{bmatrix} 2 & -3 \\ 1 & 2 \end{bmatrix}$ (b) $B = \begin{bmatrix} 2 & 1 \\ 4 & 2 \end{bmatrix}$ (c) $C = \begin{bmatrix} 0 & 3 \\ 2 & 4 \end{bmatrix}$

Solution:

(a) $|A| = \begin{vmatrix} 2 & -3 \\ 1 & 2 \end{vmatrix} = 2(2) - 1(-3) = 4 + 3 = 7$

(b) $|B| = \begin{vmatrix} 2 & 1 \\ 4 & 2 \end{vmatrix} = 2(2) - 4(1) = 4 - 4 = 0$

(c) $|C| = \begin{vmatrix} 0 & 3 \\ 2 & 4 \end{vmatrix} = 0(4) - 2(3) = 0 - 6 = -6$

The determinant of a 3×3 matrix can be found as follows.

THE DETERMINANT OF A MATRIX OF ORDER 3

The determinant of the matrix

$$A = \begin{bmatrix} a_{11} & a_{12} & a_{13} \\ a_{21} & a_{22} & a_{23} \\ a_{31} & a_{32} & a_{33} \end{bmatrix}$$

is given by

$$|A| = \begin{vmatrix} a_{11} & a_{12} & a_{13} \\ a_{21} & a_{22} & a_{23} \\ a_{31} & a_{32} & a_{33} \end{vmatrix} = a_{11}a_{22}a_{33} + a_{12}a_{23}a_{31} + a_{13}a_{21}a_{32} - a_{31}a_{22}a_{13} - a_{32}a_{23}a_{11} - a_{33}a_{21}a_{12}$$

The formula for the determinant of a 3×3 matrix would be difficult to memorize in terms of subscripts. A much simpler way to remember this formula is to recopy the first and second columns of the matrix (forming columns 4 and 5) and use products of diagonals, as shown in the following diagram.

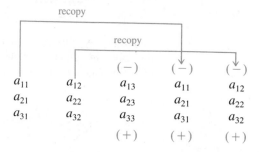

EXAMPLE 2
The Determinant
of a Matrix of Order 3

Find the determinant of the following matrix:

$$A = \begin{bmatrix} 0 & 2 & 1 \\ 3 & -1 & 2 \\ 4 & 0 & 1 \end{bmatrix}$$

Solution:
We begin by recopying the first two columns and computing the six "diagonal products," as follows:

$$-4(-1)(1) - 0(2)(0) - 1(3)(2) \qquad \text{Negative diagonals}$$

$$\begin{vmatrix} 0 & 2 & 1 & 0 & 2 \\ 3 & -1 & 2 & 3 & -1 \\ 4 & 0 & 1 & 4 & 0 \end{vmatrix} = 0 + 16 + 0 - (-4) - 0 - 6$$

$$= 16 + 4 - 6 = 14$$

$$+0(-1)(1) + 2(2)(4) + 1(3)(0) \qquad \text{Positive diagonals}$$

The diagonal product diagram used in Examples 1 and 2 is quite useful for finding the determinant for matrices of order 2 or 3. However, this scheme does not generalize to matrices of higher order. A more general method for finding determinants of matrices involves the concepts of **minors** and **cofactors**.

DEFINITION OF THE MINORS AND COFACTORS OF A MATRIX

$$M_{21} = \begin{bmatrix} a_{11} & a_{12} & a_{13} \\ a_{21} & a_{22} & a_{23} \\ a_{31} & a_{32} & a_{33} \end{bmatrix} = \begin{bmatrix} a_{12} & a_{13} \\ a_{32} & a_{33} \end{bmatrix}$$

FIGURE 8.2
Minor of a_{21}

If A is a square matrix (of order 3 or greater), then the **minor** M_{ij} of the element a_{ij} is the determinant of the matrix obtained by deleting the ith row and jth column of A. (See Figure 8.2.)

The **cofactor** C_{ij} is given by

$$C_{ij} = (-1)^{i+j} M_{ij}$$

Remark: Note that the minors and cofactors of a matrix differ at most in sign. To obtain the cofactors of a matrix, first find the minors and then apply the following "checkerboard" pattern of $+$'s and $-$'s.

Sign Pattern for Cofactors

3 × 3 Matrix

$$\begin{bmatrix} + & - & + \\ - & + & - \\ + & - & + \end{bmatrix}$$

4 × 4 Matrix

$$\begin{bmatrix} + & - & + & - \\ - & + & - & + \\ + & - & + & - \\ - & + & - & + \end{bmatrix}$$

Note that *odd* positions of a_{ij} (where $i + j$ is odd) have negative signs, and *even* positions of a_{ij} (where $i + j$ is even) have positive signs.

EXAMPLE 3
Finding the Minors and Cofactors of a Matrix

Find all the minors and cofactors of the following matrix:

$$A = \begin{bmatrix} 0 & 2 & 1 \\ 3 & -1 & 2 \\ 4 & 0 & 1 \end{bmatrix}$$

Solution:

To find the minor M_{ij}, we delete the *i*th row and *j*th column and evaluate the resulting determinant. To find M_{11}:

$$M_{11} = \begin{vmatrix} 0 & 2 & 1 \\ 3 & -1 & 2 \\ 4 & 0 & 1 \end{vmatrix} = \begin{vmatrix} -1 & 2 \\ 0 & 1 \end{vmatrix} = -1(1) - 0(2) = -1$$

To find M_{12}:

$$M_{12} = \begin{vmatrix} 0 & 2 & 1 \\ 3 & -1 & 2 \\ 4 & 0 & 1 \end{vmatrix} = \begin{vmatrix} 3 & 2 \\ 4 & 1 \end{vmatrix} = 3(1) - 4(2) = -5$$

By continuing this pattern, we can obtain the following *matrix of minors:*

$$\begin{bmatrix} M_{11} & M_{12} & M_{13} \\ M_{21} & M_{22} & M_{23} \\ M_{31} & M_{32} & M_{33} \end{bmatrix} = \begin{bmatrix} \begin{vmatrix} -1 & 2 \\ 0 & 1 \end{vmatrix} & \begin{vmatrix} 3 & 2 \\ 4 & 1 \end{vmatrix} & \begin{vmatrix} 3 & -1 \\ 4 & 0 \end{vmatrix} \\ \begin{vmatrix} 2 & 1 \\ 0 & 1 \end{vmatrix} & \begin{vmatrix} 0 & 1 \\ 4 & 1 \end{vmatrix} & \begin{vmatrix} 0 & 2 \\ 4 & 0 \end{vmatrix} \\ \begin{vmatrix} 2 & 1 \\ -1 & 2 \end{vmatrix} & \begin{vmatrix} 0 & 1 \\ 3 & 2 \end{vmatrix} & \begin{vmatrix} 0 & 2 \\ 3 & -1 \end{vmatrix} \end{bmatrix}$$

$$= \begin{bmatrix} -1 & -5 & 4 \\ 2 & -4 & -8 \\ 5 & -3 & -6 \end{bmatrix}$$

Now to find the cofactors, we combine our checkerboard pattern with these minors to obtain the *matrix of cofactors.*

$$\begin{bmatrix} C_{11} & C_{12} & C_{13} \\ C_{21} & C_{22} & C_{23} \\ C_{31} & C_{32} & C_{33} \end{bmatrix} = \begin{bmatrix} M_{11} & -M_{12} & M_{13} \\ -M_{21} & M_{22} & -M_{23} \\ M_{31} & -M_{32} & M_{33} \end{bmatrix}$$

$$= \begin{bmatrix} -1 & -(-5) & 4 \\ -(2) & -4 & -(-8) \\ 5 & -(-3) & -6 \end{bmatrix}$$

$$= \begin{bmatrix} -1 & 5 & 4 \\ -2 & -4 & 8 \\ 5 & 3 & -6 \end{bmatrix}$$

Remark: The matrix of cofactors for a given matrix A is useful for finding the inverse of A, as we will see in Section 8.5.

Now that we can find the minors and cofactors of a matrix, here is a general method for finding the determinant of a matrix of order 3 or greater. This method is referred to as **expansion by cofactors.**

FINDING DETERMINANTS BY EXPANSION BY COFACTORS

> If A is a matrix of order 3 or greater, then the determinant of A is found by adding the products of the elements in *any* row (or column) of A with its corresponding cofactors.

This method gives us many options to choose from when evaluating the determinant of a given matrix. As a case in point, let's return to the matrix shown in Examples 2 and 3 and find its determinant with this new method.

EXAMPLE 4
Expansion by Cofactors

Use expansion by cofactors to find the determinant of

$$A = \begin{bmatrix} 0 & 2 & 1 \\ 3 & -1 & 2 \\ 4 & 0 & 1 \end{bmatrix}$$

Solution:
From Example 3, the matrix of cofactors is

$$\begin{bmatrix} -1 & 5 & 4 \\ -2 & -4 & 8 \\ 5 & 3 & -6 \end{bmatrix}$$

(1) Using the *first row*, we have

$$|A| = 0(-1) + 2(5) + 1(4) = 10 + 4 = 14$$

(2) Using the *second row*, we have

$$|A| = 3(-2) + (-1)(-4) + 2(8) = -6 + 4 + 16 = 14$$

(3) Using the *first column*, we have

$$|A| = 0(-1) + 3(-2) + 4(5) = -6 + 20 = 14$$

You should try some other possibilities to see that the determinant of A can be evaluated by expanding by *any* row or column of A.

Remark: Note that in expansion by cofactors we are not using matrix multiplication, but are multiplying corresponding elements. For instance, in the expansion

$$|A| = a_{11}C_{11} + a_{21}C_{21} + \cdots + a_{n1}C_{n1}$$

along Row 1 of an $n \times n$ matrix A, we multiply each row entry by its corresponding cofactor (entry) in the matrix of cofactors of A.

In practice, the row (or column) containing the most zeros is the best choice for expansion by cofactors. Watch how this works in our next example.

EXAMPLE 5
Finding the Determinant
of a Matrix of Order 4

Use the method of expansion by cofactors to find the determinant of the following matrix:

$$A = \begin{bmatrix} 1 & -2 & 3 & 0 \\ -1 & 1 & 0 & 2 \\ 0 & 2 & 0 & 3 \\ 3 & 4 & 1 & -2 \end{bmatrix}$$

Solution:
After inspecting this matrix, we see that two of the elements in the third column are zero. Thus, we can eliminate some of the work in the expansion by using the third column.

$$|A| = a_{13}C_{13} + a_{23}C_{23} + a_{33}C_{33} + a_{43}C_{43}$$
$$= 3(C_{13}) + 0(C_{23}) + 0(C_{33}) + 1(C_{43})$$

Since two of the coefficients in this expansion are zero, we need only find the two cofactors C_{13} and C_{43}. For C_{13}, we have

$$C_{13} = (-1)^{1+3} \begin{vmatrix} 1 & -2 & 3 & 0 \\ -1 & 1 & 0 & 2 \\ 0 & 2 & 0 & 3 \\ 3 & 4 & 1 & -2 \end{vmatrix} = (1) \begin{vmatrix} -1 & 1 & 2 \\ 0 & 2 & 3 \\ 3 & 4 & -2 \end{vmatrix}$$

The diagonal procedure then yields

$$C_{13} = 4 + 0 + 9 - 12 + 12 + 0 = 13$$

For C_{43}, we have

$$C_{43} = (-1)^{4+3} \begin{vmatrix} 1 & -2 & 3 & 0 \\ -1 & 1 & 0 & 2 \\ 0 & 2 & 0 & 3 \\ 3 & 4 & 1 & -2 \end{vmatrix} = (-1) \begin{vmatrix} 1 & -2 & 0 \\ -1 & 1 & 2 \\ 0 & 2 & 3 \end{vmatrix}$$

Expanding the 3×3 determinant by minors along the first column yields

$$C_{43} = (-1)\left((1)(-1)^2 \begin{vmatrix} 1 & 2 \\ 2 & 3 \end{vmatrix} + (-1)(-1)^3 \begin{vmatrix} -2 & 0 \\ 2 & 3 \end{vmatrix} \right)$$
$$= -(3 - 4) - (-6 - 0) = 7$$

Thus, we have

$$|A| = 3(C_{13}) + 1(C_{43}) = 3(13) + 1(7) = 39 + 7 = 46$$

Section Exercises 8.4 ————————————————————————

In Exercises 1–20, find the determinant of the matrix.

1. $\begin{bmatrix} 2 & 1 \\ 3 & 4 \end{bmatrix}$ 2. $\begin{bmatrix} -3 & 1 \\ 5 & 2 \end{bmatrix}$ 3. $\begin{bmatrix} 5 & 2 \\ -6 & 3 \end{bmatrix}$

4. $\begin{bmatrix} 2 & -2 \\ 4 & 3 \end{bmatrix}$ 5. $\begin{bmatrix} -7 & 6 \\ \frac{1}{2} & 3 \end{bmatrix}$ 6. $\begin{bmatrix} 4 & -3 \\ 0 & 0 \end{bmatrix}$

7. $\begin{bmatrix} 2 & 6 \\ 0 & 3 \end{bmatrix}$ 8. $\begin{bmatrix} 2 & -3 \\ -6 & 9 \end{bmatrix}$

9. $\begin{bmatrix} 2 & -1 & 0 \\ 4 & 2 & 1 \\ 4 & 2 & 1 \end{bmatrix}$ 10. $\begin{bmatrix} -2 & 2 & 3 \\ 1 & -1 & 0 \\ 0 & 1 & 4 \end{bmatrix}$

11. $\begin{bmatrix} 3 & 2 & 2 \\ 2 & 2 & 2 \\ -4 & 4 & 3 \end{bmatrix}$ 12. $\begin{bmatrix} 0.1 & 0.2 & 0.3 \\ -0.3 & 0.2 & 0.2 \\ 0.5 & 0.4 & 0.4 \end{bmatrix}$

13. $\begin{bmatrix} 1 & 4 & -2 \\ 3 & 6 & -6 \\ -2 & 1 & 4 \end{bmatrix}$ 14. $\begin{bmatrix} 2 & 3 & 1 \\ 0 & 5 & -2 \\ 0 & 0 & -2 \end{bmatrix}$

15. $\begin{bmatrix} 6 & 3 & -7 \\ 0 & 0 & 0 \\ 4 & -6 & 3 \end{bmatrix}$ 16. $\begin{bmatrix} 1 & 1 & 2 \\ 3 & 1 & 0 \\ -2 & 0 & 3 \end{bmatrix}$

17. $\begin{bmatrix} -0.4 & 0.4 & 0.3 \\ 0.2 & 0.2 & 0.2 \\ 0.3 & 0.2 & 0.2 \end{bmatrix}$ 18. $\begin{bmatrix} x & y & 1 \\ 2 & 3 & 1 \\ 0 & -1 & 1 \end{bmatrix}$

19. $\begin{bmatrix} x & y & 1 \\ -2 & -2 & 1 \\ 1 & 5 & 1 \end{bmatrix}$ 20. $\begin{bmatrix} 3 - \lambda & 2 \\ 4 & 1 - \lambda \end{bmatrix}$

In Exercises 21–24, find (a) the matrix of minors, and (b) the matrix of cofactors of the given matrix (see Example 3).

21. $\begin{bmatrix} -3 & 2 & 1 \\ 4 & 5 & 6 \\ 2 & -3 & 1 \end{bmatrix}$ 22. $\begin{bmatrix} 10 & 8 & 3 & -7 \\ 4 & 0 & 5 & -6 \\ 0 & 3 & 2 & 7 \\ 1 & 0 & -3 & 2 \end{bmatrix}$

23. $\begin{bmatrix} 6 & 0 & -3 & 5 \\ 4 & 13 & 6 & -8 \\ -1 & 0 & 7 & 4 \\ 8 & 6 & 0 & 2 \end{bmatrix}$

24. $\begin{bmatrix} -3 & 4 & 2 \\ 6 & 3 & 1 \\ 4 & -7 & -8 \end{bmatrix}$

25. Find the determinant of the matrix in Exercise 21 by the method of expansion by cofactors using (a) the first row and (b) the second column.

26. Find the determinant of the matrix in Exercise 22 by the method of expansion by cofactors using (a) the second column and (b) the first row.

27. Find the determinant of the matrix in Exercise 23 by the method of expansion by cofactors using (a) the second row and (b) the second column.

28. Find the determinant of the matrix in Exercise 24 by the method of expansion by cofactors using (a) the third row and (b) the first column.

In Exercises 29–40, find the determinant of the matrix by the method of expansion by cofactors. Choose the row or column that may simplify the calculations.

29. $\begin{bmatrix} 1 & 4 & -2 \\ 3 & 2 & 0 \\ -1 & 4 & 3 \end{bmatrix}$ 30. $\begin{bmatrix} 2 & -1 & 3 \\ 1 & 4 & 4 \\ 1 & 0 & 2 \end{bmatrix}$

31. $\begin{bmatrix} 2 & 4 & 6 \\ 0 & 3 & 1 \\ 0 & 0 & -5 \end{bmatrix}$ 32. $\begin{bmatrix} -3 & 0 & 0 \\ 7 & 11 & 0 \\ 1 & 2 & 2 \end{bmatrix}$

33. $\begin{bmatrix} 3 & 6 & -5 & 4 \\ -2 & 0 & 6 & 0 \\ 1 & 1 & 2 & 2 \\ 0 & 3 & -1 & -1 \end{bmatrix}$ 34. $\begin{bmatrix} 2 & 6 & 6 & 2 \\ 2 & 7 & 3 & 6 \\ 1 & 5 & 0 & 1 \\ 3 & 7 & 0 & 7 \end{bmatrix}$

35. $\begin{bmatrix} 5 & 3 & 0 & 6 \\ 4 & 6 & 4 & 12 \\ 0 & 2 & -3 & 4 \\ 0 & 1 & -2 & 2 \end{bmatrix}$ 36. $\begin{bmatrix} 1 & 4 & 3 & 2 \\ -5 & 6 & 2 & 1 \\ 0 & 0 & 0 & 0 \\ 3 & -2 & 1 & 5 \end{bmatrix}$

37. $\begin{bmatrix} 3 & 2 & 4 & -1 & 5 \\ -2 & 0 & 1 & 3 & 2 \\ 1 & 0 & 0 & 4 & 0 \\ 6 & 0 & 2 & -1 & 0 \\ 3 & 0 & 5 & 1 & 0 \end{bmatrix}$

38. $\begin{bmatrix} 5 & 2 & 0 & 0 & -2 \\ 0 & 1 & 4 & 3 & 2 \\ 0 & 0 & 2 & 6 & 3 \\ 0 & 0 & 3 & 4 & 1 \\ 0 & 0 & 0 & 0 & 2 \end{bmatrix}$

39. $\begin{bmatrix} 4 & 3 & -2 & 1 & 2 \\ 0 & 0 & 0 & 0 & 0 \\ 1 & 2 & -7 & 13 & 12 \\ 6 & -2 & 5 & 6 & 7 \\ 1 & 4 & 2 & 0 & 9 \end{bmatrix}$

40. $\begin{bmatrix} 3 & 0 & 7 & 0 \\ 2 & 6 & 11 & 12 \\ 4 & 1 & -1 & 2 \\ 1 & 5 & 2 & 10 \end{bmatrix}$

The equation of a line through the points (x_1, y_1) and (x_2, y_2) can be written

$$\begin{vmatrix} x & y & 1 \\ x_1 & y_1 & 1 \\ x_2 & y_2 & 1 \end{vmatrix} = 0$$

In Exercises 41–45, use this method to find an equation of the line through the given points (x_1, y_1) and (x_2, y_2).

41. $(0, 0), (5, 3)$ 42. $(0, 0), (-2, 2)$
43. $(-4, 3), (2, 1)$ 44. $(10, 7), (-2, -7)$
45. $(-\frac{1}{2}, 3), (\frac{5}{2}, 1)$

The area of a triangle with vertices (x_1, y_1), (x_2, y_2), and (x_3, y_3) is the absolute value of

$$\frac{1}{2}\begin{vmatrix} x_1 & y_1 & 1 \\ x_2 & y_2 & 1 \\ x_3 & y_3 & 1 \end{vmatrix}$$

In Exercises 46–50, use this method to find the area of the triangle with the given vertices.

46. $(0, 0), (3, 1), (1, 5)$ 47. $(0, 0), (5, -2), (4, 5)$
48. $(-2, -3), (2, -3), (0, 4)$
49. $(-2, 1), (3, -1), (1, 6)$
50. $(0, \frac{1}{2}), (\frac{5}{2}, 0), (4, 3)$

Properties of Determinants 8.5

Which of the following determinants is easier to evaluate?

$$|A| = \begin{vmatrix} 4 & 1 & 3 & 2 \\ -2 & 2 & 1 & -3 \\ 1 & -2 & 3 & 1 \\ 3 & 1 & 3 & 2 \end{vmatrix} \qquad |B| = \begin{vmatrix} 0 & 9 & -9 & -2 \\ 0 & -2 & 7 & -1 \\ 1 & -2 & 3 & 1 \\ 0 & 7 & -6 & -1 \end{vmatrix}$$

From what we know about expansion by cofactors, it is clear that the second determinant is much simpler to evaluate, since three of the elements in the first column are zero. If we expand these two determinants by cofactors (using the first column), we get

$$|A| = (4)\begin{vmatrix} 2 & 1 & -3 \\ -2 & 3 & 1 \\ 1 & 3 & 2 \end{vmatrix} - (-2)\begin{vmatrix} 1 & 3 & 2 \\ -2 & 3 & 1 \\ 1 & 3 & 2 \end{vmatrix}$$

$$+ (1)\begin{vmatrix} 1 & 3 & 2 \\ 2 & 1 & -3 \\ 1 & 3 & 2 \end{vmatrix} - (3)\begin{vmatrix} 1 & 3 & 2 \\ 2 & 1 & -3 \\ -2 & 3 & 1 \end{vmatrix}$$

$$= 4(38) + 2(0) + 1(0) - 3(38) = 38$$

$$|B| = (0)C_{11} + (0)C_{12} + (1)\begin{vmatrix} 9 & -9 & -2 \\ -2 & 7 & -1 \\ 7 & -6 & -1 \end{vmatrix} + (0)C_{14}$$

$$= 1(38) = 38$$

It is not mere coincidence that these two determinants have the same value. In fact, we obtained the second determinant from the first by performing *elementary row operations* on the matrix *A*.

In this section, we identify the effect of elementary row operations on the value of a determinant. In addition, we discuss some special properties of determinants and their relationships to the existence and calculation of the inverse of a matrix.

ELEMENTARY ROW (COLUMN) OPERATIONS AND DETERMINANTS

If *A* is a square matrix of order 2 or greater, then the following elementary row (or column) operations produce the indicated change in $|A|$:

1. The interchange of two rows results in a change of signs of the determinant.

2. Adding a multiple of the elements of one row to the elements of another row leaves the determinant unchanged.

3. Multiplying the elements of one row by *k* produces a determinant that is *k* times the original.

Note that each of these three statements is still valid if the word ''row'' is changed to the word ''column.''

EXAMPLE 1
Applying Elementary Row Operations

Use elementary row (or column) operations to evaluate the following determinant:

$$|A| = \begin{vmatrix} 4 & 1 & 3 & 2 \\ -2 & 2 & 1 & -3 \\ 1 & -2 & 3 & 1 \\ 3 & 1 & 3 & 2 \end{vmatrix}$$

Solution:
There are many ways to approach this problem, but we begin by adding multiples of the third row to rows 1, 2, and 4, to create three zeros in the first column.

$$|A| = \begin{vmatrix} 4 & 1 & 3 & 2 \\ -2 & 2 & 1 & -3 \\ 1 & -2 & 3 & 1 \\ 3 & 1 & 3 & 2 \end{vmatrix} \begin{matrix} -4R_3 + R_1 \\ 2R_3 + R_2 \\ \\ -3R_3 + R_4 \end{matrix} \begin{vmatrix} 0 & 9 & -9 & -2 \\ 0 & -2 & 7 & -1 \\ 1 & -2 & 3 & 1 \\ 0 & 7 & -6 & -1 \end{vmatrix}$$

Now, we can expand by cofactors (as shown in the beginning of this section) to obtain

$$|A| = 38$$

EXAMPLE 2
Applying Elementary
Row Operations

Use elementary row (or column) operations to evaluate the following determinants:

(a) $|A| = \begin{vmatrix} 1 & 4 & -2 \\ 2 & 5 & -4 \\ -3 & 0 & 6 \end{vmatrix}$ (b) $|B| = \begin{vmatrix} 2 & 0 & 1 & 3 & -2 \\ -2 & 1 & 0 & 1 & -1 \\ 1 & 0 & -1 & 2 & 3 \\ 3 & -1 & 2 & 4 & -3 \\ 1 & 1 & -1 & 2 & 0 \end{vmatrix}$

Solution:

(a) Notice that the third column is a multiple of the first column. This implies that the determinant is zero. To see this, note how we can use an elementary *column* operation to create *all* zeros in the third column, which then yields zero when expanded by cofactors.

$$|A| = \begin{vmatrix} 1 & 4 & -2 \\ 2 & 5 & -4 \\ -3 & 0 & 6 \end{vmatrix} = \begin{vmatrix} 1 & 4 & 0 \\ 2 & 5 & 0 \\ -3 & 0 & 0 \end{vmatrix} = 0$$

$$2C_1 + C_3$$

(b) Since the second column of this matrix already has two zeros, we choose it for the expansion.

$$|B| = \begin{vmatrix} 2 & 0 & 1 & 3 & -2 \\ -2 & 1 & 0 & 1 & -1 \\ 1 & 0 & -1 & 2 & 3 \\ 3 & -1 & 2 & 4 & -3 \\ 1 & 1 & -1 & 2 & 0 \end{vmatrix}$$

$$\begin{matrix} \\ \\ \\ 2R_2 + R_4 \\ -R_2 + R_5 \end{matrix} \begin{vmatrix} 2 & 0 & 1 & 3 & -2 \\ -2 & 1 & 0 & 1 & -1 \\ 1 & 0 & -1 & 2 & 3 \\ 1 & 0 & 2 & 5 & -4 \\ 3 & 0 & -1 & 1 & 1 \end{vmatrix}$$

$$= (-1)^4 \begin{vmatrix} 2 & 1 & 3 & -2 \\ 1 & -1 & 2 & 3 \\ 1 & 2 & 5 & -4 \\ 3 & -1 & 1 & 1 \end{vmatrix}$$

Now, suppose we choose to expand by the fourth row. We can create three zeros in the fourth row, as follows:

$$|B| = \begin{vmatrix} 2 & 1 & 3 & -2 \\ 1 & -1 & 2 & 3 \\ 1 & 2 & 5 & -4 \\ 3 & -1 & 1 & 1 \end{vmatrix} = \begin{vmatrix} 5 & 1 & 4 & -1 \\ -2 & -1 & 1 & 2 \\ 7 & 2 & 7 & -2 \\ 0 & -1 & 0 & 0 \end{vmatrix}$$

$$3C_2 + C_1 \text{——}$$
$$C_2 + C_3 \text{——}$$
$$C_2 + C_4 \text{——}$$

Consequently,

$$|B| = (-1)(-1)^{4+2} \begin{vmatrix} 5 & 4 & -1 \\ -2 & 1 & 2 \\ 7 & 7 & -2 \end{vmatrix}$$

$$= -[-10 + 14 + 56 - (-7) - 70 - 16] = 19$$

Note in Example 2, part (a), that if one column of the matrix is a multiple of another column, we can immediately conclude that the determinant is zero. This is one of three conditions, listed next, that yield a determinant of zero. Each condition is easily verified, using elementary row operations and expansion by cofactors.

CONDITIONS THAT YIELD A DETERMINANT OF ZERO

If A is a matrix of order n and any one of the following conditions is true, then $|A| = 0$.

1. An entire row (or column) is zero.
2. Two rows (or columns) have the same elements.
3. One row (or column) is a multiple of another row (or column).

When you are evaluating determinants, you can save considerable work by recognizing these conditions. Of course, these are not the only conditions that yield zero determinants. This fact is illustrated in the following example.

EXAMPLE 3
Evaluating Determinants

Show that each of the following determinants is zero:

(a) $|A| = \begin{vmatrix} 0 & 0 & 0 \\ 2 & 4 & -5 \\ 3 & -5 & 2 \end{vmatrix}$

(b) $|B| = \begin{vmatrix} 1 & -2 & 4 \\ 0 & 1 & 2 \\ 1 & -2 & 4 \end{vmatrix}$

(c) $|C| = \begin{vmatrix} 1 & 2 & -3 \\ 2 & -1 & -6 \\ -2 & 0 & 6 \end{vmatrix}$

(d) $|D| = \begin{vmatrix} 1 & 4 & 1 \\ 2 & -1 & 0 \\ 0 & -9 & -2 \end{vmatrix}$

Solution:

(a) $|A| = \begin{vmatrix} 0 & 0 & 0 \\ 2 & 4 & -5 \\ 3 & -5 & 2 \end{vmatrix} = 0$ *(Since the first row has all zero elements.)*

(b) $|B| = \begin{vmatrix} 1 & -2 & 4 \\ 0 & 1 & 2 \\ 1 & -2 & 4 \end{vmatrix} = 0$ *(Since the first and third rows have the same elements.)*

(c) $|C| = \begin{vmatrix} 1 & 2 & -3 \\ 2 & -1 & -6 \\ -2 & 0 & 6 \end{vmatrix} = 0$ *(Since the third column is a multiple of the first column.)*

(d) $|D| = \begin{vmatrix} 1 & 4 & 1 \\ 2 & -1 & 0 \\ 0 & -9 & -2 \end{vmatrix}$

It is not immediately evident that this determinant is zero. However, by adding (-2) times the first row to the second row, we have the following:

$$|D| = \begin{vmatrix} 1 & 4 & 1 \\ 2 & -1 & 0 \\ 0 & -9 & -2 \end{vmatrix}$$

$$= \begin{vmatrix} 1 & 4 & 1 \\ 0 & -9 & -2 \\ 0 & -9 & -2 \end{vmatrix}$$

Now, since the second and third rows have the same elements, the determinant is zero.

In Section 8.3 we saw that some square matrices have (multiplicative) inverses and some do not. One very useful application of determinants is that the matrices with zero determinants do not have inverses.

CLASSIFYING SINGULAR AND NONSINGULAR MATRICES

> If A is a square matrix of order n, then it has an inverse if and only if its determinant is nonzero.

From this result we can see that none of the matrices in Example 3 have inverses, since they all have zero determinants.

Another application of determinants is that they provide us with a convenient alternative procedure for finding the inverse of a nonsingular matrix, especially those of order 2 or 3.

COFACTOR FORM FOR INVERSE OF A MATRIX

If A is a nonsingular matrix of order n, then its inverse is given by

$$A^{-1} = \frac{1}{|A|} \begin{bmatrix} C_{11} & C_{21} & C_{31} & \cdots & C_{n1} \\ C_{12} & C_{22} & C_{32} & \cdots & C_{n2} \\ C_{13} & C_{23} & C_{33} & \cdots & C_{n3} \\ \cdot & \cdot & \cdot & & \cdot \\ \cdot & \cdot & \cdot & & \cdot \\ \cdot & \cdot & \cdot & & \cdot \\ C_{1n} & C_{2n} & C_{3n} & \cdots & C_{nn} \end{bmatrix}$$

Remark: Be sure you notice that the cofactor of a_{ij} appears in the jth row and ith column (rather than the ith row and jth column). That is, the rows and columns have been interchanged. This is called *transposing* a matrix. For instance, if

$$A = \begin{bmatrix} 1 & 2 & 3 \\ 4 & 3 & 2 \\ 1 & 1 & 3 \end{bmatrix} \qquad \text{then} \qquad A^T = \begin{bmatrix} 1 & 4 & 1 \\ 2 & 3 & 1 \\ 3 & 2 & 3 \end{bmatrix}$$

If $A = \begin{bmatrix} a & b \\ c & d \end{bmatrix}$, then the inverse of A is given by

$$A^{-1} = \frac{1}{|A|} \begin{bmatrix} C_{11} & C_{21} \\ C_{12} & C_{22} \end{bmatrix} = \frac{1}{|A|} \begin{bmatrix} d & -b \\ -c & a \end{bmatrix}$$

EXAMPLE 4
Finding the Inverse
of a Matrix by Cofactors

Use cofactors to find the inverses of the following matrices:

(a) $A = \begin{bmatrix} -1 & 2 \\ 3 & -8 \end{bmatrix}$
(b) $B = \begin{bmatrix} -1 & 3 & 2 \\ 0 & 1 & -2 \\ 1 & 4 & 1 \end{bmatrix}$

Solution:

(a) Since $|A| = 8 - 6 = 2$, the inverse of A is given by

$$A^{-1} = \frac{1}{2} \begin{bmatrix} -8 & -2 \\ -3 & -1 \end{bmatrix} = \begin{bmatrix} -4 & -1 \\ -\frac{3}{2} & -\frac{1}{2} \end{bmatrix}$$

(b) Since $|B| = -1 - 6 + 0 - 2 - 8 - 0 = -17$, the inverse of B is given by

$$B^{-1} = \frac{1}{-17} \begin{bmatrix} \begin{vmatrix} 1 & -2 \\ 4 & 1 \end{vmatrix} & -\begin{vmatrix} 3 & 2 \\ 4 & 1 \end{vmatrix} & \begin{vmatrix} 3 & 2 \\ 1 & -2 \end{vmatrix} \\ -\begin{vmatrix} 0 & -2 \\ 1 & 1 \end{vmatrix} & \begin{vmatrix} -1 & 2 \\ 1 & 1 \end{vmatrix} & -\begin{vmatrix} -1 & 2 \\ 0 & -2 \end{vmatrix} \\ \begin{vmatrix} 0 & 1 \\ 1 & 4 \end{vmatrix} & -\begin{vmatrix} -1 & 3 \\ 1 & 4 \end{vmatrix} & \begin{vmatrix} -1 & 3 \\ 0 & 1 \end{vmatrix} \end{bmatrix}$$

$$B^{-1} = \frac{1}{-17} \begin{bmatrix} 9 & 5 & -8 \\ -2 & -3 & -2 \\ -1 & 7 & -1 \end{bmatrix}$$

Section Exercises 8.5

In Exercises 1–15, state the property of determinants that allows you to determine that the equation is true without expanding the determinants.

1. $\begin{vmatrix} 2 & -6 \\ 1 & -3 \end{vmatrix} = 0$

2. $\begin{vmatrix} -4 & 5 \\ 12 & -15 \end{vmatrix} = 0$

3. $\begin{vmatrix} 1 & 4 & 2 \\ 0 & 0 & 0 \\ 5 & 6 & -7 \end{vmatrix} = 0$

4. $\begin{vmatrix} -4 & 3 & 2 \\ 8 & 0 & 0 \\ -4 & 3 & 2 \end{vmatrix} = 0$

5. $\begin{vmatrix} 1 & 3 & 4 \\ -7 & 2 & -5 \\ 6 & 1 & 2 \end{vmatrix} = - \begin{vmatrix} 1 & 4 & 3 \\ -7 & -5 & 2 \\ 6 & 2 & 1 \end{vmatrix}$

6. $\begin{vmatrix} 1 & 3 & 4 \\ -2 & 2 & 0 \\ 1 & 6 & 2 \end{vmatrix} = - \begin{vmatrix} 1 & 6 & 2 \\ -2 & 2 & 0 \\ 1 & 3 & 4 \end{vmatrix}$

7. $\begin{vmatrix} 5 & 10 \\ 2 & -7 \end{vmatrix} = 5 \begin{vmatrix} 1 & 2 \\ 2 & -7 \end{vmatrix}$

8. $\begin{vmatrix} 1 & 8 & -3 \\ 3 & -12 & 6 \\ 7 & 4 & 9 \end{vmatrix} = 12 \begin{vmatrix} 1 & 2 & -1 \\ 3 & -3 & 2 \\ 7 & 1 & 3 \end{vmatrix}$

9. $\begin{vmatrix} 5 & 0 & 10 \\ 25 & -30 & 40 \\ -15 & 5 & 20 \end{vmatrix} = 5^3 \begin{vmatrix} 1 & 0 & 2 \\ 5 & -6 & 8 \\ -3 & 1 & 4 \end{vmatrix}$

10. $\begin{vmatrix} 6 & 0 & 0 & 0 \\ 0 & 6 & 0 & 0 \\ 0 & 0 & 6 & 0 \\ 0 & 0 & 0 & 6 \end{vmatrix} = 6^4 \begin{vmatrix} 1 & 0 & 0 & 0 \\ 0 & 1 & 0 & 0 \\ 0 & 0 & 1 & 0 \\ 0 & 0 & 0 & 1 \end{vmatrix}$

11. $\begin{vmatrix} 2 & -3 \\ 8 & 7 \end{vmatrix} = \begin{vmatrix} 2 & -3 \\ 0 & 19 \end{vmatrix}$

12. $\begin{vmatrix} 1 & -3 & 2 \\ 5 & 2 & -1 \\ -1 & 0 & 6 \end{vmatrix} = \begin{vmatrix} 1 & -3 & 2 \\ 0 & 17 & -11 \\ 0 & -3 & 8 \end{vmatrix}$

13. $\begin{vmatrix} 3 & 2 & 4 \\ -2 & 1 & 5 \\ 5 & -7 & -20 \end{vmatrix} = \begin{vmatrix} 7 & 2 & -6 \\ 0 & 1 & 0 \\ -9 & -7 & 15 \end{vmatrix}$

14. $\begin{vmatrix} 5 & 4 & 2 \\ 2 & -3 & 4 \\ 7 & 6 & 3 \end{vmatrix} = \begin{vmatrix} 1 & 10 & -6 \\ 2 & -3 & 4 \\ 7 & 6 & 3 \end{vmatrix}$

15. $\begin{vmatrix} 5 & 4 & 2 \\ 2 & -3 & 4 \\ 7 & 6 & 3 \end{vmatrix} = \begin{vmatrix} 2 & 4 & 5 \\ 3 & 6 & 7 \\ 4 & -3 & 2 \end{vmatrix}$

In Exercises 16–30, use elementary row (or column) operations as aids to evaluating the determinant.

16. $\begin{vmatrix} 1 & 7 & -3 \\ 1 & 3 & 1 \\ 4 & 8 & 1 \end{vmatrix}$

17. $\begin{vmatrix} 1 & 1 & 1 \\ 2 & -1 & -2 \\ 1 & -2 & -1 \end{vmatrix}$

18. $\begin{vmatrix} 2 & -1 & -1 \\ 1 & 3 & 2 \\ 1 & 1 & 3 \end{vmatrix}$

19. $\begin{vmatrix} 3 & -1 & -3 \\ -1 & -4 & -2 \\ 3 & -1 & -1 \end{vmatrix}$

20. $\begin{vmatrix} 4 & 3 & -2 \\ 5 & 4 & 1 \\ -2 & 3 & 4 \end{vmatrix}$

21. $\begin{vmatrix} 3 & 8 & -7 \\ 0 & -5 & 4 \\ 8 & 1 & 6 \end{vmatrix}$

22. $\begin{vmatrix} 5 & -8 & 0 \\ 9 & 7 & 4 \\ -8 & 7 & 1 \end{vmatrix}$

23. $\begin{vmatrix} 4 & -8 & 5 \\ 8 & -5 & 3 \\ 8 & 5 & 2 \end{vmatrix}$

24. $\begin{vmatrix} 4 & -7 & 9 & 1 \\ 6 & 2 & 7 & 0 \\ 3 & 6 & -3 & 3 \\ 0 & 7 & 4 & -1 \end{vmatrix}$

25. $\begin{vmatrix} 9 & -4 & 2 & 5 \\ 2 & 7 & 6 & -5 \\ 4 & 1 & -2 & 0 \\ 7 & 3 & 4 & 10 \end{vmatrix}$

26. $\begin{vmatrix} 1 & -2 & 7 & 9 \\ 3 & -4 & 5 & 5 \\ 3 & 6 & 1 & -1 \\ 4 & 5 & 3 & 2 \end{vmatrix}$

27. $\begin{vmatrix} 0 & -3 & 8 & 2 \\ 8 & 1 & -1 & 6 \\ -4 & 6 & 0 & 9 \\ -7 & 0 & 0 & 14 \end{vmatrix}$

28. $\begin{vmatrix} 1 & -1 & 8 & 4 & 2 \\ 2 & 6 & 0 & -4 & 3 \\ 2 & 0 & 2 & 6 & 2 \\ 0 & 2 & 8 & 0 & 0 \\ 0 & 1 & 1 & 2 & 2 \end{vmatrix}$

29. $\begin{vmatrix} 3 & -2 & 4 & 3 & 1 \\ -1 & 0 & 2 & 1 & 0 \\ 5 & -1 & 0 & 3 & 2 \\ 4 & 7 & -8 & 0 & 0 \\ 1 & 2 & 3 & 0 & 2 \end{vmatrix}$

30. $\begin{vmatrix} 4 & 2 & -1 & 0 & 3 \\ 0 & 1 & 1 & 2 & -3 \\ 0 & 0 & -2 & 8 & 12 \\ 0 & 0 & 0 & 5 & 13 \\ 0 & 0 & 0 & 0 & 3 \end{vmatrix}$

In Exercises 31–40, use cofactors to find the inverse (if it exists) of the matrix.

31. $\begin{bmatrix} 2 & 4 \\ -1 & 2 \end{bmatrix}$

32. $\begin{bmatrix} -3 & 4 \\ 2 & -4 \end{bmatrix}$

33. $\begin{bmatrix} 3 & 6 \\ 2 & 4 \end{bmatrix}$

34. $\begin{bmatrix} 2 & 7 \\ -3 & -9 \end{bmatrix}$

35. $\begin{bmatrix} 1 & 1 & 2 \\ 3 & 1 & 0 \\ -2 & 0 & 3 \end{bmatrix}$

36. $\begin{bmatrix} 3 & 2 & 2 \\ 2 & 2 & 2 \\ -4 & 4 & 3 \end{bmatrix}$

37. $\begin{bmatrix} 0.1 & 0.2 & 0.3 \\ -0.3 & 0.2 & 0.2 \\ 0.5 & 0.4 & 0.4 \end{bmatrix}$

38. $\begin{bmatrix} 1 & 0 & 0 \\ 3 & 4 & 0 \\ 2 & 5 & 5 \end{bmatrix}$

39. $\begin{bmatrix} 1 & 0 & -2 & 2 \\ 0 & 1 & 1 & -1 \\ -2 & 3 & 4 & 0 \\ 3 & 1 & -2 & -1 \end{bmatrix}$

40. $\begin{bmatrix} 1 & 0 & 3 & 0 \\ 0 & 2 & 0 & 4 \\ 1 & 0 & 3 & 0 \\ 0 & 2 & 0 & 4 \end{bmatrix}$

41. Show that $\begin{vmatrix} 1 & x & x^2 \\ 1 & y & y^2 \\ 1 & z & z^2 \end{vmatrix} = (y - x)(z - x)(z - y)$.

42. Show that $\begin{vmatrix} a_{11} & 0 & 0 & \cdots & 0 \\ a_{21} & a_{22} & 0 & \cdots & 0 \\ \vdots & \vdots & & & \vdots \\ \vdots & \vdots & & & 0 \\ a_{n1} & a_{n2} & a_{n3} & \cdots & a_{nn} \end{vmatrix} = a_{11}a_{22}a_{23} \cdots a_{nn}$.

Cramer's Rule   8.6

In Chapter 7 we looked at two general algebraic methods for solving systems of linear equations: substitution and elimination. Then, in the beginning of this chapter, we studied some matrix methods for solving systems of linear equations. In this section we study Cramer's Rule, a method to use if the system of linear equations has a unique solution *and* the same number of equations as variables. This method can be used to find the *complete* solution or to single out a particular variable and solve for that variable alone.

To see how Cramer's Rule arises, let's look at the solution of a general system involving two linear equations in two unknowns.

$$a_{11}x + a_{12}y = c_1$$
$$a_{21}x + a_{22}y = c_2$$

If we solve this system by elimination, we obtain the following. Multiply the first equation by $-a_{21}$:

$$-a_{21}a_{11}x - a_{21}a_{12}y = -a_{21}c_1$$

Multiply the second equation by a_{11}:

$$a_{21}a_{11}x + a_{11}a_{22}y = a_{11}c_2$$

Add equations to eliminate x:

$$(a_{11}a_{22} - a_{21}a_{12})y = a_{11}c_2 - a_{21}c_1$$

Solve for y (provided $a_{11}a_{22} - a_{21}a_{12} \neq 0$):

$$y = \frac{a_{11}c_2 - a_{21}c_1}{a_{11}a_{22} - a_{21}a_{12}}$$

In a similar way, we can solve for x, to obtain

$$x = \frac{a_{22}c_1 - a_{12}c_2}{a_{11}a_{22} - a_{21}a_{12}}$$

Finally, by recognizing that the numerator and denominator for both x and y can be represented as determinants, we have

$$x = \frac{\begin{vmatrix} c_1 & a_{12} \\ c_2 & a_{22} \end{vmatrix}}{\begin{vmatrix} a_{11} & a_{12} \\ a_{21} & a_{22} \end{vmatrix}}, \quad y = \frac{\begin{vmatrix} a_{11} & c_1 \\ a_{21} & c_2 \end{vmatrix}}{\begin{vmatrix} a_{11} & a_{12} \\ a_{21} & a_{22} \end{vmatrix}}, \quad a_{11}a_{22} - a_{21}a_{12} \neq 0$$

The denominator for both x and y is simply the determinant of the coefficient matrix of the system, and we denote this determinant by

$$D = \begin{vmatrix} a_{11} & a_{12} \\ a_{21} & a_{22} \end{vmatrix}$$

The determinants that form the numerators for x and y can be obtained from D by replacing the first or second column by a column representing the constants of the system. We denote these two determinants by D_x and D_y, as follows:

$$D_x = \begin{vmatrix} c_1 & a_{12} \\ c_2 & a_{22} \end{vmatrix}, \quad D_y = \begin{vmatrix} a_{11} & c_1 \\ a_{21} & c_2 \end{vmatrix}$$

This determinant form of the solution for x and y is called **Cramer's Rule.**

CRAMER'S RULE FOR TWO EQUATIONS IN TWO VARIABLES

The solution of a system of two linear equations in two variables is given by

$$x = \frac{|D_x|}{|D|} \quad \text{and} \quad y = \frac{|D_y|}{|D|}$$

provided $|D| \neq 0$.

Recall from our earlier discussion of systems of linear equations that a system of two equations in two variables can have infinitely many solutions, no solution, or precisely one solution. Our development of Cramer's Rule gives us the following quick test for determining if a system of n linear equations in n variables has a unique solution.

TEST FOR UNIQUE SOLUTION

A system of n linear equations in n variables has a unique solution if and only if the coefficient matrix has a nonzero determinant

$$|D| \neq 0$$

EXAMPLE 1
Solving a Linear System
with Cramer's Rule

Use Cramer's Rule to solve the following system:

$$4x - 2y = 10$$
$$3x - 5y = 11$$

Solution:

We have

$$|D| = \begin{vmatrix} 4 & -2 \\ 3 & -5 \end{vmatrix} = -20 + 6 = -14$$

Since $|D| \neq 0$, we know that the system has a unique solution. Using Cramer's Rule, we have

$$x = \frac{|D_x|}{|D|} = \frac{\begin{vmatrix} 10 & -2 \\ 11 & -5 \end{vmatrix}}{-14} = \frac{-50 + 22}{-14} = \frac{-28}{-14} = 2$$

$$y = \frac{|D_y|}{|D|} = \frac{\begin{vmatrix} 4 & 10 \\ 3 & 11 \end{vmatrix}}{-14} = \frac{44 - 30}{-14} = \frac{14}{-14} = -1$$

Cramer's Rule generalizes easily to systems of n linear equations in n variables. The denominator of the quotient is the determinant of the coefficient matrix, and the numerator is the determinant formed by replacing the column corresponding to a given variable (the one being solved for) with the column representing the constants in the system.

CRAMER'S RULE FOR n EQUATIONS IN n VARIABLES

The solution for a system of n linear equations in the n variables x_1, $x_2, \ldots, x_n$ is given by

$$x_1 = \frac{|D_{x_1}|}{|D|}, \qquad x_2 = \frac{|D_{x_2}|}{|D|}, \qquad \ldots, \qquad x_n = \frac{|D_{x_n}|}{|D|}$$

provided $|D| \neq 0$.

Remark: Before applying Cramer's Rule to any system of linear equations, you must write the system in the standard form:

$$a_{11}x_1 + a_{12}x_2 + a_{13}x_3 + \cdots + a_{1n}x_n = c_1$$
$$a_{21}x_1 + a_{22}x_2 + a_{23}x_3 + \cdots + a_{2n}x_n = c_2$$
$$a_{31}x_1 + a_{32}x_2 + a_{33}x_3 + \cdots + a_{3n}x_n = c_3$$
$$\vdots \qquad \vdots \qquad \vdots \qquad \qquad \vdots \qquad \vdots$$
$$a_{n1}x_1 + a_{n2}x_2 + a_{n3}x_3 + \cdots + a_{nn}x_n = c_n$$

EXAMPLE 2
Applying Cramer's Rule to a
System of Three Equations

Use Cramer's Rule to solve the following system:

$$-x + 2y - 3z = 1$$
$$2x \qquad + z = 0$$
$$3x - 4y + 4z = 2$$

Solution:
We begin by evaluating the determinant of the coefficient matrix:

$$|D| = \begin{vmatrix} -1 & 2 & -3 \\ 2 & 0 & 1 \\ 3 & -4 & 4 \end{vmatrix} = 0 + 6 + 24 - 0 - 4 - 16 = 10$$

Since $|D| \neq 0$, we know that the solution is unique, and we obtain

$$x = \frac{\begin{vmatrix} 1 & 2 & -3 \\ 0 & 0 & 1 \\ 2 & -4 & 4 \end{vmatrix}}{10} = \frac{(-1)^5 \begin{vmatrix} 1 & 2 \\ 2 & -4 \end{vmatrix}}{10}$$

$$= \frac{(-1)(-4 - 4)}{10} = \frac{4}{5}$$

$$y = \frac{\begin{vmatrix} -1 & 1 & -3 \\ 2 & 0 & 1 \\ 3 & 2 & 4 \end{vmatrix}}{10} = \frac{0 + 3 - 12 - 0 + 2 - 8}{10}$$

$$= -\frac{15}{10} = -\frac{3}{2}$$

$$z = \frac{\begin{vmatrix} -1 & 2 & 1 \\ 2 & 0 & 0 \\ 3 & -4 & 2 \end{vmatrix}}{10} = \frac{2(-1)^3 \begin{vmatrix} 2 & 1 \\ -4 & 2 \end{vmatrix}}{10}$$

$$= \frac{-2(4+4)}{10} = -\frac{8}{5}$$

You should check this solution by substituting these values into *all* three equations.

EXAMPLE 3
Applying Cramer's Rule to a System of Four Equations

Solve for x in the following system:

$$1.3x - 2.7y + 0.3w = 2.5$$
$$-4.7x + 3.2y - 0.8z + 2.5w = -3.8$$
$$5.8x + 3.5z - 1.7w = 5.8$$
$$2.4x - 1.2y + 2.3w = -3.3$$

Solution:
Expanding by cofactors along Column 3 yields

$$|D| = \begin{vmatrix} 1.3 & -2.7 & 0 & 0.3 \\ -4.7 & 3.2 & -0.8 & 2.5 \\ 5.8 & 0 & 3.5 & -1.7 \\ 2.4 & -1.2 & 0 & 2.3 \end{vmatrix} = -80.0233$$

(You should check this using your calculator.) Now, to solve for x, we again expand by cofactors, using Column 3:

$$|D_x| = \begin{vmatrix} 2.5 & -2.7 & 0 & 0.3 \\ -3.8 & 3.2 & -0.8 & 2.5 \\ 5.8 & 0 & 3.5 & -1.7 \\ -3.3 & -1.2 & 0 & 2.3 \end{vmatrix} = 112.8419$$

Finally, we have

$$x = \frac{112.8419}{-80.0233} \approx -1.410$$

We have now completed our discussion of *five* methods for solving systems of linear equations. To be efficient in solving systems of equations, you should keep each method fresh in your memory so that for any given system of equations you can make a wise choice of solution method. For your convenience, we list a summary of the methods studied, along with their unique advantages and/or disadvantages.

METHODS OF SOLVING SYSTEMS OF LINEAR EQUATIONS

Method	Advantage of Method
Substitution (Section 7.1)	General method (can be applied to linear and nonlinear systems)
Elimination (Sections 7.2 and 7.3)	General linear method (can be used with any number of equations and variables)
Matrix solution with elementary row operations (Section 8.1)	Generalization of elimination method (easily adapted to computer techniques)
Solution by inverse matrix (Section 8.3)	Efficient method for several systems with *same* coefficient matrix (only applies to systems having n equations, n variables, and a unique solution)
Cramer's Rule (Section 8.6)	Efficient method for finding the value of a single variable in a solution (only applies to systems having n equations, n variables, and a unique solution)

Section Exercises 8.6

In Exercises 1–20, use Cramer's Rule to solve (if possible) the system of equations.

1. $\quad x + 2y = 5$
   $-x + y = 1$

2. $2x - y = -10$
   $3x + 2y = -1$

3. $3x + 4y = -2$
   $5x + 3y = 4$

4. $18x + 12y = 13$
   $30x + 24y = 23$

5. $20x + 8y = 11$
   $12x - 24y = 21$

6. $13x - 6y = 17$
   $26x - 12y = 8$

7. $-0.4x + 0.8y = 1.6$
   $2x - 4y = 5$

8. $-0.4x + 0.8y = 1.6$
   $0.2x + 0.3y = 2.2$

9. $3x + 6y = 5$
   $6x + 14y = 11$

10. $3x + 2y = 1$
    $2x + 10y = 6$

11. $4x - y + z = -5$
    $2x + 2y + 3z = 10$
    $5x - 2y + 6z = 1$

12. $4x - 2y + 3z = -2$
    $2x + 2y + 5z = 16$
    $8x - 5y - 2z = 4$

13. $3x + 4y + 4z = 11$
    $4x - 4y + 6z = 11$
    $6x - 6y = 3$

14. $14x - 21y - 7z = 10$
    $-4x + 2y - 2z = 4$
    $56x - 21y + 7z = 5$

15. $3x + 3y + 5z = 1$
    $3x + 5y + 9z = 2$
    $5x + 9y + 17z = 4$

16. $2x + 3y + 5z = 4$
    $3x + 5y + 9z = 7$
    $5x + 9y + 17z = 13$

17. $7x - 3y + 2w = 41$
    $-2x + y - w = -13$
    $4x + z - 2w = 12$
    $-x + y - w = -8$

18. $2x + 5y + w = 11$
    $x + 4y + 2z - 2w = -7$
    $2x - 2y + 5z + w = 3$
    $x - 3w = 1$

19. $5x - 3y + 2z = 2$
    $2x + 2y - 3z = 3$
    $x - 7y + 8z = -4$

20. $3x + 2y + 5z = 4$
    $4x - 3y - 4z = 1$
    $-8x + 2y + 3z = 0$

In Exercises 21–24, find the least squares regression line for the given points. Use Cramer's Rule and the system of equations given for Exercises 39–42 in Section 7.2.

21. $(1, 1), (2, 1), (2, 2), (4, 5)$

22. $(0, -3), (1, -2), (1, 1), (2, 4)$
23. $(0, 5), (0, 3), (1, 4), (2, 2), (3, 0)$
24. $(-2, 3), (-1, 3), (0, 1), (2, 0), (3, -1)$

In Exercises 25 and 26, find the least squares regression parabola

for the given points. Use Cramer's Rule and the system of equations given for Exercises 45–48 in Section 7.3.

25. $(0, 1), (1, 0), (2, 3), (3, 5)$
26. $(-1, 1), (0, 4), (2, 2), (3, -1)$

Review Exercises / Chapter 8

In Exercises 1–12, use matrices and elementary row operations to solve the systems of equations.

1. $5x + 4y = 2$
    $-x + y = -22$

2. $2x - 5y = 2$
    $3x - 7y = 1$

3. $0.2x - 0.1y = 0.07$
    $0.4x - 0.5y = -0.01$

4. $2x + y = 0.3$
    $3x - y = -1.3$

5. $-x + y + 2z = 1$
    $2x + 3y + z = -2$
    $5x + 4y + 2z = 4$

6. $2x + 3y + z = 10$
    $2x - 3y - 3z = 22$
    $4x - 2y + 3z = -2$

7. $2x + 3y + 3z = 3$
    $6x + 6y + 12z = 13$
    $12x + 9y - z = 2$

8. $4x + 4y + 4z = 5$
    $4x - 2y - 8z = 1$
    $5x + 3y + 8z = 6$

9. $2x + y + 2z = 4$
    $2x + 2y = 5$
    $2x - y + 6z = 2$

10. $3x + 21y - 29z = -1$
    $2x + 15y - 21z = 0$

11. $x + 2y + 6z = 1$
    $2x + 5y + 15z = 4$
    $3x + y + 3z = -6$

12. $x_1 + 5x_2 + 3x_3 = 14$
    $4x_2 + 2x_3 + 5x_4 = 3$
    $3x_3 + 8x_4 + 6x_5 = 16$
    $2x_1 + 4x_2 - 2x_5 = 0$
    $2x_1 - x_3 = 0$

In Exercises 13–20, perform (if possible) the indicated matrix operations.

13. $\begin{bmatrix} 2 & 1 & 0 \\ 0 & 5 & -4 \end{bmatrix} - 3\begin{bmatrix} 5 & 3 & -6 \\ 0 & -2 & 5 \end{bmatrix}$

14. $-2\begin{bmatrix} 1 & 2 \\ 5 & -4 \\ 6 & 0 \end{bmatrix} + 8\begin{bmatrix} 7 & 1 \\ 1 & 2 \\ 1 & 4 \end{bmatrix}$

15. $\begin{bmatrix} 1 & 2 \\ 5 & -4 \\ 6 & 0 \end{bmatrix}\begin{bmatrix} 6 & -2 & 8 \\ 4 & 0 & 0 \end{bmatrix}$

16. $\begin{bmatrix} 1 & 5 & 6 \\ 2 & -4 & 0 \end{bmatrix}\begin{bmatrix} 6 & -2 & 8 \\ 4 & 0 & 0 \end{bmatrix}$

17. $\begin{bmatrix} 1 & 5 & 6 \\ 2 & -4 & 0 \end{bmatrix}\begin{bmatrix} 6 & 4 \\ -2 & 0 \\ 8 & 0 \end{bmatrix}$

18. $\begin{bmatrix} 4 \\ 6 \end{bmatrix}[6 \quad -2]$

19. $\begin{bmatrix} 1 & 3 & 2 \\ 0 & 2 & -4 \\ 0 & 0 & 3 \end{bmatrix}\begin{bmatrix} 4 & -3 & 2 \\ 0 & 3 & -1 \\ 0 & 0 & 2 \end{bmatrix}$

20. $\begin{bmatrix} 2 & 1 \\ 6 & 0 \end{bmatrix}\left\{\begin{bmatrix} 4 & 2 \\ -3 & 1 \end{bmatrix} + \begin{bmatrix} -2 & 4 \\ 0 & 4 \end{bmatrix}\right\}$

21. Write out the system of linear equations represented by

$$\begin{bmatrix} 5 & 4 \\ -1 & 1 \end{bmatrix}\begin{bmatrix} x \\ y \end{bmatrix} = \begin{bmatrix} 2 \\ -22 \end{bmatrix}$$

22. Express the following system of linear equations in matrix form:

$$2x + 3y + z = 10$$
$$2x - 3y - 3z = 22$$
$$4x - 2y + 3z = -2$$

In Exercises 23–28, solve the system of linear equations using (a) the inverse of the coefficient matrix and (b) Cramer's Rule.

23. The system of equations in Exercise 1.
24. The system of equations in Exercise 2.
25. The system of equations in Exercise 5.
26. The system of equations in Exercise 6.
27. The system of equations in Exercise 7.
28. The system of equations in Exercise 8.
29. If A is a 3×3 matrix and $|A| = 2$, then what is the value of $|4A|$? Give the reason for your answer.
30. Prove that
$$\begin{vmatrix} a_{11} & a_{12} & a_{13} \\ a_{21} & a_{22} & a_{23} \\ a_{31} + x & a_{32} + y & a_{33} + z \end{vmatrix} =$$
$$\begin{vmatrix} a_{11} & a_{12} & a_{13} \\ a_{21} & a_{22} & a_{23} \\ a_{31} & a_{32} & a_{33} \end{vmatrix} + \begin{vmatrix} a_{11} & a_{12} & a_{13} \\ a_{21} & a_{22} & a_{23} \\ x & y & z \end{vmatrix}$$

The *characteristic equation* of a square matrix A is the equation

$$|A - \lambda I| = 0$$

Given the matrix

$$A = \begin{bmatrix} 3 & 2 \\ -1 & 4 \end{bmatrix}$$

the characteristic equation is

$$|A - \lambda I| = \begin{vmatrix} 3 - \lambda & 2 \\ -1 & 4 - \lambda \end{vmatrix} = \lambda^2 - 2\lambda + 14 = 0$$

It can be shown that a square matrix always satisfies its characteristic equation, so that in this case, $A^2 - 2A + 14I = 0$. In Exercises 31–34, find the characteristic equation and demonstrate that the matrix satisfies the equation.

31. $A = \begin{bmatrix} 2 & 1 \\ -3 & -1 \end{bmatrix}$ 32. $A = \begin{bmatrix} 10 & 4 \\ 2 & 1 \end{bmatrix}$

33. $A = \begin{bmatrix} 2 & 1 & -1 \\ 0 & 1 & 5 \\ -1 & 5 & 2 \end{bmatrix}$ 34. $A = \begin{bmatrix} 6 & 0 & 4 \\ -2 & 1 & 3 \\ -9 & 0 & 4 \end{bmatrix}$

CHAPTER

9 SEQUENCES, SERIES, AND PROBABILITY

Sequences and Summation Notation

9.1

In this chapter, we look at several special topics in algebra that are extensions or variations of concepts studied earlier, in the basic algebra portion of the text. We hope that you will find these special topics (many of which are used in calculus) to be both interesting and challenging. We begin with the notion of a **sequence.**

Infinite Sequences

In mathematics, the word "sequence" is used in much the same way as it is in ordinary English. When we say that a collection of objects is listed "in sequence," we usually mean that the collection is ordered in that it has an identified first member, second member, third member, and so on. Mathematically, we define a sequence as a *function* whose domain is the set of positive integers. For instance, the equation

$$a(n) = \frac{1}{2^n}$$

defines the sequence

$$\left\{ \frac{1}{2}, \frac{1}{4}, \frac{1}{8}, \frac{1}{16}, \ldots , \frac{1}{2^n}, \ldots \right\}$$

the terms of which correspond respectively to

$$\{a(1), a(2), a(3), a(4), \ldots , a(n), \ldots\}$$

We will generally write a sequence in the *subscript form,*

$$\{a_1, a_2, a_3, a_4, \ldots, a_n, \ldots\}$$

or we will denote it by $\{a_n\}$, where a_n is the **nth term** of the sequence.

DEFINITION OF A SEQUENCE

> A **sequence** $\{a_n\}$ is a function whose domain is the set of positive integers. The functional values $\{a_1, a_2, a_3, \ldots, a_n, \ldots\}$ are called the **terms** of the sequence.

Remark: Because the terms of a sequence comprise an infinite set, it is common to refer to a sequence as an **infinite sequence.** On occasion it is convenient to begin subscripting a sequence with zero instead of one.

EXAMPLE 1
Finding Terms in a Sequence

List the first four terms of the sequences with the following *n*th terms (in each case, assume *n* begins with 1):

(a) $a_n = 3 + (-1)^n$ (b) $b_n = \dfrac{2n}{1 + n}$

Solution:

(a) $\{a_n\} = \{3 + (-1)^n\}$

$$= \{3 + (-1)^1, 3 + (-1)^2, 3 + (-1)^3, 3 + (-1)^4, \ldots\}$$

$$= \{3 - 1, 3 + 1, 3 - 1, 3 + 1, \ldots\}$$

$$= \{2, 4, 2, 4, \ldots\}$$

(b) $\{b_n\} = \left\{\dfrac{2n}{1 + n}\right\}$

$$= \left\{\dfrac{2(1)}{1 + 1}, \dfrac{2(2)}{1 + 2}, \dfrac{2(3)}{1 + 3}, \dfrac{2(4)}{1 + 4}, \cdots\right\}$$

$$= \left\{\dfrac{2}{2}, \dfrac{4}{3}, \dfrac{6}{4}, \dfrac{8}{5}, \cdots\right\}$$

Some very important sequences in mathematics involve terms that are defined with special types of products, called **factorials.**

DEFINITION OF FACTORIAL

> If *n* is a positive integer, then **n factorial** is defined by
>
> $$n! = 1 \cdot 2 \cdot 3 \cdot 4 \cdots (n - 1) \cdot n$$
>
> If $n = 0$, then $0!$ is defined to be
>
> $$0! = 1$$

It is helpful to know the value of $n!$ for the following values of n:

$0! = 1$	$3! = 1 \cdot 2 \cdot 3 = 6$
$1! = 1$	$4! = 1 \cdot 2 \cdot 3 \cdot 4 = 24$
$2! = 1 \cdot 2 = 2$	$5! = 1 \cdot 2 \cdot 3 \cdot 4 \cdot 5 = 120$

Factorials follow the same conventions for order of operation as do exponents.

$$2n! = 2(n!) = 2(1 \cdot 2 \cdot 3 \cdot 4 \cdots n)$$
$$(2n)! = 1 \cdot 2 \cdot 3 \cdot 4 \cdots n \cdot (n + 1) \cdots (2n)$$

EXAMPLE 2
Finding Terms of a Sequence

List the first four terms of the following sequences:

(a) $\{a_n\} = \left\{\dfrac{2^n}{n!}\right\}$ (Begin at $n = 0$)

(b) $\{b_n\} = \left\{\dfrac{(2n)!}{2^n n!}\right\}$ (Begin at $n = 1$)

Solution:

(a) $\{a_n\} = \left\{\dfrac{2^n}{n!}\right\} = \left\{\dfrac{2^0}{0!}, \dfrac{2^1}{1!}, \dfrac{2^2}{2!}, \dfrac{2^3}{3!}, \cdots\right\} = \left\{\dfrac{1}{1}, \dfrac{2}{1}, \dfrac{4}{2}, \dfrac{8}{6}, \cdots\right\}$

(b) $\{b_n\} = \left\{\dfrac{(2n)!}{2^n n!}\right\}$

$$b_1 = \frac{2!}{2^1(1!)} = \frac{1 \cdot 2}{2(1)} = \frac{1 \cdot \cancel{2}}{\cancel{2}} = 1$$

$$b_2 = \frac{4!}{2^2(2!)} = \frac{1 \cdot 2 \cdot 3 \cdot 4}{2 \cdot 2(1 \cdot 2)}$$
$$= \frac{1 \cdot \cancel{2} \cdot 3 \cdot \cancel{4}}{\cancel{2} \cdot \cancel{4}} = 1 \cdot 3$$

$$b_3 = \frac{6!}{2^3(3!)} = \frac{1 \cdot 2 \cdot 3 \cdot 4 \cdot 5 \cdot 6}{2 \cdot 2 \cdot 2(1 \cdot 2 \cdot 3)}$$
$$= \frac{1 \cdot \cancel{2} \cdot 3 \cdot \cancel{4} \cdot 5 \cdot \cancel{6}}{\cancel{2} \cdot \cancel{4} \cdot \cancel{6}} = 1 \cdot 3 \cdot 5$$

$$b_4 = \frac{8!}{2^4(4!)} = \frac{1 \cdot 2 \cdot 3 \cdot 4 \cdot 5 \cdot 6 \cdot 7 \cdot 8}{2 \cdot 2 \cdot 2 \cdot 2(1 \cdot 2 \cdot 3 \cdot 4)}$$
$$= \frac{1 \cdot \cancel{2} \cdot 3 \cdot \cancel{4} \cdot 5 \cdot \cancel{6} \cdot 7 \cdot \cancel{8}}{\cancel{2} \cdot \cancel{4} \cdot \cancel{6} \cdot \cancel{8}} = 1 \cdot 3 \cdot 5 \cdot 7$$

$$\{b_n\} = \{1, (1 \cdot 3), (1 \cdot 3 \cdot 5), (1 \cdot 3 \cdot 5 \cdot 7), \ldots\}$$

Remark: Notice the cancellation that took place in part (b) of Example 2. It often happens in work with factorials that appropriate cancellation will greatly simplify an expression. For instance, the following types of cancellations are common:

$$\frac{n!}{(n-1)!} = \frac{1 \cdot 2 \cdot 3 \cdots (n-1) \cdot n}{1 \cdot 2 \cdot 3 \cdots (n-1)} = n$$

$$\frac{(2n-1)!}{(2n+1)!} = \frac{1 \cdot 2 \cdot 3 \cdots (2n-2)(2n-1)}{1 \cdot 2 \cdot 3 \cdots (2n-2)(2n-1)(2n)(2n+1)}$$

$$= \frac{1}{2n(2n+1)}$$

It is important to realize that simply listing the first few terms is not sufficient to define a sequence—the *n*th term *must* be given. To see this, consider the following sequences, both of which have the same first three terms:

$$\{a_n\} = \left\{ \frac{1}{2}, \frac{1}{4}, \frac{1}{8}, \cdots, \frac{1}{2^n}, \cdots \right\}$$

$$\{b_n\} = \left\{ \frac{1}{2}, \frac{1}{4}, \frac{1}{8}, \cdots, \frac{6}{(n+1)(n^2-n+6)}, \cdots \right\}$$

Alternating Signs

The terms of a sequence need not all be positive. One rule used to *alternate* the signs of terms of a sequence is the following:

$$\{(-1)^{n+1} a_n\} = \{a_1, -a_2, a_3, -a_4, \ldots\}, \qquad \text{where } n = 1, 2, 3, \ldots$$

Using $n+1$ as the power of (-1) causes the sequence to begin with a positive sign. If you want to begin a sequence with a negative sign, you can use n as the power of (-1).

EXAMPLE 3
Sequence with Alternating Signs

Find the first four terms and the thirty-seventh term of the sequence

$$\{a_n\} = \left\{ \frac{(-1)^n}{2n-1} \right\}$$

Solution:
We have

$$a_1 = \frac{(-1)^1}{2(1)-1} = \frac{-1}{1} = -1$$

$$a_2 = \frac{(-1)^2}{2(2)-1} = \frac{1}{3}$$

$$a_3 = \frac{(-1)^3}{2(3) - 1} = \frac{-1}{5}$$

$$a_4 = \frac{(-1)^4}{2(4) - 1} = \frac{1}{7}$$

.

.

.

$$a_{37} = \frac{(-1)^{37}}{2(37) - 1} = \frac{-1}{73}$$

One of the most important types of sequences used in mathematics is the sequence obtained by taking successive sums of terms of a given sequence. Consider the sequence

$$\{a_n\} = \{a_1, a_2, a_3, a_4, \ldots, a_n, \ldots\}$$

By finding the successive sums

$$S_1 = a_1$$
$$S_2 = a_1 + a_2$$
$$S_3 = a_1 + a_2 + a_3$$

.

.

.

$$S_n = a_1 + a_2 + a_3 + a_4 + \cdots + a_n$$

we can form a new sequence $\{S_n\}$, which is called the **sequence of partial sums**. In particular, S_n is called the ***n*th partial sum** of the sequence $\{a_n\}$.

A convenient shorthand notation for representing sums is called **sigma notation**. The name for this notation comes from the use of the uppercase Greek letter sigma, written as Σ.

DEFINITION OF SIGMA NOTATION

If $\{a_n\}$ is a sequence, then the sum of the first n terms of the sequence is represented by

$$\sum_{i=1}^{n} a_i = a_1 + a_2 + a_3 + a_4 + \cdots + a_n$$

where i is called the **index of summation**, n is the **upper limit of summation**, and 1 is the **lower limit of summation**.

In calculus, we refer to the summation

$$\sum_{i=1}^{n} a_i = a_1 + a_2 + a_3 + a_4 + \cdots + a_n$$

as a **series.** And if the summation has infinitely many terms, we call it an **infinite series.** Consider the sequence of partial sums

$$\{S_n\} = \{S_1, S_2, S_3, S_4, \ldots, S_n, \ldots\}$$

If there is a number S such that the value of S_n approaches S as n increases without bound (symbolically we write $S_n \to S$ as $n \to \infty$), then we call S the **sum** of the infinite series and write

$$S = a_1 + a_2 + a_3 + a_4 + \cdots + a_n + \cdots = \sum_{i=1}^{\infty} a_i$$

Remark: This infinite summation process involves the concept of *convergence,* a topic that can be treated more precisely in calculus.

EXAMPLE 4
Finding a Sequence
of Partial Sums

For the sequence

$$\{a_n\} = \left\{ \frac{3}{10^n} \right\}$$

find the first five partial sums, and use these results to try to evaluate the infinite summation

$$\sum_{i=1}^{\infty} \frac{3}{10^n}$$

Solution:
Since the terms of the given sequence are

$$\{a_n\} = \left\{ \frac{3}{10^1}, \frac{3}{10^2}, \frac{3}{10^3}, \frac{3}{10^4}, \frac{3}{10^5}, \cdots \right\}$$
$$= \{0.3, 0.03, 0.003, 0.0003, 0.00003, \ldots\}$$

it follows that

$$S_1 = 0.3$$
$$S_2 = 0.3 + 0.03 = 0.33$$
$$S_3 = 0.3 + 0.03 + 0.003 = 0.333$$
$$S_4 = 0.3 + 0.03 + 0.003 + 0.0003 = 0.3333$$
$$S_5 = 0.3 + 0.03 + 0.003 + 0.0003 + 0.00003 = 0.33333$$

From these results it appears that

$$S_n \to 0.3\overline{3} \quad \text{as} \quad n \to \infty \qquad \textit{(Recall that } 0.3\overline{3} = 0.3333 \ldots)$$

which suggests that

$$\sum_{i=1}^{\infty} \frac{3}{10^n} = 0.3\overline{3} = \frac{1}{3}$$

EXAMPLE 5
Sigma Notation for a Sum

Write out the terms and evaluate each of the following sums:

(a) $\displaystyle\sum_{i=1}^{5} i$ (b) $\displaystyle\sum_{k=3}^{6} (1 + k^2)$ (c) $\displaystyle\sum_{i=0}^{8} \frac{1}{i!}$

Solution:

(a) $\displaystyle\sum_{i=1}^{5} i = 1 + 2 + 3 + 4 + 5 = 15$

(b) $\displaystyle\sum_{k=3}^{6} (1 + k^2) = (1 + 3^2) + (1 + 4^2) + (1 + 5^2) + (1 + 6^2)$

$$= 10 + 17 + 26 + 37 = 90$$

(c) $\displaystyle\sum_{i=0}^{8} \frac{1}{i!} = \frac{1}{0!} + \frac{1}{1!} + \frac{1}{2!} + \frac{1}{3!} + \frac{1}{4!} + \frac{1}{5!} + \frac{1}{6!} + \frac{1}{7!} + \frac{1}{8!}$

$$= 1 + 1 + \frac{1}{2} + \frac{1}{6} + \frac{1}{24} + \frac{1}{120} + \frac{1}{720} + \frac{1}{5040} + \frac{1}{40320}$$

$$\approx 2.71828$$

Remark: Note that the lower index of a summation does not have to be 1. Also note that we can use any variable as the index. For instance, in part (b), the letter k is used.

The following properties of sigma notation are useful.

PROPERTIES OF SIGMA NOTATION

1. $\displaystyle\sum_{i=1}^{n} ca_i = c \sum_{i=1}^{n} a_i,$ c is any constant

2. $\displaystyle\sum_{i=1}^{n} (a_i + b_i) = \sum_{i=1}^{n} a_i + \sum_{i=1}^{n} b_i$

3. $\displaystyle\sum_{i=1}^{n} (a_i - b_i) = \sum_{i=1}^{n} a_i - \sum_{i=1}^{n} b_i$

Proof:
Each of these properties follows directly from the Associative, Commutative, and Distributive Properties of arithmetic. For example, note the use of the Distributive Property in the proof of Property 1.

$$\sum_{i=1}^{n} ca_i = ca_1 + ca_2 + ca_3 + \cdots + ca_n$$

$$= c(a_1 + a_2 + a_3 + \cdots + a_n)$$

$$= c \sum_{i=1}^{n} a_i$$

Be aware that variations in the upper and lower limits of summation can produce quite different-looking sigma notations for the same sum. For example,

$$\sum_{i=1}^{5} 3(2^i) = 3 \sum_{i=1}^{5} 2^i = 3(2^1 + 2^2 + 2^3 + 2^4 + 2^5)$$

$$\sum_{i=0}^{4} 3(2^{i+1}) = 3 \sum_{i=0}^{4} 2^{i+1} = 3(2^1 + 2^2 + 2^3 + 2^4 + 2^5)$$

Section Exercises 9.1

In Exercises 1–20, write out the first five terms of the specified sequence. (In each case, assume n starts with 1.)

1. $a_n = 2^n$

2. $a_n = \dfrac{n}{n+1}$

3. $a_n = (-\frac{1}{2})^n$

4. $a_n = \dfrac{n^2 - 1}{n^2 + 2}$

5. $a_n = \dfrac{3^n}{n!}$

6. $a_n = 5 - \dfrac{1}{n} + \dfrac{1}{n^2}$

7. $a_n = \dfrac{(-1)^n}{n^2}$

8. $a_n = \dfrac{3n!}{(n-1)!}$

9. $a_n = \dfrac{n+1}{n}$

10. $a_n = \dfrac{1}{n^{3/2}}$

11. $a_n = \dfrac{n^2 - 1}{n + 1}$

12. $a_n = 1 + (-1)^n$

13. $a_n = \dfrac{1 + (-1)^n}{n}$

14. $a_n = \dfrac{n!}{n}$

15. $a_n = 3 - \dfrac{1}{2^n}$

16. $a_n = \dfrac{3^n}{4^n}$

17. $a_n = (-1)^n \left(\dfrac{n}{n+1} \right)$

18. $a_n = \dfrac{3n^2 - n + 4}{2n^2 + 1}$

19. $a_1 = 3$ and $a_{k+1} = 2(a_k - 1)$

20. $a_1 = 4$ and $a_{k+1} = a_k \left(\dfrac{k+1}{2} \right)$

In Exercises 21–36, write an expression for the nth term of the sequence.

21. $\{1, 4, 7, 10, 13, \ldots\}$

22. $\{3, 7, 11, 15, 19, \ldots\}$

23. $\{-1, 2, 7, 14, 23, \ldots\}$

24. $\left\{ 1, \dfrac{1}{4}, \dfrac{1}{9}, \dfrac{1}{16}, \dfrac{1}{25}, \cdots \right\}$

25. $\left\{ \dfrac{2}{3}, \dfrac{3}{4}, \dfrac{4}{5}, \dfrac{5}{6}, \dfrac{6}{7}, \cdots \right\}$

26. $\left\{ \dfrac{2}{1}, \dfrac{3}{3}, \dfrac{4}{5}, \dfrac{5}{7}, \dfrac{6}{9}, \cdots \right\}$

27. $\left\{ \dfrac{-1}{1}, \dfrac{1}{2}, \dfrac{-1}{4}, \dfrac{1}{8}, \dfrac{-1}{16}, \cdots \right\}$

28. $\left\{ \dfrac{1}{2}, \dfrac{1}{3}, \dfrac{2}{9}, \dfrac{4}{27}, \dfrac{8}{81}, \cdots \right\}$

29. $\left\{ 2, \left(1 + \dfrac{1}{2}\right), \left(1 + \dfrac{1}{3}\right), \left(1 + \dfrac{1}{4}\right), \left(1 + \dfrac{1}{5}\right), \cdots \right\}$

30. $\left\{ \left(1 + \dfrac{1}{2}\right), \left(1 + \dfrac{3}{4}\right), \left(1 + \dfrac{7}{8}\right), \left(1 + \dfrac{15}{16}\right), \right.$
$\left. \left(1 + \dfrac{31}{32}\right), \cdots \right\}$

31. $\left\{ \dfrac{1}{2 \cdot 3}, \dfrac{2}{3 \cdot 4}, \dfrac{3}{4 \cdot 5}, \dfrac{4}{5 \cdot 6}, \dfrac{5}{6 \cdot 7}, \cdots \right\}$

32. $\left\{ 1, \dfrac{1}{2}, \dfrac{1}{6}, \dfrac{1}{24}, \dfrac{1}{120}, \cdots \right\}$

33. $\left\{1, \dfrac{1}{1 \cdot 3}, \dfrac{1}{1 \cdot 3 \cdot 5}, \dfrac{1}{1 \cdot 3 \cdot 5 \cdot 7}, \right.$

$\left. \dfrac{1}{1 \cdot 3 \cdot 5 \cdot 7 \cdot 9}, \ldots \right\}$

34. $\{2, -4, 6, -8, 10, \ldots\}$

35. $\{1, -1, 1, -1, 1, \ldots\}$

36. $\left\{1, x, \dfrac{x^2}{2}, \dfrac{x^3}{6}, \dfrac{x^4}{24}, \dfrac{x^5}{120}, \ldots \right\}$

In Exercises 37–50, find the given sum.

37. $\displaystyle\sum_{i=1}^{5} (2i + 1)$

38. $\displaystyle\sum_{i=1}^{6} 2i$

39. $\displaystyle\sum_{k=0}^{4} \dfrac{1}{k^2 + 1}$

40. $\displaystyle\sum_{j=3}^{5} \dfrac{1}{j}$

41. $\displaystyle\sum_{k=1}^{4} 10$

42. $\displaystyle\sum_{n=1}^{10} \dfrac{3}{n + 1}$

43. $\displaystyle\sum_{k=1}^{5} C,\ C$ is constant

44. $\displaystyle\sum_{n=1}^{5} \dfrac{C}{n + 1},\ C$ is constant

45. $\displaystyle\sum_{i=1}^{4} [(i - 1)^2 + (i + 1)^3]$

46. $\displaystyle\sum_{k=2}^{5} (k + 1)(k - 3)$

47. $\displaystyle\sum_{i=1}^{4} (x^2 + 2i)$

48. $\displaystyle\sum_{y=0}^{5} (x^2 + y^2)$

49. $\displaystyle\sum_{x=1}^{3} (x^2 + 4ix)$

50. $\displaystyle\sum_{j=1}^{4} (-2)^{j-1}$

In Exercises 51–60, use sigma notation to write the given sum.

51. $\dfrac{1}{3(1)} + \dfrac{1}{3(2)} + \dfrac{1}{3(3)} + \cdots + \dfrac{1}{3(9)}$

52. $\dfrac{5}{1 + 1} + \dfrac{5}{1 + 2} + \dfrac{5}{1 + 3} + \cdots + \dfrac{5}{1 + 15}$

53. $\left[2\left(\dfrac{1}{8}\right) + 3 \right] + \left[2\left(\dfrac{2}{8}\right) + 3 \right]$

$+ \left[2\left(\dfrac{3}{8}\right) + 3 \right] + \cdots + \left[2\left(\dfrac{8}{8}\right) + 3 \right]$

54. $\left[1 - \left(\dfrac{1}{4}\right)^2 \right] + \left[1 - \left(\dfrac{2}{4}\right)^2 \right]$

$+ \left[1 - \left(\dfrac{3}{4}\right)^2 \right] + \left[1 - \left(\dfrac{4}{4}\right)^2 \right]$

55. $3 - 9 + 27 - 81 + 243 - 729$

56. $1 - \dfrac{1}{2} + \dfrac{1}{4} - \dfrac{1}{8} + \cdots - \dfrac{1}{128}$

57. $\dfrac{1}{1^2} - \dfrac{1}{2^2} + \dfrac{1}{3^2} - \dfrac{1}{4^2} + \cdots - \dfrac{1}{20^2}$

58. $\dfrac{1}{1 \cdot 3} + \dfrac{1}{2 \cdot 4} + \dfrac{1}{3 \cdot 5} + \cdots + \dfrac{1}{10 \cdot 12}$

59. $\dfrac{1}{4} + \dfrac{3}{8} + \dfrac{7}{16} + \dfrac{15}{32} + \dfrac{31}{64}$

60. $\dfrac{1}{2} + \dfrac{2}{4} + \dfrac{6}{8} + \dfrac{24}{16} + \dfrac{120}{32} + \dfrac{720}{64}$

In Exercises 61 and 62, use the following definition of the arithmetic mean $\bar{x}$ of a set of n measurements $\{x_1, x_2, x_3, \ldots, x_n\}$:

$$\bar{x} = \dfrac{1}{n} \sum_{i=1}^{n} x_i$$

61. Prove that $\displaystyle\sum_{i=1}^{n} (x_i - \bar{x}) = 0$.

62. Prove that $\displaystyle\sum_{i=1}^{n} (x_i - \bar{x})^2 = \sum_{i=1}^{n} x_i^2 - \dfrac{1}{n}\left(\sum_{i=1}^{n} x_i\right)^2$.

63. A deposit of $5000 is made in an account that earns 8% interest compounded quarterly. The balance in the account after n quarters is given by

$$A_n = 5000\left(1 + \dfrac{0.08}{4}\right)^n, \qquad n = 1, 2, 3, \ldots$$

(a) Compute the first eight terms of this sequence.

(b) Find the balance in this account after 10 years by computing the fortieth term of the sequence.

64. A deposit of $100 is made *each* month in an account that earns 12% interest compounded monthly. The balance in the account after n months is given by

$$A_n = 100(101)[(1.01)^n - 1], \qquad n = 1, 2, 3, \ldots$$

(a) Compute the first six terms of this sequence.

(b) Find the balance after 5 years by computing the sixtieth term of the sequence.

(c) Find the balance after 20 years by computing the two hundred fortieth term of the sequence.

65. Simplify the following ratios of factorials:

(a) $\dfrac{10!}{8!}$

(b) $\dfrac{25!}{23!}$

(c) $\dfrac{n!}{(n - 2)!}$

Arithmetic Sequences and Series

9.2

In this section we look at a special type of sequence called an **arithmetic sequence** (or **arithmetic progression**). In an arithmetic sequence, consecutive terms have a common difference. For instance, the sequence

$$\{a_n\} = \{2 + 3n\}$$
$$= \{(2 + 3), (2 + 6), (2 + 9), (2 + 12), \ldots\}$$
$$= \{5, 8, 11, 14, \ldots\}$$

$$8 - 5 = 3 \qquad 11 - 8 = 3$$

has a common difference of 3 between consecutive terms.

DEFINITION OF AN ARITHMETIC SEQUENCE

> The sequence $\{a_n\}$ is called an **arithmetic sequence** if each pair of consecutive terms differs by the same amount
>
> $$d = a_{i+1} - a_i$$
>
> The number d is called the **common difference** of the sequence.

EXAMPLE 1
Finding the Common Difference of an Arithmetic Sequence

Which of the following sequences are arithmetic? If arithmetic, find the common difference.

(a) $\{a_n\} = \{3 - 5n\}$ (b) $\{b_n\} = \{-2 + 3^n\}$

Solution:

(a) For the sequence

$$\{a_n\} = \{3 - 5n\} = \{3, -2, -7, -12, -17, \ldots\}$$

the difference between two consecutive terms is

$$a_{i+1} - a_i = [3 - 5(i + 1)] - [3 - 5(i)]$$
$$= 3 - 5i - 5 - 3 + 5i = -5$$

Thus, the sequence *is* arithmetic, and the common difference is $d = -5$.

(b) For the sequence

$$\{b_n\} = \{-2 + 3^n\} = \{(-2 + 3), (-2 + 9),$$
$$(-2 + 27), (-2 + 81), \ldots\}$$
$$= \{1, 7, 25, 79, \ldots\}$$

we see that the difference between the first two terms is

$$b_2 - b_1 = 7 - 1 = 6$$

and the difference between the second and third terms is

$$b_3 - b_2 = 25 - 7 = 18$$

Since these two differences are not the same, we conclude that the sequence is *not* arithmetic.

Remark: In Example 1, part (a), be sure you see that the common difference of an arithmetic sequence can be negative.

There is a simple method for writing out the terms of an arithmetic sequence. If you know the first term a_1 and the common difference d, then you can form the other terms by repeatedly adding d. For instance,

$$a_2 = a_1 + d, \qquad a_3 = a_2 + d, \qquad a_4 = a_3 + d$$

and, in general, we have

$$a_{i+1} = a_i + d$$

This is called a **recursive formula,** since it defines a given term by reference to the preceding term.

Using this recursive idea, it is possible to obtain a formula for finding the nth term of an arithmetic sequence *without* having to identify all the $n - 1$ preceding terms. Consider an arithmetic sequence with a_1 as its first term and d as the common difference. Then,

$$
\begin{aligned}
a_1 &= a_1 & &= a_1 + (0)d \\
a_2 &= a_1 + d & &= a_1 + (1)d \\
a_3 &= a_1 + d + d & &= a_1 + (2)d \\
a_4 &= a_1 + d + d + d & &= a_1 + (3)d \\
a_5 &= a_1 + d + d + d + d & &= a_1 + (4)d
\end{aligned}
$$

This leads to the following rule.

THE nTH TERM OF AN ARITHMETIC SEQUENCE

> The nth term of an arithmetic sequence whose first term is a_1 and whose common difference is d is given by
>
> $$a_n = a_1 + (n - 1)d$$

EXAMPLE 2
Finding the nth Term of an
Arithmetic Sequence

Find the indicated term of the following arithmetic sequences:
(a) The sixty-fourth term of the sequence whose first three terms are

$$\{a_n\} = \{-1, 11, 23, \ldots\}$$

(b) The thirty-eighth term of the sequence whose first term is 8 and whose ith term is given by the recursive formula

$$a_{i+1} = a_i - 7$$

Solution:

(a) For the sequence

$$\{a_n\} = \{-1, 11, 23, \ldots\}$$

we find that $a_1 = -1$ and $d = 12$. Therefore, the sixty-fourth term of the sequence is

$$a_{64} = a_1 + (n - 1)d = -1 + 63(12) = -1 + 756 = 755$$

(b) From the recursive formula $a_{i+1} = a_i - 7$, we can see that $d = -7$. Since we are given that $a_1 = 8$, we find that the thirty-eighth term of the sequence is

$$a_{38} = a_1 + (n - 1)d = 8 + 37(-7) = 8 - 259 = -251$$

EXAMPLE 3
Finding the nth Term of an
Arithmetic Sequence

The seventh term of a certain arithmetic sequence is 55, and the twenty-second term is 145. Find the eighteenth term.

Solution:

Using the nth term formula, we have

$$a_7 = a_1 + 6d \quad \text{and} \quad a_{22} = a_1 + 21d$$
$$55 = a_1 + 6d \qquad\qquad 145 = a_1 + 21d$$

Solving this system of linear equations, we have

$$a_1 + 21d = 145$$
$$\underline{-a_1 - \ \ 6d = -55}$$
$$\qquad\quad 15d = 90 \quad \text{or} \quad d = 6$$

Backsubstitution yields $a_1 = 19$, and thus the eighteenth term is

$$a_{18} = a_1 + (n - 1)d = 19 + 17(6) = 121$$

Arithmetic Mean

Recall that $(a + b)/2$ is the midpoint between the two numbers a and b on the real number line. As a result, the terms

$$a, \frac{a + b}{2}, b$$

have a common difference. We call $(a + b)/2$ the **arithmetic mean** of the numbers a and b. We can generalize this concept by finding k numbers m_1, m_2, m_3, . . ., m_k between a and b such that the terms

$$a, m_1, m_2, m_3, \ldots, m_k, b$$

have a common difference. This process is referred to as **inserting k arithmetic means** between a and b.

EXAMPLE 4
*Inserting k Arithmetic Means
Between Two Numbers*

Insert three arithmetic means between 4 and 15.

Solution:
We need to find three numbers m_1, m_2, and m_3 such that the terms

$$4, m_1, m_2, m_3, 15$$

have a common difference. In this case we have $a_1 = 4$, $n = 5$, and $a_5 = 15$. Therefore,

$$a_5 = 15 = a_1 + (n - 1)d = 4 + 4d$$

Hence, $d = \frac{11}{4}$, and the three arithmetic means are

$$m_1 = a_1 + d = 4 + \tfrac{11}{4} = \tfrac{27}{4}$$
$$m_2 = m_1 + d = \tfrac{27}{4} + \tfrac{11}{4} = \tfrac{38}{4}$$
$$m_3 = m_2 + d = \tfrac{38}{4} + \tfrac{11}{4} = \tfrac{49}{4}$$

Arithmetic Series

A series that is formed from an arithmetic sequence is called an **arithmetic series.** For example, the arithmetic sequence

$$\{a_n\} = \{1, 3, 5, 7, 9, \ldots, 2n - 1, \ldots\}$$

gives rise to the following sequence of partial sums:

$$S_1 = 1 = 1 = 1^2$$
$$S_2 = 1 + 3 = 4 = 2^2$$
$$S_3 = 1 + 3 + 5 = 9 = 3^2$$
$$S_4 = 1 + 3 + 5 + 7 = 16 = 4^2$$
$$S_5 = 1 + 3 + 5 + 7 + 9 = 25 = 5^2$$

Judging from the pattern formed by these first five partial sums, the nth partial sum appears to be

$$S_n = 1 + 3 + 5 + 7 + 9 + \cdots + 2n - 1 = \sum_{i=1}^{n} (2i - 1) = n^2$$

We can verify this observation using the following formula for finding the nth partial sum of an arithmetic sequence.

THE nTH PARTIAL SUM OF AN ARITHMETIC SEQUENCE

The nth partial sum of the arithmetic sequence $\{a_n\}$ with common difference d is given by either of the following formulas:

1. $S_n = \dfrac{n}{2}(a_1 + a_n)$

2. $S_n = \dfrac{n}{2}[2a_1 + (n - 1)d]$

Proof:

There are two different ways to generate the terms of an arithmetic sequence. One way is to repeatedly add d to the first term, and another way is to repeatedly subtract d from the nth term. Thus, we have

$$S_n = a_1 + a_2 + a_3 + \cdots + a_{n-2} + a_{n-1} + a_n$$
$$= a_1 + [a_1 + d] + [a_1 + 2d] + \cdots + [a_1 + (n - 3)d]$$
$$+ [a_1 + (n - 2)d] + [a_1 + (n - 1)d]$$

and by adding terms in the opposite order, we have

$$S_n = a_n + a_{n-1} + a_{n-2} + \cdots + a_3 + a_2 + a_1$$
$$= a_n + [a_n - d] + [a_n - 2d] + \cdots + [a_n - (n - 3)d]$$
$$+ [a_n - (n - 2)d] + [a_n - (n - 1)d]$$

Now, the multiples of d cancel when these two versions of S_n are added, and we have

$$2S_n = \underbrace{(a_1 + a_n) + (a_1 + a_n) + (a_1 + a_n) + \cdots + (a_1 + a_n)}_{n \text{ terms}}$$

$$= n(a_1 + a_n)$$

Thus, we have

$$S_n = \dfrac{n}{2}(a_1 + a_n)$$

But since $a_n = a_1 + (n - 1)d$,

$$S_n = \dfrac{n}{2}[2a_1 + (n - 1)d]$$

Using this formula, we verify our observation that the sum of the first n odd integers is n^2.

$$S_n = 1 + 3 + 5 + \cdots + 2n - 1 = \frac{n}{2}[a_1 + a_n]$$

$$= \frac{n}{2}[1 + (2n - 1)] = \frac{n}{2}(2n) = n^2$$

EXAMPLE 5

Finding the *n*th Partial Sum of an Arithmetic Sequence

Find the following partial sums:

(a) The sum of the first 99 terms of the sequence given by

$$\{a_n\} = \left\{ 7 + \frac{n}{2} \right\}$$

(b) The sum of the first 150 terms of the sequence given by

$$\{5, 16, 27, 38, 49, \ldots\}$$

Solution:

(a) The sum of the first 99 terms of this sequence is

$$S_{99} = \frac{n}{2}(a_1 + a_n) = \frac{99}{2}\left(\left[7 + \frac{1}{2} \right] + \left[7 + \frac{99}{2} \right] \right)$$

$$= \frac{99}{2}\left(\frac{15}{2} + \frac{113}{2} \right) = \frac{99}{2}\left(\frac{128}{2} \right) = 3168$$

(b) For the sequence

$$\{5, 16, 27, 38, 49, \ldots\}$$

we have $a_1 = 5$ and $d = 11$. Therefore, the sum of the first 150 terms is

$$S_{150} = \frac{n}{2}[2a_1 + (n - 1)d] = \frac{150}{2}[2(5) + 149(11)]$$

$$= 75(10 + 1639) = 123{,}675$$

EXAMPLE 6

An Application to Business

A small business sells \$10,000 worth of products during its first year. The owner of the business has set a goal of increasing annual sales by \$7500 each year for 9 years. Assuming that this goal is met, find the total sales during the first 10 years this business is in operation.

Solution:

The annual sales form an arithmetic sequence (since they increase by the same amount each year), and we can find the total sales for the first 10 years as follows:

$$S_{10} = \frac{n}{2}[2a_1 + (n - 1)d] = \frac{10}{2}[20{,}000 + 9(7500)]$$

$$= 5(87{,}500) = \$437{,}500$$

Section Exercises 9.2

In Exercises 1–10, determine if the sequence is arithmetic. If it is, find the common difference.

1. $\{4, 7, 10, 13, 16, \ldots\}$ 2. $\{10, 8, 6, 4, 2, \ldots\}$
3. $\{1, 2, 4, 8, 16, \ldots\}$ 4. $\{3, \frac{5}{2}, 2, \frac{3}{2}, 1, \ldots\}$
5. $\{\frac{9}{4}, 2, \frac{7}{4}, \frac{3}{2}, \frac{5}{4}, \ldots\}$
6. $\{-12, -8, -4, 0, 4, \ldots\}$
7. $\{\frac{1}{3}, \frac{2}{3}, \frac{4}{3}, \frac{8}{3}, \frac{16}{3}, \ldots\}$
8. $\{\ln 1, \ln 2, \ln 3, \ln 4, \ln 5, \ldots\}$
9. $\{5.3, 5.7, 6.1, 6.5, 6.9, \ldots\}$
10. $\{1^2, 2^2, 3^2, 4^2, 5^2, \ldots\}$

In Exercises 11–18, write the first five terms of the arithmetic sequence.

11. $a_1 = 5, d = 6$ 12. $a_1 = 5, d = -\frac{3}{4}$
13. $a_1 = -2.6, d = -0.4$ 14. $a_1 = 6, a_{k+1} = a_k + 12$
15. $a_1 = \frac{3}{2}, a_{k+1} = a_k - \frac{1}{4}$ 16. $a_5 = 28, a_{10} = 53$
17. $a_8 = 26, a_{12} = 42$ 18. $a_3 = 1.9, a_{15} = -1.7$

In Exercises 19–26, find a_n for the given arithmetic sequence.

19. $a_1 = 1, d = 3, n = 10$
20. $a_1 = 15, d = 4, n = 25$
21. $a_1 = 100, d = -8, n = 8$
22. $a_1 = 0, d = -\frac{2}{3}, n = 12$
23. $a_1 = x, d = 2x, n = 50$
24. $a_1 = -y, d = 5y, n = 10$
25. $\{4, \frac{3}{2}, -1, -\frac{7}{2}, \ldots\}, n = 10$
26. $\{10, 5, 0, -5, -10, \ldots\}, n = 50$

In Exercises 27–34, find the nth partial sum of the arithmetic sequence.

27. $\{8, 20, 32, 44, \ldots\}, n = 10$
28. $\{2, 8, 14, 20, \ldots\}, n = 25$
29. $\{-6, -2, 2, 6, \ldots\}, n = 50$
30. $\{0.5, 0.9, 1.3, 1.7, \ldots\}, n = 10$
31. $\{40, 37, 34, 31, \ldots\}, n = 10$
32. $\{1.50, 1.45, 1.40, 1.35, \ldots\}, n = 20$
33. $a_1 = 100, a_{25} = 220, n = 25$
34. $a_1 = 15, a_{100} = 307, n = 100$

35. Find the sum of the first 50 positive integers.
36. Find the sum of the first 100 positive even integers.
37. Find the sum of the first 100 positive multiples of 5.
38. Find the sum of the integers from 50 to 100 (inclusive).
39. A person accepts a position with a company and will receive a salary of $27,500 for the first year. The person is guaranteed a raise of $1500 per year for the first five years.
    (a) What will the salary be during the sixth year of employment?
    (b) How much will the company have paid the person by the end of the six years?
40. A freely falling object will fall 16 feet during the first second, 48 feet during the second second, 80 feet during the third second, 112 feet during the fourth second, and so on. What is the total distance the object will fall in 10 seconds if this pattern continues?
41. Determine the seating capacity of an auditorium with 30 rows of seats if there are 20 seats in the first row, 24 seats in the second row, 28 seats in the third row, and so on.
42. As a farmer bales a field of hay, each trip around the field gets shorter, since he is getting closer to the center. Suppose on the first round there were 267 bales and on the second round there were 253 bales. If you assume that the decrease will be the same on each round and there are 11 more trips, how many bales of hay will he get from the field?

In Exercises 43–46, find the given sum.

43. $\displaystyle\sum_{n=1}^{20} (2n + 5)$ 44. $\displaystyle\sum_{n=1}^{100} \left(\frac{n+4}{2}\right)$

45. $\displaystyle\sum_{n=0}^{50} (1000 - 5n)$ 46. $\displaystyle\sum_{n=0}^{100} \left(\frac{8-3n}{16}\right)$

In Exercises 47–50, insert k arithmetic means between the given pair of numbers.

47. $5, 17; k = 2$ 48. $24, 56; k = 3$
49. $3, 6; k = 3$ 50. $2, 5; k = 4$

Geometric Sequences and Series   9.3

In this section we will study another important type of sequence called a **geometric sequence** (or **geometric progression**), characterized by the fact that consecutive terms have a common *ratio*. For instance, the sequence

$$\{a_n\} = \{3(2^{n-1})\}$$
$$= \{3(2^0), 3(2^1), 3(2^2), 3(2^3), \ldots\}$$
$$= \{3, 6, 12, 24, \ldots\}$$

$$\frac{6}{3} = 2 \qquad \frac{12}{6} = 2$$

has a common ratio of 2 between consecutive terms.

DEFINITION OF A GEOMETRIC SEQUENCE

The sequence $\{a_n\}$ is called a **geometric sequence** if each pair of consecutive terms has the same (nonzero) ratio

$$r = \frac{a_{i+1}}{a_i}, \qquad r \neq 0$$

The number r is called the **common ratio** of the sequence.

EXAMPLE 1
Finding the Common Ratio of a Geometric Sequence

Determine which of the following sequences are geometric. For those that are geometric, find the common ratio.

(a) $\{a_n\} = \left\{\dfrac{2}{3^n}\right\}$ (b) $\{b_n\} = \{1, -1, 1, -1, 1, \ldots\}$

(c) $\{c_n\} = \{n!\}$

Solution:

(a) For the sequence

$$\{a_n\} = \left\{\frac{2}{3^n}\right\} = \left\{\frac{2}{3}, \frac{2}{9}, \frac{2}{27}, \frac{2}{81}, \ldots\right\}$$

the ratio of two consecutive terms is

$$r = \frac{a_{i+1}}{a_i} = \frac{2/3^{i+1}}{2/3^i} = \frac{2}{3^{i+1}}\left(\frac{3^i}{2}\right) = \frac{1}{3}$$

Thus, the sequence *is* geometric, and the common ratio is $r = \frac{1}{3}$.

(b) For the sequence

$$\{b_n\} = \{1, -1, 1, -1, 1, \ldots\}$$

the nth term is given by $b_n = (-1)^{n+1}$, and therefore the ratio of consecutive terms is

$$r = \frac{b_{i+1}}{b_i} = \frac{(-1)^{i+2}}{(-1)^{i+1}} = -1$$

Thus, the sequence *is* geometric, and the common ratio is $r = -1$.

(c) For the sequence

$$\{c_n\} = \{n!\} = \{1, 2, 6, 24, 120, 720, \ldots\}$$

the ratio of the first two terms is $2/1 = 2$ and the ratio of the third term to the second is $6/2 = 3$. Since these two ratios differ, we conclude that the sequence is *not* geometric.

By rewriting the equation for the common ratio r in the form

$$a_{i+1} = ra_i$$

we see that each successive term in a geometric sequence is an r-multiple of the preceding term. Consequently, starting with a_i and multiplying by r repeatedly, we obtain the geometric sequence

$$\{a_1, \ a_1r, \ a_1r^2, \ a_1r^3, \ a_1r^4, \ \ldots\}$$
$$\downarrow \quad \downarrow \quad \downarrow \quad \downarrow \quad \downarrow$$
$$\{a_1, \ a_2, \quad a_3, \quad a_4, \quad a_5, \ \ldots\}$$

From this, it appears that the nth term of a geometric sequence has the following form.

THE nTH TERM OF A GEOMETRIC SEQUENCE

The nth term of a geometric sequence whose first term is a_1 and whose common ratio is r is given by

$$a_n = a_1r^{n-1}$$

We postpone the formal proof of this formula until the next section on mathematical induction.

EXAMPLE 2
Finding the nth Term of a Geometric Sequence

Find the indicated term of the following geometric sequences:
(a) The fifteenth term of the sequence whose first term is 20 and whose common ratio is 1.05.
(b) The twelfth term of the sequence whose first three terms are

$$\{5, \ -15, \ 45, \ \ldots\}$$

Solution:

(a) Since $a_1 = 20$ and $r = 1.05$, the fifteenth term is

$$a_{15} = a_1r^{n-1} = 20(1.05)^{14} \approx 39.599$$

(b) For the geometric sequence

$$\{a_n\} = \{5, \ -15, \ 45, \ \ldots\}$$

the common ratio is $r = -3$. Therefore, since the first term is 5, we determine the twelfth term to be

$$a_{12} = a_1r^{n-1} = 5(-3)^{11} = 5(-177{,}147) = -885{,}735$$

EXAMPLE 3
Finding the *n*th Term
of a Geometric Sequence

The fourth term of a geometric sequence is 125, and the tenth term is 125/64. Find the fourteenth term.

Solution:
Using the *n*th term formula for a geometric sequence, we have

$$a_4 = a_1 r^3 \quad \text{and} \quad a_{10} = a_1 r^9$$

$$125 = a_1 r^3 \qquad \qquad \frac{125}{64} = a_1 r^9$$

We use substitution to solve this system of (nonlinear) equations. Solving for a_1 in the first equation, we have

$$125 = a_1 r^3 \quad \Longrightarrow \quad a_1 = \frac{125}{r^3}$$

Then by substituting in the second equation, we obtain

$$\frac{125}{64} = a_1 r^9 \quad \Longrightarrow \quad \frac{125}{64} = \left(\frac{125}{r^3}\right) r^9$$

$$\frac{1}{64} = r^6 \quad \Longrightarrow \quad r = \sqrt[6]{\frac{1}{64}} = \frac{1}{2}$$

By backsubstituting we find a_1 to be

$$125 = a_1 \left(\frac{1}{2}\right)^3$$

$$125(2^3) = a_1$$

$$1000 = a_1$$

Finally, using the *n*th term formula, we find the fourteenth term to be

$$a_{14} = a_1 r^{n-1} = 1000 \left(\frac{1}{2}\right)^{13} = \frac{1000}{8192} = \frac{125}{1024}$$

The following formula gives us a simple method of calculating the *n*th partial sum of a geometric sequence.

THE *n*TH PARTIAL SUM OF A GEOMETRIC SEQUENCE

The *n*th partial sum of the geometric sequence $\{a_n\}$ with common ratio r is given by the following formula:

$$S_n = \frac{a_1(1 - r^n)}{1 - r}, \qquad r \neq 1$$

Proof:
We begin by writing out the *n*th partial sum, as follows:

$$S_n = a_1 + a_1 r + a_1 r^2 + \cdots + a_1 r^{n-2} + a_1 r^{n-1}$$

Multiplication by r gives us

$$rS_n = a_1r + a_1r^2 + a_1r^3 + \cdots + a_1r^{n-1} + a_1r^n$$

By subtracting the second equation from the first, we have

$$
\begin{aligned}
S_n &= a_1 + a_1r + a_1r^2 + \cdots + a_1r^{n-2} + a_1r^{n-1} \\
-rS_n &= \quad\quad\; - a_1r - a_1r^2 - a_1r^3 - \cdots - a_1r^{n-1} - a_1r^n \\
\hline
S_n - rS_n &= a_1 - a_1r^n
\end{aligned}
$$

Therefore, $S_n(1 - r) = a_1(1 - r^n)$, and since $r \neq 1$, we have

$$S_n = \frac{a_1(1 - r^n)}{1 - r}$$

Remark: When using this formula for the nth partial sum of a geometric sequence, be careful to check that the index begins at $i = 1$. It is often convenient to have the index of a geometric sequence begin at $i = 0$, and in such cases you must adjust the formula for the nth partial sum, as shown in the following example.

EXAMPLE 4
Finding the nth Partial Sum of a Geometric Sequence

Find the following sums:

(a) $\displaystyle\sum_{i=1}^{12} 4(0.3)^{i-1}$ (b) $\displaystyle\sum_{i=0}^{10} 10\left(-\frac{1}{2}\right)^i$

Solution:

(a) By writing out a few of the terms, we have

$$\sum_{i=1}^{12} 4(0.3)^{i-1} = 4 + 4(0.3) + 4(0.3)^2$$
$$+ 4(0.3)^3 + \cdots + 4(0.3)^{11}$$

Now, since $a_1 = 4$ and $r = 0.3$,

$$\sum_{i=1}^{12} 4(0.3)^{i-1} = \frac{a_1(1 - r^n)}{1 - r} = \frac{4[1 - (0.3)^{12}]}{1 - 0.3} \approx 5.714$$

(b) By writing out a few of the terms, we have

$$\sum_{i=0}^{10} 10\left(-\frac{1}{2}\right)^i = 10 - 10\left(\frac{1}{2}\right) + 10\left(\frac{1}{2}\right)^2$$
$$- 10\left(\frac{1}{2}\right)^3 + \cdots + 10\left(\frac{1}{2}\right)^{10}$$

Now, we see that the *first term* is $a_1 = 10$ and $r = -\frac{1}{2}$. Moreover, by starting with $i = 0$ and ending with $i = 10$, we are summing $n = 11$ terms, which means that the partial sum is

$$\sum_{i=0}^{10} 10\left(-\frac{1}{2}\right)^i = \frac{a_1(1 - r^n)}{1 - r} = \frac{10[1 - (-\frac{1}{2})^{11}]}{1 + (\frac{1}{2})} \approx 6.670$$

EXAMPLE 5
An Application:
Compound Interest

A deposit of $50.00 is made the first day of each month in a savings account that pays 12% compounded monthly. What is the balance at the end of two years?

Solution:
The formula for compound interest is

$$A = P\left(1 + \frac{r}{n}\right)^{tn}$$

where A is amount (balance), P is initial deposit (principal), r is annual percentage rate (in decimal form), n is number of compoundings per year, and t is time (in years). To find the balance in the account after 24 months, it is helpful to consider each of the 24 deposits separately. For example, the first deposit will gain interest for a full 24 months, and its balance will be

$$A_{24} = 50\left(1 + \frac{0.12}{12}\right)^{24}$$

The second deposit will gain interest for 23 months, and its balance will be

$$A_{23} = 50\left(1 + \frac{0.12}{12}\right)^{23}$$

The last (24th) deposit will gain interest for only 1 month, and its balance will be

$$A_1 = 50\left(1 + \frac{0.12}{12}\right)^1 = 50(1.01)$$

Finally, the total balance in the account will be the sum of the balances of the 24 deposits:

$$S_{24} = A_1 + A_2 + A_3 + \cdots + A_{23} + A_{24}$$

Using the formula for the nth partial sum of a geometric sequence, with $A_1 = 50(1.01)$ and $r = 1.01$, we have

$$S_{24} = 50\,\frac{1.01[1 - (1.01)^{24}]}{1 - 1.01} = \$1362.16$$

Infinite Series

We now look at a general method for finding the sum of an infinite geometric series. For the nth partial sum

$$S_n = \frac{a_1(1 - r^n)}{1 - r}$$

The following statements are true as n increases without bound ($n \to \infty$):

1. If $r > 1$, then r^n becomes large without bound. This in turn means that

the absolute value of $(1 - r^n)/(1 - r)$ increases without bound. Hence, S_n approaches no limiting value.

2. If $r < -1$, then r^n alternates between positive and negative numbers that are unbounded in absolute value. Again, S_n approaches no limiting value.

3. If $|r| < 1$, then r^n becomes arbitrarily close to zero, which means that

$$S_n \to \frac{a_1(1 - 0)}{1 - r} = \frac{a_1}{1 - r} \quad \text{as} \quad n \to \infty$$

The number $a_1/(1 - r)$ is called the **sum of the infinite geometric series**

$$\sum_{i=1}^{\infty} a_1 r^{n-1} = a_1 + a_1 r + a_1 r^2 + a_1 r^3 + \cdots + a_1 r^{n-1} + \cdots$$

as summarized in the following statement.

SUM OF AN INFINITE GEOMETRIC SERIES

If $|r| < 1$, then the infinite geometric series

$$a_1 + a_1 r + a_1 r^2 + a_1 r^3 + \cdots + a_1 r^{n-1} + \cdots$$

has the sum

$$\frac{a_1}{1 - r}$$

EXAMPLE 6
Finding the Sum of an
Infinite Geometric Series

Find the sum of the following infinite geometric series:

$$\sum_{n=1}^{\infty} 4(-0.6)^{n-1}$$

Solution:
Since $a_1 = 4$ and $r = -0.6$, we have

$$\sum_{n=1}^{\infty} 4(-0.6)^{n-1} = \frac{a_1}{1 - r} = \frac{4}{1 - (-0.6)} = 2.5$$

When we introduced the real number system in Section 1.2, we noted that if a real number has a decimal representation that (indefinitely) repeats a block of digits, then the real number must be rational. For example, the repeating decimals $0.1111 \ldots = 0.\overline{1}$ and $5.2135135135 \ldots = 5.2\overline{135}$ represent rational numbers. Irrational numbers must have nonrepeating (non-terminating) decimal representations. For example,

$$\pi = 3.141592653589793 \ldots \quad \text{and} \quad e = 2.718281828459045 \ldots$$

In order to write a repeating decimal as the ratio of two integers, we use a geometric series, as shown in the following example.

EXAMPLE 7
Writing a Repeating Decimal as a Ratio of Two Integers

Write the following repeating decimals as the ratio of two integers:

(a) $0.2\overline{2}$ (b) $2.1\overline{425}$

Solution:

(a) As an infinite geometric sequence, we have

$$0.222\ldots = 0.2 + 0.02 + 0.002 + \cdots$$
$$= 0.2 + 0.2(0.1) + 0.2(0.1)^2 + \cdots$$
$$+ 0.2(0.1)^{n-1} + \cdots$$

Thus, $a_1 = 0.2$ and $r = 0.1$, and by the formula for the sum of an infinite geometric series, we have

$$0.2\overline{2} = \frac{a_1}{1-r} = \frac{0.2}{1-0.1} = \frac{0.2}{0.9} = \frac{2}{9}$$

(b) Since the first two digits are not part of the repeating pattern, we pull them out of the representation, as follows:

$$2.1425425425\ldots = 2.1 + 0.0425425425\ldots$$

Now for the repeating portion, we write

$$0.0\overline{425} = 0.0425 + 0.0425(0.001) + 0.0425(0.001)^2 + \cdots$$
$$+ 0.0425(0.001)^{n-1} + \cdots$$
$$= \frac{a_1}{1-r} = \frac{0.0425}{1-0.001} = \frac{0.0425}{0.999} = \frac{425}{9990}$$

Finally, by adding 2.1 to this result, we have

$$2.1\overline{425} = \frac{21}{10} + \frac{425}{9990} = \frac{20979}{9990} + \frac{425}{9990}$$
$$= \frac{21404}{9990} = \frac{10702}{4995}$$

To conclude this section, we look at a technique for determining the sum of a nongeometric infinite series.

EXAMPLE 8
Finding the Sum of an Infinite Series

Use the nth partial sum to show that the infinite series

$$\frac{2}{1\cdot3} + \frac{2}{3\cdot5} + \frac{2}{5\cdot7} + \frac{2}{7\cdot9} + \cdots + \frac{2}{(2n-1)(2n+1)} + \cdots$$

has a sum.

Solution:

Notice that the series is *not* geometric. The nth term has the *partial fraction* representation

$$a_n = \frac{2}{(2n-1)(2n+1)} = \frac{1}{2n-1} - \frac{1}{2n+1}$$

Consequently, we may write the terms as

$$a_1 = 1 - \frac{1}{3}, \qquad a_2 = \frac{1}{3} - \frac{1}{5}, \qquad a_3 = \frac{1}{5} - \frac{1}{7}, \qquad \cdots$$

Therefore, the nth partial sum is

$$S_n = a_1 + a_2 + a_3 + \cdots + a_n$$

$$= 1 - \frac{1}{3} + \frac{1}{3} - \frac{1}{5} + \frac{1}{5} - \frac{1}{7} + \cdots + \frac{1}{2n-1} - \frac{1}{2n+1}$$

$$= 1 - \frac{1}{2n+1}$$

Note that all but the first and last terms cancel. Now, since $1/(2n+1) \to 0$ as $n \to \infty$, it follows that $S_n \to 1$. Thus, we conclude that the sum is

$$1 = \frac{2}{1 \cdot 3} + \frac{2}{3 \cdot 5} + \frac{2}{5 \cdot 7} + \cdots + \frac{2}{(2n-1)(2n+1)} + \cdots$$

Section Exercises 9.3

In Exercises 1–10, determine whether or not the sequence is geometric. If it is, find its common ratio.

1. $\{5, 15, 45, 135, \ldots\}$
2. $\{3, 12, 48, 192, \ldots\}$
3. $\{3, 12, 21, 30, \ldots\}$
4. $\{1, -2, 4, -8, \ldots\}$
5. $\{1, -\frac{1}{2}, \frac{1}{4}, -\frac{1}{8}, \ldots\}$
6. $\{5, 1, 0.2, 0.04, \ldots\}$
7. $\{\frac{1}{2}, \frac{2}{3}, \frac{3}{4}, \frac{4}{5}, \ldots\}$
8. $\{9, -6, 4, -\frac{8}{3}, \ldots\}$
9. $\{1, \frac{1}{2}, \frac{1}{3}, \frac{1}{4}, \ldots\}$
10. $\{\frac{1}{5}, \frac{2}{3}, \frac{3}{9}, \frac{4}{11}, \ldots\}$

In Exercises 11–18, write the first five terms of the geometric sequence.

11. $a_1 = 2, r = 3$
12. $a_1 = 6, r = 2$
13. $a_1 = 1, r = \frac{1}{2}$
14. $a_1 = 1, r = \frac{1}{3}$
15. $a_1 = 5, r = -\frac{1}{10}$
16. $a_1 = 1, r = -x$
17. $a_1 = 1, r = x/2$
18. $a_1 = 2, r = \sqrt{3}$

In Exercises 19–26, find the nth term of the geometric sequence.

19. $a_1 = 4, r = \frac{1}{2}, n = 10$
20. $a_1 = 5, r = \frac{3}{2}, n = 8$
21. $a_1 = 6, r = -\frac{1}{3}, n = 12$
22. $a_1 = 1, r = -x/3, n = 7$
23. $a_1 = 100, r = e^x, n = 9$
24. $a_1 = 8, r = \sqrt{5}, n = 9$
25. $a_1 = 500, r = 1.02, n = 40$
26. $a_1 = 1000, r = 1.005, n = 60$

27. If $1000 is invested at 10% interest, find the amount after 10 years if the interest is compounded
    (a) annually   (b) semiannually   (c) quarterly
    (d) monthly   (e) daily

28. If $2500 is invested at 12% interest, find the amount after 20 years if the interest is compounded
    (a) annually   (b) semiannually   (c) quarterly
    (d) monthly   (e) daily

29. A company buys a machine for $135,000, and it depreciates at the rate of 30% per year. (In other words, at the end of each year its depreciated value is 70% of what it was at the beginning of the year.) Find the depreciated value of the machine after it has been used five full years.

30. A city of 250,000 people is growing at a rate of 1.3% per year. Estimate the population of the city 30 years from now.

In Exercises 31–36, use the formula for the nth partial sum of a geometric series to evaluate the given sum.

31. $\displaystyle\sum_{n=0}^{20} 3\left(\frac{3}{2}\right)^n$

32. $\displaystyle\sum_{n=0}^{15} 2\left(\frac{4}{3}\right)^n$

33. $\displaystyle\sum_{i=1}^{10} 8\left(\frac{-1}{4}\right)^{i-1}$

34. $\displaystyle\sum_{i=1}^{10} 5\left(\frac{-1}{3}\right)^{i-1}$

35. $\displaystyle\sum_{n=0}^{8} 2^n$

36. $\displaystyle\sum_{n=0}^{8} (-2)^n$

37. A deposit of \$100 is made each month for five years in an account that pays 10%, compounded monthly. What is the balance A in the account at the end of five years?

$$A = 100\left(1 + \frac{0.10}{12}\right) + 100\left(1 + \frac{0.10}{12}\right)^2 + \cdots$$
$$+ 100\left(1 + \frac{0.10}{12}\right)^{60}$$

38. A deposit of \$50 is made each month for five years in an account that pays 12%, compounded monthly. What is the balance A in the account at the end of five years?

$$A = 50\left(1 + \frac{0.12}{12}\right) + 50\left(1 + \frac{0.12}{12}\right)^2 + \cdots$$
$$+ 50\left(1 + \frac{0.12}{12}\right)^{120}$$

39. A deposit of P dollars is made every month for T years in an account that pays R percent interest compounded monthly. Let $N = 12T$ be the total number of deposits. The balance A after T years is

$$A = P\left(1 + \frac{R}{12}\right) + P\left(1 + \frac{R}{12}\right)^2 + \cdots$$
$$+ P\left(1 + \frac{R}{12}\right)^N$$

Show that the balance is given by

$$A = P\left[\left(1 + \frac{R}{12}\right)^N - 1\right]\left(1 + \frac{12}{R}\right)$$

40. Use the formula in Exercise 39 to find the amount in an account earning 9% compounded monthly, after monthly deposits of \$50 have been made for 40 years.

41. Use the formula in Exercise 39 to find the amount in an account earning 12% compounded monthly, after monthly deposits of \$50 have been made for 40 years.

42. Suppose that you went to work at a company that pays \$0.01 for the first day, \$0.02 for the second day, \$0.04 for the third day, and so on. If the daily wage keeps doubling, what would your total income be for working 30 days?

In Exercises 43–52, find the sum of the infinite geometric series.

43. $1 + \frac{1}{2} + \frac{1}{4} + \frac{1}{8} + \cdots$

44. $2 + \frac{4}{3} + \frac{8}{9} + \frac{16}{27} + \cdots$

45. $1 - \frac{1}{2} + \frac{1}{4} - \frac{1}{8} + \cdots$

46. $2 - \frac{4}{3} + \frac{8}{9} - \frac{16}{27} + \cdots$

47. $4 + 1 + \frac{1}{4} + \frac{1}{16} + \cdots$

48. $1 + 0.1 + 0.01 + 0.001 + \cdots$

49. $8 + 6 + \frac{9}{2} + \frac{27}{8} + \cdots$

50. $3 - 1 + \frac{1}{3} - \frac{1}{9} + \cdots$

51. $4 - 2 + 1 - \frac{1}{2} + \cdots$

52. $2 + \sqrt{2} + 1 + 1/\sqrt{2} + \cdots$

In Exercises 53 and 54, find the sum of the infinite nongeometric series.

53. $\displaystyle\sum_{n=0}^{\infty} \left(\frac{1}{2^n} - \frac{1}{3^n}\right)$

54. $\displaystyle\sum_{n=0}^{\infty} [(0.7)^n + (0.9)^n]$

55. A ball is dropped from a height of 16 feet. Each time it drops h feet, it rebounds $0.81h$ feet. Find the total distance traveled by the ball.

56. The ball in Exercise 55 takes the following times for each fall:

$$s_1 = -16t^2 + 16, \qquad s_1 = 0 \text{ if } t = 1$$
$$s_2 = -16t^2 + 16(0.81), \qquad s_2 = 0 \text{ if } t = 0.9$$
$$s_3 = -16t^2 + 16(0.81)^2, \qquad s_3 = 0 \text{ if } t = (0.9)^2$$
$$s_4 = -16t^2 + 16(0.81)^3, \qquad s_4 = 0 \text{ if } t = (0.9)^3$$

$$\vdots \qquad\qquad\qquad \vdots$$

$$s_n = -16t^2 + 16(0.81)^{n-1}, \qquad s_n = 0 \text{ if } t = (0.9)^{n-1}$$

Beginning with s_2, the ball takes the same amount of time to bounce up as it takes to fall, and thus the total time elapsed before it comes to rest is

$$t = 1 + 2\sum_{n=1}^{\infty} (0.9)^n$$

Find this total.

In Exercises 57–64, write the repeating decimal as a rational number by considering it to be the sum of an infinite geometric series.

57. $0.1111\ldots$

58. $0.4444\ldots$

59. $0.363636\ldots$

60. $0.212121\ldots$

61. $0.432432432\ldots$

62. $0.46666\ldots$

63. $1.363636\ldots$

64. $1.185185185\ldots$

Mathematical Induction

9.4

In this section we look at a very important form of mathematical proof, called the principle of **mathematical induction.** It is important that you clearly see the logical need for it, so let's take a closer look at a problem we discussed earlier.

In Section 9.2 we looked at the following pattern for the sum of the first n odd integers:

$$S_1 = 1 = 1 = 1^2$$
$$S_2 = 1 + 3 = 4 = 2^2$$
$$S_3 = 1 + 3 + 5 = 9 = 3^2$$
$$S_4 = 1 + 3 + 5 + 7 = 16 = 4^2$$
$$S_5 = 1 + 3 + 5 + 7 + 9 = 25 = 5^2$$

Judging from the pattern formed by these first five partial sums, we decided that the nth partial sum was of the form

$$S_n = 1 + 3 + 5 + 7 + 9 + \cdots + 2n - 1 = n^2$$

Recognizing a pattern and then simply *jumping to the conclusion* that the pattern must be true for all values of n is *not* a logically valid method of proof. There are many examples in which a pattern appears to be developing for small values of n and then at some point the pattern fails. One of the most famous cases of this was the conjecture by a well-known mathematician named Pierre de Fermat, who speculated that all numbers of the form

$$F_n = 2^{2^n} + 1$$

are prime. For $n = 0, 1, 2, 3,$ and 4, the conjecture is true.

$$F_0 = 3$$
$$F_1 = 5$$
$$F_2 = 17$$
$$F_3 = 257$$
$$F_4 = 65537$$

The size of the next Fermat number ($F_5 = 4{,}294{,}967{,}297$) is so great that it was difficult for Fermat to determine whether it was prime or not. However, another well-known mathematician, Leonhard Euler, later found a factorization

$$F_5 = 4{,}294{,}967{,}297 = 641(6{,}700{,}417)$$

which proved Fermat's conjecture to be false.

Here is the point: Just because a rule, pattern, or formula seems to work for several values of n, we cannot simply decide that it is valid for all values of n without going through a *legitimate proof*. Let's see how we can prove such statements by the principle of **mathematical induction.**

THE PRINCIPLE OF MATHEMATICAL INDUCTION

Let P_n be a statement involving the positive integer n. If
1. P_1 is true and
2. the truth of P_k implies the truth of P_{k+1}

then the statement must be true for all positive integers n.

EXAMPLE 1
Using Mathematical Induction

Use mathematical induction to prove the following formula:

$$S_n = 1 + 3 + 5 + 7 + 9 + \cdots + 2n - 1 = n^2$$

Solution:
Mathematical induction consists of two distinct parts. First, we must show that the formula is true when $n = 1$.
(1) When $n = 1$, the formula is valid, since

$$S_1 = 1 = 1^2$$

The second part of mathematical induction has two steps. The first step is to assume that the formula is valid for *some* integer k. The second step is to use this assumption to prove that the formula is valid for the integer $k + 1$.

(2) Assume that

$$S_k = 1 + 3 + 5 + 7 + 9 + \cdots + 2k - 1 = k^2$$

Now, we verify the formula for $k + 1$, as follows:

$$
\begin{aligned}
S_{k+1} &= 1 + 3 + 5 + 7 + 9 + \cdots \\
&\quad + 2k - 1 + 2(k + 1) - 1 \\
&= [1 + 3 + 5 + 7 + 9 + \cdots + 2k - 1] \\
&\quad + 2(k + 1) - 1 \\
&= [S_k] + 2k + 2 - 1 \qquad \textit{Substitute } S_k \\
&= [k^2] + 2k + 1 \qquad \textit{Substitute } k^2 \\
&= (k + 1)^2
\end{aligned}
$$

Combining the result of parts (1) and (2), we conclude by mathematical induction that the formula is valid for *all* positive integer values of n.

Remark: When using mathematical induction to prove a *summation* formula (like the one in Example 1), it is helpful to think of S_{k+1} as

$$S_{k+1} = S_k + a_{k+1}$$

where a_{k+1} is the $(k + 1)$ term of the original sum.

A well-known illustration used to explain why the principle of mathematical induction works is that of an unending line of dominos, as shown in

Figure 9.1. If the line actually contains infinitely many dominos, then it is clear that we could not knock the entire line down by knocking down only *one domino* at a time. However, suppose it were true that each domino would knock down the next one as it fell. Then we could knock them all down simply by pushing the first one and starting a chain reaction. This is the same way mathematical induction works. If the truth of P_k implies the truth of P_{k+1} and if P_1 is true, then the chain reaction proceeds as follows:

P_1 implies P_2

P_2 implies P_3

P_3 implies P_4

and so on.

The first domino knocks over the second which knocks over
the third which knocks over the fourth and so on.

FIGURE 9.1

It might happen that a statement involving natural numbers is not true for the first $k - 1$ positive integers but is true for all values of $n \geq k$. In these instances, we use a slight variation of the principle of mathematical induction in which we verify $P(k)$ rather than $P(1)$. To see the validity of this, note from Figure 9.1 that all but the first $k - 1$ dominos can be knocked down by knocking over the kth domino. This suggests that we can prove a statement $P(n)$ to be true for $n \geq k$ by showing that $P(k)$ is true and that $P(k)$ implies $P(k + 1)$. In Exercises 21-23 you are asked to apply this variation of mathematical induction.

EXAMPLE 2
Using Mathematical Induction

Use mathematical induction to prove the following formula:

$$S_n = 1^2 + 2^2 + 3^2 + 4^2 + \cdots + n^2 = \frac{n(n + 1)(2n + 1)}{6}$$

Solution:

(1) When $n = 1$, the formula is valid, since

$$S_1 = 1^2 = \frac{1(2)(3)}{6}$$

(2) Assume that

$$S_k = 1^2 + 2^2 + 3^2 + 4^2 + \cdots + k^2 = \frac{k(k + 1)(2k + 1)}{6}$$

Now, we verify the formula for $k + 1$, as follows:

$$S_{k+1} = 1^2 + 2^2 + 3^2 + 4^2 + \cdots + k^2 + (k + 1)^2$$

$$= [S_k] + a_{k+1}$$

$$= [1^2 + 2^2 + 3^2 + 4^2 + \cdots + k^2] + (k + 1)^2$$

$$= \left[\frac{k(k + 1)(2k + 1)}{6}\right] + (k + 1)^2 \qquad \textit{By assumption}$$

$$= \frac{k(k + 1)(2k + 1) + 6(k + 1)^2}{6}$$

$$= \frac{(k + 1)[k(2k + 1) + 6(k + 1)]}{6}$$

$$= \frac{(k + 1)[2k^2 + 7k + 6]}{6}$$

$$= \frac{(k + 1)(k + 2)(2k + 3)}{6}$$

$$= \frac{(k + 1)[(k + 1) + 1][2(k + 1) + 1]}{6}$$

Combining the result of parts (1) and (2), we conclude by mathematical induction that the formula is valid for *all* $n \geq 1$.

The formula in Example 2 is one of a collection of formulas that prove to be quite useful in calculus. We summarize this and other formulas dealing with the sum of various powers of the first n positive integers as follows.

SUMS OF POWERS OF INTEGERS

1. $1 + 2 + 3 + 4 + \cdots + n = \dfrac{n(n + 1)}{2}$

2. $1^2 + 2^2 + 3^2 + 4^2 + \cdots + n^2 = \dfrac{n(n + 1)(2n + 1)}{6}$

3. $1^3 + 2^3 + 3^3 + 4^3 + \cdots + n^3 = \dfrac{n^2(n + 1)^2}{4}$

4. $1^4 + 2^4 + 3^4 + 4^4 + \cdots + n^4$

$$= \frac{n(n + 1)(2n + 1)(3n^2 + 3n - 1)}{30}$$

5. $1^5 + 2^5 + 3^5 + 4^5 + \cdots + n^5 = \dfrac{n^2(n + 1)^2(2n^2 + 2n - 1)}{12}$

Remark: Each of these formulas for sums can be proved by mathematical induction. (See Exercises 7–11.)

Although choosing a pattern on the basis of a few observations does *not* guarantee the validity of the pattern, recognition *is* important. Once you have a pattern or formula that you think works, you can try using mathematical induction to prove your formula. We outline these steps as follows.

FINDING A FORMULA FOR THE nTH TERM OF A SEQUENCE

1. Calculate the first several terms of the sequence. (It is often a good idea to leave the terms in factored form *without* simplifying.)
2. Try to find a recognizable pattern from these terms, and write down a formula for the nth term of the sequence. (This is your hypothesis. You might try computing one or two more terms in the sequence to test your hypothesis.)
3. Use mathematical induction to prove (or disprove) your hypothesis.

EXAMPLE 3'
Finding a Formula for the
nth Term of a Sequence

Find a formula for the nth partial sum of the sequence

$$\frac{1}{1 \cdot 2}, \frac{1}{2 \cdot 3}, \frac{1}{3 \cdot 4}, \frac{1}{4 \cdot 5}, \cdots, \frac{1}{n(n+1)}, \cdots$$

Solution:

We begin by writing out a few of the partial sums:

$$S_1 = \frac{1}{1 \cdot 2} = \frac{1}{2}$$

$$S_2 = \frac{1}{1 \cdot 2} + \frac{1}{2 \cdot 3} = \frac{4}{6} = \frac{2}{3}$$

$$S_3 = \frac{1}{1 \cdot 2} + \frac{1}{2 \cdot 3} + \frac{1}{3 \cdot 4} = \frac{9}{12} = \frac{3}{4}$$

$$S_4 = \frac{1}{1 \cdot 2} + \frac{1}{2 \cdot 3} + \frac{1}{3 \cdot 4} + \frac{1}{4 \cdot 5} = \frac{48}{60} = \frac{4}{5}$$

Now, from these first four partial sums, it appears that the formula for the kth partial sum is

$$S_k = \frac{1}{1 \cdot 2} + \frac{1}{2 \cdot 3} + \frac{1}{3 \cdot 4} + \frac{1}{4 \cdot 5} + \cdots + \frac{1}{k(k+1)} = \frac{k}{k+1}$$

To prove the validity of this formula, we use mathematical induction, as follows. Note that we have already verified the formula for $n = 1$, so we begin by assuming that the formula is valid for $n = k$ and try to show that it is valid for $n = k + 1$.

$$S_{k+1} = \left[\frac{1}{1 \cdot 2} + \frac{1}{2 \cdot 3} + \cdots + \frac{1}{k(k+1)} \right] + \frac{1}{(k+1)(k+2)}$$

$$= \left[\frac{k}{k+1}\right] + \frac{1}{(k+1)(k+2)} \qquad \text{\textit{By assumption}}$$

$$= \frac{k(k+2)+1}{(k+1)(k+2)} = \frac{k^2+2k+1}{(k+1)(k+2)}$$

$$= \frac{(k+1)^2}{(k+1)(k+2)} = \frac{k+1}{k+2}$$

Thus, the formula is valid.

EXAMPLE 4
Proving an Inequality by
Mathematical Induction

Prove that $n < 2^n$ for all positive integers n.

Solution:

(1) For $n = 1$, the formula is true, since

$$1 < 2^1$$

(2) Assume that

$$k < 2^k$$

Now, for $n = k + 1$, we have

$$2^{k+1} = 2\,\underline{2^k\,]} > 2\,\underline{k\,]} = 2k \qquad \text{\textit{By assumption}}$$

Since $2k = k + k > k + 1$ for all $k > 1$, it follows that

$$2^{k+1} > 2k > k + 1$$

or

$$k + 1 < 2^{k+1}$$

Hence, $n < 2^n$ for all integers $n \geq 1$.

Section Exercises 9.4

In Exercises 1–14, use mathematical induction to prove the given formula for every positive integer n.

1. $2 + 4 + 6 + 8 + \cdots + 2n = n(n+1)$
2. $3 + 7 + 11 + 15 + \cdots + (4n - 1) = n(2n + 1)$
3. $2 + 7 + 12 + 17 + \cdots + (5n - 3) = \frac{n}{2}(5n - 1)$
4. $1 + 4 + 7 + 10 + \cdots + (3n - 2) = \frac{n}{2}(3n - 1)$
5. $1 + 2 + 2^2 + 2^3 + \cdots + 2^{n-1} = 2^n - 1$
6. $2(1 + 3 + 3^2 + 3^3 + \cdots + 3^{n-1}) = 3^n - 1$
7. $1 + 2 + 3 + 4 + \cdots + n = \frac{n(n+1)}{2}$

8. $1^2 + 2^2 + 3^2 + 4^2 + \cdots + n^2 = \frac{n(n+1)(2n+1)}{6}$
9. $1^3 + 2^3 + 3^3 + 4^3 + \cdots + n^3 = \frac{n^2(n+1)^2}{4}$
10. $\displaystyle\sum_{i=1}^{n} i^4 = \frac{n(n+1)(2n+1)(3n^2+3n-1)}{30}$
11. $\displaystyle\sum_{i=1}^{n} i^5 = \frac{n^2(n+1)^2(2n^2+2n-1)}{12}$
12. $\left(1 + \frac{1}{1}\right)\left(1 + \frac{1}{2}\right)\left(1 + \frac{1}{3}\right) \cdots \left(1 + \frac{1}{n}\right) = n + 1$

13. $\sum\limits_{i=1}^{n} i(i + 1) = \dfrac{n(n + 1)(n + 2)}{3}$

14. $\sum\limits_{i=1}^{n} \dfrac{1}{(2i - 1)(2i + 1)} = \dfrac{n}{2n + 1}$

In Exercises 15–20, find a formula for the *n*th partial sum of the sequence.

15. $3, 7, 11, 15, \ldots$

16. $25, 22, 19, 16, \ldots$

17. $1, \dfrac{9}{10}, \dfrac{81}{100}, \dfrac{729}{1000}, \ldots$

18. $3, -\dfrac{9}{2}, \dfrac{27}{4}, -\dfrac{81}{8}, \ldots$

19. $\dfrac{1}{4}, \dfrac{1}{12}, \dfrac{1}{24}, \dfrac{1}{40}, \ldots, \dfrac{1}{2n(n + 1)}, \ldots$

20. $\dfrac{1}{2 \cdot 3}, \dfrac{1}{3 \cdot 4}, \dfrac{1}{4 \cdot 5}, \dfrac{1}{5 \cdot 6}, \ldots, \dfrac{1}{(n + 1)(n + 2)}, \ldots$

In Exercises 21–24, use mathematical induction to prove the given inequality for the indicated integer values of *n*.

21. $(\frac{4}{3})^n > n, \qquad n \geq 7$

22. $\dfrac{1}{\sqrt{1}} + \dfrac{1}{\sqrt{2}} + \dfrac{1}{\sqrt{3}} + \cdots + \dfrac{1}{\sqrt{n}} > \sqrt{n}, \qquad n \geq 2$

23. $n! > 2^n, \qquad n \geq 4$

24. If $0 < x < y$, then $\left(\dfrac{x}{y}\right)^{n+1} < \left(\dfrac{x}{y}\right)^n, \qquad n \geq 1$

In Exercises 25–32, use mathematical induction to prove the given property for all positive integers *n*.

25. $(ab)^n = a^n b^n$

26. $\left(\dfrac{a}{b}\right)^n = \dfrac{a^n}{b^n}$

27. If $x_1 \neq 0, x_2 \neq 0, x_3 \neq 0, \ldots, x_n \neq 0$, then
$$(x_1 x_2 x_3 \cdots x_n)^{-1} = x_1^{-1} x_2^{-1} x_3^{-1} \cdots x_n^{-1}$$

28. If $x_1 > 0, x_2 > 0, x_3 > 0, \ldots, x_n > 0$, then
$$\log_b(x_1 x_2 x_3 \cdots x_n) = \log_b(x_1) + \log_b(x_2) + \log_b(x_3) + \cdots + \log_b(x_n)$$

29. Generalized Distributive Law:
$$x(y_1 + y_2 + y_3 + \cdots + y_n)$$
$$= xy_1 + xy_2 + xy_3 + \cdots + xy_n$$

30. $x^n - y^n =$
$$(x - y)(x^{n-1} + x^{n-2}y + \cdots + xy^{n-2} + y^{n-1})$$
[Hint: $x^{n+1} - y^{n+1} = x^n(x - y) + y(x^n - y^n)$.]

31. Prove that 3 is a factor of $(n^3 + 3n^2 + 2n)$ for all $n \geq 1$.

32. Prove that 5 is a factor of $(2^{2n-1} + 3^{2n-1})$ for all $n \geq 1$.

The Binomial Theorem   9.5

Recall that a **binomial** is an expression that has two terms. In this section we will look at a formula that gives us a quick method of raising a binomial to a power.

Let's look at the expansion of $(x + y)^n$ for a few values of *n*.

$(x + y)^0 = 1$

$(x + y)^1 = x + y$

$(x + y)^2 = x^2 + 2xy + y^2$

$(x + y)^3 = x^3 + 3x^2y + 3xy^2 + y^3$

$(x + y)^4 = x^4 + 4x^3y + 6x^2y^2 + 4xy^3 + y^4$

$(x + y)^5 = x^5 + 5x^4y + 10x^3y^2 + 10x^2y^3 + 5xy^4 + y^5$

There are several observations we can make about these expansions of $(x + y)^n$:

1. In each expansion, there are $n + 1$ terms.

2. In each expansion, x and y have symmetrical roles. The powers of x decrease by 1 in each term, whereas the powers of y increase by 1.

3. The sum of the powers of each term in a binomial expansion is n. For example,

$$\overset{4+1=5 \qquad\ 3+2=5}{(x+y)^5 = x^5 + 5x^4y^1 + 10x^3y^2 + 10x^2y^3 + 5xy^4 + y^5}$$

4. The first term is x^n. The last term is y^n.

5. The second term is $nx^{n-1}y$. The next to last term is nxy^{n-1}.

The coefficients of the interior terms of the binomial expansion of $(x + y)^n$ are given by a well-known theorem called the **Binomial Theorem.**

THE BINOMIAL THEOREM

$$(x+y)^n = x^n + nx^{n-1}y + \frac{n(n-1)}{2!}x^{n-2}y^2 + \cdots$$
$$+ C_r^n x^{n-r}y^r + \cdots + y^n$$

where the coefficient of $x^{n-r}y^r$ is given by

$$C_r^n = \frac{n!}{(n-r)!r!} = \frac{n(n-1)(n-2)\cdots(n-r+1)}{r!}$$

Proof:

The Binomial Theorem can be proved quite nicely using mathematical induction. The steps are straightforward but look a little complex, so we will only present an outline of the proof.

(1) If $n = 1$, then we have

$$(x+y)^1 = x^1 + y^1 = C_0^1 x + C_1^1 y$$

and the formula is valid.

(2) Assuming the formula is true for $n = k$, then the coefficient of $x^{k-r}y^r$ is given by

$$C_r^k = \frac{k!}{(k-r)!r!} = \frac{k(k-1)(k-2)\cdots(k-r+1)}{r!}$$

To show that the formula is true for $n = k + 1$, we look at the coefficient of $x^{k+1-r}y^r$ in the expansion of

$$(x+y)^{k+1} = (x+y)^k(x+y)$$

From the right-hand side, we can determine that the term involving $x^{k+1-r}y^r$ is the sum of two products, as follows:

$$(C_r^k x^{k-r} y^r)(x) + (C_{r-1}^k x^{k+1-r} y^{r-1})(y)$$

$$= \left[\frac{k!}{(k-r)!r!} + \frac{k!}{(k-r+1)!(r-1)!} \right] x^{k+1-r} y^r$$

$$= \left[\frac{(k+1-r)k!}{(k+1-r)!r!} + \frac{k!r}{(k+1-r)!r!} \right] x^{k+1-r} y^r$$

$$= \left[\frac{k!(k+1-r+r)}{(k+1-r)!r!} \right] x^{k+1-r} y^r$$

$$= \left[\frac{(k+1)!}{(k+1-r)!r!} \right] x^{k+1-r} y^r$$

$$= C_r^{k+1} x^{k+1-r} y^r$$

Thus, by mathematical induction, we can assume that the Binomial Theorem is valid for all positive integers n.

The number C_r^n is called a **binomial coefficient.** Be sure you see that the expansion of $(x+y)^n$ has $n+1$ coefficients. They are

$$C_0^n, \quad C_1^n, \quad C_2^n, \quad \cdots, \qquad C_r^n, \qquad \cdots, \quad C_{n-1}^n, \quad C_n^n$$

$$1, n, \quad \frac{n(n-1)}{2}, \ldots, \quad \frac{n(n-1)(n-2)\cdots(n-r+1)}{r!}, \ldots, n, 1$$

Remark: A common alternative notation for binomial coefficients is

$$C_r^n = \binom{n}{r}$$

EXAMPLE 1
Finding Binomial Coefficients

Find the following binomial coefficients:

(a) C_3^7 (b) C_5^{12} (c) C_7^{12} (d) C_0^8

Solution:

(a) $C_3^7 = \dfrac{7 \cdot 6 \cdot 5}{3 \cdot 2 \cdot 1} = 7(5) = 35$

Note that the denominator of a binomial coefficient is $r!$ and the numerator has r factors.

(b) $C_5^{12} = \dfrac{12 \cdot 11 \cdot 10 \cdot 9 \cdot 8}{5 \cdot 4 \cdot 3 \cdot 2 \cdot 1} = 11(9)(8) = 792$

(c) $C_7^{12} = \dfrac{12 \cdot 11 \cdot 10 \cdot 9 \cdot 8 \cdot 7 \cdot 6}{7 \cdot 6 \cdot 5 \cdot 4 \cdot 3 \cdot 2 \cdot 1} = 11(9)(8) = 792$

(d) $C_0^8 = \dfrac{8!}{8!(0!)} = 1$

Remark: In Example 1, note that the coefficients in parts (b) and (c) are the same. This is not a coincidence, since it is true in general that

$$C_r^n = C_{n-r}^n$$

This is in keeping with our earlier observation that the coefficients of a binomial expansion occur in a symmetrical pattern.

EXAMPLE 2
Using the Binomial Theorem

Use the Binomial Theorem to expand the following:

(a) $(x + 2)^5$ (b) $(1 - a^2)^4$

Solution:

(a) $(x + 2)^5 = C_0^5 x^5 + C_1^5 x^4(2) + C_2^5 x^3(2^2) + C_3^5 x^2(2^3)$

$$+ C_4^5 x(2^4) + C_5^5(2^5)$$

$$= x^5 + 5(2)x^4 + 10(4)x^3 + 10(8)x^2 + 5(16)x + 32$$

$$= x^5 + 10x^4 + 40x^3 + 80x^2 + 80x + 32$$

(b) Since this binomial is a difference, rather than a sum, we write

$$(1 - a^2)^4 = [1 + (-a^2)]^4$$

From this we see that the signs of the terms will alternate, as follows:

$$(1 - a^2)^4 = C_0^4(1^4) - C_1^4(1^3)(a^2) + C_2^4(1^2)(a^2)^2$$

$$- C_3^4(1)(a^2)^3 + C_4^4(a^2)^4$$

$$= 1 - 4a^2 + 6a^4 - 4a^6 + a^8$$

Pascal's Triangle

There is an interesting way to remember the patterns for binomial coefficients. By arranging the coefficients in a triangular pattern, we have the following array, which is called **Pascal's Triangle:**

$$
\begin{array}{ccccccccccccc}
&&&&&&1&&&&&& \\
&&&&&1&&1&&&&& \\
&&&&1&&2&&1&&&& \\
&&&1&&3&&3&&1&&& \\
&&1&&4&&6&&4&&1&& \\
&1&&5&&10&&10&&5&&1& \\
1&&6&&15&&20&&15&&6&&1 \\
\end{array}
$$

$$1 \quad 7 \quad 21 \quad 35 \quad 35 \quad 21 \quad 7 \quad 1$$

The first and last number in each row is 1, and every other number in the triangle is formed by adding the two numbers immediately above that number. For example, the two numbers above 35 are 15 and 20:

$$15 + 20 = 35$$

Since the top row corresponds to the binomial expansion $(x + y)^0 = 1$, we call it the **zero row.** Similarly, the next row corresponds to the binomial expansion $(x + y)^1 = 1(x) + 1(y)$, and we call it the **first row.** In general, the ***n*th row** in Pascal's Triangle gives us the coefficients of $(x + y)^n$.

EXAMPLE 3
Using Pascal's Triangle
in a Binomial Expansion

Use Pascal's Triangle to expand the following:

$$\left(x - \frac{1}{2}\right)^6$$

Solution:

$$\left(x - \frac{1}{2}\right)^6 = x^6 - 6x^5\left(\frac{1}{2}\right) + 15x^4\left(\frac{1}{2}\right)^2 - 20x^3\left(\frac{1}{2}\right)^3$$

$$+ 15x^2\left(\frac{1}{2}\right)^4 - 6x\left(\frac{1}{2}\right)^5 + \left(\frac{1}{2}\right)^6$$

$$= x^6 - \frac{6x^5}{2} + \frac{15x^4}{4} - \frac{20x^3}{8} + \frac{15x^2}{16} - \frac{6x}{32} + \frac{1}{64}$$

$$= x^6 - 3x^5 + \frac{15x^4}{4} - \frac{5x^3}{2} + \frac{15x^2}{16} - \frac{3x}{16} + \frac{1}{64}$$

Remark: Remember that when expanding a binomial *difference* you must alternate signs, as shown in Example 3.

EXAMPLE 4
Finding a Specified Term
in a Binomial Expansion

Find the sixth term in the expansion of $(3a + 2b)^{12}$.

Solution:
Using the Binomial Theorem, we let $x = 3a$ and $y = 2b$ and note that in the *sixth* term the exponent of y is $r = 5$ and the exponent of x is $12 - 5 = 7$. Consequently, the sixth term of the expansion is

$$C_5^{12}x^7y^5 = \frac{12 \cdot 11 \cdot 10 \cdot 9 \cdot 8}{5!}\,(3a)^7(2b)^5$$

Section Exercises 9.5

In Exercises 1–10, evaluate C_r^n.

1. C_3^5

2. C_6^8

3. C_0^{12}

4. C_{20}^{20}

5. C_{15}^{20}

6. C_5^{12}

7. C_{98}^{100}

8. C_4^{10}

9. C_2^{100}

10. C_6^{10}

In Exercises 11–24, use the Binomial Theorem to expand the given binomial. Simplify your answer.

11. $(x + y)^5$
12. $(x + y)^6$
13. $(a + 2)^4$
14. $(s + 3)^5$
15. $(r + 3s)^6$
16. $(x + 2y)^4$
17. $(x - y)^5$
18. $(2x - y)^5$
19. $(1 - 2x)^3$
20. $\left(\dfrac{x}{2} - 3y\right)^3$
21. $(x^2 + 5)^4$
22. $(x^2 + y^2)^6$
23. $\left(\dfrac{1}{x} + y\right)^5$
24. $\left(\dfrac{1}{x} + 2y\right)^6$

In Exercises 25–30, use the Binomial Theorem to expand the given complex number. Simplify your answer by using the fact that $i^2 = -1$.

25. $(1 + i)^4$
26. $(2 - i)^5$
27. $(2 - 3i)^6$
28. $(5 + \sqrt{-9})^3$
29. $\left(\dfrac{-1}{2} + \dfrac{\sqrt{3}}{2}i\right)^3$
30. $(5 - \sqrt{3}\,i)^4$

31. Find the term involving x^5 in the expansion of $(x + 3)^{12}$.
32. Find the term involving x^8 in the expansion of $(x^2 + 3)^{12}$.
33. Find the term involving x^8 in the expansion of $(x - 2y)^{10}$.
34. What is the coefficient of $x^2 y^8$ in the expansion of $(4x - y)^{10}$?
35. What is the coefficient of $x^4 y^{11}$ in the expansion of $(3x - 2y)^{15}$?

36. What is the middle term in the expansion of $(x^2 - 5)^8$?
37. What is the middle term in the expansion of $(\sqrt{x} + \sqrt{y})^{12}$?
38. What is the middle term in the expansion of $\left(\dfrac{2}{x} - \dfrac{3}{y}\right)^{10}$?
39. Use Pascal's Triangle to expand $(2t - s)^5$.
40. Use Pascal's Triangle to expand $\left(\dfrac{x}{2} + 2y\right)^6$.

In the study of probability, it is sometimes necessary to use the expansion of $(p + q)^n$, where $p + q = 1$. In Exercises 41–46, use the Binomial Theorem to expand the given expression.

41. $(\tfrac{1}{2} + \tfrac{1}{2})^7$
42. $(\tfrac{1}{4} + \tfrac{3}{4})^{10}$
43. $(\tfrac{1}{3} + \tfrac{2}{3})^8$
44. $(0.3 + 0.7)^{12}$
45. $(0.6 + 0.4)^5$
46. $(0.35 + 0.65)^6$

In Exercises 47–50, prove the given property for all integers r and n, $1 \le r \le n$.

47. $C_r^n = C_{n-r}^n$
48. $C_0^n - C_1^n + C_2^n - \cdots \pm C_n^n = 0$
49. $C_r^{n+1} = C_r^n + C_{r-1}^n$
50. $C_n^{2n} = (C_0^n)^2 + (C_1^n)^2 + (C_2^n)^2 + \cdots + (C_n^n)^2$
51. Prove that the sum of the numbers in the nth row of Pascal's Triangle is 2^n. [Hint: Consider $2^n = (1 + 1)^n$.]

Counting Principles, Permutations, and Combinations   9.6

In the last two sections of this chapter, we give a brief introduction to some fundamental counting principles and their application to probability. Much of probability has to do with counting the number of ways an event can occur. The following example describes some simple cases.

EXAMPLE 1
Some Simple Counting Problems

A random number generator (on a computer) selects an integer from 1 to 40. Find the number of ways the following events can occur:
(a) An even integer is selected.
(b) A number less than 10 is selected.
(c) A square number is selected.
(d) A prime number is selected.

Solution:

To solve each of these questions, we simply count the outcomes in each event.

(a) Since half the numbers between 1 and 40 (inclusive) are even, this event can occur in 20 ways.

(b) The numbers that are less than 10 are

$$\{1, 2, 3, 4, 5, 6, 7, 8, 9\}$$

Therefore, there are 9 ways this event can happen.

(c) The square numbers from 1 to 40 are

$$\{1, 4, 9, 16, 25, 36\}$$

Therefore, there are 6 ways this event can happen.

(d) The prime numbers from 1 to 40 are

$$\{2, 3, 5, 7, 11, 13, 17, 19, 23, 29, 31, 37\}$$

Therefore, there are 12 ways this event can happen.

Each of the parts of Example 1 consists of a single event. The situation becomes somewhat more complicated when we try to count the number of ways two or more events can occur in succession (or in order). To do this, we make use of the Fundamental Counting Principle.

FUNDAMENTAL COUNTING PRINCIPLE

If E_1, E_2, E_3, . . ., E_n is a sequence of events such that E_1 can occur in m_1 ways, and after E_1 has occurred E_2 can occur in m_2 ways, and after E_2 has occurred E_3 can occur in m_3 ways, and so forth, then the number of ways the sequence can occur is

$$m_1 \cdot m_2 \cdot m_3 \cdots m_n$$

Remark: Be sure you see that the Fundamental Counting Principle applies to a *sequence* of events. This means that the order of the events is important. It also means that when you are counting the number of ways the second event can occur, you must take into consideration the fact that the first event has already occurred.

The following example describes two similar sequences of events that differ in a subtle, yet important, way.

EXAMPLE 2
Applying the Fundamental
Counting Principle

(a) A certain auto license number is made using three digits. How many different numbers are possible? (Leading zeros such as 001 or 027 are legitimate.)

(b) The digits from 0 to 9 are written on slips of paper and placed in a box. Three of the slips of paper are drawn and placed in order. How many different outcomes are possible?

Solution:

(a) By considering the selection of each digit as a separate event, we have

> Event 1 = Choice of first digit
>
> Event 2 = Choice of second digit
>
> Event 3 = Choice of third digit

Since there are 10 choices for each position, we can apply the Fundamental Counting Principle to conclude that there are

$$10 \cdot 10 \cdot 10 = 1000 \text{ different license numbers}$$

(b) This problem is quite like the first, except for one very important distinction. Once the first slip of paper has been drawn, there are only 9 slips left for the second draw. Moreover, once the second slip has been drawn, there are only 8 slips left for the third draw. Hence, the Fundamental Counting Principle tells us that there are

$$10 \cdot 9 \cdot 8 = 720 \text{ different outcomes}$$

Remark: The distinction between the two parts of Example 2 is that in part (a) we considered different **arrangements with repetition** and in part (b) we considered different **arrangements without repetition.**

EXAMPLE 3
Applying the Fundamental Counting Principle

(a) Some versions of the BASIC programming language allow variable names that can be two characters long. If the first character can be any letter and the second character can be any letter *or* digit, how many different variable names are possible?

(b) Most versions of the FORTRAN programming language allow variable names that can be up to six characters long. The first character can be any letter, and the other characters can be any alphanumeric character (the letters, a dollar sign, or a digit). How many different variable names are possible?

Solution:

(a) Since a variable name can have one or two characters, we look at these two cases separately.

> *One-character length:*   Since the first character must be a letter of the alphabet, there are
>
> 26 different one-character names

Two-character length: There are 26 choices for the first character, and 36 (any letter or digit) for the second, which gives a total of

$$26 \cdot 36 = 936 \text{ different two-character names}$$

Thus the total number of variable names is

$$26 + 936 = 962 \text{ different names}$$

(b) The solution is similar to that in part (a), except that the dollar sign ($) gives us 37 choices for the second through sixth characters. The total number of variable names is given by

1-character length:	26	$=$	26
2-character length:	$26 \cdot 37$	$=$	962
3-character length:	$26 \cdot 37 \cdot 37$	$=$	$35{,}594$
4-character length:	$26 \cdot 37 \cdot 37 \cdot 37$	$=$	$1{,}316{,}978$
5-character length:	$26 \cdot 37 \cdot 37 \cdot 37 \cdot 37$	$=$	$48{,}728{,}186$
6-character length:	$26 \cdot 37 \cdot 37 \cdot 37 \cdot 37 \cdot 37$	$=$	$1{,}802{,}942{,}882$
	Total:		$1{,}853{,}024{,}628$

One important application of the Fundamental Counting Principle is in determining the number of ways that n elements can be arranged (in order). We call an ordering of n elements a **permutation** of elements.

DEFINITION OF PERMUTATION

A **permutation** of n distinct elements is an ordering of the elements such that one element is first, one is second, and so on:

$$(a_1, a_2, a_3, a_4, \ldots , a_n)$$

EXAMPLE 4
Finding the Number of
Permutations on n Elements

A horse race has five entries. In how many different orders can the horses finish? (Assume that there are no ties.)

Solution:
We have

1st place:	Any of the five horses.
2nd place:	Any of the remaining four horses.
3rd place:	Any of the remaining three horses.
4th place:	Any of the remaining two horses.
5th place:	The one remaining horse.

Multiplying these five numbers together, we obtain the total number of orders for the horses.

$$5 \cdot 4 \cdot 3 \cdot 2 \cdot 1 = 5! = 120 \text{ orders}$$

We can generalize the result obtained in Example 4 to conclude that the number of permutations (orderings) of n distinct elements is $n!$.

NUMBER OF PERMUTATIONS OF n ELEMENTS

There are $n!$ different permutations of n elements.

Proof:
We can use the Fundamental Counting Principle as follows:

1st position:	Any of the n elements.
2nd position:	Any of the remaining $n - 1$ elements.
3rd position:	Any of the remaining $n - 2$ elements.
4th position:	Any of the remaining $n - 3$ elements.
. .	.
. .	.
. .	.
$n - 1$ position:	Any of the remaining 2 elements.
nth position:	The remaining 1 element.

By multiplying these n numbers together, we find the total number of permutations to be

$$n(n - 1)(n - 2)(n - 3) \cdots 3 \cdot 2 \cdot 1 = n!$$

Occasionally, we are interested in ordering a subset of a collection of elements rather than ordering the entire collection. For example, we might want to choose (and order) r elements out of a collection of n elements. We call such an ordering a **permutation of n elements taken r at a time.** The following example illustrates such a case.

EXAMPLE 5
Permutations of n Elements Taken r at a Time

Eight horses are running in a race. In how many different ways can these horses come in first, second, and third place? (Assume that there are no ties.)

Solution:
We have

Win (1st position):	Eight choices
Place (2nd position):	Seven choices
Show (3rd position):	Six choices

Using the Fundamental Counting Principle, we multiply these three numbers together to obtain

$8 \cdot 7 \cdot 6 = 336$ orders

We can generalize the result of Example 5 as follows. The proof of this result closely parallels that given for the number of permutations of n elements.

NUMBER OF PERMUTATIONS OF n ELEMENTS TAKEN r AT A TIME

The number of permutations of n elements taken r at a time is

$$P_r^n = \frac{n!}{(n-r)!} = n(n-1)(n-2) \cdots (n-r+1)$$

To help visualize the Fundamental Counting Principle, we can make a **tree diagram** which actually lists the various permutations on three elements. In the following diagram (Figure 9.2), we can see in the right-hand branches of the tree the six possible permutations of the letters A, B, and C.

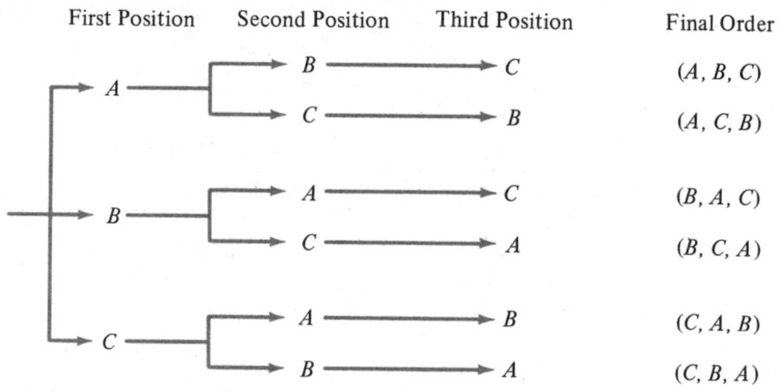

FIGURE 9.2

EXAMPLE 6
Listing Permutations

List the 24 different permutations of the letters A, B, C, and D.

Solution:
The various permutations are shown in the table on page 419.

Remember that for permutations order is important. Thus, if we are looking at the possible permutations of the letters A, B, C, and D taken three at a time, the permutations

(A, B, D) and (B, A, D)

are different, since the order of the elements is different.

A in 1st Place	B in 1st Place	C in 1st Place	D in 1st Place
(A, B, C, D)	(B, A, C, D)	(C, A, B, D)	(D, A, B, C)
(A, B, D, C)	(B, A, D, C)	(C, A, D, B)	(D, A, C, B)
(A, C, B, D)	(B, C, A, D)	(C, B, A, D)	(D, B, A, C)
(A, C, D, B)	(B, C, D, A)	(C, B, D, A)	(D, B, C, A)
(A, D, B, C)	(B, D, A, C)	(C, D, A, B)	(D, C, A, B)
(A, D, C, B)	(B, D, C, A)	(C, D, B, A)	(D, C, B, A)

Distinguishable Permutations

In Example 6, each letter is distinguishable from the other. But suppose there were four A's and one each of B, C, and D. The *total* number of permutations of the seven letters would be

$$P_7^7 = 7!$$

However, not all of these arrangements would be *distinguishable*, since the four A's are alike. In other words, any rearrangement of the A's, keeping the B, C, and D in the same locations, would not be a distinguishable permutation. Thus, the total number of *distinguishable* permutations of the seven letters would be less than 7!. To be more precise, let P be the number of distinguishable permutations of the seven letters. Then, since the four A's have $4! = 24$ permutations, the total number of permutations would be $24P$. Consequently, we have

$$24P = 7!$$

$$P = \frac{7!}{24} = \frac{7!}{4!}$$

This argument can be generalized to obtain the following result.

DISTINGUISHABLE PERMUTATIONS

Suppose a set of n objects has n_1 alike of one kind, n_2 alike of a second kind, n_3 alike of a third kind, and so on, with

$$n = n_1 + n_2 + n_3 + \cdots + n_k$$

Then the number of **distinguishable permutations** of the n objects is

$$\frac{n!}{n_1! \cdot n_2! \cdot n_3! \cdots n_k!}$$

EXAMPLE 7
Distinguishable Permutations

In how many distinguishable ways can the product x^2y^3z be written without using exponents?

Solution:

Without exponents, this product can be written as

$$x \cdot x \cdot y \cdot y \cdot y \cdot z$$

Since there are two x's, three y's, and only one z, the number of distinguishable permutations is

$$\frac{6!}{2!(3!)(1!)} = \frac{6 \cdot 5 \cdot 4}{2} = 6 \cdot 5 \cdot 2 = 60$$

Circular Arrangements

Circular arrangements also produce a slight variation in the permutation rule. For instance, when arranging the four letters A, B, C, and D around a circle, we cannot distinguish between the four arrangements shown in Figure 9.3.

$$(A, B, C, D), \qquad (B, C, D, A), \qquad (C, D, A, B), \qquad (D, A, B, C)$$

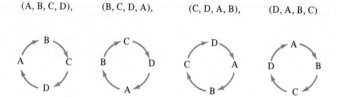

Four Equivalent Circular Arrangements of Four Letters

FIGURE 9.3

One way of thinking about circular arrangements is that the placement of the first object has no bearing on the total number of distinguishable permutations. This leads us to the following rule for determining the number of circular arrangements.

NUMBER OF CIRCULAR ARRANGEMENTS

The number of distinguishable circular arrangements of n objects is

$$(n - 1)!$$

Combinations

As a final topic in this section, we look at a related method of selecting subsets of a larger set in which order is *not* counted. We call such subsets **combinations of n elements taken r at a time.** We denote combinations with braces (rather than parentheses). Thus, the combinations

$$\{A,\ B,\ D\} \qquad \text{and} \qquad \{B,\ A,\ D\}$$

are equivalent, since both sets contain the same elements and order is not important. A classic example of how a combination occurs is in card games in which the player is free to reorder the cards in his or her hand after they have been dealt.

NUMBER OF COMBINATIONS OF n ELEMENTS TAKEN r AT A TIME

The number of combinations of n elements taken r at a time is

$$C_r^n = \frac{n!}{(n-r)!r!} = \frac{n(n-1)(n-2)\cdots(n-r+1)}{r!}$$

Remark: Note that the formula for C_r^n is the same one given for binomial coefficients. This means that we can use Pascal's Triangle as an alternative way of computing the number of combinations of n elements taken r at a time.

EXAMPLE 8
Combinations of n Elements Taken r at a Time

A standard poker hand consists of 5 cards dealt from a deck of 52 cards. How many different poker hands are possible? (After the cards are dealt, the player may reorder them, and therefore order is not important.)

Solution:
We use the formula for the number of combinations of 52 elements taken 5 at a time to obtain

$$\frac{52!}{47!(5!)} = \frac{52 \cdot 51 \cdot 50 \cdot 49 \cdot 48}{5 \cdot 4 \cdot 3 \cdot 2 \cdot 1}$$

$$= 2{,}598{,}960 \text{ different hands}$$

We can see from Example 8 that it is easily possible for the number of combinations to be quite large. Even though the formula for C_r^n is given in fractional form, we know that the fraction must always reduce to an integer. (Remember that Pascal's Triangle consists entirely of integers.) For large values of C_r^n, we suggest that you lessen your chance of calculator overflow by reducing the fraction before multiplying the factors of the numerator. For example, the calculation in Example 8 could be simplified as follows:

$$\frac{52 \cdot 51 \cdot 50 \cdot 49 \cdot 48}{5 \cdot 4 \cdot 3 \cdot 2 \cdot 1} = \left(\frac{52}{4}\right)\left(\frac{51}{3}\right)\left(\frac{50}{5 \cdot 2}\right)(49 \cdot 48)$$

$$= 13(17)(5)(49)(48)$$

In deciding whether a problem calls for the number of permutations or the number of combinations, you will find the following guidelines helpful.

GUIDELINES FOR USE OF PERMUTATIONS AND COMBINATIONS

1. Use **permutations** if a problem calls for the number of *arrangements* of objects and different orders are to be counted.

2. Use **combinations** if a problem calls for the number of ways of *selecting* objects and the order of selection is not to be counted.

EXAMPLE 9
Using Combinations with the Fundamental Counting Principle

The traveling squad for a college basketball team consists of two centers, five forwards, and four guards. In how many ways can the coach select a starting team of one center, two forwards, and two guards?

Solution:
The number of ways to select one center is

$$C_1^2 = \frac{2!}{1!(1!)} = 2$$

The number of ways to select two forwards from among five is

$$C_2^5 = \frac{5!}{3!(2!)} = 10$$

The number of ways to select two guards from among four is

$$C_2^4 = \frac{4!}{2!(2!)} = 6$$

Therefore, the total number of ways to select a starting team is

$$C_1^2 \cdot C_2^5 \cdot C_2^4 = 2 \cdot 10 \cdot 6 = 120$$

Section Exercises 9.6

In Exercises 1–10, use the Fundamental Counting Principle.

1. A small college is in need of two additional faculty. One is to be a chemist, and the other a statistician. In how many ways can these positions be filled if there are three applicants for the chemistry position and four for the position in statistics?

2. A traveler can take one of three departure times from New York to Chicago, one of two departure times from Chicago to Lincoln, and one of two from Lincoln to San Francisco. How many choices of departure times exist for a trip from New York to San Francisco?

3. Four people are lining up for a ride on a toboggan, but only two of the four are willing to take the first position. With that constraint, how many ways can the four be seated on the toboggan?

4. A college student is preparing her course schedule for the next semester. She may select one of two mathematics courses, one of three science courses, and one of five courses from the social sciences and humanities. In how many ways can she select her schedule?

5. In a certain state the automobile license plates consist of two letters followed by a four-digit number. How many distinct license plate numbers can be formed?

6. In how many ways can a six-question true-false exam be answered? (Assume that no questions are omitted.)

7. How many three-digit numbers (leading digits cannot be zero) can be formed under the following conditions?

    (a) There are no restrictions.
    (b) No repetition of digits is allowed.
    (c) The number must be a multiple of 5.

8. A combination lock will open when the right choice of three numbers is selected. If the numbers range from 1 to 40 (inclusive), how many different lock combinations are possible?

9. Three couples have reserved seats in a given row for a concert. In how many different ways can they be seated, given the following conditions?
   (a) There are no seating restrictions.
   (b) Each couple wishes to sit together.

10. In how many orders can three girls and two boys walk through a doorway single-file, given the following conditions?
    (a) There are no restrictions.
    (b) The boys go before the girls.
    (c) The girls go before the boys.

In Exercises 11–16, evaluate P_r^n.

11. P_4^4
12. P_5^5
13. P_3^8
14. P_2^{20}
15. P_5^{20}
16. P_1^{100}

In Exercises 17–20, use permutations to solve the given counting problem.

17. In how many ways can five children line up in one row to have their picture taken?

18. In how many ways can six people sit in a six-passenger car?

19. From a pool of 12 candidates, the offices of president, vice-president, secretary, and treasurer will be selected. In how many different ways can the offices be filled, if each of the 12 candidates can hold any office?

20. There are four processes involved in assembling a certain product, and these can be performed in any order. The management wants to test each order to determine which is the least time-consuming. How many different orders will have to be tested?

In Exercises 21–24, use combinations to solve the given counting problem.

21. In order to conduct a certain experiment, 4 students are randomly selected from a class of 20 students. How many different groups of 4 students are possible?

22. A student may answer any 10 questions from a total of 12 questions on an exam. How many different ways can the student select the questions?

23. There are 40 numbers in a particular state lottery. In how many ways can a player select 6 of the numbers? (The order of selection is not important.)

24. How many subsets of 4 elements can be formed from a set of 100 elements?

In Exercises 25–35, use any of the counting methods discussed in this section. More than one method may be required to solve a given question.

25. A committee composed of three graduate students and two undergraduate students is to be selected from a group of eight graduates and five undergraduates. How many different committees can be formed?

26. A shipment of 12 microwave ovens contains 3 defective units. In how many ways can a vending company purchase 4 of these units and receive
    (a) all good units?
    (b) two good units?
    (c) at least two good units?

27. An employer interviews eight people for four openings in the company. Three of the eight people are from a minority group. If all eight are qualified, in how many ways could the employer fill the four positions if
    (a) the selection is random?
    (b) exactly two are selected from the minority group?

28. Five cards are selected from an ordinary deck of 52 playing cards. In how many ways can you get a full house? (A full house consists of 3 of one kind and 2 of another. For example, A-A-A-5-5 and K-K-K-10-10 are full houses.)

29. Four people are to be selected at random from a group of four couples. In how many ways can this be done, given the following conditions?
    (a) There are no restrictions.
    (b) There is to be at least one couple in the group of four.
    (c) The selection must include one member from each couple.

30. Repeat Exercise 29 (a) and (b) if there are five couples from which to make the selection. (Assume that there are still only four to be selected.)

31. A Scrabble game player has the following seven letters: A, A, G, E, E, E, M.
    (a) How many different seven-letter "words" can the player spell? (The words don't have to make sense in English.)
    (b) How many different six-letter "words" can the player spell?

32. Find the number of distinguishable permutations of the following groups of letters:
    (a) B, B, B, T, T, T, T, T
    (b) A, A, Y, Y, Y, Y, X, X, X
    (c) K, K, M, M, L, L, N, N

33. Find the number of distinguishable arrangements of eight people seated around a circular table.

34. Find the number of distinguishable arrangements of six keys on a key ring.

35. Find the number of distinguishable arrangements of six keys on a key ring if two of the keys are alike.

Probability   9.7

As a member of a complex society, you are used to living with varying amounts of uncertainty. For example, you may be questioning the likelihood of getting a good job after graduation, the likelihood of winning a state lottery, the likelihood of an accident on your next trip home, or any of countless other possibilities. In assigning measurements to uncertainties in everyday life, we often use ambiguous terminology, such as "fairly certain," "probable," or "highly unlikely." In mathematics, we attempt to remove this ambiguity by assigning a number to the likelihood of the occurrence of an event. We call this numerical measurement the **probability** that the event will occur. For example, if we toss a coin, we say the probability that it will land heads up is one-half, or 50%.

In the study of probability, we call any happening whose result is uncertain an **experiment.** The various possible results of the experiment are called **outcomes,** and the collection of all possible outcomes of an experiment is called the **sample space** of the experiment. Finally, any subcollection of a sample space is called an **event.** In this section we will deal only with sample spaces in which each outcome is equally likely.

EXAMPLE 1
Finding the Sample Space

An experiment consists of tossing a six-sided die.
(a) What is the sample space?
(b) Describe the event corresponding to a number greater than 2 turning up.

Solution:

(a) The sample space consists of six outcomes, which we represent by the numbers 1 through 6. That is,

$$S = \{1, 2, 3, 4, 5, 6\}$$

Note that each of the outcomes in the sample space is equally likely (assuming that the die is balanced).
(b) The event corresponding to a number greater than 2 turning up is

$$A = \{3, 4, 5, 6\}$$

To describe sample spaces in such a way that each outcome is equally likely, we must sometimes distinguish between various outcomes in ways that appear artificial. The next example illustrates such a situation.

EXAMPLE 2
Finding the Sample Space

Find the sample spaces for the following:
(a) One coin is tossed.
(b) Two coins are tossed.
(c) Three coins are tossed.

Solution:

(a) Since the coin will land either heads up (denoted by H) or tails up (denoted by T), the sample space is

$$S = \{H, T\}$$

(b) Since either coin can land heads up or tails up, the possible outcomes are

HH = heads up on both coins

HT = heads up on first coin and tails up on second coin

TH = tails up on first coin and heads up on second coin

TT = tails up on both coins

Thus, the sample space is

$$S = \{HH, HT, TH, TT\}$$

Note that we must distinguish between the two cases HT and TH, even though these two outcomes appear to be similar.

(c) Following the notation of part (b), the sample space is

$$S = \{HHH, HHT, HTH, HTT, THH, THT, TTH, TTT\}$$

To calculate the probability of an event, we must count the number of outcomes in the event and in the sample space. We denote the *number of outcomes* in Event A by $n(A)$ and the number of outcomes in the sample space S by $n(S)$. In this section we will assume that every sample space has a finite number of outcomes.

THE PROBABILITY OF AN EVENT

> If an event A has $n(A)$ equally likely outcomes and its sample space has $n(S)$ equally likely outcomes, then the **probability** of event A is
>
> $$P(A) = \frac{n(A)}{n(S)}$$

Remark: Since the number of outcomes in an event must be less than or equal to the number of outcomes in the sample space, we can see that the probability of an event must be a number between 0 and 1. That is, for any event A, it must be true that

$$0 \leq P(A) \leq 1$$

The statement $P(A) = 0$ means that event A *cannot occur*, whereas the statement $P(A) = 1$ means that event A *must occur*.

Historically, a great deal of interest in probability stemmed from its applications to gambling.

EXAMPLE 3
Finding the Probability
of an Event

Find the probability of the following events:

(a) Two coins are tossed. What is the probability that both land heads up?
(b) A card is drawn from a standard deck of playing cards. What is the probability that it is an ace?
(c) Two six-sided dice are tossed. What is the probability that the total of the two dice is 7?

Solution:

(a) Following Example 2, we let

$$A = \{HH\} \quad \text{and} \quad S = \{HH, HT, TH, TT\}$$

The probability of getting two heads is

$$P(A) = \frac{n(A)}{n(S)} = \frac{1}{4}$$

(b) Since there are 52 cards in a standard deck of playing cards and there are 4 aces (one in each suit), the probability of drawing an ace is

$$P(A) = \frac{n(A)}{n(S)} = \frac{4}{52} = \frac{1}{13}$$

(c) Since there are six possible outcomes on each die, we use the Fundamental Counting Principle to conclude that there are

$$6 \cdot 6 = 36 \text{ different outcomes}$$

when two dice are tossed. To find the probability of rolling a total of 7, we must first count the number of ways this can be done.

	Total of 7
First die	1 2 3 4 5 6
Second die	6 5 4 3 2 1

Since there are six ways, the probability is

$$P(A) = \frac{n(A)}{n(S)} = \frac{6}{36} = \frac{1}{6}$$

We could have written out each sample space in Example 3 and simply counted the outcomes in the desired events. For larger sample spaces, however, we must make more use of the counting principles discussed in the previous section.

EXAMPLE 4
Comparing Probabilities

Twelve-sided dice can be constructed (in the shape of regular dodecahedrons) so that each of the numbers from one to six appears twice on each die, as shown in Figure 12.4. Prove that these dice can be used in any game using ordinary six-sided dice without changing the probabilities for different outcomes.

Solution:
For an ordinary six-sided die, the probability of any particular number coming up is

$$P(A) = \frac{n(A)}{n(S)} = \frac{1}{6}$$

For one of the twelve-sided dice, the probability of any particular number coming up is

$$P(A) = \frac{n(A)}{n(S)} = \frac{2}{12} = \frac{1}{6}$$

FIGURE 9.4

Thus, the probabilities for any outcome using the two different types of dice are the same.

EXAMPLE 5
Finding the Probability of an Event

A state lottery is set up so that each player chooses six numbers from 1 to 40. If these six numbers match the six numbers drawn by the lottery commission, the player wins (or shares) the top prize. What is the probability of winning the top prize in this game?

Solution:
Since the order of the numbers is not important, we use our formula for the number of combinations of 40 elements taken 6 at a time to determine the size of the sample space.

$$n(S) = C_6^{40} = \frac{40 \cdot 39 \cdot 38 \cdot 37 \cdot 36 \cdot 35}{6 \cdot 5 \cdot 4 \cdot 3 \cdot 2 \cdot 1} = 3{,}838{,}380$$

If a person buys only one ticket, the probability of winning is

$$P(A) = \frac{n(A)}{n(S)} = \frac{1}{3{,}838{,}380}$$

We say that two events are **independent** if the occurrence of one has no effect on the occurrence of the other. To find the probability that two independent events will both occur, we multiply the probabilities of each.

PROBABILITY OF INDEPENDENT EVENTS BOTH OCCURRING

> If A and B are independent events, then the probability that both A and B will occur is
>
> $$P(A) \cdot P(B)$$

Remark: This result can be generalized to any number of events, provided they are all independent of each other. Thus, the probability of n independent events all occurring is given by the product of their individual probabilities, as follows:

$$P(A_1) \cdot P(A_2) \cdot P(A_3) \cdots P(A_n)$$

EXAMPLE 6
Probability of Independent Events

A random number generator on a computer selects three integers from 1 to 20. What is the probability that all three numbers are less than (or equal to) 5?

Solution:
If the random number generator is truly random, then we can conclude that the selection of any given number will not affect the selection of the next number. This means that the three choices represent independent events. Furthermore, since the probability of selecting a number from 1 to 5 is

$$P(A) = \frac{5}{20} = \frac{1}{4}$$

we can conclude that the probability of selecting all three numbers less than or equal to 5 is

$$P(A) \cdot P(A) \cdot P(A) = \left(\frac{1}{4}\right)\left(\frac{1}{4}\right)\left(\frac{1}{4}\right) = \left(\frac{1}{64}\right)$$

The **complement of an event** A is the collection of all outcomes in the sample space that are not in A. We denote the complement of event A as A', and the probability of A' is given by

$$P(A') = 1 - P(A)$$

For instance, if the probability of winning a certain game is

$$P(A) = \frac{1}{4}$$

then the probability of losing the game is

$$P(A') = 1 - \frac{1}{4} = \frac{3}{4}$$

EXAMPLE 7
Finding the Probability of the
Complement of an Event

A manufacturer has determined that a certain machine averages 1 faulty unit for every 1,000 it produces. What is the probability that an order of 200 units will have 1 or more faulty units?

Solution:
To solve this problem as stated, we would need to find the probability of having exactly 1 faulty unit, exactly 2 faulty units, exactly 3 faulty units, and so on. However, using complements, we can simply find the probability that all units are perfect and then subtract this value from 1. Since the probability that any given unit is perfect is $\frac{999}{1000}$, the probability that all 200 units are perfect is

$$P(A) = \left(\frac{999}{1000}\right)^{200} \approx 0.8186$$

Therefore, the probability that at least 1 unit is faulty is

$$P(A') = 1 - P(A) \approx 0.1814$$

Section Exercises 9.7

1. Consider the experiment of tossing a coin three times.
   (a) Find the sample space.
   (b) Find the probability of getting exactly one tail.
   (c) Find the probability of getting a head on first toss.
   (d) Find the probability of getting at least one head.
2. Consider the experiment of selecting 1 card from a standard deck of 52 playing cards.
   (a) Find the probability of getting a face card.
   (b) Find the probability of not getting a face card.
   (c) Find the probability of getting a black card that is not a face card.
   (d) Find the probability that the card will be a 6 or less.
3. Consider the experiment of tossing a six-sided die twice.
   (a) Find the probability that the total is 4.
   (b) Find the probability that the total is less than 11.
   (c) Find the probability that the total is 7 or 11.
   (d) Find the probability that the total is 2, 3, or 12.
4. Three people have been nominated for president of a college class. From a small poll, it is estimated that the probability of Jane's winning the election is 0.37 and the probability of Larry's winning the election is 0.44. What is the probability of the third candidate's winning the election?
5. Taylor, Moore, and Jenkins are candidates for public office. It is estimated that Moore and Jenkins have about the same probability of winning, and Taylor is believed to be twice

as likely to win as either of the others. Find the probability of each candidate's winning the election.
6. In a high school graduating class of 72 students, 28 were on the honor roll; of these, 18 are going on to college. Of the other 44 students, 12 are going on to college. If a student is selected at random from the class, what is the probability that the person chosen is
   (a) going to college?
   (b) not going to college?
   (c) on the honor roll, but not going to college?
7. Two integers (from 1 to 30 inclusive) are chosen by a random number generator on a computer. What is the probability that
   (a) the numbers are both even?
   (b) one number is even and one is odd?
   (c) both numbers are less than 10?
   (d) the same number is chosen twice?
8. An instructor gives her class a list of eight study problems, from which she will select five to be answered on an exam. If a given student knows how to solve six of the problems, find the probability that the student will be able to answer
   (a) all five questions on the exam
   (b) exactly four questions on the exam
   (c) at least four questions on the exam
9. Two cards are selected at random from an ordinary deck

of 52 playing cards. Find the probability that two aces are selected, given the following conditions:

(a) The cards are drawn in sequence, with the first card being replaced and the deck reshuffled prior to the second drawing.

(b) The two cards are drawn consecutively, without replacement.

10. On a certain game show you are given four different digits, which are to be arranged in the proper order to give the price of a car. If you are correct, you win the car. What is the probability of winning, given the following conditions?

(a) You guess at the position of each digit.

(b) You know the first digit, but must guess at the remaining three.

11. Four letters and envelopes are addressed to four different people. If the letters are randomly inserted into the envelopes, what is the probability that

(a) exactly one will be inserted in the correct envelope?

(b) at least one will be inserted in the correct envelope?

12. Five cards are drawn from an ordinary deck of 52 playing cards. What is the probability of getting a full house? (See Exercise 28 in Section 9.6.)

13. A shipment of 12 microwave ovens contains 3 defective units. A vending company has ordered 4 of these units, and since each is identically packaged, the selection will be random. (See Exercise 26 in Section 9.6.) What is the probability that

(a) all 4 units are good?

(b) exactly 2 units are good?

(c) at least 2 units are good?

14. A fire company keeps two rescue vehicles to serve the community. Because of the demand on their time and the chance of mechanical failure, the probability that a specific vehicle is available when needed is 90%. If the availability of one vehicle is independent of the other, find the probability that

(a) both vehicles are available at a given time

(b) neither vehicle is available at a given time

(c) at least one is available at a given time

15. One component of a space vehicle contains a main system and one independent backup system. The probability that a specific system will function satisfactorily for the duration of a flight is 0.985. What is the probability that during a given flight

(a) both systems function satisfactorily?

(b) at least one system functions satisfactorily?

(c) both systems fail?

16. A door-to-door sales representative makes a sale at approximately one-third of the homes she calls on. If on a given day she goes to four homes, what is the probability that she will make a sale at

(a) all four homes?

(b) none of the homes?

(c) at least one home?

17. Assume that the probability of the birth of a child of a particular sex is 50%. In a family with six children, what is the probability that

(a) all the children are girls?

(b) all the children are the same sex?

(c) there is at least one girl?

In Exercises 18–21, use the following information about binomial experiments. A **binomial experiment** is distinguished by the following four conditions:

1. There are n identical trials.
2. There are two possible outcomes to each trial, one of which is called success and the other failure.
3. The probability of a success is the same for every trial and is denoted by p; the probability of failure is denoted by $q = 1 - p$.
4. The n trials are independent.

Each term in the expansion of $(p + q)^n$ gives the probability of a certain number of successes. For example, if there are 10 trials, then the probability of 7 successes is given by $C_7^{10} p^7 q^3$.

18. Use the information in Exercise 16 ($p = \frac{1}{3}$, $q = \frac{2}{3}$, $n = 4$) to expand $(p + q)^n$ and find the probability of

(a) four sales   (b) three sales   (c) two sales

(d) one sale   (e) no sale

19. A particular binomial experiment has 10 trials. The probability of success is 10%. Find the following:

(a) the probability that all 10 trials fail

(b) the probability that all 10 trials succeed

(c) the probability that at least one trial succeeds

(d) the probability that exactly one trial succeeds

20. A basketball player makes 65% of his shots from the floor. Use the expansion of $(p + q)^n$ to find the probability that he makes x of his next five shots.

x	0	1	2	3	4	5
Probability						

21. Use the information in Exercise 17 ($p = q = \frac{1}{2}$, $n = 6$) to expand $(p + q)^n$ and find the probability for each of the following:

Number of Girls	0	1	2	3	4	5	6
Probability							

Review Exercises / Chapter 9

In Exercises 1–14, find the sum.

1. $\displaystyle\sum_{i=1}^{6} 5$

2. $\displaystyle\sum_{k=2}^{5} 4k$

3. $\displaystyle\sum_{j=3}^{10} (2j - 3)$

4. $\displaystyle\sum_{j=1}^{8} (20 - 3j)$

5. $\displaystyle\sum_{i=0}^{6} 2^i$

6. $\displaystyle\sum_{i=0}^{4} 3^i$

7. $\displaystyle\sum_{i=0}^{\infty} (\tfrac{7}{8})^i$

8. $\displaystyle\sum_{i=0}^{\infty} (\tfrac{1}{3})^i$

9. $\displaystyle\sum_{k=0}^{\infty} 4(\tfrac{2}{3})^k$

10. $\displaystyle\sum_{k=0}^{\infty} 1.3(\tfrac{1}{10})^k$

11. $\displaystyle\sum_{k=1}^{11} (\tfrac{2}{3}k + 4)$

12. $\displaystyle\sum_{k=1}^{25} \left(\frac{3k + 1}{4}\right)$

13. $\displaystyle\sum_{n=0}^{10} (n^2 + 3)$

14. $\displaystyle\sum_{n=1}^{100} \left(\frac{1}{n} - \frac{1}{n+1}\right)$

In Exercises 15–18, use mathematical induction to prove the given formula for every positive integer n.

15. $1 + 4 + 7 + \cdots + (3n - 2) = \dfrac{n}{2}(3n - 1)$

16. $1 + \dfrac{3}{2} + 2 + \dfrac{5}{2} + \cdots + \dfrac{1}{2}(n + 1) = \dfrac{n}{4}(n + 3)$

17. $\displaystyle\sum_{j=0}^{n-1} ar^j = \frac{a(1 - r^n)}{1 - r}$

18. $\displaystyle\sum_{k=0}^{n-1} [a + kd] = \frac{n}{2}[2a + (n - 1)d]$

In Exercises 19–24, write the repeating decimal as a rational number by considering each to be the sum of an infinite geometric series.

19. $0.454545\ldots$

20. $0.151515\ldots$

21. $1.0666\ldots$

22. $0.0222\ldots$

23. $0.01333\ldots$

24. $2.7333\ldots$

In Exercises 25–30, use the Binomial Theorem to expand the binomial. Simplify your answer.

25. $\left(\dfrac{x}{2} + y\right)^4$

26. $(a - 3b)^5$

27. $\left(\dfrac{2}{x} + 3x\right)^6$

28. $(3x - y^2)^7$

29. $(5 + 2i)^4$

30. $(4 - 5i)^3$

31. As a family increases in number, the number of different interpersonal relationships that exist increases at an even faster rate. Find the number of different interpersonal relationships that exist (between two people) if the number of members in the family is (a) 2, (b) 4, and (c) 6.

32. In Morse Code, all characters are transmitted using a sequence of dits and dahs. How many different characters can be formed by using a sequence of three dits and dahs? (These can be repeated. For example, dit-dit-dit represents the letter s.)

33. A Novice Amateur Radio license consists of two letters, one digit, and then three more letters. How many different licenses could be issued if no restrictions were placed on the letters or digits that could be used?

34. How many different straight lines are determined by (a) 5 points and (b) 10 points? (Assume that no 3 of the given points are collinear and that 2 points determine a line.)

35. A man has five pairs of socks (no two pairs are the same color). If he randomly selects two socks from the drawer, what is the probability that he gets a matching pair?

36. A child carries a five-volume set of books to a bookshelf. She is not able to read and hence cannot distinguish one volume from another. What is the probability that she puts the books on the shelf in the correct order?

37. Are the chances of rolling a 3 with one die the same as that of rolling a total of 6 with two dice? If not, which has the higher probability?

38. A die is rolled six times. What is the probability that each side of the die will appear exactly once?

39. Five cards are drawn from an ordinary deck of 52 playing cards. Find the probability of getting two pairs. (For example, the hand could be A-A-5-5-Q or 4-4-7-7-K.)

40. (a) What is the probability that in a group of 10 people at least 2 people have the same birthday?

    (b) How large must the group be before the probability that at least 2 people have the same birthday is at least 50%?

APPENDIX A

EXPONENTIAL TABLES

x	e^x	e^{-x}	x	e^x	e^{-x}
0.0	1.0000	1.0000	3.0	20.086	0.0498
0.1	1.1052	0.9048	3.1	22.198	0.0450
0.2	1.2214	0.8187	3.2	24.533	0.0408
0.3	1.3499	0.7408	3.3	27.113	0.0369
0.4	1.4918	0.6703	3.4	29.964	0.0334
0.5	1.6487	0.6065	3.5	33.115	0.0302
0.6	1.8221	0.5488	3.6	36.598	0.0273
0.7	2.0138	0.4966	3.7	40.447	0.0247
0.8	2.2255	0.4493	3.8	44.701	0.0224
0.9	2.4596	0.4066	3.9	49.402	0.0202
1.0	2.7183	0.3679	4.0	54.598	0.0183
1.1	3.0042	0.3329	4.1	60.340	0.0166
1.2	3.3201	0.3012	4.2	66.686	0.0150
1.3	3.6693	0.2725	4.3	73.700	0.0136
1.4	4.0552	0.2466	4.4	81.451	0.0123
1.5	4.4817	0.2231	4.5	90.017	0.0111
1.6	4.9530	0.2019	4.6	99.484	0.0101
1.7	5.4739	0.1827	4.7	109.95	0.0091
1.8	6.0496	0.1653	4.8	121.51	0.0082
1.9	6.6859	0.1496	4.9	134.29	0.0074
2.0	7.3891	0.1353	5.0	148.41	0.0067
2.1	8.1662	0.1225	5.1	164.02	0.0061
2.2	9.0250	0.1108	5.2	181.27	0.0055
2.3	9.9742	0.1003	5.3	200.34	0.0050
2.4	11.023	0.0907	5.4	221.41	0.0045
2.5	12.182	0.0821	5.5	244.69	0.0041
2.6	13.464	0.0743	5.6	270.43	0.0037
2.7	14.880	0.0672	5.7	298.87	0.0033
2.8	16.445	0.0608	5.8	330.30	0.0030
2.9	18.174	0.0550	5.9	365.04	0.0027

EXPONENTIAL TABLES (Continued)

x	e^x	e^{-x}	x	e^x	e^{-x}
6.0	403.43	0.0025	8.0	2980.96	0.0003
6.1	445.86	0.0022	8.1	3294.47	0.0003
6.2	492.75	0.0020	8.2	3640.95	0.0003
6.3	544.57	0.0018	8.3	4023.87	0.0002
6.4	601.85	0.0017	8.4	4447.07	0.0002
6.5	665.14	0.0015	8.5	4914.77	0.0002
6.6	735.10	0.0014	8.6	5431.66	0.0002
6.7	812.41	0.0012	8.7	6002.91	0.0002
6.8	897.85	0.0011	8.8	6634.24	0.0002
6.9	992.27	0.0010	8.9	7331.97	0.0001
7.0	1096.63	0.0009	9.0	8103.08	0.0001
7.1	1211.97	0.0008	9.1	8955.29	0.0001
7.2	1339.43	0.0007	9.2	9897.13	0.0001
7.3	1480.30	0.0007	9.3	10938.02	0.0001
7.4	1635.98	0.0006	9.4	12088.38	0.0001
7.5	1808.04	0.0006	9.5	13359.73	0.0001
7.6	1998.20	0.0005	9.6	14764.78	0.0001
7.7	2208.35	0.0005	9.7	16317.61	0.0001
7.8	2440.60	0.0004	9.8	18033.74	0.0001
7.9	2697.28	0.0004	9.9	19930.37	0.0001
			10.0	22026.47	0.0000

APPENDIX B

NATURAL LOGARITHMIC TABLES

	0.00	0.01	0.02	0.03	0.04	0.05	0.06	0.07	0.08	0.09
1.0	0.0000	0.0100	0.0198	0.0296	0.0392	0.0488	0.0583	0.0677	0.0770	0.0862
1.1	0.0953	0.1044	0.1133	0.1222	0.1310	0.1398	0.1484	0.1570	0.1655	0.1740
1.2	0.1823	0.1906	0.1989	0.2070	0.2151	0.2231	0.2311	0.2390	0.2469	0.2546
1.3	0.2624	0.2700	0.2776	0.2852	0.2927	0.3001	0.3075	0.3148	0.3221	0.3293
1.4	0.3365	0.3436	0.3507	0.3577	0.3646	0.3716	0.3784	0.3853	0.3920	0.3988
1.5	0.4055	0.4121	0.4187	0.4253	0.4318	0.4383	0.4447	0.4511	0.4574	0.4637
1.6	0.4700	0.4762	0.4824	0.4886	0.4947	0.5008	0.5068	0.5128	0.5188	0.5247
1.7	0.5306	0.5365	0.5423	0.5481	0.5539	0.5596	0.5653	0.5710	0.5766	0.5822
1.8	0.5878	0.5933	0.5988	0.6043	0.6098	0.6152	0.6206	0.6259	0.6313	0.6366
1.9	0.6419	0.6471	0.6523	0.6575	0.6627	0.6678	0.6729	0.6780	0.6831	0.6881
2.0	0.6931	0.6981	0.7031	0.7080	0.7129	0.7178	0.7227	0.7275	0.7324	0.7372
2.1	0.7419	0.7467	0.7514	0.7561	0.7608	0.7655	0.7701	0.7747	0.7793	0.7839
2.2	0.7885	0.7930	0.7975	0.8020	0.8065	0.8109	0.8154	0.8198	0.8242	0.8286
2.3	0.8329	0.8372	0.8416	0.8459	0.8502	0.8544	0.8587	0.8629	0.8671	0.8713
2.4	0.8755	0.8796	0.8838	0.8879	0.8920	0.8961	0.9002	0.9042	0.9083	0.9123
2.5	0.9163	0.9203	0.9243	0.9282	0.9322	0.9361	0.9400	0.9439	0.9478	0.9517
2.6	0.9555	0.9594	0.9632	0.9670	0.9708	0.9746	0.9783	0.9821	0.9858	0.9895
2.7	0.9933	0.9969	1.0006	1.0043	1.0080	1.0116	1.0152	1.0188	1.0225	1.0260
2.8	1.0296	1.0332	1.0367	1.0403	1.0438	1.0473	1.0508	1.0543	1.0578	1.0613
2.9	1.0647	1.0682	1.0716	1.0750	1.0784	1.0818	1.0852	1.0886	1.0919	1.0953
3.0	1.0986	1.1019	1.1053	1.1086	1.1119	1.1151	1.1184	1.1217	1.1249	1.1282
3.1	1.1314	1.1346	1.1378	1.1410	1.1442	1.1474	1.1506	1.1537	1.1569	1.1600
3.2	1.1632	1.1663	1.1694	1.1725	1.1756	1.1787	1.1817	1.1848	1.1878	1.1909
3.3	1.1939	1.1969	1.2000	1.2030	1.2060	1.2090	1.2119	1.2149	1.2179	1.2208
3.4	1.2238	1.2267	1.2296	1.2326	1.2355	1.2384	1.2413	1.2442	1.2470	1.2499
3.5	1.2528	1.2556	1.2585	1.2613	1.2641	1.2669	1.2698	1.2726	1.2754	1.2782
3.6	1.2809	1.2837	1.2865	1.2892	1.2920	1.2947	1.2975	1.3002	1.3029	1.3056
3.7	1.3083	1.3110	1.3137	1.3164	1.3191	1.3218	1.3244	1.3271	1.3297	1.3324
3.8	1.3350	1.3376	1.3403	1.3429	1.3455	1.3481	1.3507	1.3533	1.3558	1.3584
3.9	1.3610	1.3635	1.3661	1.3686	1.3712	1.3737	1.3762	1.3788	1.3813	1.3838

NATURAL LOGARITHMIC TABLES (Continued)

	0.00	0.01	0.02	0.03	0.04	0.05	0.06	0.07	0.08	0.09
4.0	1.3863	1.3888	1.3913	1.3938	1.3962	1.3987	1.4012	1.4036	1.4061	1.4085
4.1	1.4110	1.4134	1.4159	1.4183	1.4207	1.4231	1.4255	1.4279	1.4303	1.4327
4.2	1.4351	1.4375	1.4398	1.4422	1.4446	1.4469	1.4493	1.4516	1.4540	1.4563
4.3	1.4586	1.4609	1.4633	1.4656	1.4679	1.4702	1.4725	1.4748	1.4770	1.4793
4.4	1.4816	1.4839	1.4861	1.4884	1.4907	1.4929	1.4951	1.4974	1.4996	1.5019
4.5	1.5041	1.5063	1.5085	1.5107	1.5129	1.5151	1.5173	1.5195	1.5217	1.5239
4.6	1.5261	1.5282	1.5304	1.5326	1.5347	1.5369	1.5390	1.5412	1.5433	1.5454
4.7	1.5476	1.5497	1.5518	1.5539	1.5560	1.5581	1.5602	1.5623	1.5644	1.5665
4.8	1.5686	1.5707	1.5728	1.5748	1.5769	1.5790	1.5810	1.5831	1.5851	1.5872
4.9	1.5892	1.5913	1.5933	1.5953	1.5974	1.5994	1.6014	1.6034	1.6054	1.6074
5.0	1.6094	1.6114	1.6134	1.6154	1.6174	1.6194	1.6214	1.6233	1.6253	1.6273
5.1	1.6292	1.6312	1.6332	1.6351	1.6371	1.6390	1.6409	1.6429	1.6448	1.6467
5.2	1.6487	1.6506	1.6525	1.6544	1.6563	1.6582	1.6601	1.6620	1.6639	1.6658
5.3	1.6677	1.6696	1.6715	1.6734	1.6752	1.6771	1.6790	1.6808	1.6827	1.6845
5.4	1.6864	1.6882	1.6901	1.6919	1.6938	1.6956	1.6974	1.6993	1.7011	1.7029
5.5	1.7047	1.7066	1.7084	1.7102	1.7120	1.7138	1.7156	1.7174	1.7192	1.7210
5.6	1.7228	1.7246	1.7263	1.7281	1.7299	1.7317	1.7334	1.7352	1.7370	1.7387
5.7	1.7405	1.7422	1.7440	1.7457	1.7475	1.7492	1.7509	1.7527	1.7544	1.7561
5.8	1.7579	1.7596	1.7613	1.7630	1.7647	1.7664	1.7681	1.7699	1.7716	1.7733
5.9	1.7750	1.7766	1.7783	1.7800	1.7817	1.7834	1.7851	1.7867	1.7884	1.7901
6.0	1.7918	1.7934	1.7951	1.7967	1.7984	1.8001	1.8017	1.8034	1.8050	1.8066
6.1	1.8083	1.8099	1.8116	1.8132	1.8148	1.8165	1.8181	1.8197	1.8213	1.8229
6.2	1.8245	1.8262	1.8278	1.8294	1.8310	1.8326	1.8342	1.8358	1.8374	1.8390
6.3	1.8405	1.8421	1.8437	1.8453	1.8469	1.8485	1.8500	1.8516	1.8532	1.8547
6.4	1.8563	1.8579	1.8594	1.8610	1.8625	1.8641	1.8656	1.8672	1.8687	1.8703
6.5	1.8718	1.8733	1.8749	1.8764	1.8779	1.8795	1.8810	1.8825	1.8840	1.8856
6.6	1.8871	1.8886	1.8901	1.8916	1.8931	1.8946	1.8961	1.8976	1.8991	1.9006
6.7	1.9021	1.9036	1.9051	1.9066	1.9081	1.9095	1.9110	1.9125	1.9140	1.9155
6.8	1.9169	1.9184	1.9199	1.9213	1.9228	1.9242	1.9257	1.9272	1.9286	1.9301
6.9	1.9315	1.9330	1.9344	1.9359	1.9373	1.9387	1.9402	1.9416	1.9430	1.9445
7.0	1.9459	1.9473	1.9488	1.9502	1.9516	1.9530	1.9544	1.9559	1.9573	1.9587
7.1	1.9601	1.9615	1.9629	1.9643	1.9657	1.9671	1.9685	1.9699	1.9713	1.9727
7.2	1.9741	1.9755	1.9769	1.9782	1.9796	1.9810	1.9824	1.9838	1.9851	1.9865
7.3	1.9879	1.9892	1.9906	1.9920	1.9933	1.9947	1.9961	1.9974	1.9988	2.0001
7.4	2.0015	2.0028	2.0042	2.0055	2.0069	2.0082	2.0096	2.0109	2.0122	2.0136
7.5	2.0149	2.0162	2.0176	2.0189	2.0202	2.0215	2.0229	2.0242	2.0255	2.0268
7.6	2.0281	2.0295	2.0308	2.0321	2.0334	2.0347	2.0360	2.0373	2.0386	2.0399
7.7	2.0412	2.0425	2.0438	2.0451	2.0464	2.0477	2.0490	2.0503	2.0516	2.0528
7.8	2.0541	2.0554	2.0567	2.0580	2.0592	2.0605	2.0618	2.0631	2.0643	2.0656
7.9	2.0669	2.0681	2.0694	2.0707	2.0719	2.0732	2.0744	2.0757	2.0769	2.0782

NATURAL LOGARITHMIC TABLES (Continued)

	0.00	0.01	0.02	0.03	0.04	0.05	0.06	0.07	0.08	0.09
8.0	2.0794	2.0807	2.0819	2.0832	2.0844	2.0857	2.0869	2.0882	2.0894	2.0906
8.1	2.0919	2.0931	2.0943	2.0956	2.0968	2.0980	2.0992	2.1005	2.1017	2.1029
8.2	2.1041	2.1054	2.1066	2.1078	2.1090	2.1102	2.1114	2.1126	2.1138	2.1150
8.3	2.1163	2.1175	2.1187	2.1199	2.1211	2.1223	2.1235	2.1247	2.1258	2.1270
8.4	2.1282	2.1294	2.1306	2.1318	2.1330	2.1342	2.1353	2.1365	2.1377	2.1389
8.5	2.1401	2.1412	2.1424	2.1436	2.1448	2.1459	2.1471	2.1483	2.1494	2.1506
8.6	2.1518	2.1529	2.1541	2.1552	2.1564	2.1576	2.1587	2.1599	2.1610	2.1622
8.7	2.1633	2.1645	2.1656	2.1668	2.1679	2.1691	2.1702	2.1713	2.1725	2.1736
8.8	2.1748	2.1759	2.1770	2.1782	2.1793	2.1804	2.1815	2.1827	2.1838	2.1849
8.9	2.1861	2.1872	2.1883	2.1894	2.1905	2.1917	2.1928	2.1939	2.1950	2.1961
9.0	2.1972	2.1983	2.1994	2.2006	2.2017	2.2028	2.2039	2.2050	2.2061	2.2072
9.1	2.2083	2.2094	2.2105	2.2116	2.2127	2.2138	2.2148	2.2159	2.2170	2.2181
9.2	2.2192	2.2203	2.2214	2.2225	2.2235	2.2246	2.2257	2.2268	2.2279	2.2289
9.3	2.2300	2.2311	2.2322	2.2332	2.2343	2.2354	2.2364	2.2375	2.2386	2.2396
9.4	2.2407	2.2418	2.2428	2.2439	2.2450	2.2460	2.2471	2.2481	2.2492	2.2502
9.5	2.2513	2.2523	2.2534	2.2544	2.2555	2.2565	2.2576	2.2586	2.2597	2.2607
9.6	2.2618	2.2628	2.2638	2.2649	2.2659	2.2670	2.2680	2.2690	2.2701	2.2711
9.7	2.2721	2.2732	2.2742	2.2752	2.2762	2.2773	2.2783	2.2793	2.2803	2.2814
9.8	2.2824	2.2834	2.2844	2.2854	2.2865	2.2875	2.2885	2.2895	2.2905	2.2915
9.9	2.2925	2.2935	2.2946	2.2956	2.2966	2.2976	2.2986	2.2996	2.3006	2.3016

APPENDIX C

COMMON LOGARITHMIC TABLES

	0.00	0.01	0.02	0.03	0.04	0.05	0.06	0.07	0.08	0.09
1.0	0.0000	0.0043	0.0086	0.0128	0.0170	0.0212	0.0253	0.0294	0.0334	0.0374
1.1	0.0414	0.0453	0.0492	0.0531	0.0569	0.0607	0.0645	0.0682	0.0719	0.0755
1.2	0.0792	0.0828	0.0864	0.0899	0.0934	0.0969	0.1004	0.1038	0.1072	0.1106
1.3	0.1139	0.1173	0.1206	0.1239	0.1271	0.1303	0.1335	0.1367	0.1399	0.1430
1.4	0.1461	0.1492	0.1523	0.1553	0.1584	0.1614	0.1644	0.1673	0.1703	0.1732
1.5	0.1761	0.1790	0.1818	0.1847	0.1875	0.1903	0.1931	0.1959	0.1987	0.2014
1.6	0.2041	0.2068	0.2095	0.2122	0.2148	0.2175	0.2201	0.2227	0.2253	0.2279
1.7	0.2304	0.2330	0.2355	0.2380	0.2405	0.2430	0.2455	0.2480	0.2504	0.2529
1.8	0.2553	0.2577	0.2601	0.2625	0.2648	0.2672	0.2695	0.2718	0.2742	0.2765
1.9	0.2788	0.2810	0.2833	0.2856	0.2878	0.2900	0.2923	0.2945	0.2967	0.2989
2.0	0.3010	0.3032	0.3054	0.3075	0.3096	0.3118	0.3139	0.3160	0.3181	0.3201
2.1	0.3222	0.3243	0.3263	0.3284	0.3304	0.3324	0.3345	0.3365	0.3385	0.3404
2.2	0.3424	0.3444	0.3464	0.3483	0.3502	0.3522	0.3541	0.3560	0.3579	0.3598
2.3	0.3617	0.3636	0.3655	0.3674	0.3692	0.3711	0.3729	0.3747	0.3766	0.3784
2.4	0.3802	0.3820	0.3838	0.3856	0.3874	0.3892	0.3909	0.3927	0.3945	0.3962
2.5	0.3979	0.3997	0.4014	0.4031	0.4048	0.4065	0.4082	0.4099	0.4116	0.4133
2.6	0.4150	0.4166	0.4183	0.4200	0.4216	0.4232	0.4249	0.4265	0.4281	0.4298
2.7	0.4314	0.4330	0.4346	0.4362	0.4378	0.4393	0.4409	0.4425	0.4440	0.4456
2.8	0.4472	0.4487	0.4502	0.4518	0.4533	0.4548	0.4564	0.4579	0.4594	0.4609
2.9	0.4624	0.4639	0.4654	0.4669	0.4683	0.4698	0.4713	0.4728	0.4742	0.4757
3.0	0.4771	0.4786	0.4800	0.4814	0.4829	0.4843	0.4857	0.4871	0.4886	0.4900
3.1	0.4914	0.4928	0.4942	0.4955	0.4969	0.4983	0.4997	0.5011	0.5024	0.5038
3.2	0.5052	0.5065	0.5079	0.5092	0.5105	0.5119	0.5132	0.5145	0.5159	0.5172
3.3	0.5185	0.5198	0.5211	0.5224	0.5237	0.5250	0.5263	0.5276	0.5289	0.5302
3.4	0.5315	0.5328	0.5340	0.5353	0.5366	0.5378	0.5391	0.5403	0.5416	0.5428
3.5	0.5441	0.5453	0.5465	0.5478	0.5490	0.5502	0.5514	0.5527	0.5539	0.5551
3.6	0.5563	0.5575	0.5587	0.5599	0.5611	0.5623	0.5635	0.5647	0.5658	0.5670
3.7	0.5682	0.5694	0.5705	0.5717	0.5729	0.5740	0.5752	0.5763	0.5775	0.5786
3.8	0.5798	0.5809	0.5821	0.5832	0.5843	0.5855	0.5866	0.5877	0.5888	0.5899
3.9	0.5911	0.5922	0.5933	0.5944	0.5955	0.5966	0.5977	0.5988	0.5999	0.6010

COMMON LOGARITHMIC TABLES (Continued)

	0.00	0.01	0.02	0.03	0.04	0.05	0.06	0.07	0.08	0.09
4.0	0.6021	0.6031	0.6042	0.6053	0.6064	0.6075	0.6085	0.6096	0.6107	0.6117
4.1	0.6128	0.6138	0.6149	0.6160	0.6170	0.6180	0.6191	0.6201	0.6212	0.6222
4.2	0.6232	0.6243	0.6253	0.6263	0.6274	0.6284	0.6294	0.6304	0.6314	0.6325
4.3	0.6335	0.6345	0.6355	0.6365	0.6375	0.6385	0.6395	0.6405	0.6415	0.6425
4.4	0.6435	0.6444	0.6454	0.6464	0.6474	0.6484	0.6493	0.6503	0.6513	0.6522
4.5	0.6532	0.6542	0.6551	0.6561	0.6571	0.6580	0.6590	0.6599	0.6609	0.6618
4.6	0.6628	0.6637	0.6646	0.6656	0.6665	0.6675	0.6684	0.6693	0.6702	0.6712
4.7	0.6721	0.6730	0.6739	0.6749	0.6758	0.6767	0.6776	0.6785	0.6794	0.6803
4.8	0.6812	0.6821	0.6830	0.6839	0.6848	0.6857	0.6866	0.6875	0.6884	0.6893
4.9	0.6902	0.6911	0.6920	0.6928	0.6937	0.6946	0.6955	0.6964	0.6972	0.6981
5.0	0.6990	0.6998	0.7007	0.7016	0.7024	0.7033	0.7042	0.7050	0.7059	0.7067
5.1	0.7076	0.7084	0.7093	0.7101	0.7110	0.7118	0.7126	0.7135	0.7143	0.7152
5.2	0.7160	0.7168	0.7177	0.7185	0.7193	0.7202	0.7210	0.7218	0.7226	0.7235
5.3	0.7243	0.7251	0.7259	0.7267	0.7275	0.7284	0.7292	0.7300	0.7308	0.7316
5.4	0.7324	0.7332	0.7340	0.7348	0.7356	0.7364	0.7372	0.7380	0.7388	0.7396
5.5	0.7404	0.7412	0.7419	0.7427	0.7435	0.7443	0.7451	0.7459	0.7466	0.7474
5.6	0.7482	0.7490	0.7497	0.7505	0.7513	0.7520	0.7528	0.7536	0.7543	0.7551
5.7	0.7559	0.7566	0.7574	0.7582	0.7589	0.7597	0.7604	0.7612	0.7619	0.7627
5.8	0.7634	0.7642	0.7649	0.7657	0.7664	0.7672	0.7679	0.7686	0.7694	0.7701
5.9	0.7709	0.7716	0.7723	0.7731	0.7738	0.7745	0.7752	0.7760	0.7767	0.7774
6.0	0.7782	0.7789	0.7796	0.7803	0.7810	0.7818	0.7825	0.7832	0.7839	0.7846
6.1	0.7853	0.7860	0.7868	0.7875	0.7882	0.7889	0.7896	0.7903	0.7910	0.7917
6.2	0.7924	0.7931	0.7938	0.7945	0.7952	0.7959	0.7966	0.7973	0.7980	0.7987
6.3	0.7993	0.8000	0.8007	0.8014	0.8021	0.8028	0.8035	0.8041	0.8048	0.8055
6.4	0.8062	0.8069	0.8075	0.8082	0.8089	0.8096	0.8102	0.8109	0.8116	0.8122
6.5	0.8129	0.8136	0.8142	0.8149	0.8156	0.8162	0.8169	0.8176	0.8182	0.8189
6.6	0.8195	0.8202	0.8209	0.8215	0.8222	0.8228	0.8235	0.8241	0.8248	0.8254
6.7	0.8261	0.8267	0.8274	0.8280	0.8287	0.8293	0.8299	0.8306	0.8312	0.8319
6.8	0.8325	0.8331	0.8338	0.8344	0.8351	0.8357	0.8363	0.8370	0.8376	0.8382
6.9	0.8388	0.8395	0.8401	0.8407	0.8414	0.8420	0.8426	0.8432	0.8439	0.8445
7.0	0.8451	0.8457	0.8463	0.8470	0.8476	0.8482	0.8488	0.8494	0.8500	0.8506
7.1	0.8513	0.8519	0.8525	0.8531	0.8537	0.8543	0.8549	0.8555	0.8561	0.8567
7.2	0.8573	0.8579	0.8585	0.8591	0.8597	0.8603	0.8609	0.8615	0.8621	0.8627
7.3	0.8633	0.8639	0.8645	0.8651	0.8657	0.8663	0.8669	0.8675	0.8681	0.8686
7.4	0.8692	0.8698	0.8704	0.8710	0.8716	0.8722	0.8727	0.8733	0.8739	0.8745
7.5	0.8751	0.8756	0.8762	0.8768	0.8774	0.8779	0.8785	0.8791	0.8797	0.8802
7.6	0.8808	0.8814	0.8820	0.8825	0.8831	0.8837	0.8842	0.8848	0.8854	0.8859
7.7	0.8865	0.8871	0.8876	0.8882	0.8887	0.8893	0.8899	0.8904	0.8910	0.8915
7.8	0.8921	0.8927	0.8932	0.8938	0.8943	0.8949	0.8954	0.8960	0.8965	0.8971
7.9	0.8976	0.8982	0.8987	0.8993	0.8998	0.9004	0.9009	0.9015	0.9020	0.9025

COMMON LOGARITHMIC TABLES (Continued)

	0.00	0.01	0.02	0.03	0.04	0.05	0.06	0.07	0.08	0.09
8.0	0.9031	0.9036	0.9042	0.9047	0.9053	0.9058	0.9063	0.9069	0.9074	0.9079
8.1	0.9085	0.9090	0.9096	0.9101	0.9106	0.9112	0.9117	0.9122	0.9128	0.9133
8.2	0.9138	0.9143	0.9149	0.9154	0.9159	0.9165	0.9170	0.9175	0.9180	0.9186
8.3	0.9191	0.9196	0.9201	0.9206	0.9212	0.9217	0.9222	0.9227	0.9232	0.9238
8.4	0.9243	0.9248	0.9253	0.9258	0.9263	0.9269	0.9274	0.9279	0.9284	0.9289
8.5	0.9294	0.9299	0.9304	0.9309	0.9315	0.9320	0.9325	0.9330	0.9335	0.9340
8.6	0.9345	0.9350	0.9355	0.9360	0.9365	0.9370	0.9375	0.9380	0.9385	0.9390
8.7	0.9395	0.9400	0.9405	0.9410	0.9415	0.9420	0.9425	0.9430	0.9435	0.9440
8.8	0.9445	0.9450	0.9455	0.9460	0.9465	0.9469	0.9474	0.9479	0.9484	0.9489
8.9	0.9494	0.9499	0.9504	0.9509	0.9513	0.9518	0.9523	0.9528	0.9533	0.9538
9.0	0.9542	0.9547	0.9552	0.9557	0.9562	0.9566	0.9571	0.9576	0.9581	0.9586
9.1	0.9590	0.9595	0.9600	0.9605	0.9609	0.9614	0.9619	0.9624	0.9628	0.9633
9.2	0.9638	0.9643	0.9647	0.9652	0.9657	0.9661	0.9666	0.9671	0.9675	0.9680
9.3	0.9685	0.9689	0.9694	0.9699	0.9703	0.9708	0.9713	0.9717	0.9722	0.9727
9.4	0.9731	0.9736	0.9741	0.9745	0.9750	0.9754	0.9759	0.9764	0.9768	0.9773
9.5	0.9777	0.9782	0.9786	0.9791	0.9795	0.9800	0.9805	0.9809	0.9814	0.9818
9.6	0.9823	0.9827	0.9832	0.9836	0.9841	0.9845	0.9850	0.9854	0.9859	0.9863
9.7	0.9868	0.9872	0.9877	0.9881	0.9886	0.9890	0.9894	0.9899	0.9903	0.9908
9.8	0.9912	0.9917	0.9921	0.9926	0.9930	0.9934	0.9939	0.9943	0.9948	0.9952
9.9	0.9956	0.9961	0.9965	0.9969	0.9974	0.9978	0.9983	0.9987	0.9991	0.9996

APPENDIX D

TRIGONOMETRIC TABLES

1 degree ≈ 0.01745 radians
1 radian ≈ 57.29578 degrees

For 0 ≤ θ ≤ 45, read from upper left.
For 45 ≤ θ ≤ 90, read from lower right.
For 90 ≤ θ ≤ 360, use the identities:

θ	Quadrant II	Quadrant III	Quadrant IV
sin θ	sin(180−θ)	−sin(θ−180)	−sin(360−θ)
cos θ	−cos(180−θ)	−cos(θ−180)	cos(360−θ)
tan θ	−tan(180−θ)	tan(θ−180)	−tan(360−θ)
cot θ	−cot(180−θ)	cot(θ−180)	−cot(360−θ)

Degrees	Radians	sin	cos	tan	cot			Degrees	Radians	sin	cos	tan	cot		
0°00′	.0000	.0000	1.0000	.0000	−	1.5708	90°00′	4°00′	.0698	.0698	.9976	.0699	14.301	1.5010	86°00′
10	.0029	.0029	1.0000	.0029	343.774	1.5679	50	10	.0727	.0727	.9974	.0729	13.727	1.4981	50
20	.0058	.0058	1.0000	.0058	171.885	1.5650	40	20	.0756	.0756	.9971	.0758	13.197	1.4952	40
30	.0087	.0087	1.0000	.0087	114.589	1.5621	30	30	.0785	.0785	.9969	.0787	12.706	1.4923	30
40	.0116	.0116	.9999	.0116	85.940	1.5592	20	40	.0814	.0814	.9967	.0816	12.251	1.4893	20
50	.0145	.0145	.9999	.0145	68.750	1.5563	10	50	.0844	.0843	.9964	.0846	11.826	1.4864	10
1°00′	.0175	.0175	.9998	.0175	57.290	1.5533	89°00′	5°00′	.0873	.0872	.9962	.0875	11.430	1.4835	85°00′
10	.0204	.0204	.9998	.0204	49.104	1.5504	50	10	.0902	.0901	.9959	.0904	11.059	1.4806	50
20	.0233	.0233	.9997	.0233	42.964	1.5475	40	20	.0931	.0929	.9957	.0934	10.712	1.4777	40
30	.0262	.0262	.9997	.0262	38.188	1.5446	30	30	.0960	.0958	.9954	.0963	10.385	1.4748	30
40	.0291	.0291	.9996	.0291	34.368	1.5417	20	40	.0989	.0987	.9951	.0992	10.078	1.4719	20
50	.0320	.0320	.9995	.0320	31.242	1.5388	10	50	.1018	.1016	.9948	.1022	9.788	1.4690	10
2°00′	.0349	.0349	.9994	.0349	28.636	1.5359	88°00′	6°00′	.1047	.1045	.9945	.1051	9.514	1.4661	84°00′
10	.0378	.0378	.9993	.0378	26.432	1.5330	50	10	.1076	.1074	.9942	.1080	9.255	1.4632	50
20	.0407	.0407	.9992	.0407	24.542	1.5301	40	20	.1105	.1103	.9939	.1110	9.010	1.4603	40
30	.0436	.0436	.9990	.0437	22.904	1.5272	30	30	.1134	.1132	.9936	.1139	8.777	1.4573	30
40	.0465	.0465	.9989	.0466	21.470	1.5243	20	40	.1164	.1161	.9932	.1169	8.556	1.4544	20
50	.0495	.0494	.9988	.0495	20.206	1.5213	10	50	.1193	.1190	.9929	.1198	8.345	1.4515	10
3°00′	.0524	.0523	.9986	.0524	19.081	1.5184	87°00′	7°00′	.1222	.1219	.9925	.1228	8.144	1.4486	83°00′
10	.0553	.0552	.9985	.0553	18.075	1.5155	50	10	.1251	.1248	.9922	.1257	7.953	1.4457	50
20	.0582	.0581	.9983	.0582	17.169	1.5126	40	20	.1280	.1276	.9918	.1287	7.770	1.4428	40
30	.0611	.0610	.9981	.0612	16.350	1.5097	30	30	.1309	.1305	.9914	.1317	7.596	1.4399	30
40	.0640	.0640	.9980	.0641	15.605	1.5068	20	40	.1338	.1334	.9911	.1346	7.429	1.4370	20
50	.0669	.0669	.9978	.0670	14.924	1.5039	10	50	.1367	.1363	.9907	.1376	7.269	1.4341	10
		cos	sin	cot	tan	Radians	Degrees			cos	sin	cot	tan	Radians	Degrees

TRIGONOMETRIC TABLES (Continued)

Degrees	Radians	sin	cos	tan	cot		
8°00'	.1396	.1392	.9903	.1405	7.115	1.4312	82°00'
10	.1425	.1421	.9899	.1435	6.968	1.4283	50
20	.1454	.1449	.9894	.1465	6.827	1.4254	40
30	.1484	.1478	.9890	.1495	6.691	1.4224	30
40	.1513	.1507	.9886	.1524	6.561	1.4195	20
50	.1542	.1536	.9881	.1554	6.435	1.4166	10
9°00'	.1571	.1564	.9877	.1584	6.314	1.4137	81°00'
10	.1600	.1593	.9872	.1614	6.197	1.4108	50
20	.1629	.1622	.9868	.1644	6.084	1.4079	40
30	.1658	.1650	.9863	.1673	5.976	1.4050	30
40	.1687	.1679	.9858	.1703	5.871	1.4021	20
50	.1716	.1708	.9853	.1733	5.769	1.3992	10
10°00'	.1745	.1736	.9848	.1763	5.671	1.3963	80°00'
10	.1774	.1765	.9843	.1793	5.576	1.3934	50
20	.1804	.1794	.9838	.1823	5.485	1.3904	40
30	.1833	.1822	.9833	.1853	5.396	1.3875	30
40	.1862	.1851	.9827	.1883	5.309	1.3846	20
50	.1891	.1880	.9822	.1914	5.226	1.3817	10
11°00'	.1920	.1908	.9816	.1944	5.145	1.3788	79°00'
10	.1949	.1937	.9811	.1974	5.066	1.3759	50
20	.1978	.1965	.9805	.2004	4.989	1.3730	40
30	.2007	.1994	.9799	.2035	4.915	1.3701	30
40	.2036	.2022	.9793	.2065	4.843	1.3672	20
50	.2065	.2051	.9787	.2095	4.773	1.3643	10
12°00'	.2094	.2079	.9781	.2126	4.705	1.3614	78°00'
10	.2123	.2108	.9775	.2156	4.638	1.3584	50
20	.2153	.2136	.9769	.2186	4.574	1.3555	40
30	.2182	.2164	.9763	.2217	4.511	1.3526	30
40	.2211	.2193	.9757	.2247	4.449	1.3497	20
50	.2240	.2221	.9750	.2278	4.390	1.3468	10
13°00'	.2269	.2250	.9744	.2309	4.331	1.3439	77°00'
10	.2298	.2278	.9737	.2339	4.275	1.3410	50
20	.2327	.2306	.9730	.2370	4.219	1.3381	40
30	.2356	.2334	.9724	.2401	4.165	1.3352	30
40	.2385	.2363	.9717	.2432	4.113	1.3323	20
50	.2414	.2391	.9710	.2462	4.061	1.3294	10
14°00'	.2443	.2419	.9703	.2493	4.011	1.3265	76°00'
10	.2473	.2447	.9696	.2524	3.962	1.3235	50
20	.2502	.2476	.9689	.2555	3.914	1.3206	40
30	.2531	.2504	.9681	.2586	3.867	1.3177	30
40	.2560	.2532	.9674	.2617	3.821	1.3148	20
50	.2589	.2560	.9667	.2648	3.776	1.3119	10
15°00'	.2618	.2588	.9659	.2679	3.732	1.3090	75°00'
10	.2647	.2616	.9652	.2711	3.689	1.3061	50
20	.2676	.2644	.9644	.2742	3.647	1.3032	40
30	.2705	.2672	.9636	.2773	3.606	1.3003	30
40	.2734	.2700	.9628	.2805	3.566	1.2974	20
50	.2763	.2728	.9621	.2836	3.526	1.2945	10
16°00'	.2793	.2756	.9613	.2867	3.487	1.2915	74°00'
10	.2822	.2784	.9605	.2899	3.450	1.2886	50
20	.2851	.2812	.9596	.2931	3.412	1.2857	40
30	.2880	.2840	.9588	.2962	3.376	1.2828	30
40	.2909	.2868	.9580	.2994	3.340	1.2799	20
50	.2938	.2896	.9572	.3026	3.305	1.2770	10
17°00'	.2967	.2924	.9563	.3057	3.271	1.2741	73°00'
10	.2996	.2952	.9555	.3089	3.237	1.2712	50
20	.3025	.2979	.9546	.3121	3.204	1.2683	40
30	.3054	.3007	.9537	.3153	3.172	1.2654	30
40	.3083	.3035	.9528	.3185	3.140	1.2625	20
50	.3113	.3062	.9520	.3217	3.108	1.2595	10
		cos	sin	cot	tan	Radians	Degrees

Degrees	Radians	sin	cos	tan	cot		
18°00'	.3142	.3090	.9511	.3249	3.078	1.2566	72°00'
10	.3171	.3118	.9502	.3281	3.047	1.2537	50
20	.3200	.3145	.9492	.3314	3.018	1.2508	40
30	.3229	.3173	.9483	.3346	2.989	1.2479	30
40	.3258	.3201	.9474	.3378	2.960	1.2450	20
50	.3287	.3228	.9465	.3411	2.932	1.2421	10
19°00'	.3316	.3256	.9455	.3443	2.904	1.2392	71°00'
10	.3345	.3283	.9446	.3476	2.877	1.2363	50
20	.3374	.3311	.9436	.3508	2.850	1.2334	40
30	.3403	.3338	.9426	.3541	2.824	1.2305	30
40	.3432	.3365	.9417	.3574	2.798	1.2275	20
50	.3462	.3393	.9407	.3607	2.773	1.2246	10
20°00'	.3491	.3420	.9397	.3640	2.747	1.2217	70°00'
10	.3520	.3448	.9387	.3673	2.723	1.2188	50
20	.3549	.3475	.9377	.3706	2.699	1.2159	40
30	.3578	.3502	.9367	.3739	2.675	1.2130	30
40	.3607	.3529	.9356	.3772	2.651	1.2101	20
50	.3636	.3557	.9346	.3805	2.628	1.2072	10
21°00'	.3665	.3584	.9336	.3839	2.605	1.2043	69°00'
10	.3694	.3611	.9325	.3872	2.583	1.2014	50
20	.3723	.3638	.9315	.3906	2.560	1.1985	40
30	.3752	.3665	.9304	.3939	2.539	1.1956	30
40	.3782	.3692	.9293	.3973	2.517	1.1926	20
50	.3811	.3719	.9283	.4006	2.496	1.1897	10
22°00'	.3840	.3746	.9272	.4040	2.475	1.1868	68°00'
10	.3869	.3773	.9261	.4074	2.455	1.1839	50
20	.3898	.3800	.9250	.4108	2.434	1.1810	40
30	.3927	.3827	.9239	.4142	2.414	1.1781	30
40	.3956	.3854	.9228	.4176	2.394	1.1752	20
50	.3985	.3881	.9216	.4210	2.375	1.1723	10
23°00'	.4014	.3907	.9205	.4245	2.356	1.1694	67°00'
10	.4043	.3934	.9194	.4279	2.337	1.1665	50
20	.4072	.3961	.9182	.4314	2.318	1.1636	40
30	.4102	.3987	.9171	.4348	2.300	1.1606	30
40	.4131	.4014	.9159	.4383	2.282	1.1577	20
50	.4160	.4041	.9147	.4417	2.264	1.1548	10
24°00'	.4189	.4067	.9135	.4452	2.246	1.1519	66°00'
10	.4218	.4094	.9124	.4487	2.229	1.1490	50
20	.4247	.4120	.9112	.4522	2.211	1.1461	40
30	.4276	.4147	.9100	.4557	2.194	1.1432	30
40	.4305	.4173	.9088	.4592	2.177	1.1403	20
50	.4334	.4200	.9075	.4628	2.161	1.1374	10
25°00'	.4363	.4226	.9063	.4663	2.145	1.1345	65°00'
10	.4392	.4253	.9051	.4699	2.128	1.1316	50
20	.4422	.4279	.9038	.4734	2.112	1.1286	40
30	.4451	.4305	.9026	.4770	2.097	1.1257	30
40	.4480	.4331	.9013	.4806	2.081	1.1228	20
50	.4509	.4358	.9001	.4841	2.066	1.1199	10
26°00'	.4538	.4384	.8988	.4877	2.050	1.1170	64°00'
10	.4567	.4410	.8975	.4913	2.035	1.1141	50
20	.4596	.4436	.8962	.4950	2.020	1.1112	40
30	.4625	.4462	.8949	.4986	2.006	1.1083	30
40	.4654	.4488	.8936	.5022	1.991	1.1054	20
50	.4683	.4514	.8923	.5059	1.977	1.1025	10
27°00'	.4712	.4540	.8910	.5095	1.963	1.0996	63°00'
10	.4741	.4566	.8897	.5132	1.949	1.0966	50
20	.4771	.4592	.8884	.5169	1.935	1.0937	40
30	.4800	.4617	.8870	.5206	1.921	1.0908	30
40	.4829	.4643	.8857	.5243	1.907	1.0879	20
50	.4858	.4669	.8843	.5280	1.894	1.0850	10
		cos	sin	cot	tan	Radians	Degrees

TRIGONOMETRIC TABLES (Continued)

Degrees	Radians	sin	cos	tan	cot		Degrees
28°00′	.4887	.4695	.8829	.5317	1.881	1.0821	62°00′
10	.4916	.4720	.8816	.5354	1.868	1.0792	50
20	.4945	.4746	.8802	.5392	1.855	1.0763	40
30	.4974	.4772	.8788	.5430	1.842	1.0734	30
40	.5003	.4797	.8774	.5467	1.829	1.0705	20
50	.5032	.4823	.8760	.5505	1.816	1.0676	10
29°00′	.5061	.4848	.8746	.5543	1.804	1.0647	61°00′
10	.5091	.4874	.8732	.5581	1.792	1.0617	50
20	.5120	.4899	.8718	.5619	1.780	1.0588	40
30	.5149	.4924	.8704	.5658	1.767	1.0559	30
40	.5178	.4950	.8689	.5696	1.756	1.0530	20
50	.5207	.4975	.8675	.5735	1.744	1.0501	10
30°00′	.5236	.5000	.8660	.5774	1.732	1.0472	60°00′
10	.5265	.5025	.8646	.5812	1.720	1.0443	50
20	.5294	.5050	.8631	.5851	1.709	1.0414	40
30	.5323	.5075	.8616	.5890	1.698	1.0385	30
40	.5325	.5100	.8601	.5930	1.686	1.0356	20
50	.5381	.5125	.8587	.5969	1.675	1.0327	10
31°00′	.5411	.5150	.8572	.6009	1.664	1.0297	59°00′
10	.5440	.5175	.8557	.6048	1.653	1.0268	50
20	.5469	.5200	.8542	.6088	1.643	1.0239	40
30	.5498	.5225	.8526	.6128	1.632	1.0210	30
40	.5527	.5250	.8511	.6168	1.621	1.0181	20
50	.5556	.5275	.8496	.6208	1.611	1.0152	10
32°00′	.5585	.5299	.8480	.6249	1.600	1.0123	58°00′
10	.5614	.5324	.8465	.6289	1.590	1.0094	50
20	.5643	.5348	.8450	.6330	1.580	1.0065	40
30	.5672	.5373	.8434	.6371	1.570	1.0036	30
40	.5701	.5398	.8418	.6412	1.560	1.0007	20
50	.5730	.5422	.8403	.6453	1.550	.9977	10
33°00′	.5760	.5446	.8387	.6494	1.540	.9948	57°00′
10	.5789	.5471	.8371	.6536	1.530	.9919	50
20	.5818	.5495	.8355	.6577	1.520	.9890	40
30	.5847	.5519	.8339	.6619	1.511	.9861	30
40	.5876	.5544	.8323	.6661	1.501	.9832	20
50	.5905	.5568	.8307	.6703	1.492	.9803	10
34°00′	.5934	.5592	.8290	.6745	1.483	.9774	56°00′
10	.5963	.5616	.8274	.6787	1.473	.9745	50
20	.5992	.5640	.8258	.6830	1.464	.9716	40
30	.6021	.5664	.8241	.6873	1.455	.9687	30
40	.6050	.5688	.8225	.6916	1.446	.9657	20
50	.6080	.5712	.8208	.6959	1.437	.9628	10
35°00′	.6109	.5736	.8192	.7002	1.428	.9599	55°00′
10	.6138	.5760	.8175	.7046	1.419	.9570	50
20	.6167	.5783	.8158	.7089	1.411	.9541	40
30	.6196	.5807	.8141	.7133	1.402	.9512	30
40	.6225	.5831	.8124	.7177	1.393	.9483	20
50	.6254	.5854	.8107	.7221	1.385	.9454	10
36°00′	.6283	.5878	.8090	.7265	1.376	.9425	54°00′
10	.6312	.5901	.8073	.7310	1.368	.9396	50
20	.6341	.5925	.8056	.7355	1.360	.9367	40
30	.6370	.5948	.8039	.7400	1.351	.9338	30
40	.6400	.5972	.8021	.7445	1.343	.9308	20
50	.6429	.5995	.8004	.7490	1.335	.9279	10
		cos	sin	cot	tan	Radians	Degrees

Degrees	Radians	sin	cos	tan	cot		Degrees
37°00′	.6458	.6018	.7986	.7536	1.327	.9250	53°00′
10	.6487	.6041	.7969	.7581	1.319	.9221	50
20	.6516	.6065	.7951	.7627	1.311	.9192	40
30	.6545	.6088	.7934	.7673	1.303	.9163	30
40	.6574	.6111	.7916	.7720	1.295	.9134	20
50	.6603	.6134	.7898	.7766	1.288	.9105	10
38°00′	.6632	.6157	.7880	.7813	1.280	.9076	52°00′
10	.6661	.6180	.7862	.7860	1.272	.9047	50
20	.6690	.6202	.7844	.7907	1.265	.9018	40
30	.6720	.6225	.7826	.7954	1.257	.8988	30
40	.6749	.6248	.7808	.8002	1.250	.8959	20
50	.6778	.6271	.7790	.8050	1.242	.8930	10
39°00′	.6807	.6293	.7771	.8098	1.235	.8901	51°00′
10	.6836	.6316	.7753	.8146	1.228	.8872	50
20	.6865	.6338	.7735	.8195	1.220	.8843	40
30	.6894	.6361	.7716	.8243	1.213	.8814	30
40	.6923	.6383	.7698	.8292	1.206	.8785	20
50	.6952	.6406	.7679	.8342	1.199	.8756	10
40°00′	.6981	.6428	.7660	.8391	1.192	.8727	50°00′
10	.7010	.6450	.7642	.8441	1.185	.8698	50
20	.7039	.6472	.7623	.8491	1.178	.8668	40
30	.7069	.6494	.7604	.8541	1.171	.8639	30
40	.7098	.6517	.7585	.8591	1.164	.8610	20
50	.7127	.6539	.7566	.8642	1.157	.8581	10
41°00′	.7156	.6561	.7547	.8693	1.150	.8552	49°00′
10	.7185	.6583	.7528	.8744	1.144	.8523	50
20	.7214	.6604	.7509	.8796	1.137	.8494	40
30	.7243	.6626	.7490	.8847	1.130	.8465	30
40	.7272	.6648	.7470	.8899	1.124	.8436	20
50	.7301	.6670	.7451	.8952	1.117	.8407	10
42°00′	.7330	.6691	.7431	.9004	1.111	.8378	48°00′
10	.7359	.6713	.7412	.9057	1.104	.8348	50
20	.7389	.6734	.7392	.9110	1.098	.8319	40
30	.7418	.6756	.7373	.9163	1.091	.8290	30
40	.7447	.6777	.7353	.9217	1.085	.8261	20
50	.7476	.6799	.7333	.9271	1.079	.8232	10
43°00′	.7505	.6820	.7314	.9325	1.072	.8203	47°00′
10	.7534	.6841	.7294	.9380	1.066	.8174	50
20	.7563	.6862	.7274	.9435	1.060	.8145	40
30	.7592	.6884	.7254	.9490	1.054	.8116	30
40	.7621	.6905	.7234	.9545	1.048	.8087	20
50	.7650	.6926	.7214	.9601	1.042	.8058	10
44°00′	.7679	.6947	.7193	.9657	1.036	.8029	46°00′
10	.7709	.6967	.7173	.9713	1.030	.7999	50
20	.7738	.6988	.7153	.9770	1.024	.7970	40
30	.7767	.7009	.7133	.9827	1.018	.7941	30
40	.7796	.7030	.7112	.9884	1.012	.7912	20
50	.7825	.7050	.7092	.9942	1.006	.7883	10
45°00′	.7854	.7071	.7071	1.0000	1.000	.7854	45°00′
		cos	sin	cot	tan	Radians	Degrees

APPENDIX E

TABLE OF SQUARE ROOTS AND CUBE ROOTS

n	$\sqrt{n}$	$\sqrt[3]{n}$	n	$\sqrt{n}$	$\sqrt[3]{n}$	n	$\sqrt{n}$	$\sqrt[3]{n}$
1	1.00000	1.00000	31	5.56776	3.14138	61	7.81025	3.93650
2	1.41421	1.25992	32	5.65685	3.17480	62	7.87401	3.95789
3	1.73205	1.44225	33	5.74456	3.20753	63	7.93725	3.97906
4	2.00000	1.58740	34	5.83095	3.23961	64	8.00000	4.00000
5	2.23607	1.70998	35	5.91608	3.27107	65	8.06226	4.02073
6	2.44949	1.81712	36	6.00000	3.30193	66	8.12404	4.04124
7	2.64575	1.91293	37	6.08276	3.33222	67	8.18535	4.06155
8	2.82843	2.00000	38	6.16441	3.36198	68	8.24621	4.08166
9	3.00000	2.08008	39	6.24500	3.39121	69	8.30662	4.10157
10	3.16228	2.15443	40	6.32456	3.41995	70	8.36660	4.12129
11	3.31662	2.22398	41	6.40312	3.44822	71	8.42615	4.14082
12	3.46410	2.28943	42	6.48074	3.47603	72	8.48528	4.16017
13	3.60555	2.35133	43	6.55744	3.50340	73	8.54400	4.17934
14	3.74166	2.41014	44	6.63325	3.53035	74	8.60233	4.19834
15	3.87298	2.46621	45	6.70820	3.55689	75	8.66025	4.21716
16	4.00000	2.51984	46	6.78233	3.58305	76	8.71780	4.23582
17	4.12311	2.57128	47	6.85565	3.60883	77	8.77496	4.25432
18	4.24264	2.62074	48	6.92820	3.63424	78	8.83176	4.27266
19	4.35890	2.66840	49	7.00000	3.65931	79	8.88819	4.29084
20	4.47214	2.71442	50	7.07107	3.68403	80	8.94427	4.30887
21	4.58258	2.75892	51	7.14143	3.70843	81	9.00000	4.32675
22	4.69042	2.80204	52	7.21110	3.73251	82	9.05539	4.34448
23	4.79583	2.84387	53	7.28011	3.75629	83	9.11043	4.36207
24	4.89898	2.88450	54	7.34847	3.77976	84	9.16515	4.37952
25	5.00000	2.92402	55	7.41620	3.80295	85	9.21954	4.39683
26	5.09902	2.96250	56	7.48331	3.82586	86	9.27362	4.41400
27	5.19615	3.00000	57	7.54983	3.84850	87	9.32738	4.43105
28	5.29150	3.03659	58	7.61577	3.87088	88	9.38083	4.44796
29	5.38516	3.07232	59	7.68115	3.89300	89	9.43398	4.46475
30	5.47723	3.10723	60	7.74597	3.91487	90	9.48683	4.48140

TABLE OF SQUARE ROOTS AND CUBE ROOTS (Continued)

n	$\sqrt{n}$	$\sqrt[3]{n}$	n	$\sqrt{n}$	$\sqrt[3]{n}$	n	$\sqrt{n}$	$\sqrt[3]{n}$
91	9.53939	4.49794	128	11.3137	5.03968	165	12.8452	5.48481
92	9.59166	4.51436	129	11.3578	5.05277	166	12.8841	5.49586
93	9.64365	4.53065	130	11.4018	5.06580	167	12.9228	5.50688
94	9.69536	4.54684	131	11.4455	5.07875	168	12.9615	5.51785
95	9.74679	4.56290	132	11.4891	5.09164	169	13.0000	5.52877
96	9.79796	4.57886	133	11.5326	5.10447	170	13.0384	5.53966
97	9.84886	4.59470	134	11.5758	5.11723	171	13.0767	5.55050
98	9.89949	4.61044	135	11.6190	5.12993	172	13.1149	5.56130
99	9.94987	4.62606	136	11.6619	5.14256	173	13.1529	5.57205
100	10.0000	4.64159	137	11.7047	5.15514	174	13.1909	5.58277
101	10.0499	4.65701	138	11.7473	5.16765	175	13.2288	5.59344
102	10.0995	4.67233	139	11.7898	5.18010	176	13.2665	5.60408
103	10.1489	4.68755	140	11.8322	5.19249	177	13.3041	5.61467
104	10.1980	4.70267	141	11.8743	5.20483	178	13.3417	5.62523
105	10.2470	4.71769	142	11.9164	5.21710	179	13.3791	5.63574
106	10.2956	4.73262	143	11.9583	5.22932	180	13.4164	5.64622
107	10.3441	4.74746	144	12.0000	5.24148	181	13.4536	5.65665
108	10.3923	4.76220	145	12.0416	5.25359	182	13.4907	5.66705
109	10.4403	4.77686	146	12.0830	5.26564	183	13.5277	5.67741
110	10.4881	4.79142	147	12.1244	5.27763	184	13.5647	5.68773
111	10.5357	4.80590	148	12.1655	5.28957	185	13.6015	5.69802
112	10.5830	4.82028	149	12.2066	5.30146	186	13.6382	5.70827
113	10.6301	4.83459	150	12.2474	5.31329	187	13.6748	5.71848
114	10.6771	4.84881	151	12.2882	5.32507	188	13.7113	5.72865
115	10.7238	4.86294	152	12.3288	5.33680	189	13.7477	5.73879
116	10.7703	4.87700	153	12.3693	5.34848	190	13.7840	5.74890
117	10.8167	4.89097	154	12.4097	5.36011	191	13.8203	5.75897
118	10.8628	4.90487	155	12.4499	5.37169	192	13.8564	5.76900
119	10.9087	4.91868	156	12.4900	5.38321	193	13.8924	5.77900
120	10.9545	4.93242	157	12.5300	5.39469	194	13.9284	5.78896
121	11.0000	4.94609	158	12.5698	5.40612	195	13.9642	5.79889
122	11.0454	4.95968	159	12.6095	5.41750	196	14.0000	5.80879
123	11.0905	4.97319	160	12.6491	5.42884	197	14.0357	5.81865
124	11.1355	4.98663	161	12.6886	5.44012	198	14.0712	5.82848
125	11.1803	5.00000	162	12.7279	5.45136	199	14.1067	5.83827
126	11.2250	5.01330	163	12.7671	5.46256	200	14.1421	5.84804
127	11.2694	5.02653	164	12.8062	5.47370			

ANSWERS TO ODD-NUMBERED EXERCISES

CHAPTER 1

Section 1.2

1. Commutative   **3.** Distributive   **5.** Closure
7. Inverse   **9.** Identity   **11.** Associative
13. Closure   **15.** Distributive   **17.** Associative and
Commutative   **19.** Distributive and Identity   **21.** -6
23. 2   **25.** 6   **27.** -14 **29.** 2   **31.** $1/2$
33. $-9/4$ **35.** $7/20$ **37.** $3/10$ **39.** $\dfrac{x(2-y)}{2y}$
41. 0   **43.** 0   **47.** $24/35$ **49.** (a) 0.625   (b) $0.\overline{3}$
(c) $0.\overline{123}$ (d) $0.5\overline{4}$ (e) $0.11\overline{3}$

51.

h	10	1	0.5
$\dfrac{\sqrt{4+h}-2}{h}$	0.17417	0.23607	0.24264

h	0.01	0.0001	0.000001
$\dfrac{\sqrt{4+h}-2}{h}$	0.24984	0.25000	0.25000

Section 1.3

1.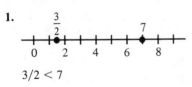

$3/2 < 7$

3.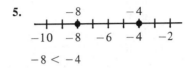

$-6 < \pi$

5.

$-8 < -4$

7.

$2/3 < 5/6$

9.

$-2.5 < -1.75$

11. $0 < 9$ **13.** $5.5 > -2.5$ **15.** $4 < x + y$
17. $-12 < 6$ **19.** $15 > -3x$ **21.** $-b/2 < 5/2$
23. $[-2, 0)$ $-2 \leq x < 0$ **25.** $[3, 11/2]$ $3 \leq x \leq 11/2$

27. $[100, \infty)$ $100 \leq x$

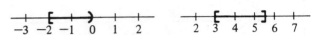

29. $(\sqrt{2}, 8]$ $\sqrt{2} < x \leq 8$

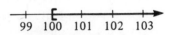

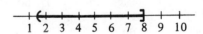

31. $x < 0$ **33.** $A \geq 30$ **35.** $1/3\% \leq R \leq 1\%$
37. 4   **39.** 23/4   **41.** 51   **43.** 14.99
45. $|x - 5| \leq 3$ **47.** $|z - (3/2)| > 1$ **49.** $|y| < |y - 8|$

51. (a) $127/90, 584/413, 7071/5000, \sqrt{2}, 47/33$
(b) $7071/5000$

Section 1.4

	Exponential Form	Base	Exponent	Factored Form
1.	4^3	4	3	$(4)(4)(4)$
3.	5^4	5	4	$(5)(5)(5)(5)$
5.	$(2x)^4$	$2x$	4	$(2x)(2x)(2x)(2x)$
7.	$(x^2 + 4)^3$	$x^2 + 4$	3	$(x^2 + 4)(x^2 + 4)(x^2 + 4)$
9.	$(x/y)^6$	x/y	6	$(x/y)(x/y)(x/y)(x/y)(x/y)(x/y)$

11. -24 **13.** $1/2$ **15.** 5   **17.** $5x^6$ **19.** $24y^{10}$
21. $10x^4$ **23.** $7x^5$ **25.** $(4/3)(x + y)^2$ **27.** $-2x^3$
29. $x^2/9z^4$ **31.** $a^6/64b^9$ **33.** 1   **35.** x^4/y^4
37. b^5/a^5 **39.** 1   **41. (a)** 9.3×10^7 **(b)** 9.0×10^8
(c) 4.35×10^{-6} **43. (a)** 1,910,000
(b) 234,500,000,000   **(c)** 6.21   **(d)** 0.00852
(e) 0.00007021   **(f)** 0.00000003798   **45. (a)** 7697.13
(b) 954.45   **(c)** 1.479   **(d)** 1.11×10^8
47. (a) 3.0981×10^6 **(b)** 3.0769×10^{10}
49. $8\frac{1}{3}$ min

51.

t	5	10	20
A	\$910.97	\$1659.73	\$5509.41

t	30	40	50
A	\$18,288.29	\$60,707.30	\$201,515.58

25. (a) $\dfrac{5x}{y^2}\sqrt{3}$ **(b)** $(x - y)\sqrt{5(x - y)}$ **27. (a)** $\sqrt{x}$

(b) $\sqrt[3]{x}$ **29. (a)** $\sqrt[4]{3x^2y^3}$ **(b)** $xy^2\sqrt{6}$ **31. (a)** $\dfrac{\sqrt{3}}{3}$

(b) $\dfrac{\sqrt{21}}{7}$ **33. (a)** $4\sqrt[3]{4}$ **(b)** $\dfrac{\sqrt[3]{18}}{6}$

35. (a) $\dfrac{x(5 + \sqrt{3})}{11}$ **(b)** $\dfrac{8(6 - \sqrt{10})}{13}$

37. (a) $\dfrac{3(\sqrt{x} - \sqrt{y})}{x - y}$ **(b)** $-(4/5)(\sqrt{2} + 2\sqrt{3})$

39. (a) $\dfrac{13}{2\sqrt{13}}$ **(b)** $\dfrac{2}{3\sqrt{2}}$ **41. (a)** $\dfrac{1}{x(\sqrt{3} + \sqrt{2})}$

(b) $\dfrac{1}{2(\sqrt{15} - 3)}$ **43. (a)** $2\sqrt{x}$ **(b)** $13\sqrt{2}$
45. (a) $8\sqrt{xy}$ **(b)** $(13/2)\sqrt{2ab}$ **47. (a)** $|x|y\sqrt{15}$
(b) $3\sqrt{3}/|a|$ **49. (a)** $2\sqrt[4]{2}$ **(b)** $a\sqrt[6]{10ab}$
51. (a) 7.5498   **(b)** 3.5477   **(c)** 12.6515   **(d)** 6.3096
(e) 0.0221   **(f)** 9609.4958   **53. (a)** $R \approx 1.72$
(b) $R \approx 1.08$

Section 1.5

	$\sqrt[n]{b^m} = a$	$b^{m/n} = a$	$a^{n/m} = b$
1.	$\sqrt{9} = 3$	$9^{1/2} = 3$	$3^2 = 9$
3.	$\sqrt[5]{32} = 2$	$32^{1/5} = 2$	$2^5 = 32$
5.	$\sqrt{196} = 14$	$196^{1/2} = 14$	$14^2 = 196$
7.	$\sqrt[3]{-216} = -6$	$(-216)^{1/3} = -6$	$(-6)^3 = -216$
9.	$\sqrt[3]{27^2} = 9$	$27^{2/3} = 9$	$9^{3/2} = 27$
11.	$\sqrt[4]{81^3} = 27$	$81^{3/4} = 27$	$27^{4/3} = 81$

13. (a) $3/2$ **(b)** $3/2$ **15. (a)** 64   **(b)** 4   **17. (a)** 326
(b) 562   **19. (a)** $1/16$ **(b)** $2/3$ **21. (a)** $2\sqrt{2}$
(b) $3\sqrt{2}$ **23. (a)** $2x\sqrt[3]{2x^2}$ **(b)** $2xz\sqrt[4]{2z}$

Section 1.6

1. $i, -1, -i, 1, i, -1, -i, 1, i, -1, -i, 1, i, -1, -i, 1,$
$i, -1, -i, 1$ **3.** $4 + 3i, 4 - 3i$
5. $-7 - 8i, -7 + 8i$ **7.** $-7 + 9i, -7 - 9i$
9. $2 - 3\sqrt{3}\,i, 2 + 3\sqrt{3}\,i$ **11.** $5\sqrt{3}\,i, -5\sqrt{3}\,i$
13. $-1 - 6i, -1 + 6i$ **15.** $8 + 7\sqrt{7}\,i, 8 - 7\sqrt{7}\,i$
17. $8, 8$ **19.** $5 - 3\sqrt{3}, 5 - 3\sqrt{3}$ **21.** $11 - i$
23. 4   **25.** $-5 + 3i$ **27.** $(3 - 5\sqrt{2}) + 2\sqrt{2}\,i$
29. $1/6 + (7/6)i$ **31.** $5 + i$ **33.** 41
35. $12 + 30i$ **37.** -90 **39.** $-40 + 16i$
41. 24   **43.** $-9 + 40i$ **45.** $-236 + 115i$

47. $16/41 + (20/41)i$　　**49.** $3/5 + (4/5)i$

51. $-7 - 6i$　　**53.** $31/5 + (2/5)i$　　**55.** $(1/8)i$

57. $-5/4 - (5/4)i$　　**59.** $35/29 + (595/29)i$

61. $a = -10, b = 6$　　**63.** $a = 6, b = 5$

Section 1.7

	Degree	Leading Coefficient	Terms	Coefficients
1.	2	2	$2x^2, -x, 1$	$2, -1, 1$
3.	5	1	$x^5, -1$	$1, -1$
5.	5	4	$4x^5, 6x^4, -3x^3, 10x^2, -x, -1$	$4, 6, -3, 10, -1, -1$

7. 4　　**9.** 4　　**11.** $-3x^2 + x - 8$

13. $8x^3 + 29x^2 + 11$

15. $9x^5 + 5x^4 + 19x^3 + 19x^2 + x - 9$

17. $25x^7 + 16x^3 - 26$　　**19.** $x^3 - 12xy - 9y^2$

21. $x^4 - 5x^3 - 2x^2 + 11x - 5$

23. $x^4 - x^3 + 5x^2 - 9x - 36$　　**25.** $x^3 + 27$

27. $x^4 - 1$　　**29.** $x^3 + 4x^2 - 5x - 20$

31. $x^2 + 7x + 12$　　**33.** $6x^2 - 7x - 5$

35. $x^2 + 12x + 36$　　**37.** $4x^2 - 20xy + 25y^2$

39. $x^2 + 2xy + y^2 - 6x - 6y + 9$　　**41.** $x^2 - 4y^2$

43. $4r^4 - 25$　　**45.** $x^3 + 3x^2 + 3x + 1$

47. $8x^3 - 12x^2y + 6xy^2 - y^3$

49. $x^3 - 3x^2y + 3xy^2 - y^3 + 3x^2 - 6xy + 3y^2 + 3x - 3y + 1$　　**51.** $x - y$　　**53.** $16r^6 - 24r^3s^2 + 9s^4$

55. $30m^2 + 31mn - 12n^2$　　**57. (a)** 647.7　**(b)** 13.5

(c) 1.4464　**(d)** 2.6111

Section 1.8

1. $3(x + 2)$　　**3.** $x(y - z)$　　**5.** $3ab(3a - 4b^2)$

7. $(x + y)(z^2 - 1)$　　**9.** $(x + 6)(x - 6)$

11. $(4y - 3)(4y + 3)$　　**13.** $(x + 1)(x - 3)$

15. $(9 + y)(9 - y)$　　**17.** $(x - 2)^2$

19. $(2x + 1)^2$　　**21.** $(x - 2y)^2$　　**23.** $(ab - c)^2$

25. $(x + 2)(x - 1)$　　**27.** $(x - 2)(x - 3)$

29. $(3x - 2)(x - 1)$　　**31.** $(2y - 3z)(y + 12z)$

33. $(x - 2)(x^2 + 2x + 4)$　　**35.** $(y + 4)(y^2 - 4y + 16)$

37. $(x - 3y)(x^2 + 3xy + 9y^2)$

39. $(x^2 + 4)(x^4 - 4x^2 + 16)$　　**41.** $(x - 1)(y + z)$

43. $(5x - 3)(y + 2)$　　**45.** $(10x - 7)(y - 1)$

47. $2x(x + 1)(x - 2)$　　**49.** $7r(3s + r)(3s - r)$

51. $xy(2x - y)^2$　　**53.** $6x(x - 2y)(x^2 + 2xy + 4y^2)$

55. $(x + 2y)(x - 2y)(x + y)(x - y)$

57. $2x(2x - 1)(4x - 1)$　　**59.** $2xz(x + 2y)(z + 1)$

61. $(x - 2)^5$

Section 1.9

1. $\dfrac{3y}{y + 1}$　　**3.** $\dfrac{x(x + 3)}{x - 2}$　　**5.** $\dfrac{(x + z)(y + 2)}{x(y + z)}$

7. $a - 2$　　**9.** $\dfrac{x + 5}{x - 1}$　　**11.** $\dfrac{4x - 3}{x^2}$

13. $\dfrac{x - 6}{(x + 2)(x - 2)}$　　**15.** $\dfrac{-(x^2 + 3)}{(x + 1)(x - 2)(x - 3)}$

17. $\dfrac{-(x - 1)^2}{x(x^2 + 1)}$　　**19.** $\dfrac{1}{5(x - 2)}$　　**21.** $\dfrac{-x(x + 7)}{x + 1}$

23. $\dfrac{r + 1}{r}$　　**25.** $\dfrac{t - 3}{(t - 2)(t + 3)}$

27. $\dfrac{x - y}{x(x + y)^2}$　　**29.** $3/2$　　**31.** $xy(x + y)$

33. $\dfrac{x^2}{5(x^2 + 4x + 16)}$　　**35.** $\dfrac{x + y}{xy}$　　**37.** -2

39. $1/x$　　**41.** $1/y$　　**43.** $\dfrac{(x + 3)^3}{2x(x - 3)}$　　**45.** $\dfrac{1}{2y + 1}$

47. $\dfrac{-(2x + h)}{x^2(x + h)^2}$　　**49.** $\dfrac{x^3 + y^2}{y^2 - xy + x^2}$

51. $(1/16)(4x^2 + 1)$　　**53.** $\dfrac{2x - 1}{2x}$　　**55.** $\dfrac{-1}{t^2(t^2 + 1)^{1/2}}$

57. $\dfrac{1}{2(x^2 + 1)^{5/4}}$　　**59.** $\dfrac{(x^2 - 2)}{x^3(1 - x^2)^{1/2}}$

Section 1.10

1. $2x - 3y - 4$　　**3.** $5z + 3x - 6$　　**5.** $\dfrac{3 - x}{x - 1}$

7. ax/y　　**9.** $16x^2$　　**11.** $\sqrt{x + 9}$　　**13.** $\dfrac{6x + y}{6x - y}$

15. $\dfrac{y}{xy + 1}$ **17.** $\dfrac{4y + xy}{x}$ **19.** $\sqrt[3]{x^2(x + 7)}$

21. $\dfrac{4x + 3y}{xy}$ **23.** $1/2y$ **25.** $(1/3)(2x^2 + x + 15)$

27. $\sqrt{x}(1 + x)$ **29.** $(1/2)x^{-1/2}(5x^2 + 1)$

31. $(1/3)(1 - 2x)^{-2/3}(3 - 8x)$ **33.** $\dfrac{-1}{x^2\sqrt{x^2 + 1}}$

35. $(1/15)(2x + 1)^{3/2}(3x - 1)$

37. $(3/28)(t + 1)^{4/3}(4t - 3)$

39. $(1/15)(2x - 1)^{1/2}(3x^2 + 2x - 13)$ **41.** $-1/4$

43. 2 **45.** $1/2$ **47.** $25/9, 49/16$ **49.** $1/(2x^2)$

51. $16x^{-1} - 5 - x$ **53.** $4x^{8/3} - 7x^{5/3} + x^{-1/3}$

55. $\dfrac{x^2}{x^4 + 1} + \dfrac{4x}{x^4 + 1} + \dfrac{8}{x^4 + 1}$

Chapter 1 Review Exercises

1. Do not add denominators. $7/16 + 3/16 = 10/16 = 5/8$
3. Multiplication is not distributive over multiplication.
$10(4 \cdot 7) = 40 \cdot 7 = 4 \cdot 70 = 280$ **5.** Only the
numerator is multiplied by 4. $4\left(\dfrac{3}{7}\right) = \dfrac{4 \cdot 3}{7} = \dfrac{12}{7}$
7. Invert the divisor before multiplying.
$15/16 \div 2/3 = 15/16 \cdot 3/2 = 45/32$
9. Multiply before adding. $12 + 8 \times 6 = 12 + 48 = 60$
11. Distribute the negative.
$2[5 - (3 - 2)] = 2[5 - 3 + 2] = 2[4] = 8$
13. Exponent applies to each factor in the base.
$(2x)^4 = 2^4x^4 = 16x^4$ **15.** Order of operations; perform
the operations within the symbol of grouping first.
$(5 + 8)^2 = 13^2 = 169$ **17.** Multiply the exponents.
$(3^4)^4 = 3^{16}$ **19.** Perform the operations on the radicand
first. $\sqrt{3^2 + 4^2} = \sqrt{25} = 5$ **21.** Radicals cannot be
multiplied if the indices are not the same. $(7x)^{1/2}(2)^{1/3} = $
$(7x)^{3/6}(2)^{2/6} = \sqrt[6]{(7x)^3(4)} = \sqrt[6]{1372x^3}$ **23.** Multiplication
by a negative number changes the direction of the inequality.
$-5 < -3,$ $-2(-5) > -2(-3),$ $10 > 6$
25. -20 **27.** -11 **29.** -11 **31.** 25
33. -144 **35.** $(5/3)^6$ **37.** 50 **39.** 9×10^8
41. $|x - 7| \geq 4$ **43.** $|y + 30| < 5$

45.

47.

49. $3 + 7i$ **51.** $-\sqrt{2}\,i$ **53.** $40 + 65i$
55. $-4 - 46i$ **57.** $1 - 6i$ **59.** $(4/3)i$
61. $x^5 - 2x^4 + x^3 - x^2 + 2x - 1$

63. $x^4 + x^2 - x - \dfrac{1}{x}$ **65.** $y^6 + y^4 - y^3 - y$

67. $\dfrac{1}{x^2(x^2 + x + 1)}$ **69.** $\dfrac{x^2(x^2 + 4x + 2)}{(x + 2)(x - 2)(x^2 + 2)^2}$

71. $\dfrac{3}{(x - 1)(x + 2)}$ **73.** $\dfrac{x^3 - x + 3}{(x - 1)(x + 2)}$

75. $\dfrac{x^4}{(x - 1)^3}$ **77.** $\dfrac{x + 1}{x(x^2 + 1)}$ **79.** $\dfrac{2x^3}{(x^2 - 4)^2}$

81. $\dfrac{x(x + 1)}{x^2 + x + 1}$ **83.** $(1/5)x(5x - 6)$ **85.** $\dfrac{-1}{xy(x + y)}$

87. $\dfrac{3ax^2}{(a^2 - x)(a - x)}$ **89.** $\dfrac{1}{x(x - 4)(x - 5)}$

91. $(x - 1)(x^2 + x + 1)$ **93.** $(x + y)^2(x - y)^2$

95. $(1/12)(9x^2 - 10x + 48)$ **97.** $\dfrac{1}{2(x + 1)^{3/2}}(x + 2)$

99. $(z - 5)(z + 2)(z^2 - 2z + 4)$ **101.** 50625
103. 0

105.

n	1	10	10^2
$5/n$	5	0.5	0.05

n	10^4	10^6	10^{10}
$5/n$	0.0005	5×10^{-6}	5×10^{-10}

CHAPTER 2

Section 2.1

1. (a) No (b) No (c) Solution (d) No
3. (a) Solution (b) Solution (c) No (d) No
5. (a) Solution (b) No (c) No (d) No   **7.** (a) No
(b) No (c) Solution (d) Solution   **9.** (a) No (b) No
(c) No (d) Solution   **11.** Conditional   **13.** Identity
15. Conditional   **17.** Conditional   **19.** Identity
21. $x = 3$ **23.** $x = -3/2$ **25.** $x = 9$
27. $x = -26$ **29.** $z = -6/5$ **31.** $x = 9$
33. $u = 10$ **35.** $x = 4$ **37.** $x = 3$ **39.** $x = 11/6$
41. $x = 5$ **43.** $x = 0$ **45.** Identity

47. No solution　**49.** $x = \dfrac{4b + 1}{a + 2}$　**51.** $x = 138.889$

53. $x = 62.372$　**55.** $x = -2.386$　**57. (a)** 6.46

(b) 6.41　**59. (a)** 1.06　**(b)** 1.06　**61. (a)** 1.00

(b) 1.01

51. $\dfrac{1}{(x + 1)^2 + 4}$　**53.** $\dfrac{1}{(x - 2)^2 - 16}$

55. $\dfrac{1}{\sqrt{9 - (x - 3)^2}}$　**57.** $\dfrac{1}{\sqrt{(x - 1/2)^2 + 1}}$

59. $\dfrac{1}{41 - (x - 5)^2}$

Section 2.2

1. $h = \dfrac{2A}{b}$　**3.** $L = \dfrac{V}{WH}$　**5.** $C = \dfrac{S}{1 + R}$

7. $R = \dfrac{A - P}{PT}$　**9.** $b = \dfrac{2A - ha}{h}$

11. $r = \dfrac{3V + \pi h^3}{3\pi h^2}$　**13.** 262, 263　**15.** 37, 185

17. 22.5 in $\times$ 15 in　**19.** 97　**21.** 13.5　**23.** 1192.5

25. 135%　**27.** 175　**29.** \$22,316.98

31. (a) 80/21 hr, 16/5 hr　**(b)** 25/23 hr　**(c)** 25.6 mi

33. $66\frac{2}{3}$ mi/hr　**35.** 1.29 sec　**37.** 27 quarters and 8

fifty-cent pieces　**39.** \$4500, \$7500　**41.** 11.43%

43. 8823 units

45.

	Solution 1	*Solution 2*
(a)	25 gal	75 gal
(b)	4 L	1 L
(c)	5 qt	5 qt
(d)	$\dfrac{75}{4}$ gal	$\dfrac{25}{4}$ gal

47. 0.48 gal　**49.** 20/9 hr　**51.** 40/13 hr

53. 20/3 hr　**55.** $C_1 = \dfrac{CC_2}{C_2 - C}$　**57.** $\alpha = \dfrac{L - L_0}{L_0(\Delta t)}$

59. $m_2 = \dfrac{Fr^2}{\alpha m_1}$　**61.** $R_1 = \dfrac{(n - 1)R_2 f}{R_2 + (n - 1)f}$

63. $n = \dfrac{L - a + d}{d}$　**65.** $r = \dfrac{S - a}{S - L}$

Section 2.3

1. $-2, 4$　**3.** -5　**5.** 3/2　**7.** $-1/2, 0$

9. $-3/4, 3/4$　**11.** $-1/2, 3$　**13.** $-10/3, 5$

15. 3, 9　**17.** $-a$　**19.** $-5, -1$　**21.** $-4, 4$

23. $-\sqrt{7}, \sqrt{7}$　**25.** $-8, 8$　**27.** $-2\sqrt{3}, 2\sqrt{3}$

29. $-12, 6$　**31.** $12 - 3\sqrt{2}, 12 + 3\sqrt{2}$

33. 2　**35.** $-1/2$　**37.** $-3 - 2i, -3 + 2i$

39. $1 - 2i, 1 + 2i$　**41.** 0, 2　**43.** 1/3, 5/3

45. $-1, 5$　**47.** $\dfrac{6 - 5\sqrt{2}}{10}, \dfrac{6 + 5\sqrt{2}}{10}$　**49.** $-1/2, 3/2$

Section 2.4

1. 1　**3.** 2　**5.** 0　**7.** 2　**9.** $-3/4, 1/4$

11. $1 \pm \sqrt{3}$　**13.** $-7 \pm \sqrt{5}$　**15.** $2(-2 \pm \sqrt{5})$

17. $\dfrac{2 \pm \sqrt{7}}{3}$　**19.** $\dfrac{-2 \pm \sqrt{11}}{6}$　**21.** $\dfrac{-1 \pm 2\sqrt{2}}{2}$

23. 2/7　**25.** $\dfrac{-8 \pm \sqrt{3}}{5}$　**27.** $\dfrac{-1 \pm \sqrt{11}\,i}{8}$

29. $6 \pm \sqrt{11}$　**31.** $-0.643, 0.976$

33. $0.343 \pm 0.661i$　**35.** $-0.488, 1.687$　**37.** 2000

39. 6　**41.** 50, 50　**43.** 1/5, 5　**45.** 8, 7

47. (a) $\dfrac{\sqrt{550}}{4} = 5.86$ sec　**(b)** $\dfrac{1 + \sqrt{17}}{2} = 2.56$ sec

49. $25(3 - \sqrt{5}) = 19.1$ ft, 9.55 times　**51.** $30\sqrt{6}$

53. 550 mi/hr, 600 mi/hr　**55.** $7 + \sqrt{65}, 9 + \sqrt{65}$

57. 14×14 in　**59.** 11%

Section 2.5

1. $-3/\sqrt{2}, 0, 3/\sqrt{2}$　**3.** $-1, 0, 3$　**5.** $-3, 3, -3i, 3i$

7. $-3, 0$　**9.** $-1, 1, 3$　**11.** $\pm 1, (1/2) \pm (\sqrt{3}/2)i$

13. 5　**15.** 26　**17.** -16　**19.** 2/3　**21.** 5, 6

23. $-5, 2$　**25.** 0　**27.** 101/4　**29.** 0

31. $\pm\sqrt{69}$　**33.** 1　**35.** $\pm 1, \pm\sqrt{33}/11$

37. $-2/3, 4/3$　**39.** 2/3　**41.** ± 2

43. $-1/3, -1/5$　**45.** $\pm 4, \pm 1/2$　**47.** $\pm\sqrt{7}/6$

49. $-2, 1$　**51.** $-\sqrt[3]{5}$　**53.** 1/4　**55.** $-125/8, 1$

57. $\pm 1.038, \pm 0.780i$　**59.** 16.756

61. 26.25 thousand　**63.** 1 or 0.262　**65.** $\sqrt{\dfrac{S^2 - \pi^2 r^4}{\pi r}}$

Section 2.6

1. (a) Yes　**(b)** No　**(c)** Yes　**(d)** No　**3. (a)** Yes

(b) No　**(c)** No　**(d)** Yes

5. $x \geq 12$　**7.** $x < -1/2$

9. $x \geq 1/2$

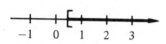

11. $x > 1/2$

13. $-9/2 < x < 15/2$

15. $-3/4 < x < -1/4$

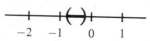

17. $-5 < x < 5$

19. $x < -6, x > 6$

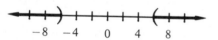

21. $-7 < x < 3$

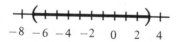

23. $x \leq -7, x \geq 13$

25. $4 < x < 5$ **27.** $-3 \leq x \leq 3$

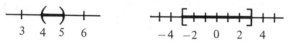

29. $x < -2, x > 2$

31. $-7 < x < 3$

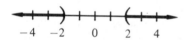

33. $x \leq -7/2, x \geq -1/2$ **35.** $-3 < x < 2$

37. $x < -1, x > 1$ **39.** $x < 0, 0 < x < 3/2$

41. $x \leq 3, x \geq 4$ **43.** $-2 \leq x \leq 2$ **45.** $-4 \leq x \leq 3$

47. $-\infty < x < \infty$ **49. (a)** $|x| \leq 2$ **(b)** $|x| > 2$

51. $x < -1, x > 4$ **53.** $5 < x < 15$

55. $-5 < x < -3/2, x > -1$

57. $-9/7 < x < 6/5, x \geq 36$ **59.** $x \geq -1/2$

61. (a) 10 sec **(b)** $4 < t < 6$ **63.** $r > 12.5\%$

65. $x > 36$ units   67. $65.8 \leq h \leq 71.2$

Chapter 2 Review Exercises

1. $0, 12/5$ **3.** $0, 2$ **5.** $1/5$ **7.** $\pm\sqrt{2}/2$

9. $2, -1 \pm \sqrt{3}i$ **11.** $0, 1, 2$ **13.** $1, 5/3$ **15.** $0, 2$

17. $3 \pm i$ **19.** $126, -124$ **21.** $2/5, 2$ **23.** $2, 6$

25. $-1, 1$ **27.** $\pm 2, \pm 2\sqrt{33}/11$ **29.** No solution

31. $x > -5/3$ **33.** $-2 \leq x \leq 2$ **35.** $x < 3, x > 5$

37. $x < -7/3, x > 1$ **39.** $-5 \leq x < -1, x > 1$

41. $r = \sqrt{3V/\pi h}$ **43.** $t = \dfrac{V_0 \pm \sqrt{V_0{}^2 - 64S}}{32}$

45. $p = \dfrac{k}{3\pi r^2 L}$ **47.** $1000(-4 + \sqrt{17}) = 123.1$ mi

49. 4

CHAPTER 3

Section 3.1

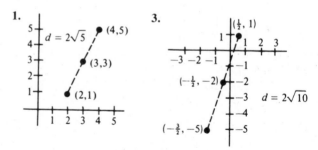

1.   **3.**

5.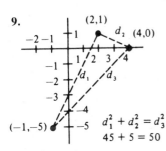
$d = 2\sqrt{37}$

7.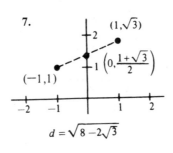
$d = \sqrt{8 - 2\sqrt{3}}$

17. $x = \pm 3$ **19.** $2y = 3x - 1$

21. $\left(\dfrac{3x_1 + x_2}{4}, \dfrac{3y_1 + y_2}{4}\right)$, $\left(\dfrac{x_1 + x_2}{2}, \dfrac{y_1 + y_2}{2}\right)$,

$\left(\dfrac{x_1 + 3x_2}{4}, \dfrac{y_1 + 3y_2}{4}\right)$ **25. (a)** 15.5%   **(b)** 14.8%

(c) 13.3%   **(d)** 16.1%   **27. (a)** 175   **(b)** 700   **(c)** 280

(d) 750   **29.** 1972, 1973, 1976   **31. (a)** (1.25, 3.6)

(b) 10.534   **33. (a)** (6, −45)   **(b)** 99.860

35. (a) (0.1880, −0.5995)   **(b)** 1.471

9.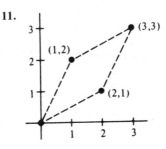
$d_1^2 + d_2^2 = d_3^2$
$45 + 5 = 50$

11.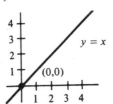
The length of each side
is $\sqrt{5}$

Section 3.2

1. c   **3.** b   **5.** a   **7.** (0, −3), (3/2, 0)

9. (0, −2), (−2, 0), (1, 0)   **11.** (0, 0), (−3, 0), (3, 0)

13. (0, 1/2), (1, 0)   **15.** (0, 0)   **17.** Symmetric with

respect to the y-axis   **19.** Symmetric with respect to the

y-axis   **21.** Symmetric with respect to the x-axis

23. Symmetric with respect to the origin   **25.** Symmetric

with respect to the origin

13.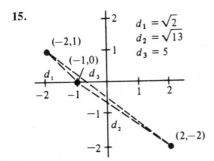

Collinear since $d_1 + d_2 = d_3$
$2\sqrt{5} + \sqrt{5} = 3\sqrt{5}$

27. $y = x$

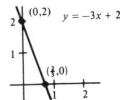

29. $y = -3x + 2$

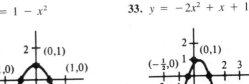

31. $y = 1 - x^2$

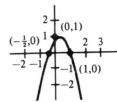

33. $y = -2x^2 + x + 1$

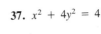
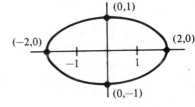

15.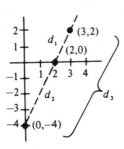
$d_1 = \sqrt{2}$
$d_2 = \sqrt{13}$
$d_3 = 5$

Not on a line since $d_1 + d_2 > d_3$

35. $y = x^3 + 2$

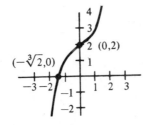

37. $x^2 + 4y^2 = 4$

39. $y = (x + 2)^2$

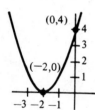

41. $y = \dfrac{1}{x}$

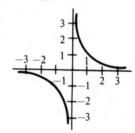

59. $\left(x + \dfrac{1}{2}\right)^2 + \left(y + \dfrac{5}{4}\right)^2 = \dfrac{9}{4}$

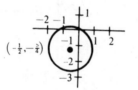

61. $(1, 2)$ is not on the graph
$(1, -1)$ is on the graph
$(4, 5)$ is on the graph

63. $(1, 1/5)$ is on the graph
$(2, 1/2)$ is on the graph
$(-1, -2)$ is not on the graph

43. $y = |x - 2|$

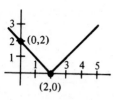

45. $y = \sqrt{x - 3}$

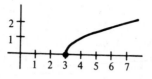

47. $x^2 + y^2 - 9 = 0$
49. $x^2 + y^2 - 4x + 2y - 11 = 0$
51. $x^2 + y^2 + 2x - 4y = 0$
53. $(x - 1)^2 + (y + 3)^2 = 4$

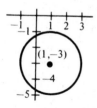

55. $(x - 1)^2 + (y + 3)^2 = 0$

57. $\left(x - \dfrac{1}{2}\right)^2 + \left(y - \dfrac{1}{2}\right)^2 = 2$

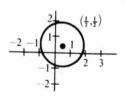

65. (a)

Year	1970	1971	1972	1973	1974
t	0	1	2	3	4
CPI	114.4	120.8	128.3	136.9	146.6

Year	1975	1976	1977	1978	1979
t	5	6	7	8	9
CPI	157.4	169.3	182.3	196.4	211.6

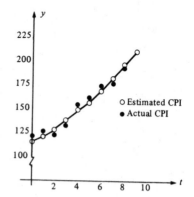

○ Estimated CPI
● Actual CPI

(b) 325.91

67. (a)

Year	1950	1955	1960	1965
t	0	5	10	15
Percentage	20.4	11.3	7.8	6.0

Year	1970	1975	1979
t	20	25	29
Percentage	4.8	4.1	3.6

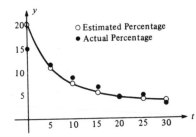

(b) 2.74

y	30	40
F(y)	2,334,527.36	4,792,320.00

47.

t	0	1	2	4	6	8	10
N(t)	500	539	574	621	640	640	633

Section 3.3

1. (a) -1 **(b)** -3 **(c)** -9 **(d)** $2b - 3$ **(e)** $2x - 5$
(f) $-5/2$ **3. (a)** 0 **(b)** 1 **(c)** $\sqrt{3}$ **(d)** 3
(e) $\sqrt{x + \Delta x + 3}$ **(f)** $\sqrt{c + 3}$ **5. (a)** 1 **(b)** -1
(c) -1 **(d)** 1 **(e)** 1 **(f)** $|x - 1|/(x - 1)$ **7.** $3 + \Delta x$

9. $3x^2 + 3x \Delta x + (\Delta x)^2$ **11.** $\dfrac{-1}{\sqrt{x - 1}(1 + \sqrt{x - 1})}$

13. ± 3 **15.** $10/7$
17. Domain: $[1, \infty)$ **19.** Domain: $(-\infty, \infty)$
    Range: $[0, \infty)$ Range: $[0, \infty)$
21. Domain: $[-3, 3]$
    Range: $[0, 3]$
23. Domain: $(-\infty, 0)$ and $(0, \infty)$
    Range: $(0, \infty)$
25. Domain: $(-\infty, 0)$ and $(0, \infty)$
    Range: -1 and 1
27. y is not a function of x **29.** y is a function of x
31. y is a function of x **33.** y is not a function of x
35. y is a function of x **37.** $V = 1750a + 500,000, \; a > 0$
39. (a) $C = 12.30x + 98,000$ **(b)** $R = 17.98x$
(c) $P = 5.68x - 98,000$ **41. (a)** $x = \dfrac{100(14.75 - p)}{p}$

(b) $x = 47.5$ **43. (a)** $p = 91 - 0.01x, \; 100 < x < 1600$
(b) $P = 31x - 0.01x^2, \; 100 < x < 1600$ **(c)** \$21,000.00

45.

y	5	10	20
F(y)	26,474.08	149.760.00	847,170.49

Section 3.4

1. y is a function of x **3.** y is not a function of x
5. y is a function of x **7.** y is a function of x
9. y is not a function of x **11. (a)** Increasing on $(-\infty, \infty)$
(b) Odd function   **13. (a)** Increasing on $(-\infty, 0)$ and
$(2, \infty)$, decreasing on $(0, 2)$ **(b)** Neither even nor odd
15. (a) Increasing on $(-1, 0)$ and $(1, \infty)$, decreasing on
$(-\infty, -1)$ and $(0, 1)$ **(b)** Even function
17. (a) Increasing on $(1, \infty)$, decreasing on $(-\infty, -1)$,
constant on $(-1, 1)$ **(b)** Even function
19. (a) Increasing on $(-2, \infty)$, decreasing on $(-3, -2)$
(b) Neither even nor odd

21.

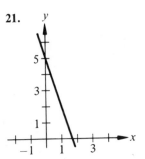

23.

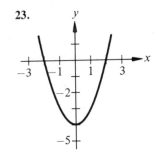

25.

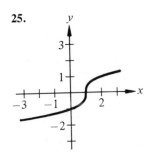

27.

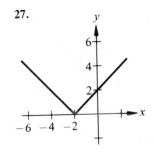

29.

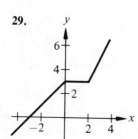

31. (a)

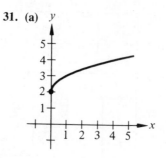

Section 3.5

1. 1   **3.** 0   **5.** −3

7.

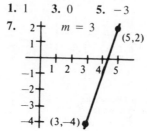

$m = 3$

9. $m = 0$

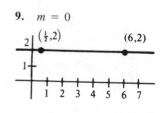

11.

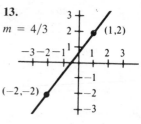

m is undefined

13. $m = 4/3$

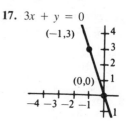

(b)

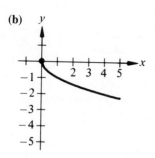

(c)

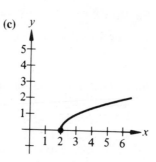

15.

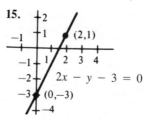

$2x - y - 3 = 0$

17. $3x + y = 0$

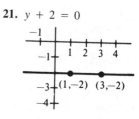

19.

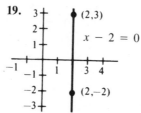

$x - 2 = 0$

21. $y + 2 = 0$

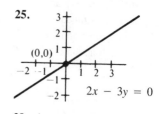

(d)

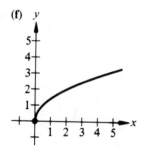

(e)

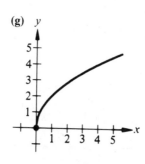

23.

$3x - 4y + 12 = 0$

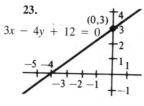

25.

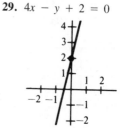

$2x - 3y = 0$

27. $2x + y - 5 = 0$

29. $4x - y + 2 = 0$

(f)

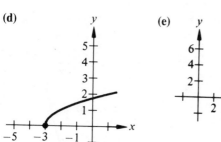

(g)

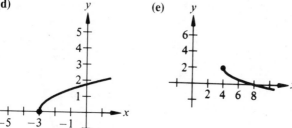

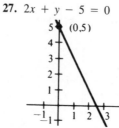

33. (a) $y = (x + 1)\sqrt{x + 4}$ **(b)** $y = x\sqrt{x + 3} + 2$
(c) $y = -x\sqrt{x + 3}$ **(d)** $y = (1 - x)\sqrt{x + 2}$

31. $9x - 12y + 8 = 0$ **33.** $x - 3 = 0$

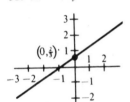

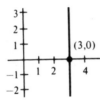

13.

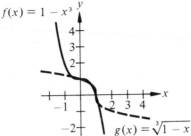

$g(x) = x^2 + 4$

$f(x) = \sqrt{x - 4}$

35. $3x + 2y = 6$ **37.** $12x + 3y = -2$
39. $x + y = 3$ **41.** $2x + 4y = 5$
43. (a) $2x - y = 3$ **(b)** $x + 2y = 4$
45. (a) $40x + 24y = 53$ **(b)** $24x - 40y = -9$
47. (a) $x = 2$ **(b)** $y = 5$ **49.** $5F - 9C = 160$
51. $W = 4.50 + 0.75x$ **53.** $V = 825{,}000 - 30{,}000t$
55. (a) $y = (76/3)t + 96$ **(b)** 1976: 121.33, 1977: 146.67
(c) 1980: 222.67 **(d)** Amount spent for energy imports was increasing at the rate of 25.33 billion dollars per year.
57. (a) $C = 26{,}500 + 14.75t$ **(b)** $R = 25t$
(c) $P = 10.25t - 26{,}500$ **(d)** $t = 2585.4$ hr

15.

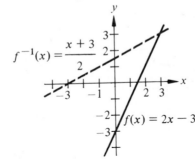

$f(x) = 1 - x^3$

$g(x) = \sqrt[3]{1 - x}$

17. $f^{-1}(x) = \dfrac{x + 3}{2}$

Section 3.6

1. (a) $2x$ **(b)** 2 **(c)** $x^2 - 1$ **(d)** $\dfrac{x + 1}{x - 1}$ **(e)** x **(f)** x

3. (a) $x^2 - x + 1$ **(b)** $x^2 + x - 1$ **(c)** $x^2 - x^3$

(d) $\dfrac{x^2}{1 - x}$ **(e)** $(1 - x)^2$ **(f)** $1 - x^2$

5. (a) $x^2 + 5 + \sqrt{1 - x}$ **(b)** $x^2 + 5 - \sqrt{1 - x}$

(c) $(x^2 + 5)\sqrt{1 - x}$ **(d)** $\dfrac{x^2 + 5}{\sqrt{1 - x}}$ **(e)** $6 - x$

(f) Undefined   **7. (a)** $\dfrac{x + 1}{x^2}$ **(b)** $\dfrac{x - 1}{x^2}$ **(c)** $\dfrac{1}{x^3}$

(d) x **(e)** x^2 **(f)** x^2

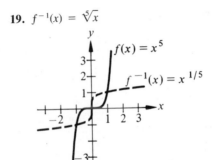

$f^{-1}(x) = \dfrac{x + 3}{2}$

$f(x) = 2x - 3$

9.

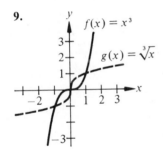

$f(x) = x^3$

$g(x) = \sqrt[3]{x}$

11.

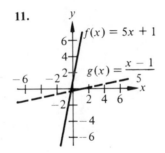

$f(x) = 5x + 1$

$g(x) = \dfrac{x - 1}{5}$

19. $f^{-1}(x) = \sqrt[5]{x}$

$f(x) = x^5$

$f^{-1}(x) = x^{1/5}$

21. $f^{-1}(x) = x^2, 0 \le x$

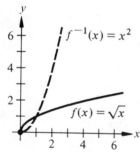

23. $f^{-1}(x) = \sqrt{4 - x^2}, 0 \le x \le 2$

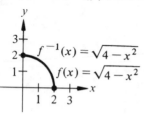

25. $f^{-1}(x) = x^3 + 1$

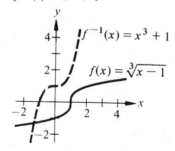

27. $f^{-1}(x) = x^{3/2}$

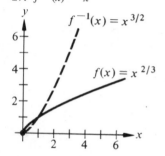

29. Yes, as long as $a \ne 0$, $f^{-1}(x) = \dfrac{x - b}{a}$ **31.** No

33. Yes, $f^{-1}(x) = \dfrac{x\sqrt{5}}{\sqrt{1 - x^2}}$ **35.** No

37. Yes, $f^{-1}(x) = \dfrac{1}{x}$ **39.** Yes, $f^{-1}(x) = \dfrac{x^2 - 3}{2}$

Section 3.7

1. $y = kx, k = \dfrac{5}{2}$ **3.** $s = kt^2, k = 16$

5. $y = \dfrac{k}{x}, k = 75$ **7.** $h = \dfrac{k}{t^3}, k = 12$

9. $z = kxy, k = 2$ **11.** $F = krs^3, k = 14$

13. $z = \dfrac{kx^2}{y}, k = \dfrac{2}{3}$ **15.** (a) 2 in (b) 15 lb

17. $\sqrt{6}/4 = 0.61$ mi/hr   **19.** 400 ft

21. The illumination is 1/4 as great.   **23.** $d \ge 0.0542$ in

Chapter 3 Review Exercises

1. (a) 6 (b) (3, 0) (c) $y = 0$ (d) $x^2 + y^2 - 6x = 0$
3. (a) 5 (b) (0, 1/2) (c) $3x - 4y + 2 = 0$
  (d) $x^2 + y^2 - y - 6 = 0$ **5.** (a) 13 (b) (8, 7/2)
  (c) $5x - 12y + 2 = 0$ (d) $x^2 + y^2 - 16x - 7y + 34 = 0$
7. (a) 8 (b) (5, 2) (c) $x = 5$
  (d) $x^2 + y^2 - 10x - 4y + 13 = 0$ **9.** (a) $\sqrt{53}$
  (b) (5/2, 1) (c) $2x - 7y + 2 = 0$
  (d) $x^2 + y^2 - 5x - 2y - 6 = 0$ **11.** 7/3
15. (a) Center (1/2, 5), radius 3/2
  (b)   **17.**

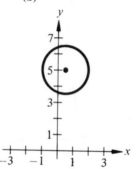

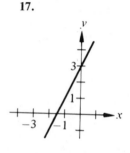

19.

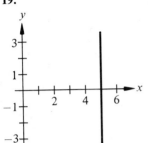

21.

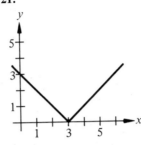

37. (a) $f^{-1}(x) = x^2 - 1$

(b)

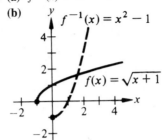

23.

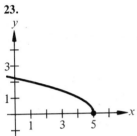

25.

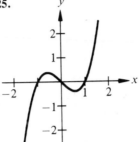

39. (a) $f^{-1}(x) = \sqrt{x + 5}$

(b)

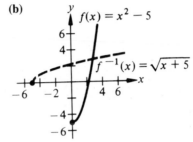

27.

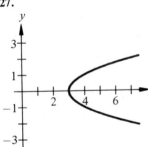

29.

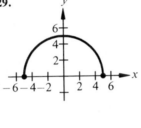

41. $x \geq 4$, $f^{-1}(x) = \dfrac{\sqrt{2x + 8}}{2}$

43. $z = kx^2/y$, $k = 32/25$

CHAPTER 4

Section 4.1

1. h   **3.** d   **5.** g   **7.** a   **9.** c

11. (a) $f(x) = (x - 2)^2$ (b) $f(x) = x^2 - 4x + 4$

13. (a) $f(x) = -(x + 2)^2 + 4$ (b) $f(x) = -x^2 - 4x$

15. (a) $f(x) = -(x + 3)^2 + 3$ (b) $f(x) = -x^2 - 6x - 6$

17.

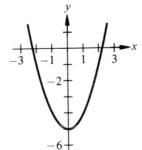

19.

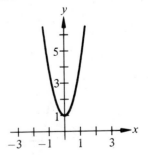

31. Domain: $[-5, 5]$, range: $[0, 5]$ **33.** (a) 5   (b) 17
(c) $t^4 + 1$ (d) $-x^2 - 1$ (e) $x^2 + 2x\Delta x + (\Delta x)^2 + 1$
35. (a) $f^{-1}(x) = 2x + 6$
(b)

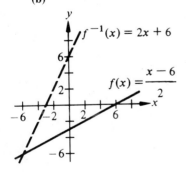

21.

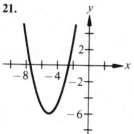

23.

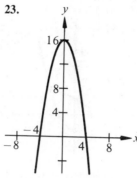

25.

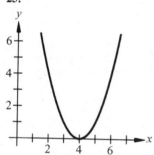

27.

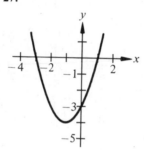

29.

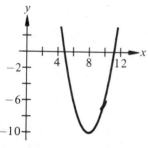

31.

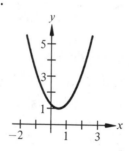

33.

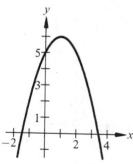

35.

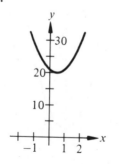

37. $x = 25$ **39.** $x = 20$ **41.** 200
43. $f(x) = (3/4)(x - 5)^2 + 12$
45. $f(x) = (1/2)(2x^2 + x - 10)$
$g(x) = (-1/2)(2x^2 + x - 10)$

Section 4.2

1. g   **3.** d   **5.** h   **7.** i   **9.** j   **11.** Up to the right and to the left   **13.** Down to the right and to the left   **15.** Up to the right and down to the left   **17.** Down to the right and up to the left   **19.** Down to the right and to the left   **21.** ± 5 **23.** 3   **25.** $-2, 1$ **27.** $2 \pm \sqrt{3}$ **29.** 0, 2   **31.** ± 1 **33.** $\pm \sqrt{5}$ **35.** No real roots
37. $f(x) = x^2 - 10x$ **39.** $f(x) = x^2 + 4x - 12$
41. $f(x) = x^3 + 5x^2 + 6x$
43. $f(x) = x^4 - 4x^3 - 9x^2 + 36x$
45. **(a)**

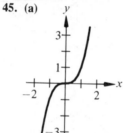

(b) (i)

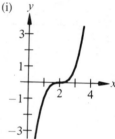

(ii)

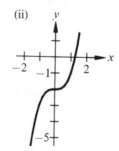

(iii)

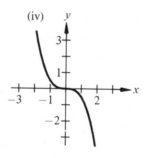

(iv)

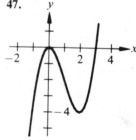

47.

49.

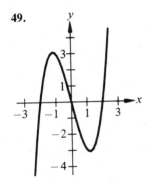

51.

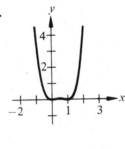

35. $-3x^3 - 6x^2 - 12x - 24 - \dfrac{48}{x - 2}$

37. $-x^2 + 3x - 6 + \dfrac{11}{x + 1}$ **39.** $4x^2 + 14x - 30$

41. (a) 1   (b) 4   (c) 4   (d) 1954

43. $(x - 2)(x + 3)(x - 1)$ **45.** $(2x - 1)(x - 5)(x - 2)$

47. $(x - 0.4)(x - 0.5)(x - 1)$
49. $[x - (1 + \sqrt{3})][x - (1 - \sqrt{3})](x - 1)$
51. $27, -274.625$ **53.** $11.705, -5646.972$

Section 4.4

1. 1, 2, 3   **3.** $-1, 1, 4$ **5.** $-10, -1$
7. 2   **9.** 1, 2   **11.** $-1, 1/2$ **13.** $-1/2, 1$
15. $-3/4, 1/2, 1$ **17.** $-3/4$ **19.** $\pm 1, \pm\sqrt{2}$
21. $-1, 2$ **23.** $\pm 1, \pm 2$ **25.** $-3, -1, 0, 4$
27. $-2, -1/2, 4$ **29.** $\pm\sqrt{2}, 0, 3, 4$
31. (a) $\pm 1, \pm 1/2, \pm 1/4, \pm 1/8, \pm 1/16, \pm 1/32, \pm 3,$
$\pm 3/2, \pm 3/4, \pm 3/8, \pm 3/16, \pm 3/32$

(b)   (c) $-1/8, 3/4, 1$

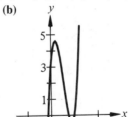

53.

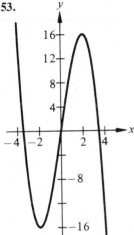

55.

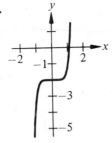

57. 0.7   **59.** 3.3

Section 4.3

1. $x^2 - 3x + 1$ **3.** $x^3 + 3x^2 - 1$ **5.** $3x^2 + 3x - 4$

7. $7 - \dfrac{11}{x + 2}$ **9.** $x + \dfrac{5x}{x^2 - 5}$ **11.** $x - \dfrac{x + 9}{x^2 + 1}$

13. $x^2 + \dfrac{x^2 + 7}{x^3 - 1}$ **15.** $2x^2 - 6x + 21 - \dfrac{144x - 68}{2x^2 + 6x - 3}$

17. $-x - 4 + \dfrac{20}{5 + 4x - x^2}$ **19.** $2x + \dfrac{x + 5}{x^2 - 2x - 8}$

21. $3x^2 - 2x + 5$ **23.** $4x^2 - 9$

25. $-x^2 + 10x - 25$ **27.** $5x^2 + 14x + 56 + \dfrac{232}{x - 4}$

29. $10x^3 + 10x^2 + 60x + 360 + \dfrac{1360}{x - 6}$

31. $x^4 + 2x^3 + 4x^2 + 8x + 16$ **33.** $x^2 - 8x + 64$

33. (a) $\pm 1, \pm 2, \pm 3, \pm 6, \pm 9, \pm 18, \pm 1/2, \pm 3/2,$
$\pm 9/2, \pm 1/4, \pm 3/4, \pm 9/4$

(b)   (c) $-2, \dfrac{1 \pm \sqrt{145}}{8}$

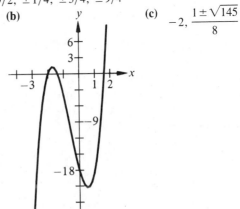

35. (a) $\pm 1,\ \pm 2,\ \pm 5,\ \pm 10,\ \pm 1/2,\ \pm 5/2,\ \pm 1/4,\ \pm 5/4$
(b)   **(c)** $-2,\ -1/2,\ 5/2$

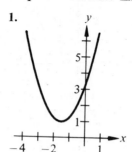

37. $-1,\ 1/4,\ 1$ **39.** $-2/3,\ 1/3,\ 1$
41. (a) Upper bound **(b)** Lower bound **(c)** Neither
43. (a) Neither **(b)** Lower bound **(c)** Upper bound

Section 4.5

1. $-11,\ -7;\ (x + 7)(x + 11)$ **3.** $2 \pm \sqrt{3}$;
$[x - (2 - \sqrt{3})][x - (2 + \sqrt{3})]$
5. $\pm 5i;\ (x - 5i)(x + 5i)$
7. $\pm 3,\ \pm 3i;\ (x + 3)(x - 3)(x + 3i)(x - 3i)$
9. $1 \pm i;\ [x - (1 + i)][x - (1 - i)]$
11. $2,\ 2 \pm i;\ (x - 2)[x - (2 + i)][x - (2 - i)]$
13. $-5,\ 4 \pm 3i;\ (x + 5)[x - (4 + 3i)][x - (4 - 3i)]$
15. $-10,\ -7 \pm 5i;\ (x + 10)[x - (-7 + 5i)]$
$[x - (-7 - 5i)]$
17. $-\dfrac{3}{4},\ 1 \pm \dfrac{1}{2}i;\ (4x + 3)(2x - 2 + i)(2x - 2 - i)$
19. $-2,\ 1 \pm \sqrt{2}i;\ (x + 2)[x - (1 + \sqrt{2}i)]$
$[x - (1 - \sqrt{2}i)]$
21. $-1/5,\ 1 \pm \sqrt{5}\,i;\ (5x + 1)[x - (1 + \sqrt{5}\,i)]$
$[x - (1 - \sqrt{5}i)]$
23. $2,\ \pm 2i;\ (x - 2)^2(x + 2i)(x - 2i)$
25. $\pm i,\ \pm 3i;\ (x + i)(x - i)(x + 3i)(x - 3i)$
27. $-1,\ \pm 2,\ \pm 3;\ (x + 1)(x + 2)(x - 2)(x + 3)(x - 3)$
29. $-2,\ 1 \pm \sqrt{2}i;\ (x + 2)^2[x - (1 + \sqrt{2}i)]$
$[x - (1 - \sqrt{2}i)]$
31. $f(x) = x^3 - x^2 + 25x - 25$
33. $f(x) = x^3 - 10x^2 + 33x - 34$
35. $f(x) = x^4 + 37x^2 + 36$
37. $f(x) = x^4 + 8x^3 + 9x^2 - 10x + 100$
39. $f(x) = 16x^4 + 36x^3 + 16x^2 + x - 30$

41. (a) $(x^2 + 9)(x^2 - 3)$ **(b)** $(x^2 + 9)(x + \sqrt{3})(x - \sqrt{3})$
(c) $(x + 3i)(x - 3i)(x + \sqrt{3})(x - \sqrt{3})$
43. (a) $(x^2 - 2x - 2)(x^2 - 2x + 3)$
(b) $[x - (1 + \sqrt{3})][x - (1 - \sqrt{3})](x^2 - 2x + 3)$
(c) $[x - (1 + \sqrt{3})][x - (1 - \sqrt{3})][x - (1 + \sqrt{2}i)]$
$[x - (1 - \sqrt{2}i)]$

Section 4.6

1. 0.682   **3.** 0.206   **5.** 2.769   **7.** $-1.164,\ 1.453$
9. 1.563

Chapter 4 Review Exercises

1.

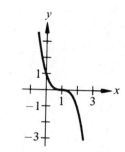

3.

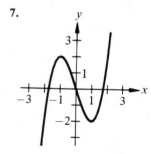

5.

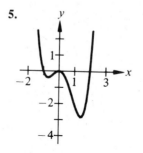

7.

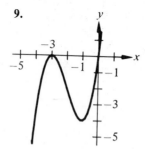

9.

11. $(1,\ -1)$ Minimum   **13.** $(3,\ 9)$ Maximum
15. $(-3/2,\ -41/4)$ Minimum   **17.** $x = 4500$ units
19. $x^2 - 2$ **21.** $3x^3 + 6x^2 - 4x - 31 - \dfrac{46x - 129}{x^2 - 2x + 4}$

23. $\dfrac{3}{2} + \dfrac{x - 2}{2(2x^3 + x)}$

25. $\dfrac{1}{4}x^3 - \dfrac{7}{2}x^2 - 7x - 14 - \dfrac{28}{x - 2}$ **27.** $6x^3 - 27x$

29. $2x^2 + (-3 + 4i)x + (1 - 2i)$ **31.** $1, 3/4$

33. $5/6, \pm 2i$ **35.** $-1, \dfrac{2}{3}, \dfrac{3}{2}, 3$

37. $6x^4 + 13x^3 + 7x^2 - x - 1$ **39.** $y_0 = 11.30$

CHAPTER 5

Section 5.1

1. f   **3.** a   **5.** j   **7.** c   **9.** h

11.

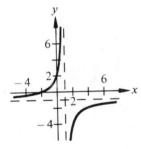

13.

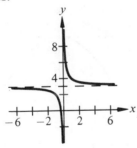

15.

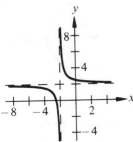

17.

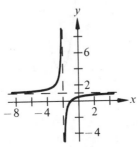

19.

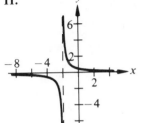

21.

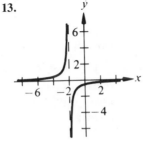

23.

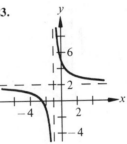

25.

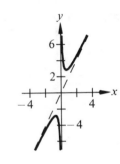

27.

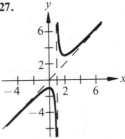

29.

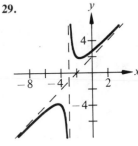

31. Horizontal asymptote: $y = 0$
    Vertical asymptote: $x = 0$

33. Horizontal asymptote: $y = -1$
    Vertical asymptote: $x = 2$

35. Vertical asymptotes: $x = \pm 1$
    Slant asymptote: $y = x$

37. Horizontal asymptote: $y = 3$

39. No asymptotes

41.

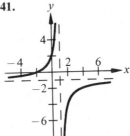

43.

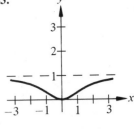

45.

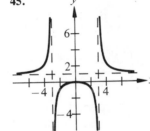

47.

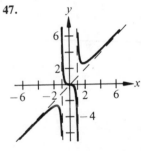

49.

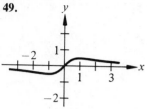

51.

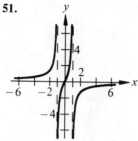

53.

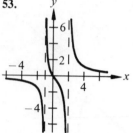

55.

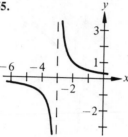

57.

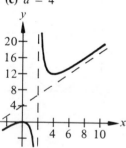

59.

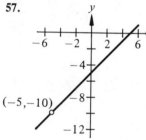

61. (c) $a = 4$

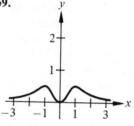

63. $8/3$

Section 5.2

1. $\dfrac{1}{2}\left[\dfrac{1}{x - 1} - \dfrac{1}{x + 1}\right]$

3. $\dfrac{1}{x - 1} - \dfrac{1}{x + 2}$

5. $\dfrac{3}{2x - 1} - \dfrac{2}{x + 1}$

7. $\dfrac{5}{x - 2} - \dfrac{1}{x + 2} - \dfrac{3}{x}$

9. $2x + \dfrac{3/2}{x - 4} - \dfrac{1/2}{x + 2}$

11. $\dfrac{3}{x} - \dfrac{1}{x^2} + \dfrac{1}{x + 1}$

13. $x + 3 + \dfrac{6}{x - 1} + \dfrac{4}{(x - 1)^2} + \dfrac{1}{(x - 1)^3}$

15. $3\left[\dfrac{1}{x - 3} + \dfrac{3}{(x - 3)^2}\right]$

17. $\dfrac{2x}{x^2 + 1} - \dfrac{1}{x}$

19. $\dfrac{1}{6}\left[\dfrac{1}{x - 2} - \dfrac{1}{x + 2} + \dfrac{2}{x^2 + 2}\right]$

21. $\dfrac{1}{8}\left[\dfrac{1}{2x - 1} + \dfrac{1}{2x + 1} - \dfrac{4x}{4x^2 + 1}\right]$

23. $\dfrac{1}{x^2 + 2} + \dfrac{x}{(x^2 + 2)^2}$

25. $\dfrac{1}{x + 1} + \dfrac{2}{x^2 - 2x + 3}$

27. $\dfrac{1}{x} + \dfrac{1 - x}{x^2 + 1}$

29. $\dfrac{1}{L}\left(\dfrac{1}{y} + \dfrac{1}{L - y}\right)$

Section 5.3

1. h   **3.** e   **5.** f   **7.** c   **9.** g

11. Vertex: $(0, 0)$
Focus: $(0, 1/16)$
Directrix: $y = -1/16$

13. Vertex: $(0, 0)$
Focus: $(-3/2, 0)$
Directrix: $x = 3/2$

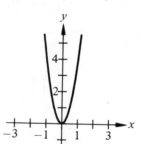

15. Vertex: $(0, 0)$
Focus: $(0, -2)$
Directrix: $y = 2$

17. Vertex: $(0, 0)$
Focus: $(2, 0)$
Directrix: $x = -2$

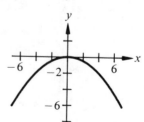

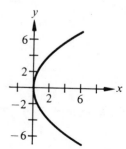

19. $x^2 + 6y = 0$ **21.** $3y^2 + 8x = 0$

23. $x^2 - 4y = 0$ **25.** $x^2 + 16y = 0$

27. $y^2 - 9x = 0$

29. Center: $(0, 0)$
    Foci: $(\pm 3, 0)$
    Vertices: $(\pm 5, 0)$

31. Center: $(0, 0)$
    Foci: $(0, \pm 3)$
    Vertices: $(0, \pm 5)$

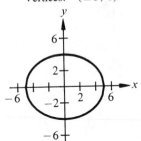

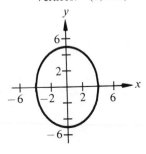

33. Center: $(0, 0)$
    Foci: $(\pm 2, 0)$
    Vertices: $(\pm 3, 0)$

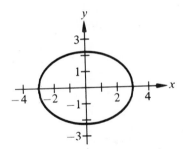

35. Center: $(0, 0)$
    Foci: $(0, \pm \sqrt{2})$
    Vertices: $(0, \pm \sqrt{5})$

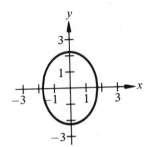

37. $\dfrac{x^2}{36} + \dfrac{y^2}{11} = 1$ **39.** $\dfrac{x^2}{25} + \dfrac{y^2}{21} = 1$

41. $\dfrac{x^2}{1} + \dfrac{y^2}{4} = 1$ **43.** $\dfrac{x^2}{24} + \dfrac{y^2}{49} = 1$

45. Center: $(0, 0)$
    Vertices: $(\pm 1, 0)$
    Foci: $(\pm \sqrt{2}, 0)$

47. Center: $(0, 0)$
    Vertices: $(0, \pm 1)$
    Foci: $(0, \pm \sqrt{5})$

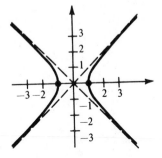

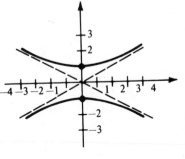

49. Center: $(0, 0)$
    Vertices: $(0, \pm 5)$
    Foci: $(0, \pm 13)$

51. Center: $(0, 0)$
    Vertices: $(\pm \sqrt{3}, 0)$
    Foci: $(\pm \sqrt{5}, 0)$

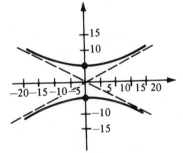

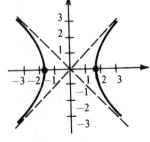

53. $\dfrac{y^2}{4} - \dfrac{x^2}{12} = 1$ **55.** $\dfrac{x^2}{1} - \dfrac{y^2}{9} = 1$

57. $\dfrac{y^2}{9} - \dfrac{x^2}{9/4} = 1$ **59.** $\dfrac{x^2}{9} - \dfrac{y^2}{7} = 1$

61. $x^2 - 800y = 0$

Section 5.4

1. Vertex: $(1, -2)$
    Focus: $(1, -4)$
    Directrix: $y = 0$

3. Vertex: $(5, -1/2)$
    Focus: $(11/2, -1/2)$
    Directrix: $x = 9/2$

5. Vertex: $(1, 1)$
    Focus: $(1, 2)$
    Directrix: $y = 0$

7. Vertex: $(8, -1)$
    Focus: $(9, -1)$
    Directrix: $x = 7$

9. Vertex: $(-2, -3)$
    Focus: $(-4, -3)$
    Directrix: $x = 0$

11. Vertex: $(-1, 2)$
    Focus: $(0, 2)$
    Directrix: $x = -2$

13. $y^2 - 4y + 8x - 20 = 0$

15. $x^2 - 8y + 32 = 0$

17. $y^2 - 4y - 8x + 4 = 0$

19. $x^2 + y - 4 = 0$

21. Center: $(1, 5)$
    Foci: $(1, 9), (1, 1)$
    Vertices: $(1, 10), (1, 0)$
    $e = \dfrac{4}{5}$

23. Center: $(-2, 3)$
    Foci: $(-2, 3 \pm \sqrt{5})$
    Vertices: $(-2, 6), (-2, 0)$
    $e = \dfrac{\sqrt{5}}{3}$

25. Center: $(1, -1)$ $e = \dfrac{3}{5}$
    Foci: $\left(1 \pm \dfrac{3\sqrt{10}}{20}, -1\right)$
    Vertices: $\left(1 \pm \dfrac{\sqrt{10}}{4}, -1\right)$

27. Center: $\left(\dfrac{1}{2}, -1\right)$ $e = \dfrac{\sqrt{10}}{5}$
    Foci: $\left(\dfrac{1}{2} \pm \sqrt{2}, -1\right)$
    Vertices: $\left(\dfrac{1}{2} \pm \sqrt{5}, -1\right)$

29. $\dfrac{(x-2)^2}{4} + \dfrac{(y-2)^2}{1} = 1$

31. $\dfrac{x^2}{48} + \dfrac{(y-4)^2}{64} = 1$

33. $\dfrac{(x-3)^2}{9} + \dfrac{(y-5)^2}{16} = 1$

35. $\dfrac{x^2}{16} + \dfrac{(y-4)^2}{12} = 1$

37. Center: $(1, -2)$
    Vertices: $(-1, -2), (3, -2)$
    Foci: $(1 \pm \sqrt{5}, -2)$

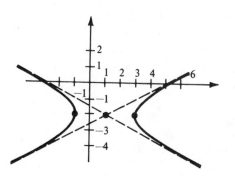

39. Center: $(2, -6)$
    Vertices: $(2, -5), (2, -7)$
    Foci: $(2, -6 \pm \sqrt{2})$

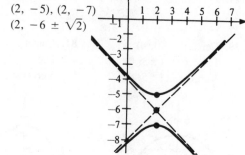

41. Center: $(2, -3)$
    Vertices: $(1, -3), (3, -3)$
    Foci: $(2 \pm \sqrt{10}, -3)$

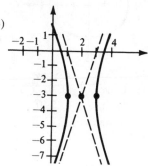

43. Center: $(1, -3)$
    Vertices: $(1, -3 \pm \sqrt{2})$
    Foci: $(1, -3 \pm 2\sqrt{5})$

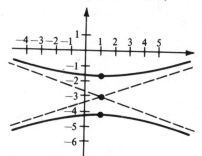

45. Degenerate hyperbola: graph is two intersecting lines with center at $(-1, -3)$

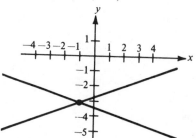

47. $\dfrac{(x-3)^2}{9} - \dfrac{(y-2)^2}{4} = 1$ **49.** $\dfrac{y^2}{9} - \dfrac{(x-2)^2}{9/4} = 1$

51. Circle   **53.** Hyperbola   **55.** Ellipse
57. Parabola   **59. (a)** $17,500\sqrt{2} = 24,750$ mi/hr
(b) $x^2 = -16,400(y - 4100)$
61. 91,419,000 mi,   94,581,000 mi

17. $1 - \dfrac{1}{8}\left[\dfrac{25}{x+5} - \dfrac{9}{x-3}\right]$ **19.** $\dfrac{1}{2}\left[\dfrac{3}{x-1} - \dfrac{x-3}{x^2+1}\right]$

21. $\dfrac{3x}{x^2+1} + \dfrac{x}{(x^2+1)^2}$ **23.** $\dfrac{y^2}{1} - \dfrac{x^2}{8} = 1$

25. $\dfrac{(x-2)^2}{25} + \dfrac{y^2}{21} = 1$ **27.** $8x + y^2 = 16$

29. $\dfrac{x^2}{16/5} - \dfrac{y^2}{64/5} = 1$ **31.** $y = 2 - 2x^2$

Chapter 5 Review Exercises

1.
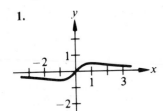

3.
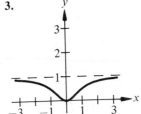

5.
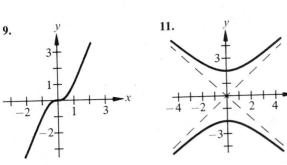

7.
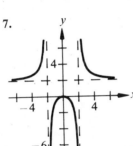

9.
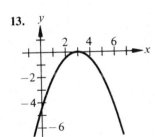

11.
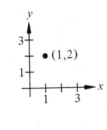

13.
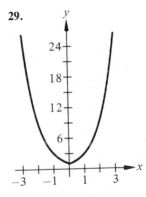

15.

CHAPTER 6

Section 6.1

1. i   **3.** c   **5.** a   **7.** f   **9.** e   **11.** $x = 4$
13. $x = -2$ **15.** $x = 2$ **17.** $x = 16$
19. $x = -2$

21.
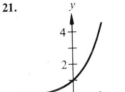

23.
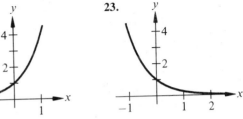

25.
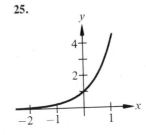

27.

29.
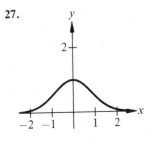

31.
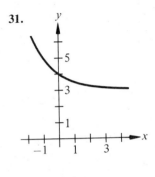

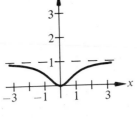

33.

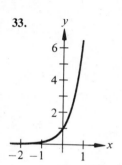

35.

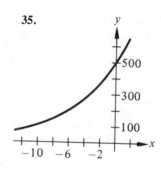

37.

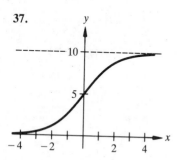

39. (a) $24,115.73 (b) $25,714.29 (c) $26,602.23
(d) $27,231.38 (e) $27,547.07 (f) $27,557.94
41. (a) $472.70 (b) $298.29 **43.** (a) 100 (b) 300
(c) 900 **45.** 96% **47.** (a) 0.15 (b) 0.49 (c) 0.81
49. 229.2 units/mL

Section 6.2

1. $2^4 = 16$ **3.** $5^{-2} = 1/25$ **5.** $16^{1/2} = 4$
7. $7^0 = 1$ **9.** $e^2 = e^2$ **11.** $3 = \log_5 125$
13. $1/4 = \log_{81} 3$ **15.** $-2 = \log_6(1/36)$
17. $3 = \ln(20.0855...)$ **19.** $v = \log_u w$ **21.** $x = 3$
23. $x = -3$ **25.** $x = 1/3$ **27.** $b = 3$ **29.** $x = 3$
31. $x = -1, x = 2$ **33.** $x = 8$ **35.** 0.9208
37. 0.2084 **39.** 1.6542 **41.** 0.1781 **43.** 1.8957
45. -0.7124 **47.** 0.91355 **49.** 2.0367
51. $\log_2 x + \log_2 y + \log_2 z$ **53.** $(1/2)\ln(a - 1)$
55. $\ln z + 2\ln(z - 1)$ **57.** $2\log_b x - 2\log_b y - 3\log_b z$
59. $(1/3)[\ln x + (1/2)\ln y]$ **61.** $\log_3\left(\dfrac{x - 2}{x + 2}\right)$
63. $\ln\sqrt[3]{\dfrac{x(x + 3)^2}{x^2 - 1}}$ **65.** $\ln\dfrac{9}{\sqrt{x^2 + 1}}$ **67.** 0.7124
69. -0.4312 **71.** 1.6117 **73.** 3.8227

75. -1.4436 **77.** 23.68 yr **79.** (a) 7.9 (b) 7.7
81. 1.6×10^{-6} **83.** 4.64 **85.** (a) 2.6201
(b) 1.6201 (c) -1.3799 **87.** (a) 3.6420 (b) 1.6420
(c) -2.3580 **89.** (a) 426 (b) 4.26 (c) 0.00426
91. 18.11 **93.** 2.079 **95.** 4.423 **97.** 0.4194
99. 0.2761

Section 6.3

1.

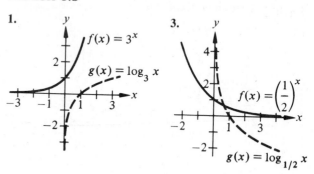

3.

5.

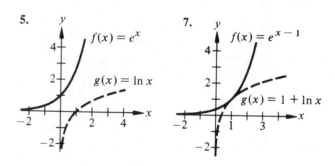

7.

9. d **11.** j **13.** g **15.** a **17.** h
19. $(1/2, \infty)$ **21.** $(-\infty, 0)$ **23.** $(-2, 2)$
25. $(-\infty, -2), (-2, \infty)$ **27.** (a) 20 (b) 70 (c) 95
(d) 120

29. (a)

K	1	2	3	4
t	0	7.30	11.56	14.59

K	6	8	10	12
t	18.86	21.89	24.24	26.16

(b)

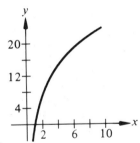

31.

x	y	$\dfrac{\ln x}{\ln y}$	$\ln \dfrac{x}{y}$	$\ln x - \ln y$
1	2	0	-0.6931	-0.6931
3	4	0.7925	-0.2877	-0.2877
10	5	1.4307	0.6931	0.6931
4	0.5	-2.0000	2.0794	2.0794

Section 6.4

1. $x = -2$ **3.** $x = 5$ **5.** $x = \dfrac{\log 12}{\log 4} = 1.7925$

7. $-\dfrac{\ln 360}{\ln 8.5} = -2.7504$ **9.** $t = \dfrac{\ln 3}{0.09} = 12.2068$

11. $t = \dfrac{\ln 2}{12 \ln(1 + 0.10/12)} = 6.9603$

13. $t = 5(\ln 19 - \ln 4) = 7.7907$

15. $N = -\dfrac{\ln 0.2247}{\ln 1.0775} = 20.0016$ **17.** $x = 3.7087$

19. $x = 3$ ($x = -1$ is not in the domain of $\log x$)

21. $x = -1, x = 3$ **23.** $x = 3$ **25.** $x = e^2, 1$

27. $x = (1/2) \ln 3$ **29.** $x = \pm 2$ **31. (a)** 6.64 yr

(b) 6.33 yr   **(c)** 6.30 yr   **(d)** 6.30 yr

33.

r	2%	4%	6%	8%	10%	12%
t	54.93	27.47	18.31	13.73	10.99	9.16

35. (a) 303 units   **(b)** 528 units
(b) 39.79 yr   **39.** 7.79 months

37. (a) 29.33 yr
41. 22.4°

Chapter 6 Review Exercises

1.

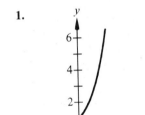

3

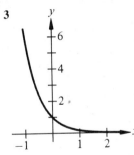

5.

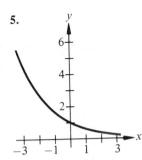

7.

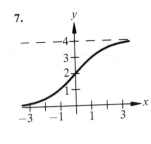

9.

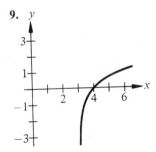

11.

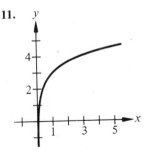

13.

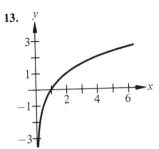

15.

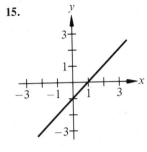

17. $\ln(x^2 + 1) - \ln |x|$ **19.** $\ln(x^2 + 1) + \ln |x - 1|$

21. $\ln \dfrac{\sqrt{2x - 1}}{(x + 1)^2}$ **23.** $\ln(x^2\sqrt[3]{y})$ **25.** $\ln \dfrac{3\sqrt[3]{4 - x^2}}{x}$

27. False   **29.** True   **31. (a)** $x = 1151.3$ units
(b) $x = 1324.6$ units   **33.** $y = 2e^{0.1014t}$

35. $y = 4e^{-0.4159t}$ **37. (a)** $r = \dfrac{\ln 2}{7.75} = 9\%$

(b) $1834.37   **39. (a)** $N = 30(1 - e^{-0.0502t})$

(b) $t = \dfrac{\ln 6}{0.0502} = 36$ days

41. (a) $S = 30(1 - e^{-0.1823t})$ **(b)** 17,944 units

(c)

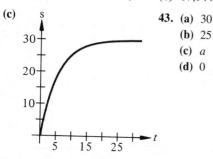

43. (a) 30
(b) 25
(c) a
(d) 0

CHAPTER 7

Section 7.1

1. $(1, 2)$ **3.** $(-1, 2), (2, 5)$ **5.** $(0, 5), (3, 4)$
7. $(0, 0), (2, 4)$ **9.** $(-1, 1), (8, 4)$ **11.** $(5, 5)$
13. $(1/2, 3)$ **15.** $(1.5, 0.3)$ **17.** $(20/3, 40/3)$
19. $(1, 2)$ **21.** $(29/10, 21/10), (-2, 0)$
23. $(-1, -2), (2, 1)$ **25.** $(-1, 0), (0, 1), (1, 0)$
27. $(1, 2, 3)$ **29.** $(1/2, 2), (-4, -1/4)$
31. $(2, 2), (4, 0)$ **33.** No points of intersection
35. $(0, 1)$ **37.** $(1, 0), (5\pi/6, -\sqrt{3}/2)$
39. 6400 units   **41.** 4, 8   **43.** $x = 5, y = 5, \lambda = -5$
45. $x = \pm\sqrt{2}/2, y = 1/2, \lambda = 1$

Section 7.2

1. $(2, 0)$ **3.** $(-1, -1)$ **5.** No solution
7. Dependent   **9.** $(-1/3, -2/3)$ **11.** $(5/2, 3/4)$
13. $(3, 4)$ **15.** $(4, -1)$ **17.** $(40, 40)$
19. No solution   **21.** Dependent   **23.** $(18/5, 3/5)$
25. $(5, -2)$ **27.** $(90/31, -67/31)$
29. $(-6/35, 43/35)$ **31.** 550 mi/hr, 50 mi/hr
33. 20/3 gal of 20% solution, 10/3 gal of 50% solution
35. 375 adults, 125 children   **37.** 12×8 ft

39. (a) $y = (3/4)x + (4/3)$
(b)

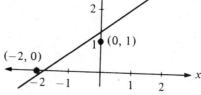

41. (a) $y = -2x + 4$

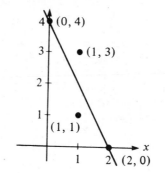

Section 7.3

1. $(1, 2, 3)$ **3.** $(2, -3, -2)$ **5.** $(5, -2, 0)$
7. Inconsistent   **9.** $(1, -3/2, 1/2)$
11. $(-3a + 10, 5a - 7, a)$
13. $\left(13 - 4a, \dfrac{45}{2} - \dfrac{15}{2}a, a\right)$
15. $(-a, 2a - 1, a)$ **17.** $\left(\dfrac{1}{2} - \dfrac{3}{2}a, 1 - \dfrac{2}{3}a, a\right)$
19. Inconsistent   **21.** $(1, 1, 1, 1)$ **23.** $(0, 0, 0)$
25. $\left(-\dfrac{3}{5}a, \dfrac{4}{5}a, a\right)$ **27.** $y = 2x^2 + 3x - 4$
29. $y = x^2 - 4x + 3$ **31.** $x^2 + y^2 - 4x = 0$
33. $x^2 + y^2 - 6x - 8y = 0$
35. $a = -32, v_0 = 0, s_0 = 144$
37. $a = -32, v_0 = -32, s_0 = 500$
39. $\dfrac{1}{2}\left(-\dfrac{2}{x} + \dfrac{1}{x-1} + \dfrac{1}{x+1}\right)$
41. $\dfrac{1}{2}\left(\dfrac{1}{x} - \dfrac{1}{x-2} + \dfrac{2}{x+3}\right)$
43. 4 medium *or* 2 large, 1 medium, 2 small

45. $y = (3/7)x^2 + (6/5)x + (26/35)$

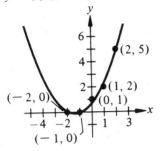

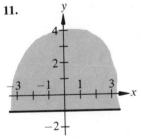

19.

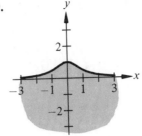

21.

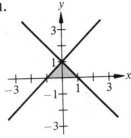

23.

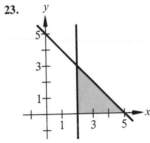

25.

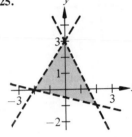

Section 7.4

1. f   **3.** g   **5.** a   **7.** b   **9.** d

11.

13.

27.

29.

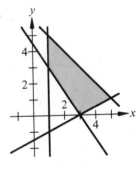

15.

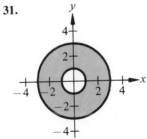

17.

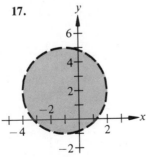

31.

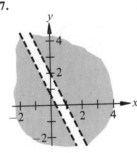

33.

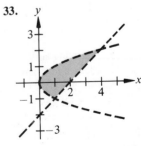

35.
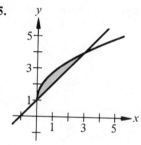

37.
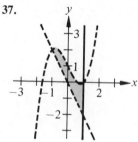

Chapter 7 Review Exercises

1. $(1, 1)$ **3.** $(5, 4)$ **5.** $(0, 0)$, $(2, 8)$, $(-2, 8)$
7. $(0, 0)$, $(-3, 3)$ **9.** $(5/2, 3)$ **11.** $(-0.5, 0.8)$
13. $(0, 0)$ **15.** $(4.8, 4.4, -1.6)$
17. $(3a + 4, 2a + 5, a)$ **19.** $(-3a + 2, 5a + 6, a)$
21.

39.
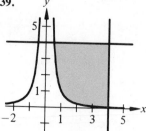

23.

21.
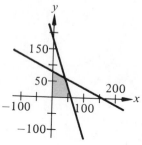

23.
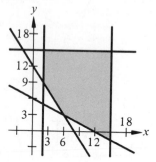

41. $x + (3/2)y \le 12$
  $(4/3)x + (3/2)y \le 15$
  $x \ge 0$
  $y \ge 0$

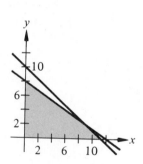

25.
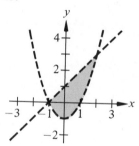

27.
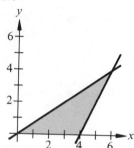

29. $(X, Y, Z) = (2, 5, 4)$

43. $x \ge 2$, $x \le 5$, $y \ge 1$, $y \le 7$
45. $y \le x + 1$, $y \le -x + 1$, $y \ge 0$
47. $x^2 + y^2 \le 16$, $y \ge x$, $x \ge 0$
49. (a) Minimum at $(0, 0)$: 0
      Maximum at $(3, 4)$: 17
   (b) Minimum at $(0, 0)$: 0
      Maximum at $(4, 0)$: 20
51. (a) Minimum at $(0, 0)$: 0
      Maximum at $(60, 20)$: 740
   (b) Minimum at $(0, 0)$: 0
      Maximum at any point along line segment connecting
      $(60, 20)$ and $(30, 45)$: 2100
53. 200 units (at \$250)
    50 units (at \$400)
55. Minimum at $(5, 3)$: 35

CHAPTER 8

Section 8.1

1. (a) $\begin{bmatrix} 1 & 2 & 3 \\ 0 & -5 & -10 \\ 3 & 1 & -1 \end{bmatrix}$ **(b)** $\begin{bmatrix} 1 & 2 & 3 \\ 0 & -5 & -10 \\ 0 & -5 & -10 \end{bmatrix}$

(c) $\begin{bmatrix} 1 & 2 & 3 \\ 0 & -5 & -10 \\ 0 & 0 & 0 \end{bmatrix}$ **(d)** $\begin{bmatrix} 1 & 2 & 3 \\ 0 & 1 & 2 \\ 0 & 0 & 0 \end{bmatrix}$

(e) $\begin{bmatrix} 1 & 0 & -1 \\ 0 & 1 & 2 \\ 0 & 0 & 0 \end{bmatrix}$ **3.** $\begin{bmatrix} 1 & 0 & 0 & 3 \\ 0 & 1 & 0 & 2 \\ 0 & 0 & 1 & -1 \end{bmatrix}$

5. $\begin{bmatrix} 1 & 0 & 5 & 4 \\ 0 & 1 & 6 & 3 \\ 0 & 0 & 0 & 0 \end{bmatrix}$ **7.** $\begin{bmatrix} 1 & 0 & 0 \\ 0 & 1 & 0 \\ 0 & 0 & 1 \\ 0 & 0 & 0 \end{bmatrix}$

9. $\begin{bmatrix} 1 & 2 & 0 & 0 \\ 0 & 0 & 1 & 0 \\ 0 & 0 & 0 & 1 \\ 0 & 0 & 0 & 0 \end{bmatrix}$ **11.** $(3, 2)$ **13.** $(4, -2)$

15. $(1/2, -3/4)$ **17.** Inconsistent **19.** $(4, -3, 2)$
21. $(2a + 1, 3a + 2, a)$ **23.** $(5a + 4, -3a + 2, a)$
25. $(0, 2 - 4a, a)$ **27.** $(1, 0, 4, -2)$
29. $(2, -2, 3, -5, 1)$ **31.** \$800,000 at 8%, \$500,000 at
9%, \$200,000 at 12% **33.** $a = 1/2, b = -3, c = 7/2$
35. $D = -5, E = -3, F = 6$

13. Not possible **15.** $\begin{bmatrix} -1 & 19 \\ 4 & -27 \\ 0 & 14 \end{bmatrix}$

17. $\begin{bmatrix} 1 & 0 & 0 \\ 0 & 1 & 0 \\ 0 & 0 & 7/2 \end{bmatrix}$ **19.** $\begin{bmatrix} 60 & 72 \\ -20 & -24 \\ 10 & 12 \\ 60 & 72 \end{bmatrix}$

21. $\begin{bmatrix} -4 & 0 \\ 8 & 2 \end{bmatrix}$ **23.** $\begin{bmatrix} 0 & 0 & 0 \\ 0 & 0 & 0 \\ 0 & 0 & 0 \end{bmatrix}$

25. $AC = BC = \begin{bmatrix} 12 & -6 & 9 \\ 16 & -8 & 12 \\ 4 & -2 & 3 \end{bmatrix}$ **27.** $AB \neq BA$

29. $AB = [\$1250 \quad \$1331.25 \quad \$981.25]$

Section 8.2

1. (a) $\begin{bmatrix} 3 & -2 \\ 1 & 7 \end{bmatrix}$ **(b)** $\begin{bmatrix} -1 & 0 \\ 3 & -9 \end{bmatrix}$ **(c)** $\begin{bmatrix} 3 & -3 \\ 6 & -3 \end{bmatrix}$

(d) $\begin{bmatrix} -1 & -1 \\ 8 & -19 \end{bmatrix}$ **3. (a)** $\begin{bmatrix} 7 & 3 \\ 1 & 9 \\ -2 & 15 \end{bmatrix}$

(b) $\begin{bmatrix} 5 & -5 \\ 3 & -1 \\ -4 & -5 \end{bmatrix}$ **(c)** $\begin{bmatrix} 18 & -3 \\ 6 & 12 \\ -9 & 15 \end{bmatrix}$ **(d)** $\begin{bmatrix} 16 & -11 \\ 8 & 2 \\ -11 & -5 \end{bmatrix}$

5. (a) $\begin{bmatrix} 3 & 3 & -2 & 1 & 1 \\ -2 & 5 & 7 & -6 & -8 \end{bmatrix}$

(b) $\begin{bmatrix} 1 & 1 & 0 & -1 & 1 \\ 4 & -3 & -11 & 6 & 6 \end{bmatrix}$

(c) $\begin{bmatrix} 6 & 6 & -3 & 0 & 3 \\ 3 & 3 & -6 & 0 & -3 \end{bmatrix}$

(d) $\begin{bmatrix} 4 & 4 & -1 & -2 & 3 \\ 9 & -5 & -24 & 12 & 11 \end{bmatrix}$ **7. (a)** $\begin{bmatrix} 0 & 15 \\ 6 & 12 \end{bmatrix}$

(b) $\begin{bmatrix} -2 & 2 \\ 31 & 14 \end{bmatrix}$ **9. (a)** $\begin{bmatrix} 0 & -10 \\ 10 & 0 \end{bmatrix}$ **(b)** $\begin{bmatrix} 0 & -10 \\ 10 & 0 \end{bmatrix}$

11. (a) $\begin{bmatrix} 6 & -21 & 15 \\ 8 & -23 & 19 \\ 4 & 7 & 5 \end{bmatrix}$ **(b)** $\begin{bmatrix} 9 & 0 & 13 \\ 7 & -2 & 21 \\ 1 & 4 & -19 \end{bmatrix}$

Section 8.3

7. $\begin{bmatrix} -3 & 2 \\ -2 & 1 \end{bmatrix}$ **9.** $\begin{bmatrix} 1 & -1 \\ 2 & -1 \end{bmatrix}$ **11.** Does not exist

13. $\frac{1}{3}\begin{bmatrix} -9 & -7 \\ 3 & 2 \end{bmatrix}$ **15.** $\begin{bmatrix} 1 & 1 & -1 \\ -3 & 2 & -1 \\ 3 & -3 & 2 \end{bmatrix}$

17. $\begin{bmatrix} -175 & 37 & -13 \\ 95 & -20 & 7 \\ 14 & -3 & 1 \end{bmatrix}$

19. $\begin{bmatrix} -24 & 7 & 1 & -2 \\ -10 & 3 & 0 & -1 \\ -29 & 7 & 3 & -2 \\ 12 & -3 & -1 & 1 \end{bmatrix}$ **21.** $\frac{1}{2}\begin{bmatrix} -3 & 3 & 2 \\ 9 & -7 & -6 \\ -2 & 2 & 2 \end{bmatrix}$

23. $\frac{5}{11}\begin{bmatrix} 0 & -4 & 2 \\ -22 & 11 & 11 \\ 22 & -6 & -8 \end{bmatrix}$

25. $\begin{bmatrix} -1/8 & 0 & 0 & 0 \\ 0 & 1 & 0 & 0 \\ 0 & 0 & 1/4 & 0 \\ 0 & 0 & 0 & -1/5 \end{bmatrix}$

27. $\begin{bmatrix} 1 & 0 & 0 \\ -0.75 & 0.25 & 0 \\ 0.35 & -0.25 & 0.2 \end{bmatrix}$ **29.** Does not exist

31. (a) $(4, 8)$ **(b)** $(-8, -11)$ **(c)** $(-7, -7)$
33. (a) $(-5, -81/2, 48)$ **(b)** $(-3, -24, 28)$
(c) $(0, 1, -1)$

Section 8.4

1. 5   **3.** 27   **5.** -24 **7.** 6   **9.** 0   **11.** -2
13. 0   **15.** 0   **17.** 0.002   **19.** $-7x + 3y - 8$

21. (a) $\begin{bmatrix} 23 & -8 & -22 \\ 5 & -5 & 5 \\ 7 & -22 & -23 \end{bmatrix}$ **(b)** $\begin{bmatrix} 23 & 8 & -22 \\ -5 & -5 & -5 \\ 7 & 22 & -23 \end{bmatrix}$

23. (a) $\begin{bmatrix} 662 & 708 & 394 & 524 \\ -282 & -298 & -174 & -234 \\ 42 & 48 & 44 & 24 \\ 611 & 674 & 377 & 507 \end{bmatrix}$

(b) $\begin{bmatrix} 662 & -708 & 394 & -524 \\ 282 & -298 & 174 & -234 \\ 42 & -48 & 44 & -24 \\ -611 & 674 & -377 & 507 \end{bmatrix}$ **25.** -75

27. 170   **29.** -58 **31.** -30 **33.** -108 **35.** 0
37. 412   **39.** 0   **41.** $3x - 5y = 0$
43. $x + 3y - 5 = 0$ **45.** $2x + 3y - 8 = 0$
47. 33/2   **49.** 31/2

Section 8.5

1. Column 2 is a multiple of Column 1.   **3.** Row 2 has only zero elements.   **5.** The interchange of Columns 2 and 3 results in a change of sign of the determinant.
7. Multiplying the elements of Row 1 by 1/5 produces a determinant that is 1/5 times the original.   **9.** Multiplying the elements of all three rows by 1/5 produces a determinant that is $(1/5)^3$ times the original.   **11.** Adding -4 times the elements of Row 1 to the elements of Row 2 leaves the determinant unchanged.   **13.** Adding multiples of Column 2 to Columns 1 and 3 leaves the determinant unchanged.

15. The interchange of two rows (or columns) results in a change of sign of the determinant. Rows 2 and 3 were interchanged, followed by an interchange of Columns 1 and 3.
17. -6 **19.** -26 **21.** -126 **23.** 236

25. -3740 **27.** 7441   **29.** 410   **31.** $\frac{1}{8}\begin{bmatrix} 2 & -4 \\ 1 & 2 \end{bmatrix}$

33. Inverse does not exist.   **35.** $\frac{1}{2}\begin{bmatrix} -3 & 3 & 2 \\ 9 & -7 & -6 \\ -2 & 2 & 2 \end{bmatrix}$

37. $\frac{1}{11}\begin{bmatrix} 0 & -20 & 10 \\ -110 & 55 & 55 \\ 110 & -30 & -40 \end{bmatrix}$

39. $\frac{1}{3}\begin{bmatrix} -9 & -26 & 6 & 8 \\ 6 & 16 & -3 & -4 \\ -9 & -25 & 6 & 7 \\ -3 & -12 & 3 & 3 \end{bmatrix}$

Section 8.6

1. $(1, 2)$ **3.** $(2, -2)$ **5.** $(3/4, -1/2)$ **7.** $|D| = 0$
9. $(2/3, 1/2)$ **11.** $(-1, 3, 2)$ **13.** $(1, 1/2, 3/2)$
15. $(0, -1/2, 1/2)$ **17.** $(5, 0, -2, 3)$ **19.** $|D| = 0$
21. $y = (27/19)x - (18/19)$
23. $y = -(22/17)x + (74/17)$
25. $y = (3/4)(x^2 - x + 1)$

Chapter 8 Review Exercises

1. $(10, -12)$ **3.** $(0.6, 0.5)$ **5.** $(2, -3, 3)$
7. $(1/2, -1/3, 1)$ **9.** $\left(-2a + \dfrac{3}{2}, 2a + 1, a\right)$

11. Inconsistent   **13.** $\begin{bmatrix} -13 & -8 & 18 \\ 0 & 11 & -19 \end{bmatrix}$

15. $\begin{bmatrix} 14 & -2 & 8 \\ 14 & -10 & 40 \\ 36 & -12 & 48 \end{bmatrix}$ **17.** $\begin{bmatrix} 44 & 4 \\ 20 & 8 \end{bmatrix}$

19. $\begin{bmatrix} 4 & 6 & 3 \\ 0 & 6 & -10 \\ 0 & 0 & 6 \end{bmatrix}$

21. $5x + 4y = 2, -x + y = -22$ **23.** $(10, -12)$
25. $(2, -3, 3)$ **27.** $(1/2, -1/3, 1)$ **29.** 128
31. $\lambda^2 - \lambda + 1 = 0$
33. $-\lambda^3 + 5\lambda^2 + 18\lambda - 52 = 0$

CHAPTER 9

Section 9.1

1. 2, 4, 8, 16, 32   **3.** $-1/2$, $1/4$, $-1/8$, $1/16$, $-1/32$
5. 3, 9/2, 9/2, 27/8, 81/40
7. -1, $1/4$, $-1/9$, $1/16$, $-1/25$
9. 2, 3/2, 4/3, 5/4, 6/5   **11.** 0, 1, 2, 3, 4
13. 0, 1, 0, 1/2, 0   **15.** 5/2, 11/4, 23/8, 47/16, 95/32
17. $-1/2$, 2/3, $-3/4$, 4/5, $-5/6$ **19.** 3, 4, 6, 10, 18
21. $a_n = 3n - 2$ **23.** $a_n = n^2 - 2$ **25.** $a_n = \dfrac{n + 1}{n + 2}$
27. $a_n = \dfrac{(-1)^n}{2^{n-1}}$ **29.** $a_n = 1 + \dfrac{1}{n}$
31. $a_n = \dfrac{n}{(n + 1)(n + 2)}$ **33.** $a_n = \dfrac{2^n n!}{(2n)!}$
35. $a_n = (-1)^{n+1}$ **37.** 35   **39.** 158/85   **41.** 40
43. $5C$ **45.** 238   **47.** $4x^2 + 20$ **49.** $14 + 24i$
51. $\displaystyle\sum_{i=1}^{9} \dfrac{1}{3i}$ **53.** $\displaystyle\sum_{i=1}^{8} \left[2\left(\dfrac{i}{8}\right) + 3 \right]$
55. $\displaystyle\sum_{i=1}^{6} (-1)^{i+1} 3^i$ **57.** $\displaystyle\sum_{i=1}^{20} \dfrac{(-1)^{i+1}}{i^2}$
59. $\displaystyle\sum_{i=1}^{5} \dfrac{2^i - 1}{2^{i+1}}$
63. (a) $A_1 = \$5100.00$, $A_2 = \$5202.00$,
$A_3 = \$5306.04$, $A_4 = \$5412.16$, $A_5 = \$5520.40$,
$A_6 = \$5630.81$, $A_7 = \$5743.43$, $A_8 = \$5858.30$
(b) $\$11,040.20$ **65. (a)** $10 \cdot 9 = 90$
(b) $25 \cdot 24 = 600$ **(c)** $n(n - 1)$

Section 9.2

1. Arithmetic sequence, $d = 3$ **3.** Not an arithmetic
sequence   **5.** Arithmetic sequence, $d = -1/4$ **7.** Not
an arithmetic sequence   **9.** Arithmetic sequence, $d = 0.4$
11. $\{5, 11, 17, 23, 29, \ldots\}$
13. $\{-2.6, -3.0, -3.4, -3.8, -4.2, \ldots\}$
15. $\{3/2, 5/4, 1, 3/4, 1/2, \ldots\}$
17. $\{-2, 2, 6, 10, 14, \ldots\}$ **19.** 28   **21.** 44
23. $99x$ **25.** $-37/2$ **27.** 620   **29.** 4600
31. 265   **33.** 4000   **35.** 1275   **37.** 25,250
39. (a) $\$35,000$ **(b)** $\$187,500$ **41.** 2340   **43.** 520
45. 44,625   **47.** 9, 13   **49.** 15/4, 9/2, 21/4

Section 9.3

1. Geometric sequence, $r = 3$ **3.** Not a geometric
sequence   **5.** Geometric sequence, $r = -1/2$ **7.** Not
a geometric sequence   **9.** Not a geometric sequence
11. $\{2, 6, 18, 54, 162, \ldots\}$
13. $\{1, 1/2, 1/4, 1/8, 1/16, \ldots\}$
15. $\{5, -1/2, 1/20, -1/200, 1/2000, \ldots\}$
17. $\{1, x/2, x^2/4, x^3/8, x^4/16, \ldots\}$ **19.** $(1/2)^7$
21. $-2(-1/3)^{10}$ **23.** $100e^{8x}$ **25.** $500(1.02)^{39}$
27. (a) $\$2593.74$ **(b)** $\$2653.30$ **(c)** $\$2685.06$
(d) $\$2707.04$ **(e)** $\$2717.91$ **29.** $\$22,689.45$
31. 29,921.31   **33.** 6.400   **35.** 511   **37.** $\$7808.24$
41. $\$594,121.01$ **43.** 2   **45.** 2/3   **47.** 16/3
49. 32   **51.** 8/3   **53.** 1/2   **55.** 152.42 ft
57. 1/9   **59.** 4/11   **61.** 16/37   **63.** 15/11

Section 9.4

15. $n(2n + 1)$ **17.** $10[1 - (0.9)^n]$ **19.** $\dfrac{n}{2(n + 1)}$

Section 9.5

1. 10   **3.** 1   **5.** 15,504   **7.** 4950   **9.** 4950
11. $x^5 + 5x^4y + 10x^3y^2 + 10x^2y^3 + 5xy^4 + y^5$
13. $a^4 + 8a^3 + 24a^2 + 32a + 16$
15. $r^6 + 18r^5s + 135r^4s^2 + 540r^3s^3 + 1215r^2s^4 + 1458rs^5 + 729s^6$
17. $x^5 - 5x^4y + 10x^3y^2 - 10x^2y^3 + 5xy^4 - y^5$
19. $1 - 6x + 12x^2 - 8x^3$
21. $x^8 + 20x^6 + 150x^4 + 500x^2 + 625$
23. $\dfrac{1}{x^5} + \dfrac{5y}{x^4} + \dfrac{10y^2}{x^3} + \dfrac{10y^3}{x^2} + \dfrac{5y^4}{x} + y^5$ **25.** -4
27. $2035 + 828i$ **29.** 1   **31.** $1{,}732{,}104x^5$
33. $180x^8y^2$ **35.** $-226{,}437{,}120$ **37.** $924x^3y^3$
39. $32t^5 - 80t^4s + 80t^3s^2 - 40t^2s^3 + 10ts^4 - s^5$
41. $\dfrac{1}{128} + \dfrac{7}{128} + \dfrac{21}{128} + \dfrac{35}{128} + \dfrac{35}{128} + \dfrac{21}{128} + \dfrac{7}{128} + \dfrac{1}{128}$
43. $\dfrac{1}{6561} + \dfrac{16}{6561} + \dfrac{112}{6561} + \dfrac{448}{6561} + \dfrac{1120}{6561} + \dfrac{1792}{6561} +$
$\dfrac{1792}{6561} + \dfrac{1024}{6561} + \dfrac{256}{6561}$ **45.** $0.07776 + 0.25920 +$
$0.34560 + 0.23040 + 0.07680 + 0.01024$

Section 9.6

1. 12   **3.** 12   **5.** 6,760,000   **7. (a)** 900 **(b)** 648
(c) 180   **9. (a)** 720 **(b)** 48   **11.** 24   **13.** 336
15. 1,860,480   **17.** 120   **19.** 11,880   **21.** 4845
23. 3,838,380   **25.** 560   **27. (a)** 70 **(b)** 30
29. (a) 70 **(b)** 54 **(c)** 16   **31. (a)** 420 **(b)** 420
33. 5040   **35.** 60

21.

Number of girls	0	1	2	3
Probability	1/64	6/64	15/64	20/64

Number of girls	4	5	6
Probability	15/64	6/64	1/64

Section 9.7

1. (a) $S = \{HHH, HHT, HTH, HTT, THH, THT, TTH, TTT\}$
(b) 3/8 **(c)** 1/2 **(d)** 7/8   **3. (a)** 1/12 **(b)** 11/12
(c) 2/9 **(d)** 1/9   **5.** $P(\{\text{Taylor wins}\}) = 1/2$,
$P(\{\text{Moore wins}\}) = P(\{\text{Jenkins wins}\}) = 1/4$ **7. (a)** 1/4
(b) 1/2 **(c)** 9/100 **(d)** 1/30   **9. (a)** 1/169 **(b)** 1/221
11. (a) 1/3 **(b)** 5/8   **13. (a)** 14/55 **(b)** 12/55
(c) 54/55   **15. (a)** 0.9702 **(b)** 0.9998 **(c)** 0.0002
17. (a) 1/64 **(b)** 1/32 **(c)** 63/64   **19. (a)** 0.9^{10}
(b) 0.1^{10} **(c)** $1 - 0.9^{10}$ **(d)** $C_1^{10}(0.1)^1(0.9)^9$

Chapter 9 Review Exercises

1. 30   **3.** 80   **5.** 127   **7.** 8   **9.** 12   **11.** 88
13. 418   **19.** 5/11   **21.** 16/15   **23.** 1/75
25. $\dfrac{x^4}{16} + \dfrac{x^3y}{2} + \dfrac{3x^2y^2}{2} + 2xy^3 + y^4$
27. $\dfrac{64}{x^6} + \dfrac{576}{x^4} + \dfrac{2160}{x^2} + 4320 + 4860x^2 + 2916x^4 + 729x^6$
29. $41 + 840i$ **31. (a)** 1 **(b)** 6 **(c)** 15
33. 118,813,760   **35.** 1/9
37. $P(\{3\}) = 1/6 > 5/36 = P(\{(1, 5), (5, 1), (2, 4), (4, 2),$
$(3, 3)\})$ **39.** 0.0475

INDEX

Abscissa, 122
Absolute value, 14
Absolute value inequality, 112
Acute angle, 295
Addition
  of complex numbers, 36
  of matrices, 339
  of real numbers, 4
Additive identity, 5
  matrix, 340
Additive inverse, 5, 7
  matrix, 340
Algebra of calculus, 2, 60, 64
Algebraic expression, 40
Algebraic fraction, 56
Alternation of signs, 380
Antilogarithm, 273
Arithmetic mean, 388
Arithmetic progression, 386
Arithmetic sequence, 386
Arithmetic series, 389
Arrangement
  circular, 420
  with repetition, 415
  without repetition, 415
Associative property, 5
  matrix addition, 340
Asymptote, 224
  of hyperbola, 224, 248
Augmented matrix, 331
Axis, 121
  ellipse, 244
  hyperbola, 246
  parabola, 184

Back substitution, 336

Base, 17
Binomial, 41, 408
Binomial coefficient, 410
Binomial difference, 412
Binomial product, 44
Binomial Theorem, 409
Bisection method, 217
Break-even analysis, 299

Calculator, xiii
Cancellation Laws, 6
Cartesian plane, 121
Change of base, 272
Characteristic, 276
Circle, 252
  general form, 134
  standard form, 133
Circular arrangement, 420
Closed interval, 13
Closure property, 5
Coefficient, 41, 179
Coefficient matrix, 331
Cofactor, 357
Cofactor form for inverse, 367
Column, matrix, 330
Combination of n elements taken r at a
  time, 420
Common difference, arithmetic se-
  quence, 386
Common logarithm, 272
Common ratio of a geometric sequence,
  393
Commutative property, 5
  matrix addition, 340
Complement of an event, 428
Completion of square, 90, 92

Complex conjugates, 38
Complex number, 35
Complex root, 97
Complex zero, 213
Composite function, 166
Compound fraction, 59
Compound interest, 266
Conditional equation, 72
Conic, 241
  equation, 251
Conjugate, 30
Consistent, 306
Constant, 40
Constant function, 142, 147
Constant of proportionality, 172
Constant term, 179
Continuity, 188
Coordinate, 11, 122
Coordinate system, 121
Cramer's Rule, 370, 372
Critical number, 113
Cross multiplication, 58
Cube root, 24

Decreasing function, 147
Degree of a polynomial function, 41,
  179
Denominator, 7
Dependent, 306
Dependent variable, 137
Determinant
  of a matrix of order 2, 355
  of a matrix of order 3, 356
  properties of, 362
Difference of two squares, 50
Direct variation, 172

Direct variation as *n*th power, 173
Directly proportional, 172
Discount, 83
Discriminant, 96
Distance between two points on the real
    line, 15
Distance Formula, 122, 123
Distinguishable permutation, 571
Distributive property, 5
Division Algorithm, 197
Division
    of complex numbers, 38
    of polynomials, 196
    of real numbers, 4, 7
    synthetic, 198
    by zero, 9
Domain, 137, 142, 145
Domain of a variable, 46

e, 264
Echelon form, 333
Economics, 299
Element of a matrix, 331
Elementary column operation, 363
Elementary row operation, 331, 332,
    363
Elimination, 303
Ellipse, 244, 252
    major axis, 244
    minor axis, 244
    standard form, 244
Equality
    of matrices, 339
    properties of, 6
Equation, 6
    quadratic, 96, 108
Equivalent equation, 72
Equivalent expression, 41
Equivalent fraction, 7
Evaluation of an algebraic expression, 46
Even function, 149
Event, 424
Expansion by minors, 359
Experiment, 424
Exponent, 17
    properties of, 18
Exponential equation, solution of, 286
Exponential form, 17
Exponential function, 258
    properties of, 261

Exponential graph, 260
Exponential key, calculator, 22
Extraneous solution, 76

Factor, 48, 90
    irreducible, 214
    of a polynomial, 190, 214
Factor Theorem, 201
Factorial, 378
Factorization
    by grouping, 54
    of a polynomial, 49
First row, 412
FOIL method, 44
Formula, 78
Fraction
    improper, 197
    proper, 197
    properties of, 7
Fractional expression, 56
Function, 137, 142
Function
    alternative definition, 145
Function notation, 139, 142
Functional value, 139
Fundamental Counting Principle, 414
Fundamental Theorem of Algebra, 211

Gauss-Jordan elimination, 334
General equation, 161
General form of the equation of a cir-
    cle, 134
General second-degree equation, 241
Geometric progression, 392
Geometric sequence, 392
Graph of an equation, 128
Greatest integer function, 152

Half-open interval, 13
Horizontal asymptote, 224
Horizontal line, 161
Horizontal line test, 170
Horner's Method, 202
Human memory model, 283
Hyperbola, 224, 246, 252

i, 34
Identity, 71
Image, 145
Imaginary number, 34, 97
Implied domain, 142

Improper fraction, 197
Inconsistent, 306, 337
Increasing function, 147
Independent, probability, 427
Independent variable, 137
Index
    of a radical, 25
    of summation, 381
Inequality
    linear, 111
    perfect square, 117
    properties of, 14
Inequality sign, 12
Infinite interval, 13
Infinite sequence, 377, 378
Infinite series, 382
Integer, 4
Intercept, 129
Interest, simple, 85
Intermediate Value Theorem, 192
Interval, 13
Inverse function, 168
Inverse matrix, 349
Inverse variation, 174
Inversely proportional, 174
Iinvertible, matrix, 349
Irrational number, 4
Irreducible, 214
Irreducible quadratic polynomial, 52
Iterative process, 217

Joint variation, 175
Jointly proportional, 175

Law of Trichotomy, 13
Leading coefficient, 41, 179
Leading coefficient test, 189
Linear equation, 73, 161
    solution guidelines, 74
Linear equation in one variable, 303
Linear inequality, 111
Linear interpolation, 278
Linear programming, 324–326
Logarithm
    common, 272
    natural, 272
    properties of, 270
Logarithm to base *b*, 269
Logarithmic equation, solution of, 286
Logarithmic expression, 274

Logarithmic function, 279
Logarithmic tables, 276
Lower bound, 209
Lower limit of summation, 381
Lowest common denominator, 8

m by n matrix, 331
Main diagonal, 331
Mantissa, 276
Mathematical induction, 402
Mathematical model, 83, 172
Matrix, 330
   augmented, 331
   coefficient, 331
   of cofactors, 358
   inverse, 349
   of minors, 358
   multiplication, 339, 343
   notation, 339
   subtraction, 340
Matrix addition, 339
Method of elimination, 303
Method of substitution, 295
Midpoint Rule, 124
Minor, 357
Missing factor model, 58
Model, 83
Monomial, 41, 180
Multiplication
   of complex numbers, 37
   of real numbers, 4
Multiplicative identity, 5
Multiplicative inverse, 5

Natural base e, 264
Natural logarithm, 272
Natural number, 4
Negation, properties of, 5
Negative direction, 11
Negative of a real number, 7
Negative slope, 157
Newton's Law of Cooling, 291
Newton's Method, 219
Noncommutativity of matrix multiplica-
   tion, 346
Nonnegative, 12
Nonsingular, 349
Nonsingular matrix, 366
nth partial sum, 381
nth root, 24

nth row, 412
nth term
   of an arithmetic sequence, 387
   of a geometric sequence, 394
   of a sequence, 406
Numerator, 7

Odd function, 149
One-to-one correspondence, 11
Open interval, 13
Operations with rational expressions, 57
Order property of real numbers, 11
Ordered pair, 121
Ordinate, 122
Origin, 11, 121
Outcome, 424

Parabola, 179, 242, 251
Parabola
   intercept, 184
   standard form, 182
Parallel lines, 162
Partial fraction, 234
   basic equation, 236
Partial fraction decomposition, 234
Partial sum
   of an arithmetic sequence, 390
   of a geometric sequence, 395
Pascal's Triangle, 411, 421
Percentage, 83
Perfect square inequality, 117
Perfect square trinomial, 50
Perimeter, 79
Permutation, 416
   of n elements, 417
   of n elements taken r at a time, 417
Perpendicular lines, 162
Point-plotting method, 129
Point-slope equation of a line, 158, 161
Points of intersection, 298
Polynomial, 41
Polynomial factor, 190, 214
Polynomial function, 179
Polynomial operations, 42
Position equation, 99
Positive direction, 11
Positive slope, 157
Prime number, 9
Principal nth root of b, 25
Priority of operations, 73

Probability, 424
   of an event, 425
   of independent events, 428
Proper fraction, 197

Quadrant, 121
Quadratic equation, 96, 108
Quadratic form, 90, 404
Quadratic formula, 95
Quadratic function, 179
Quadratic type, 108

Radical, 24
   properties of, 28
Radical equation, 106
Radicand, 25
Range, 137, 142, 145
Rational exponent, 26
Rational expression, 56
Rational function, 222, 228
Rational number, 4
Rational zero test, 205
Rationalization of a radical, 30
Real number line, 11
Real number system, 3
Real numbers, properties of, 5
Real part of a complex number, 35
Real zeros of a polynomial function,
   190, 192
Reciprocal of a real number, 7
Rectangular coordinate system, 121
Recursive formula, 387
Reduced echelon form, 333
Reflexive property, 6
Remainder Theorem, 200
Repeated root, 91
Rigid transformation, 152
Root
   of an equation, 71
   of a polynomial equation, 190
Row, matrix, 330
Rule of signs, 7

Sample space, 424
Scalar multiplication, 341
Scalar quantity, 271
Scientific notation, 20
Sequence, 378
   of partial sums, 381
Series, 382

FORMULAS FROM GEOMETRY

Triangle:

$h = a \sin\theta$

Area $= \dfrac{1}{2} bh$

(Law of Cosines)

$c^2 = a^2 + b^2 - 2ab \cos\theta$

Right Triangle:

(Pythagorean Theorem)

$c^2 = a^2 + b^2$

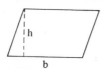

Equilateral Triangle:

$h = \dfrac{\sqrt{3}\,s}{2}$

Area $= \dfrac{\sqrt{3}\,s^2}{4}$

Parallelogram:

Area $= bh$

Trapezoid:

Area $= \dfrac{h}{2}(a+b)$

Circle:

Area $= \pi r^2$

Circumference $= 2\pi r$

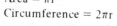

Sector of Circle:

(θ in radians)

Area $= \dfrac{\theta r^2}{2}$

$s = r\theta$

Circular Ring:

(p = average radius,

w = width of ring)

Area $= \pi(R^2 - r^2)$

$= 2\pi pw$

Sector of Circular Ring:

(p = average radius,

w = width of ring,

θ in radians)

Area $= \theta pw$

Ellipse:

Area $= \pi ab$

Circumference $\approx 2\pi \sqrt{\dfrac{a^2+b^2}{2}}$

Cone:

(A = area of base)

Volume $= \dfrac{Ah}{3}$

Right Circular Cone:

Volume $= \dfrac{\pi r^2 h}{3}$

Lateral Surface Area $= \pi r \sqrt{r^2 + h^2}$

Frustum of Right Circular Cone:

Volume $= \dfrac{\pi(r^2 + rR + R^2)h}{3}$

Lateral Surface Area $= \pi s(R+r)$

Right Circular Cylinder:

Volume $= \pi r^2 h$

Lateral Surface Area $= 2\pi rh$

Sphere:

Volume $= \dfrac{4}{3}\pi r^3$

Surface Area $= 4\pi r^2$

Wedge:

(A = area of upper face,

B = area of base)

$A = B \sec\theta$

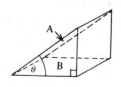

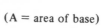